Brain Drug Delivery for the Treatment of Neurological Disorders

Although a wealth of knowledge is available on how best to reduce the burden of stroke and infectious neurological diseases, intervention and reduction of the burden of other non-communicable neurological disorders is much more complex and requires the development of new and more effective treatments and prevention measures. This book addresses this demand by collecting key elements required in developing novel treatments for non-communicable neurological disorders. Key chapters include the blood-brain barrier's important role in the treatment of neurological disorders, physiological barriers for drug delivery to brain, novel drug delivery systems, and industrial production and regulatory aspects.

Key Features

- Summarizes key *in vitro* and *in vivo* evaluations that should be considered in developing successful brain drug delivery systems.
- Addresses points that should be considered, when large-scale brain drug delivery systems are produced at the industrial level, and how batch consistencies can be met.
- Helps formulation scientists to consider important regulatory requirements when developing novel drug delivery systems to brain.
- Includes an important chapter describing *in vivo* evaluation of diffusing therapeutic agents from the blood into the brain parenchyma and how this approach would complement the current imaging techniques such as IVIS.
- Provides key information about the use of monoclonal antibodies for the treatment of neurological disorders.

Drugs and the Pharmaceutical Sciences: A Series of Textbooks and Monographs

Series Editor: Anthony J. Hickey, RTI International, Research Triangle Park, USA

The Drugs and Pharmaceutical Sciences series is designed to enable the pharmaceutical scientist to stay abreast of the changing trends, advances, and innovations associated with therapeutic drugs and that area of expertise and interest that has come to be known as the pharmaceutical sciences. The body of knowledge that those working in the pharmaceutical environment have to work with, and master, has been, and continues, to expand at a rapid pace as new scientific approaches, technologies, instrumentations, clinical advances, economic factors, and social needs arise and influence the discovery, development, manufacture, commercialization, and clinical use of new agents and devices.

Recent Titles in Series

An Introduction to Generative Drug Discovery
Sean Ekins

Viral Oncology: New Approaches to Molecular Cancer Therapeutics
Edited by Umesh Kumar, Deepak Parashar and Subodh Kumar

A Guide to Particulate Science in Pharmaceutical Product Development
Margaret Louey, Timothy Crowder and Anthony J. Hickey

Purification of Biotechnological Products: A Focus on Industrial Applications
Adalberto Pessoa, Beatriz Vahan Kilikian and Paul Frederick Long

GMP Audits in Pharmaceutical and Biotechnology Industries
Mustafa Edik

Nanomedicine and Applications in Cancer: A Complete Guide to Nanomedicine and Cancer Applications
Imran Saleem and Ahmed AH Abdellatif

Pharmaceutical Computer Systems Validation: Quality Assurance, Risk Management and Regulatory Compliance, Volume 1
Guy Wingate

Pharmaceutical Computer Systems Validation: Volume 2 - Real Life Case Studies
Guy Wingate

Good Manufacturing Practices for Pharmaceuticals, Eighth Edition
Graham P. Bunn

Pharmaceutical Process Engineering, Third Edition
Anthony J. Hickey

Computer Simulations in the Pharmaceutical Industry, Volume 1
Kiran Gangarapu

Brain Drug Delivery for the Treatment of Neurological Disorders
Edited by Touraj Ehtezazi

For more information about this series, please visit: www.crcpress.com/Drugs-and-the-Pharmaceutical-Sciences/book-series/IHCDRUPHASCI

Brain Drug Delivery for the Treatment of Neurological Disorders

Edited by
Touraj Ehtezazi

CRC Press
Taylor & Francis Group
Boca Raton London New York

CRC Press is an imprint of the
Taylor & Francis Group, an **informa** business

Designed cover image: Shutterstock

First edition published 2027
by CRC Press
2385 NW Executive Center Drive, Suite 320, Boca Raton FL 33431

and by CRC Press
4 Park Square, Milton Park, Abingdon, Oxon, OX14 4RN

CRC Press is an imprint of Taylor & Francis Group, LLC

ISBN: 978-1-032-77191-5 (hbk)
ISBN: 978-1-032-77316-2 (pbk)
ISBN: 978-1-003-48240-6 (ebk)

DOI: 10.1201/9781003482406

Typeset in Times
by KnowledgeWorks Global Ltd.

Contents

List of Contributors

Selin Akpinar Adscheid
MyBiotech GmbH
Industriestraße 1B
Überherrn, Germany
Department of Pharmacy, Biopharmaceutics and Pharmaceutical Technology
Saarland University
Saarbrücken, Germany

Saheed O. Benson
School of Pharmacy and Biomolecular Sciences, Liverpool
John Moores University
Byrom Street, Liverpool L3 3AF, UK

Shreyansh Chauhan
Institute of Pharmacy
Nirma University
Ahmedabad, Gujarat

Rhys Davies
The Walton Centre NHS Foundation Trust
Liverpool, UK

Touraj Ehtezazi
School of Pharmacy and Biomolecular Sciences, Liverpool
John Moores University
Byrom Street, Liverpool L3 3AF, UK

Mosavar Farahani
School of Life and Environmental Science
University of Northampton, UK

Amos Fatokun
School of Pharmacy and Biomolecular Sciences, Liverpool
John Moores University
Byrom Street, Liverpool L3 3AF, UK

Nazende Günday-Türeli
MyBiotech GmbH
Industriestraße 1B
Überherrn, Germany

Anam Hameed
Department of Human Nutrition and Dietetics, Faculty of Rehabilitation and Allied Health Sciences,
Riphah International University
Gulberg III, Lahore, Pakistan.

Huma Hameed
Faculty of Pharmaceutical Sciences
University of Central Punjab (UCP)
Lahore, Pakistan

Disha Joshi
L.J. Institute of Pharmacy
LJ University, LJ Campus
Ahmedabad, India

Mahtab Ahmad Khan
Faculty of Pharmaceutical Sciences
University of Central Punjab (UCP)
Lahore, Pakistan

Humaira Liaqat
Faculty of Pharmaceutical Sciences
University of Central Punjab (UCP)
Lahore, Pakistan

Jai Naik
Institute of Pharmacy
Nirma University
Ahmedabad, Gujarat

Ana Cláudia Paiva-Santos
Department of Pharmaceutical Technology, Faculty of Pharmacy of the University of Coimbra
University of Coimbra
Coimbra, Portugal

REQUIMTE/LAQV, Group of Pharmaceutical Technology, Faculty of Pharmacy of the University of Coimbra
University of Coimbra
Coimbra, Portugal

Mayur Patel
Institute of Pharmacy
Nirma University
Ahmedabad, Gujarat

Riya Patel
L.J. Institute of Pharmacy
LJ University, LJ Campus
Ahmedabad, India

Yashwant Pathak
University of South Florida
Florida, USA.

Luca Peruzzotti-Jametti
Department of Clinical Neurosciences and National Institute for Health Research (NIHR) Biomedical Research Centre
University of Cambridge
Cambridge, UK

Stefano Pluchino
Department of Clinical Neurosciences and National Institute for Health Research (NIHR) Biomedical Research Centre
University of Cambridge
Cambridge, UK

Bhupendra Prajapati
Department of Pharmaceutics, Parul Institute of Pharmacy, Faculty of Pharmacy
Parul University
Waghodia, Vadodara, Gujarat, India

Jidhin Raj
The Walton Centre NHS Foundation Trust
Liverpool, UK

Hakan Sahin
Istanbul University-Cerrahpasa
Istanbul, TR

Darren Sexton
School of Pharmacy and Biomolecular Sciences, Liverpool
John Moores University
Byrom Street, Liverpool L3 3AF, UK

Shreeraj Shah
L.J. Institute of Pharmacy
LJ University, LJ Campus
Ahmedabad, India

Changiz Taghibiglou
Dept. Anatomy, Physiology, Pharmacology, College of Medicine,
University of Saskatchewan, Health Science Bldg.
Saskatoon, SK, Canada

Akif E. Türeli
MyBiotech GmbH
Industriestraße 1B
Überherrn, Germany

Bindu Kumari Yadav
L.J. Institute of Pharmacy
LJ University, LJ Campus
Ahmedabad, India

Muhammad Zaman
Faculty of Pharmaceutical Sciences
University of Central Punjab (UCP)
Lahore, Pakistan

About the Editor

Dr. Touraj Ehtezazi is a Reader in Pharmaceutics at Liverpool John Moores University. A pharmacist by training, he graduated from the University of Tabriz, Iran, and obtained his PhD from the University of Nottingham in 1997. He conducted postdoctoral studies at the University of Nottingham in 1998 and held a lectureship at the University of Leicester from 2000.

Since joining Liverpool John Moores University in 2002, Dr. Ehtezazi has established himself as a leading researcher in drug delivery to the brain. He has published extensively in peer-reviewed journals, with particular focus on neuroinflammation, traumatic brain injury, Alzheimer's disease, and multiple sclerosis, including editing a special issue dedicated to brain drug delivery. Dr. Ehtezazi has contributed to the planning of major Horizon-2020 projects addressing the treatment of neurological disorders. His work combines expertise in pharmaceutical formulation, mathematical modelling, and targeted drug delivery systems to advance therapeutic strategies for challenging neurological conditions.

List of Abbreviations

%IA/g	The percentage of the injected activity per gram of tissue
5-FC	5-fluorocytosine
$5HT_{2A}$	5-hydroxytryptamine 2A
5xFAD	Tg(APPSwFlLon,PSEN1*M146L*L286V)6799Vas/Mmjax
6-OHDA	6-hydroxydopamine
aa	Amino acid
AAV	Adeno-associated virus
ACC	Agenesis of the Corpus Callosum
ACSD	Ataxias and Cerebellar or Spinocerebellar Degeneration
AD	Alzheimer's Disease
ADA	Anti-Drug Antibodies
ADAS-Cog 11	Alzheimer's Disease Assessment Scale-Cognitive subscale
ADAS-cog	Alzheimer's Disease Assessment Scale – cognitive
ADCC	Antibody-dependent cellular cytotoxicity
ADCOMS	Alzheimer's Disease Composite Score
ADCS-MCI-ADL	Alzheimer's Disease Cooperative Study--Activities of Daily Living Scale for Mild Cognitive Impairment
ADEM	Acute disseminated encephalomyelitis
ADHD	Attention-Deficit/Hyperactivity Disorder
AE	Adverse events
AF4	Asymmetric flow field-flow fractionation
AGS	Aicardi–Goutières Syndrome
AIS	Acute ischemic stroke
ALA	α–linolenic acid
ALS	Amyotrophic lateral sclerosis
ALSAQ40	ALS Assessment Questionnaire-40
ALSFRS-R	Amyotrophic Lateral Sclerosis Functional Rating Scale-Revised
AMPA	Alpha-amino-3-hydroxy-5-methylisoxazole-4-propionic acid
AMPKα1	AMP-activated protein kinase α1
AMT	Adsorptive-mediated endocytosis
Ang-2	Angiopoietin-2
Angpt2	Angiopoietin-2
AP2	Adaptor protein complex 2
aPD-1	antibody against the immune checkpoint inhibitor PD-1
API	Active Pharmaceutical Ingredient
ApoE	Apolipoprotein E
APP	Amyloid beta A4 Precursor Protein
APP	Amyloid precursor protein
APP/PS1	APPswe/PS1dE9
aPTT	Activated Partial Thromboplastin Time
ARCA	Autosomal Recessive Cerebellar Ataxias
ArgiNLS	Arginine-rich nuclear localization signal
ARIA	Amyloid related imaging abnormalities
ARTD	ADP-ribosyltransferase diphtheria toxin-like
ATP	Adenosine Triphosphate
ATRP	Atomic transfer radical polymerisation
ATXN3	Ataxin-3 protein

AUC	Area under the curve
AUF	Auranofin
AuNPs	Gold Nanoparticles
Aß	Amyloid-beta protein
Aβ	Amyloid-beta
BACE1	Beta-site amyloid precursor protein cleaving enzyme 1
BACE-1	Beta-site amyloid precursor protein-cleaving enzyme 1
BACE-1	β-secretase
BAM	Border-associated macrophages
BBB	Blood-brain barrier
BBBMPs	BBB-Modulating Peptides
BCECs/BMECs	Brain Microvascular Endothelial Cells
BCSFB	Blood-cerebrospinal fluid barrier
BDNF	Brain-Derived Neurotrophic Factor
BECs	Brain endothelial cells
Ben-PCL-mPEG	Benzyl-poly(ε-caprolactone)-*b*-methoxy poly(ethylene glycol)
BER	base excision repair
BID	Twice daily
BLECs	Brain-Like Endothelial Cells
BMDM	Bone marrow derived macrophage
BMEC	Brain Microvascular Endothelial Cell
BPI	Brachial Plexus Injury
BRCA1	Breast cancer-associated gene 1
BRCT	BRCA1 carboxy-terminal
BSA	Bovine serum albumin
bsAbs	Bispecific monoclonal antibodies
C3a/C5a	Complement Activation Fragments C3a and C5a
CAG	Cytosine–Adenine–Guanine repeat
CARPA	Complement Activation-Related Pseudoallergy
CAT	Carnitine acetyltransferase
CCI	Controlled cortical impact
CCL2	c-c motif chemokine ligand 2
CCL3	c-c motif chemokine ligand 3
CCPM	Core-crosslinked polymeric micelles
CD36	Cluster of differentiation 36
CD44	Cluster of Differentiation 44
CDC	Complement-dependent cytotoxicity
CDK12	Revealed cyclin-dependent kinase 12
CDRs	Complementarity-determining regions
CDR-SB	Clinical Dementia Rating-Sum-of-Boxes
CED	Convection enhanced delivery
CFF	Cross-Flow Filtration
CFZ	Carfilzomib
CGRP	Calcitonin gene-related peptide
CMC	Critical micelle concentration
CMT	Carrier-mediated transport
CMV	Cytomegalovirus
CNS	Central nervous system
COVID-19	Coronavirus Disease 2019
CP	Cerebral palsy
CPPs	Cell-Penetrating Peptides

CQA	critical quality attribute
CQAs	Critical Quality Attributes
CRISPR	Clustered Regularly Interspaced Short Palindromic Repeats
CRISPR-Cas9	Clustered Regularly Interspaced Short Palindromic Repeats / CRISPR-associated protein 9
Cryo-EM	Cryo-electron microscopy
CSF	Cerebrospinal fluid
CSF-1R	Colony-Stimulating Factor 1R
CT	computerised tomography
CT	X-ray computed tomography
CTD	Common technical document
CTT	C-terminal tail
CUMS	Chronic unpredictable mild stress
CVO	Circumventricular organs
CXCL1	CXC ligand 1
CXCL10	C-X-C motif chemokine ligand 10
CXCL12	C-X-C motif chemokine ligand-12
CXCL2	C-X-C motif chemokine ligand 2
CXCL3	C-X-C motif chemokine ligand 3
CXCR4	C-X-C motif chemokine receptor-4
CypD	Cyclophilin D
DALYs	Disability adjusted life year
DAM	Disease-associated microglia
DAPK1	Death-associated protein kinase 1
DBS	Deep brain stimulation
DDS	Drug Delivery System
DHA	Docosahexaenoic acid
DLG	Discs large
DLS	Dynamic light scattering
DLT	Dose-limiting toxicities
DM	DNA-binding motif
DMI	Dimethyl itaconate
DMT	Disease modifying therapy
DNA	Deoxyribonucleic Acid
DoE	Design of Experiments
DOX	Doxorubicin
DPX	Dapoxetine
DSB	Double-strand DNA break
DTI	Diffusion Tensor Imaging
DTPA	Diethylenetriaminepentaacetic
DTX	Docetaxel
EAAT 1	Excitatory amino-acid transporter 1
EAE	Experimental autoimmune encephalomyelitis
ECG	Electrocardiograph
ECM	Extracellular matrix
ECT	Electroconvulsive Therapy
EDSS	Expanded Disability Status Scale
EEG	Electroencephalogram
EEG	Electroencephalography
EGFR	Epidermal growth factor receptor
ELIXIR	engineered late-current inhibitor X by inactivation-gate release

EMA	European Medicines Agency
EMG	Electromyography
EO	Ethylene oxide
EPO	Erythropoietin
ER	Endoplasmic reticulum
ERK	Extracellular regulated kinase
ERK	Extracellular Signal-Regulated Kinase
ESCRT	Endosomal sorting complex required for transport
ESR	Erythrocyte Sedimentation Rate
EU	Endotoxin unit
EV	Extracellular vesicle
EVT	Endovascular therapy
FAD	Familial AD
FcRn	Fc receptor
FcRn	Neonatal Fc Receptor
FcγRs	Fc gamma receptors
FD4	FITC-dextran Mw 4000
FDA	U.S. Food and Drug Administration
Fe3O4	Magnetite Iron Oxide
FMT	Fluorescence molecular tomography
FND	Functional neurological disorder
FNDC5	Fibronectin type III domain-containing 5
FNP	Flash Nanoprecipitation
FPI	Fluid percussion injury
FRET	Förster resonance energy transfer
FUS	Focused ultrasound
F-waves	Late Motor Conduction Responses
GAM	Glioma associated microglia
GB	Glioblastoma
GBM	Glioblastoma
GBS	Guillain–Barré Syndrome
GFAP	Glial Fibrillary Acidic Protein
GI	Gastrointestinal Tract
GLUT1	Glucose Transporter 1
GMH	Germinal matrix haemorrhage
GMP	Good Manufacturing Practice
GSH	Glutathione
H&E	Hematoxylin & Eosin
HAMAs	Human anti-mouse antibodies
HD	Huntington's disease
HIE	Hypoxic ischemic encephalopathy
HIR	Human insulin receptor
HIRMAb	Humanized anti-insulin receptor monoclonal antibody
HIV	Human Immunodeficiency Virus
HPLC	High-pressure liquid chromatography
HR	Homologous recombination
HTS	High-Throughput Screening
HTT	Huntingtin
i.p.	Intraperitoneal
i.v.	Intravenous
I/R	Ischemic/Reperfusion

Iba1/Iba-1	Ionized Calcium Binding Adapter Molecule 1
IBA-1	Ionised calcium binding adaptor 1
ICAD	Intracranial Atherosclerotic Disease
ICAM-1	Intercellular Adhesion Molecule-1
ICH	Intracranial hypertension
ICH	The International Council for Harmonisation of Technical Requirements for Pharmaceuticals for Human Use
ICH	Intracerebral Haemorrhage
ICV	Intracerebroventricular
ID	Intellectual Disability
IgG	Immunoglobulin G
IL-1β	Interleukin-1 beta
IL-6	Interleukin-6
IL-6R	Interleukin-6 receptor
ILV	Intraluminal vesicle
IMP	Investigational medicinal product
IMPD	Investigational medicinal product dossier
IN	Intranasal
iNOS	Inducible nitric oxide synthase
Insig1	Insulin-induced gene 1
IP	Intraperitoneal
IR	Insulin receptor
IRE1α	Inositol requiring enzyme 1 alpha
Irg1	Immune responsive gene 1
IRR	Infusion related reactions
IS	Ischemic stroke
IT	Intrathecal injection
ITP	Insulin receptor transmembrane domain-binding peptide
ITZ-PM	A mixed polymeric micelle formulation of itraconazole
IV	Intravenous
IVIG	Intravenous Immunoglobulin
JAK	Janus Kinase
JC-1/TMRE	Mitochondrial Membrane Potential Dyes
LAL	Limulus amoebocyte lysate
LAM	Leukocyte adhesion molecules
LDH	Lactate Dehydrogenase
LDL	Low-Density Lipoprotein
LH	Lurasidone hydrochloride
LI	Latent inhibition
lncRNA	Long non-coding RNA
LP	Lumbar Puncture
LPS	Lipopolysaccharide
LRP1	Low density lipoprotein receptor-related protein 1
LTD	Long-term depression
mAb	Monoclonal antibody
MAPK	Mitogen-activated protein kinases
MBP	Myelin Basic Protein
MBs	Microbubbles
MCAO	Middle cerebral artery occlusion
MCI	Mild cognitive impairment
M-CSF	Macrophage-Colony Stimulating Factor

MD	Molecular dynamics
MDD	Major depressive disorder
ME	Median eminence
MePEG-*b*-PCL	Poly(ethylene glycol)-block-poly(ε-caprolactone)
MET	Magnetoencephalography
MHLW	Ministry of Health, Labour and Welfare (Japan)
mHTT	Mutant Huntingtin
MIVM	Multi-Inlet Vortex Mixer
MMP	Metalloproteinases
MMSE	Mini-Mental State Examination
MNPs	Metallic nanoparticles
MOG	Myelin oligodendrocyte glycoprotein
MOGAD	MOG Antibody-Associated Disease
mPEG-*b*-p(HPMAm-Bz	Methoxy poly(ethylene glycol)-*b*-(N-(2-benzoyloxypropyl) methacrylamide
MPO	Myeloperoxidase
MPTP	(1-methyl-4-phenyl-1,2,3,6-tetrahydropyridine)
MRI	Magnetic Resonance Imaging
mRNA	Messenger Ribonucleic Acid
MRS	Magnetic Resonance Spectroscopy
MRT	Mean residence time
MS	Multiple Sclerosis
MSA	Multiple System Atrophy
MSC	Mesenchymal stem cell
mTBI	mild traumatic brain injury
MTD	Maximum tolerated dose
mTHPG	Meta-tetra(hydroxyphenyl)chlorin
MTS/WST/XTT	Water-soluble Tetrazolium-based Assays
MTT	Methyl Thiazolyl Tetrazolium Assay
MVBs	Multivesicular bodies
MVD	Maximum valid dilution
N2B	Nose-to-Brain Delivery
NAD	Nicotinamide adenine dinucleotide
NC	Nanocapsules
NCS	Nerve Conduction Studies
ND	Not determined
NFAT	Nuclear factor of activated T cells
NF-κB	Nuclear factor kappa B
NHEJ	Non-homologous end-joining
NI	Neuroinflammation
NIH	National Institutes of Health
NINDS	National Institute of Neurological Disorders and Stroke
NIR	Near-Infrared
NIRF	Near-Infrared Fluorescence
NLC	Nanostructured lipid carrier
NLRP3	Nucleotide-binding oligomerization domain-like receptor family pyrin domain containing 3
NLS	Nuclear localisation signal
NMDA	N-methyl-D-aspartate
NMDAR	N-methyl-D-aspartate glutamate receptor
NMR	Nuclear Magnetic Resonance

NMU	Neuropeptide Neuromedin U
nNOS	Neuronal nitric oxide synthase
NO	Nitric oxide
NOAEL	No-observed-adverse-effect-level
NOX	NADPH oxidase
NP	Nanoparticle
NVA	Neurovascular array
NVU	Neurovascular unit
OBBBO	Osmotic BBB Opening
OCN	oxidized carbon nanozymes
OCT-4	Octamer-binding transcription factor 4
OGD	Oxygen glucose deprivation
OI	4-octyl itaconate
OTS	One-touch stockings of Cambridge
OVLT	Organum vasculosum of the lamina terminalis
OXOPHOS	Oxidative phosphorylation
p(CL-co-DTC)-PEG	Poly (ε-caprolactone)-*co*-poly (1,2-dithiolane-carbonate)-*b*-poly (ethylene glycol)
$P(LP_{10}$-co-$ZLL_{11})$	Poly(L-phenylalanine-co-γ-benzyl-L-lysine)
p53DM	DNA-binding motif of the tumour suppressor protein p53
PACAP38	Pituitary adenylate cyclase-activating polypeptide
PAMAM	Polyamidoamine
PAMAM	Polyamidoamine
PAN	1-(2-pyridylazo)-2-naphthol
PANX	Pannexin
PARP	Poly (ADP-ribose) polymerase
PARPi	PARP inhibitor
PBBI	Penetrating ballistic-like brain injury
PBS	Phosphate buffer saline
PCL	Poly (ε-caprolactone)
PD	Pharmacodynamic
PD	Parkinson's Disease
PDA	Polydopamine
PDGF-BB	Platelet-derived growth factor BB
PDGFR-β	Platelet-derived growth factor ß receptor
PDGF-β	Platelet-derived growth factor β
PDH	Pyruvate dehydrogenase
PDI	Polydispersity Index
pDNA	Plasmid DNA
PE	Phosphatidylethanolamine
PEG	Polyethylene Glycol
PEG	Polyethylene Glycol / PEGylation
PEG-*b*-PCC-g-DC-g-TEPA	Poly(ethylene glycol)-block-poly(2-methyl-2-carboxyl-propylene carbonate-graft-dodecanol-graft-tetraethylenepentamine)
PEG-DSPE	1,2-distearoyl-*sn*-glycero-3-phosphoethanolamine-N-[amino(polyethylene glycol)-5000
PEG-PCL	Polyethylene Glycol–Polycaprolactone
PEG-PHDCA	Methoxy polyethylene glycol cyanoacrylate-co-*n*-hexadecyl cyanoacrylate
PEG-PLA	Poly(ethylene glycol)–poly(lactic acid)
PEG-PLL	PEG-poly-L-lysine

PEI	Polyethylenimine
PEO-*b*-PBLA/PEO-PBLA	Poly(ethylene oxide)-block-poly(β-benzyl-L-aspartate)
PEO-PBLA	Poly-(ethylene oxide)-poly(β-benzyl L-aspartate)
PET	Positron emission tomography
P-gp	P-glycoprotein
PI	Propidium Iodide
PIN1	Peptidyl-prolyl isomerase
PINK1	PTEN-induced kinase 1
PK	Pharmacokinetics
PKC	Protein kinase C
PLGA	Poly(lactic-co-glycolic acid)
PM	Plasma membrane
PMAG-*b*-PAA	Poly(2-deoxy-2-methacrylamido-D-glucose) [PMAG]-*b*- poly amino acid [PAA]
PMS	Progressive MS
PNS	Peripheral Nervous System
POI	Protein of interest
PON1	Paraoxonase 1
PPDF-Hyd-DOX	Polyethylene glycol-poly (aspartate-hydrazone-DOX/ phenylalanine)~n~
PPMS	Primary progressive MS
PRP	Platelet-rich plasma
PS	Phosphatidylserine
PS80 / P188	Polysorbate-80 / Poloxamer 188
PSD-95	Postsynaptic density protein 95
PT	Prothrombin Time
PTX	Paclitaxel
PVA	Polyvinyl Alcohol
Qalb	CSF/Serum albumin
QbD	Quality-by-Design
QMS	Quality Management System
RA	Retinoic Acid
RAGE	Advanced glycation end products
RBCs	Red Blood Cells
RMT	Receptor-mediated transcytosis
RNA	Ribonucleic Acid
RNAi	RNA Interference
ROS	Reactive oxygen species
RTI	Reverse Transcriptase Inhibitor
rt-PA	Recombinant tissue plasminogen activator
RVG	Rabies virus glycoprotein
RVG	Rabies Virus Glycoprotein Peptide
SAL	Sterility assurance level
SARA	Scale for Assessment and Rating of Ataxia
SARS-CoV-2	Severe Acute Respiratory Syndrome Coronavirus 2
SAS	Subarachnoid space
SBI	Secondary brain injury
SC	Subcutaneous
SCA	Spinocerebellar Ataxia
SCA3	Spinocerebellar ataxia type 3
SCAP	SREBP cleavage-activating protein

scFv	Single-chain variable fragments
scFvs	Single-chain antibody fragment
SCI	Spinal cord injury
SDH	Succinate dehydrogenase
Sema4D	Semaphorin4D
sEV	Small extracellular vesicle
SFK	Src family kinases
SFO	Subfornical organ
SHRSP	Stroke-prone spontaneously hypertensive
siRNA	small-interfering RNA
SLN	Solid Lipid Nanoparticle
SLNP	Solid lipid nanoparticle
SMA	Spinal muscular atrophy
SNaP	Sequential Flash Nanoprecipitation
SNL	Spinal nerve ligation
SOD1	Superoxide dismutase 1
SPECT	Single-photon emission computed tomography
SPION	Superparamagnetic iron oxide NPs
SPK2	Sphingosine kinase 2
SPMS	Secondary progressive MS
SREBP	Sterol regulatory element binding protein
SSB	Single-Strand Breaks
SUS	Scanning Ultrasound
SV-40	Simian virus 40
TAM	Tumour-associated macrophages
TBI	Traumatic brain injury
TDAR	T-cell dependent antibody response
tDCS	Transcranial Direct Current Stimulation
TEER	Transepithelial/endothelial electrical resistance
TF	Transferrin protein
TfR	Transferrin receptor
TGF-β	Transforming growth factor β
TGF-β1	Transforming growth factor β1
Th17	T-helper type 17
THF	Tetrahydrofuran
TIA	Transient ischaemic attack
TLR	Toll-like receptor
tMCAO	transient middle cerebral artery occlusion
TMP	Transmembrane Pressure
TMZ	Temozolomide
TNF-α	Tumour necrosis factor-α
tPA	Tissue plasminogen activator
TREM2	Triggering receptor expressed on myeloid cells 2
TREX1	Three Prime Repair Exonuclease 1 gene
TUNEL	Terminal deoxynucleotidyl transferase dUTP Nick End Labeling
TVs	Transport vehicles
UV	Ultraviolet
VCAM-1	Vascular cells adhering to molecule-1
VEGF	Vascular endothelial growth factor
VEGF165	Vascular endothelial growth factor 165
VGEF	Vascular endothelial growth factor

VLA-4	Very late antigen-4
VLPs	Virus like particles
WHO	World Health Organization
WMH	White Matter Hyperintensities
WT	Wild type
yCD::UPRT	Yeast *cytosine deaminase–uracil phosphoribosyl transferase*
Zn	Zinc finger
ZO	Zonula occludens
α-syn	α-synuclein
γ-PGA	Poly(γ-glutamic acid)

1 Neurological Disorders and Their Impact

Jidhin Raj and Rhys Davies

1.1 INTRODUCTION

Neurology as a medical science explores the complex architecture and function of the nervous system and its interplay with human experience, following an instance of damage, or during any state of disease or disorder. This chapter provides a comprehensive outlook of Neurology's scope and limitations, emphasizing its distinctive features as a clinical discipline. As the brain is a conduit for all human experience and action, and because of the nervous system's complexity has been a major challenge to therapeutic development, diagnosis continues to be the dominant task. In short, we are concerned with where in the neural pathway damage or disease is located. What is its nature, and might it be better defined as a disordered state rather than a traditional histopathological entity?

In previous years, a small number of highly specialized neurologists were engaged largely in identifying histopathologically defined conditions of the nervous system. Especially in the United Kingdom, common presentations were managed, or perhaps we should say supported, by that very strong group of medical practitioners, the general physicians/internists. By now, proactive management decisions have become more specialized, to the point that even the commonest conditions may require specialist input. Furthermore, neural symptoms arising in the absence of discernible histopathology were often recognized as such and managed or merely accepted in community settings. The latter includes the hugely pervasive primary headache disorders and the broad range of presentations increasingly discussed under the rubric of functional neurological disorder (FND).

As in all domains of clinical practice, neurological diagnosis rests on the patient's history, the clinical examination in the consulting room and followed by investigations. The greatest of the three is history, followed by examination. However, we explore the evolving landscape of Neurology including the shift from traditional diagnostic work to interventional approaches driven by advances in imaging, laboratory data and therapeutics. This chapter aims to characterize neurological conditions while discussing diagnostic challenges and therapeutic strategies. Emphasis is placed on modern developments such as disease-modifying therapies (DMTs), neurostimulation and integration of technology with therapy. Key challenges, including drug delivery across the blood-brain barrier (BBB) and individualized therapy, are addressed. We also underscore the importance of holistic and patient-centred care, highlighting neurology's progressive role in medical practice; it may be that we are concerned with the strangest of human organs, but it is also the organ most intimately linked with human experience and identity.

1.1.1 Clinical Neuroscience and Where Neurology Fits in

Neuroscience is a relatively recent term as a title from a branch of science and title for courses of higher study. It encompasses all the fundamental biosciences as applied to the nervous system and, notably, large sections of Psychology (although of course many important subdisciplines of Psychology focus on phenomena of mind and behaviour without being concerned with the neural substrates). Clinical Neuroscience then refers to those branches of Medicine and allied

DOI: 10.1201/9781003482406-1

health professions directed towards the nervous system. This necessarily includes Neurology and Neurosurgery, concerned with the skill of undertaking surgical procedures on the nervous system and the complex decision-making that must underpin this work. Clinical Neuroscience also encompasses a range of other specialty names able to accept the 'neuro' prefix – radiology, pathology, paediatrics, genetics and rehabilitation. Special mention is required for Clinical Neurophysiology, with its expertise in electroencephalography (EEG), electromyography (EMG) and the like. Furthermore, specialists in Ophthalmology may develop into neuro-ophthalmologists, some Ear, Nose and Throat surgeons become experts in skull-base surgery and some Orthopaedic surgeons are specialists in surgical conditions of the spinal column or the peripheral nerves. Of course, psychiatrists are experts in conditions more usefully analysed in terms of mental process than neural structure, and experts in potential medical therapies that may benefit mental disorders. However, its basis in Neuroscience gains increasing importance, not least in the subspecialty of Neuropsychiatry. Clinical psychologists vie with medics to be the clinical professionals with the longest minimum period of university study and include the neuropsychologists, to conclude this long but not exhaustive list.

Neurology as a subspeciality of internal medicine, has a particular emphasis on the relationship between structure and function. Understanding the localization of function is the first step to localizing disease, to understand and manage it. The range of conditions neurologists encounter in their General Neurology practice is remarkably broad, from common disorders like migraine and stroke to chronic diseases like Parkinson's disease and multiple sclerosis (MS). Each of these diseases has distinctive presentations, yet also shares many overlapping features which require careful pattern recognition and critical thinking.

The specialty continues to evolve at a rapid pace. Advances in genetics, neuroimaging, immunology, nanotechnology have transformed both diagnosis and management. Once untreatable conditions now have DMTs and/or may benefit from specialized neurorehabilitation treatment, gene therapy or functional neurosurgical intervention. Inevitably, such specialist advances demand multidisciplinary working, with other medical specialists and allied clinical professionals. Yet the heart, and the brain, of Neurology still lie in the understanding of patients' experience of their symptoms and correlating with the map of neural wiring.

Neurology engages in a multitude of treatment strategies including prevention, rehabilitation and research. With advancements in medical science, the role of neurologists has evolved to include (Figure 1.1):

1. Preventive care: Identification of risk factors, stratification and modulation strategies for conditions such as stroke, epilepsy and migraines
2. Neurorehabilitation: Collaboration with specialists, physiotherapists and occupational therapists to help patients improve function and quality of life.
3. Research: Neurology has evolved from couch neurology to hot clinics and rapid access care for many diseases including stroke. It is in the forefront of groundbreaking research including genetic research, advanced drug therapies and innovative therapies including adaptive deep brain stimulation (DBS).
4. Technological integration: Present-day Neurology takes considerable help from advanced technology including positron emission tomography (PET), magnetoencephalography (MEG), magnetic resonance imaging (MRI) to precisely diagnose and monitor conditions.

Neurological conditions top the list as the leading cause of disability adjusted life years (DALYs) in 2021 (1). Disease of the brain's blood supply, that is, stroke is the third major contributor, and dementias are the seventh major contributor of overall death as per the WHO report in August 2024 (2). MS and migraine are the leading causes of disability in working-age adults in Western countries (3). These studies emphasize the role of neurology in public health. Indeed, public health awareness

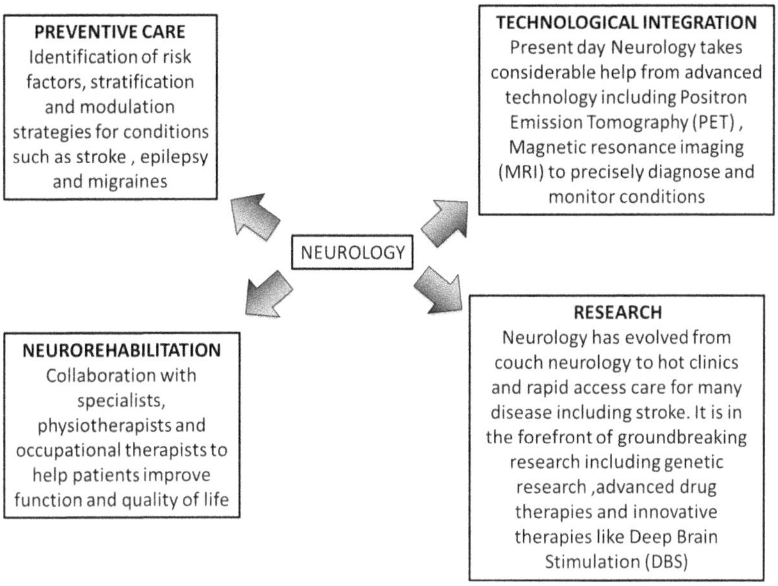

FIGURE 1.1 A wheel and spoke diagram demonstrating multidisciplinary nature of modern neurology science comprising technological advancements, multiple research fields, behavioural practices and prevention of developing neurological disorders.

is of utmost importance by promoting awareness, advocating for early interventions and developing guidelines to help to improve outcomes in these conditions.

1.1.2 Clinical Anatomy of the Nervous System

The Nervous system is the body's most complex organized system, responsible for coordinating and controlling all functions. It may be divided into central and peripheral parts, both characterized by extraordinary spatial complexity, which makes targeted pharmacotherapy very challenging. Further complexity arises from the chemistry of myriad neurotransmitters in central nervous system (CNS). Meanwhile, the number of neurotransmitters in the somatic peripheral nervous system (PNS) is limited to one, yet the pharmacological complexity of its receptor at the neuromuscular junction is a notorious complex interaction.

The nervous system has two main components: the CNS, comprising of the brain and spinal cord, and the PNS, which includes cranial nerves, spinal roots and nerves, the plexuses and their branches to the limbs and autonomic ganglia.

The CNS primarily acts as an information hub and command centre, processing sensory input and planning responses, while the PNS serves as a network to relay signals between the centre and the body as a whole. CNS and PNS may be further subdivided into functional systems for each aspect of human experience and action, including autonomic functions critical for homeostasis maintenance. Functional systems involving sensorimotor and homeostasis straddle both CNS and PNS, while the so-called higher systems of function such as memory, language, planning and awareness are confined to the CNS.

The CNS is distinctive for its surrounding structures that provide protection, including the axial skeleton of the skull and spine and the soft tissue meningeal linings. It also has protected status in the circulation, on account of autoregulation of perfusion pressure and the BBB.

In broad terms, the brain's cerebral hemispheres are the destination of sensory pathways from the periphery and the origin of motor pathways; they also form the neural substrate of most higher functions. The other main elements of the brain are the cerebellum and brainstem. The cerebellum plays a key role in coordinating and calibrating neural functions, while the brainstem connects the upper brain to spinal cord and cerebellum. It is also the origin of the cranial nerve as well as being the seat of core functions such as wakefulness and respiratory control. The human spinal cord is the conduit between brain and the limbs and trunk. While PNS includes the cranial and spinal nerves, and their branches that subserve sensory, motor and autonomic/homeostatic functions. While the molecular characteristics of the cranial nerves (especially nerves III to XII) make them vulnerable to diseases and damage that also affect the spinal and limb nerves. However, their location makes them vulnerable to diseases of the cranium, and brainstem.

1.1.3 The Process of Neurology

Neurology involves analysis of symptoms through targeted history taking focusing on unravelling the anatomical aspects of a problem. As always, history taking initiates the process, characterizing the symptoms with onset, progression and associated symptoms taking into consideration the overall context. These are then localized anatomically using structured examination techniques which either challenge or corroborate the initial hypothesis. Localization – the 'where' of the problem – is central to neurology, which then helps to clarify the 'what', 'why' and 'how' of the underlying condition.

The anatomical distribution of conditions within the nervous system can be broadly categorized into three groups – focal conditions such as tumour or trauma, which involve damage to particular structures of the nervous system. For example, a patient presenting with right-side hemiparesis and facial weakness may have a lesion localized to the corticospinal tract on the left side of the brain. Multifocal or dispersed conditions, on the other hand, often arise from a humoral or infiltrative (primary inflammatory/immune disease, infection or neoplasia). Commoner still are degenerative conditions, often sporadic but sometimes with a specific genetic or metabolic cause. Thirdly, various reactive disorders (migraine perhaps being the commonest) are not associated with significant histopathological abnormality but involve dysfunction of neural pathways. The Art of Neurology consists of understanding these three broad spatial categories of conditions, which map onto the wiring of the nervous system together with an awareness of the tempo inherent to various forms of pathology – trauma fast, tumours slower, degenerative disease slowest, membrane disorders intermittent.

There must also be an element of expecting the unexpected, as dispersed disease processes may have patchy effects and chronic processes may present through acute decompensation. Symptoms and signs in Neurology typically reflect the specific site of involvement within the nervous system, which can lead to similar clinical features irrespective of the underlying pathology. For instance, polyneuropathy refers to disease affecting the peripheral nerves in a diffuse manner and often has sensory disturbance at the feet as its initial symptom. This may occur from the microvascular effects of diabetes, the toxicity of ethanol, deficiency of vitamin B12, HIV infection, autoimmunity, genetic predisposition or from presumed multifactorial causation – idiopathic peripheral neuropathy. Moreover, similar symptoms can arise from damage at differing anatomical locations such as foot drop resulting from injury to the fibular near the knee or impingement of the fifth lumbar nerve root in the low back. An extreme version of this challenge is seen in FND, where symptoms that mimic effects of histopathological neural damage occur in cases where no damage is present, as a result of aberration in conscious interpretation of body sensations and or of initiating actions. Of course, FND is complex, but it may be viewed in continuity with the reality that nearly all patients' clinical presentations include some details that do not contribute directly to the diagnosis.

Finally, in listing distinct concepts in pathophysiology, we must recognize that several concepts may co-exist in an individual patient's experience. Taking MS as an example, which includes focal damage from inflammation, but also includes dispersed effects. It includes relapses, with inflammatory pathology arising over a period of days, but also a propensity for the neural pathways of the CNS to deteriorate faster than in normal aging. It is a notoriously difficult but important challenge for people with MS, and their doctors, to define which neural symptoms indicate an inflammatory relapse and which are general effects of background MS, altered vigilance in the presence of a serious diagnosis and so on.

1.1.4 The Changing Landscape of Neurology in Clinical Practice

The importance of Neurology within Medicine or Internal Medicine varies significantly across healthcare systems. In the UK, Neurology has historically been a relatively small specialty, with most neurologists based in large hospitals alongside neurosurgeons. A hub and spoke model have been an efficient system whereby Neurology expertise is provided two or three times per week to district hospital by visiting specialists.

Until recently, the responsibility of managing common and acute neurological presentations, Stroke disease especially, fell on senior physicians in peripheral hospitals who had undergone extensive experience in Internal Medicine generally. This was partly due to the limited therapeutic options available for neurological patients, where diagnosis rather than intervention remained the primary focus. In parallel, geographically widespread memory clinics led predominantly by psychiatrists have historically taken charge of both biomedical and mental health aspects of dementia. This division of responsibility reflected the way neurological services developed in the UK, distinguishing it from other healthcare models where a larger number of neurologists participate more directly in acute and community settings.

We now stand at the threshold of a major shift in the approach of Neurology, driven by significant advances in pharmacotherapy for conditions ranging from acute stroke to MS, to neurodegenerative dementia. The increasing availability of disease-modifying treatments is pushing neurology beyond its traditional diagnostic role to an active interventional one. The shift is already underway, with neurologists participating in rapid access or 'hot' clinics and ward rounds of emergency departments, while still maintaining their outpatient practice for long term, and sometimes rare, conditions. The expansion of mechanical thrombectomy services for occlusive cerebrovascular disease, the use of disease-modifying treatments in MS, monoclonal antibodies in Alzheimer's dementia and the emergence of new genetic and personalized medicines will continue to reshape the specialty. As evolution continues, the profession must adapt to new challenges including workforce expansion, training reforms and redefining its liaison with other specialties in an increasingly intricate healthcare landscape.

1.1.5 Clinical Conditions Affecting the Nervous System

The nervous system with all its complexities is susceptible to a wide range of disorders. Diseases can arise from myriad causes including genetic mutations, infections, metabolic imbalances, vascular events or degenerative causes. Due to the complex nature of neurological structures and their intricate functioning, patients often present with overlapping symptoms which require a meticulous approach to diagnosis and management. Understanding these classifications helps in narrowing down differential diagnoses and guiding appropriate therapeutic interventions.

1.2 CATEGORIES OF NEUROLOGICAL DISORDERS

1.2.1 Vascular Disorders

Vascular disorders arise either from impaired blood flow or loss of vessel wall integrity in the CNS. The most common forms are ischemic and haemorrhagic strokes. In ischaemic strokes, a clot or a atheromatous plague blocks an artery, causing disruption of blood flow leading to infarction (4).

A transient ischaemic attack (TIA) is a short ischaemic event without permanent damage, heralding future stroke. Haemorrhagic strokes include intracerebral haemorrhage (~10% strokes) and sub arachnoid haemorrhage (~3%) resulting from rupture of vessels or aneurysms (4). Both of the stroke types, even though varied in their presentation, share common risk factors such as hypertension, diabetes and hypercholesterolemia. Clinical presentation depends on the vascular territory (e.g. brainstem nuclei, descending motor pathways, language centres of the cerebrum) and the diagnosis is clinched by imaging. Non-contrast CT is used acutely to distinguish haemorrhage and most common imaging used in stroke intervention protocols. MRI (diffusion weighted imaging) is sensitive for early ischaemia. Vascular imaging is also undertaken to evaluate for plaques, emboli or aneurysms (in cases of hemorrhage). The treatment depends on the type of stroke. Acute ischaemic stroke may be treated with intravenous tissue plasminogen activator (tPA) within a time window. Latterly, the large vessel occlusions may be treated with endovascular intervention, permitting mechanical thrombectomy. Haemorrhagic stroke management includes intensive blood pressure management and if raised intracranial pressure occurs surgical evacuation or decompression may be considered.

Long-term management of vascular disorders includes risk factor modification, antiplatelet and anticoagulant therapy as indicated and statins.

1.2.2 Inflammatory and Autoimmune Disorders

This category is the umbrella term for diseases that cause immune-mediated damage, often presenting with relapsing or progressive neurological symptoms. The classical example is MS, a chronic inflammatory disease of the CNS. The typical presentations of MS include optic neuritis, transverse myelitis (of the spinal cord) and focal lesions of the brain separated in time and space. Clinical diagnosis is based on consensus criteria (like those of McDonald) and we cannot rely on one single data point (5). Acute relapses are treated with high-dose corticosteroids, while disease-modifying drugs (interferon beta, glatiramer acetate, ocrelizumab, natalizumab) reduce relapse rates and may attenuate progression. More recently a group of patients has been separated from the main category of MS as we now know that specific autoimmune processes are at play-Neuromyelitis Optica spectrum disorder. Other autoimmune encephalopathies include limbic encephalitis (anti-NMDA receptor, LGI1 and other autoantibodies). Their presentation may include seizures, memory loss, psychotic behaviour. They also show MRI and EEG abnormalities. Diagnosis requires cerebrospinal fluid analysis, serological testing and imaging including MRI and PET imaging. Acute disseminated encephalomyelitis (ADEM) is a monophasic postinfectious demyelinating encephalopathy seen in children, which can present with multifocal lesions and responds to steroids. Guillain-Barré syndrome is the best-known autoimmune disease of the PNS (6).

1.2.3 Neurodegenerative Disorders

It is characterized by progressive neuronal loss and abnormal protein accumulation, leading to progressive neurological decline. Alzheimer's disease most often presents with progressive impairment of recall with MRI features of brain atrophy especially in the hippocampal formation. Diagnosis is clinical with supporting biomarkers like MRI features, PET imaging and CSF assays for amyloid-beta A4 peptide and tau protein. Current treatment for Alzheimer's is predominantly symptomatic, to increasing brain acetylcholine levels. A range of monoclonal antibodies targeting protein aggregation is in the pipeline. Parkinson's disease involves neuronal loss in the deep cerebral nuclei and accumulation of Lewy Body deposits (7). Parkinson's disease presents with resting tremor, rigidity, bradykinesia and postural instability. Diagnosis is mainly clinical. Therapies focus on boosting dopamine (levodopa, peripheral dopa decarboxylase inhibitors, MAO-B inhibitors, COMT inhibitors, dopamine agonists) and other medications. Advanced cases may benefit from newer therapies like DBS, or drugs delivered by novel means, for example, subcutaneous apomorphine infusion or jejunal administration of levodopa. Amyotrophic lateral sclerosis is a degenerative disease that targets the motor neurons (8), leading to weakness and other motor features. Diagnosis is based on clinical assessment often supplemented by EMG. Treatment options are more disappointing than for Parkinson's disease but may include riluzole, which gives modest benefits. Unfortunately, treatment options for other neurodegenerative diseases, sporadic or familial, are even more sparse. Such incurable disorders include Frontotemporal dementia (young onset personality disorder and language disorder), Progressive supranuclear palsy, multiple system atrophy, corticobasal degeneration (atypical Parkinsonism syndromes with poor L-dopa response) and spinocerebellar ataxias. Management of these conditions is generally supportive with physical therapy, medications for symptomatic relief and multidisciplinary care.

1.2.4 Infectious Disorders

Infectious neurological disorders maybe caused by viruses, bacteria, parasites or prions that invade the system. The clinical deficits correlate with the structures involved. For example, meningitis causes headache and fever whereas encephalitis is associated with altered mental status and seizures. Herpes simplex virus is the most common cause of encephalitis (9). CSF PCR is diagnostic, and treatment of choice is IV acyclovir. Bacterial meningitis is treated with antibiotics and steroids. Tuberculosis meningitis should be considered in a subacute course in endemic areas. It is treated with anti-tuberculous therapy with steroids (10). Chronic granulomatous infections caused by fungi (cryptococcal, histoplasma) can cause space-occupying lesions and meningeal disease. Its detection is by CSF studies (India Ink, PCR) or tissue biopsy. Brain abscesses (from otitis, sinusitis or hematogenous spread) are diagnosed by MRI or CT showing ring-enhancing lesions. Management is through surgical drainage with antibiotics. Prion diseases cause rapidly progressive dementia with myoclonus, CSF markers (14-3-3 protein, RT-QUIC assay) and EEG with periodic complexes.

1.2.5 Neoplastic Disorders

Neoplastic diseases include primary tumours of the brain or spinal cord and metastatic lesions. Primary brain tumours range from relatively benign (e.g. meningioma, schwannoma) to malignant (glioblastoma multiforme, high-grade glioma). Secondary metastasis (from lung, breast, melanoma, etc.) is common in adults (11). The common symptoms include headaches, seizures, focal deficits or signs of raised intracranial pressure. Diagnosis is mainly by MRI with contrast followed by biopsy. Treatment usually involves surgical resection when feasible followed by radiotherapy/chemotherapy. Meningeal carcinomatosis and paraneoplastic syndromes are other tumour-related disorders, respectively, from infiltrative metastases or by triggering immune damage. With the advent of serology, our precision in diagnosing paraneoplastic syndromes have improved but the primary tumour is often elusive. The tumour most classically associated with paraneoplasia is small cell lung carcinoma and chemotherapy for the underlying neoplasm is the most definitive treatment where possible.

1.2.6 Metabolic and Toxic Disorders

Metabolic and toxic disorders form a diverse group. They include hepatic encephalopathy (failure of the liver to clear nitrogenous toxins) (12), Wernicke-Korsakoff syndrome encephalopathy from thiamine deficiency sometimes from malnutrition in the context of high ethanol intake (13) and osmotic demyelination syndrome (usually iatrogenic when electrolytes are corrected too rapidly). Neurologic toxins include heavy metals (e.g. lead encephalopathy, wrist drop), organophosphates leading to cholinergic crisis and chronic ethanol exposure (leading to a range of adverse effects in the central and peripheral systems). Metabolic disorders can also result from inborn errors of metabolism. These include Wilson's disease – (copper accumulation causing hepatic and neuropsychiatric symptoms) and mitochondrial disorders affecting different parts of the neural axis.

1.2.7 Functional Neurological Disorders

Functional neurological disorders are a new subspeciality of neurology concerned with disorders which produces neurological symptoms not associated with structural disease (14). Patients experience symptoms such as disability (weakness, tremor, sensory disturbances, altered responsiveness) but diagnostic testing (imaging, EEG, MRI) show no lesion. The key to diagnosis is the pattern recognition and presence of clinical signs that are not compatible with known pathology. For example, non-anatomical sensory loss, which does not correspond to any dermatomal distribution is usually functional in nature. Treatment emphasizes education to understand that symptoms may occur without disease, usually involving a multidisciplinary approach. Addressing coexisting anxiety, depression or trauma is often important. Engagement in therapy is required for beneficial effect though some cases are chronic and require long-term support.

1.2.8 Genetic and Hereditary Disorders

Genetic disorders of the nervous system are caused by inherited mutations. Inheritance patterns include autosomal dominant, autosomal recessive and X-linked. Huntington's disease (HD) is an AD trinucleotide repeat disorder (HTT gene CAG expansion) causing progressive chorea, psychiatric manifestations and dementia in adulthood (15). Spinal muscular atrophy (SMA) is AR and results from SMN1 gene deletions on chromosome 5. It causes damage to anterior horn cells leading to hypotonia and weakness in infants and children. Therapies include SMN-enhancing drugs (nusinersen, risdiplam) and gene therapy (onasemnogene) (16).

Genetic neuropathies include Charcot-Marie-Tooth disease. This is an umbrella term that encompasses many separate mutations, all generating a similar clinical picture (CMT1a, PMP22 duplication, is inherited as a dominant trait, while other forms are AR). Mitochondrial disease and metabolic gene disease (MELAS, Leigh Syndrome) are examples of metabolic disease with genetic basis.

Genetic disorders are identified by intricate family history, clinical pattern recognition (early onset, chronicity and multi-system involvement, family involvement) and confirmed by molecular testing. Genetic counselling is essential and some now have treatments (enzyme replacement in certain lysosomal storage diseases).

1.2.9 Traumatic Neurological Disorders

Traumatic neurological disorders occur as a result of mechanical injury to the nervous system and comprises of traumatic brain injury (TBI), spinal cord injury (SCI) and concussion. TBI ranges from mild (concussion) to severe (diffuse axonal injury, intracranial haemorrhage). Diffuse axonal injury is characterized by widespread axonal shearing injury. SCI can result in tetraplegia or paraplegia. Mechanisms include fracture dislocation or hyperflexion/hyperextension injuries. Initial management follows spine stabilization or surgery to decompress the cord. Chronic SCI complications include autonomic dysreflexia, spasticity and central pain. Rehabilitation is critical to maximize function and involves physiotherapy, occupational therapy, mobility aids and management of neurogenic bladder/bowel. Pharmacotherapy has a limited role at the moment of trauma but may contribute to neuroprotection in the period of hyperacute care, as well as in rehabilitation.

1.2.10 Peripheral Nervous System Disorders

PNS disorders affect the motor neuron in the cord or brainstem, the nerve root, plexus, peripheral nerve, neuromuscular junction or muscle.

Neuropathies are diseases of the peripheral nerves themselves and can be acute or chronic, generalized or focal. Guillain-Barré syndrome (GBS) is an acute inflammatory condition affecting the peripheral nerve myelin sheet in which patients develop rapidly progressive symmetric ascending weakness, areflexia, sensory disturbances and aetiology is likely to be post-infective. Treatment options include IVIG or plasma exchange. Common metabolic neuropathies include diabetic polyneuropathy with variable presentations including stocking-glove pattern of sensorimotor neuropathy with pain, tingling and loss of ankle reflexes. Poor glycaemic control drives its development. Treatment options include glucose control and neuropathic pain relief.

Autoimmune neuromuscular junction disorders (e.g. Myasthenia gravis) cause fatigable weakness including ptosis, diplopia and bulbar symptoms. It is treated with acetylcholinesterase inhibitors, immunotherapy and thymectomy.

Muscle disease also comprises a part of the disease spectrum with aetiologies varied from autoimmune, genetic and drug-associated. Strong clinical suspicion with supportive laboratory and electrophysiological results clinches the diagnosis in these cases. Sometimes biopsy of the affected part is undertaken when diagnosis is uncertain.

1.2.11 Developmental and Paediatric Neurological Disorders

This category includes disorders arising from infancy and childhood. Epileptic syndromes are the prime examples. West syndrome is defined by triad of epileptic spasms, hypsarrhythmia on EEG and developmental regression or arrest. Treatment with high-dose ACTH or vigabatrin (for tuberous sclerosis-related cases) can improve outcomes.

Cerebral palsy (CP) refers to a heterogeneous group of non-progressive motor deficits due to brain injury or malformation in the perinatal period that subsequently affects developmental milestones. Children with CP have spasticity, dyskinesia, hyperreflexia and often cognitive impairment depending on the areas of affection (17). MRI typically involves white matter injury and management is multi-disciplinary. Early interventional programmes (physical, speech, occupational therapies) capitalize on developmental plasticity to maximize function.

Metabolic and genetic conditions affecting children (e.g. phenylketonuria, mitochondrial disorders, leukodystrophies) can also present in development and some are identified by newborn screening or early imaging. Multidisciplinary care (neurology, genetics, rehabilitation, education) is the need for management. These disorders also underscore the need for a family-centred approach in combining medical, educational and social care for neurological wellbeing.

1.3 THERAPEUTIC APPROACHES IN NEUROLOGY

Neurology has undergone a huge transformation in recent years shifting the focus from a diagnostic specialty to a more therapeutic one with an expanding arsenal of therapies. Neurologic treatments range from symptom management to disease-modifying interventions. Breakthroughs in neuropharmacology, especially in immunology, genetics and intervention have greatly shaped this field providing patients with improved outcomes. The neurologists today need to integrate these diverse therapeutic approaches along with being up to date with new therapies to optimize patient care.

Treatment in neurology can be broadly divided into two categories: symptomatic treatments, which focus on modification of the manifestations whereas the disease process goes unaltered. For example, levodopa helps to control the motor symptoms in Parkinson's disease and lamotrigine helps to control seizures in focal epilepsy. In contrast, DMTs aim to alter progression of the disease. In MS, the introduction of monoclonal antibodies such as ocrelizumab has revolutionized treatment by reducing relapse rate and delaying disease progression. Similarly, gene silencing therapies are emerging as breakthrough in genetic diseases like SMA.

Interventional treatments in neurology have become a major part of therapeutics, especially in the subspecialties of stroke and vascular neurology. These techniques have provided solutions where pharmacological agents alone are insufficient. Mechanical thrombectomy for large vessel occlusions in stroke have drastically improved stroke outcomes when performed within a time window. DBS, used in conditions like Parkinson's disease and dystonia has revolutionized motor symptom management and improved quality of life for affected individuals. Similarly intrathecal drug delivery systems such as baclofen pumps for spasticity management highlight modern neurological interventions. With the advancements in therapeutics, the integration of neurosurgical, radiological and neurophysiological expertise is expected to further its horizons.

Neurology has always embraced a holistic approach, recognizing the interplay between neurological disorders, psychological and lifestyle factors. Cognitive rehabilitation, physiotherapy, occupational therapy and psychotherapy play crucial roles in managing chronic neurological conditions, for example, multidisciplinary approaches in neurodegenerative disorders address various mobility, cognitive and other concerns of the patient. Dietary modifications in epilepsy further highlight the importance of lifestyle interventions. Additionally digital health technologies, including wearable devices and artificial intelligence (AI)-driven support, have expanded personalized care. The future of neurology will likely see integration of these multimodal approaches tailoring care to patient-centred needs.

1.3.1 Challenges in Neurological Pharmacotherapy

Despite the advances in therapeutics, the treatment of neurological disorders continues to pose complex challenges. The variable efficacy of many drugs across individual patients often compels a process of trial and error to identify the most appropriate treatment regimen. Conditions such as treatment-resistant epilepsy and refractory migraine draw attention to the need for a more personalized approach to therapy. The integration of genetic and biomarker testing tends to promise a more personalized approach and predicting individual responses to specific therapy thereby optimizing treatment strategies. Another significant obstacle is the BBB, a physiological barrier that restricts the movement of many substances including therapeutic agents into the brain. Researchers are actively pursuing novel drug delivery systems including nanotechnology and targeted delivery to tide over this challenge. While further research is warranted, the ongoing progress in neuropharmacology provides increasing hope for effective management, reducing the rate of progression and even prevention of neurological disorders.

1.3.2 Delivery of Drugs in the Nervous System

Drug delivery to the nervous system presents unique challenges due to the distinctive nature of the BBB. The BBB serves as a highly selective membrane preventing harmful substances from entering the brain but at the same time making it difficult for the passage of many medications to reach the target. This system poses a significant hurdle in treating many neurological conditions like neurological infections, neurodegenerative conditions and tumours. Traditional oral and intravenous drugs struggle often to cross the BBB effectively thereby increasing the need for larger doses and frequent dosing. This impairs the tolerability for the patients and increases the side effects. Alternative drug delivery methods including nanoparticle-based carriers, prodrug strategies and direct administration have found a footing in drug research for this reason. Researchers are continually exploring ways to bypass the barrier such as using focused ultrasound to temporally open the BBB, allowing more seepage leading to increased efficiency.

In some cases, local delivery of medicines is the best way to achieve desired therapeutic effects while minimizing systemic side effects. For example, intrathecal drug administration, where the medication is injected directly into CSF, is commonly used for conditions like spasticity (e.g. intrathecal baclofen) or cancer-related pain. Similarly, DBS provides targets for Parkinson's disease and dystonia. Another route which is considered to bypass the BBB barrier is intranasal drug delivery where the medication is absorbed through the olfactory pathway. Such localized approaches help to improve drug efficiency, giving focused therapy and thereby reducing side effects.

A major challenge in drug delivery for neurological patients is ensuring proper medication intake particularly in patients with difficulty in swallowing. Many neurological disorders including stroke and Parkinson's disease impair the ability to swallow, making oral medications difficult or even dangerous. This necessitates alternative formulations like liquid suspensions, transdermal patches or sublingual medications to ensure treatment is received safely. Poor adherence to treatment due to dysphagia remains a major issue, highlighting the need for accessible drug formulations for patients with neurological diseases.

Another challenge is regarding treatment compliance, especially in conditions with sporadic symptoms such as epilepsy and migraine. People may be reluctant to take daily medication when they feel well, leading to inconsistent levels and reduced efficacy. To address this, long-acting formulations have been considered like CGRP inhibitors for migraine and long-term antipsychotics. Additionally, patient education and digital health interventions such as reminder apps and alarms play an important role in improving adherence. As neurological advances a key focus appears to be refining drug delivery methods to enhance accessibility, effectiveness and patient compliance.

1.3.3 Newer Modalities of Treatment

Normally, small molecule therapies are designed to cross BBB, allowing for modulation of neurotransmitter activity, enzyme function or receptor signalling. Examples include levodopa for Parkinson's disease, acetylcholinesterase inhibitors for Alzheimer's and riluzole for amyotrophic lateral sclerosis. These therapies aim to restore neurotransmitter balance of reduce protein accumulation. Advancements continue to refine formulations for improved efficacy and lesser side effects.

Monoclonal antibodies have revolutionized neurological treatment by targeting specific pathological pathways. In MS, Ocrelizumab and Natalizumab modulate immune response to reduce inflammation. Anti-amyloid antibodies like lecanemab and aducanumab have been developed for Alzheimer's although their efficiency has been a topic of discussion. These therapies target specific molecules or antibodies thus representing the shift towards precision medicine.

Gene therapy is another area of focused medicine. It holds great promise for hereditary and degenerative diseases by correcting the underlying genetic defects. This therapy requires a functional gene to be placed into cells using viral vectors such as adeno-associated viruses (AAVs). In SMA, gene therapy with onasemnogene replaces the defective SMN1 gene. CRISPR is also a new gene editing technique that holds promise for the future especially in gene-related neurological diseases.

Neurostimulation therapies offer treatment options by directly modulating neural circuits. DBS is widely used for various movement disorders like Parkinson's disease. Essential tremor and dystonia where controlled impulses are given to specific targets. Other forms of neuromodulation include transcranial magnetic stimulation for depression and epilepsy as well as vagus nerve stimulation for refractory epilepsy. Advances in adaptive and closed-loop neuromodulation are areas of ongoing research and continue to improve precision and effectiveness.

Stem cell therapy is a promising avenue in the field of neurodegeneration. It aims to replace the defective neurons and restore lost function. Mesenchymal stem cells and induced pluripotent stem cells are being investigated for Parkinson's disease, ALS and stroke recovery. There are still various challenges and hurdles to cross including techniques to ensure cell survival, integration and avoid rejection.

Advancements in genomics, biomarkers and AI are paving the way for personalized treatments. Pharmacogenomics allows tailoring drug choices based on individual genetic makeup and optimizing efficacy. Biomarker-driven approaches such as CSF or PET imaging markers for Alzheimer's disease help refine diagnosis and treatment selection. The future of neurology is shifting towards precision medicine offering individualized strategies for better patient outcomes.

1.4 CONCLUSION

As Medicine has evolved, the complexity of the nervous system has limited the range of treatments available to treat those affected by neural diseases. While other medical specialties have long pursued interventions and cures, Neurology traditionally has had to focus on diagnosis. All safe neurologists required sharp analytical skills to spot the underlying condition, the best achieved a holistic approach to communicate diagnosis in as affirmative a manner as the truth allowed, with respect for the uncertainties of disease and of life itself.

Drug development and other advances have widened our therapeutic horizons. However, there remain many neurological disorders without effective pharmacotherapy or any other targeted treatment. Indeed, a fundamental aspect of Neurology is judiciously to avoid inappropriate use of drugs and to guide patients by other means through the challenges of their disease.

The future of Neurology is promising, with expanding treatment options that target different aspects of the pathology. However, new therapies bring with them new challenges – most especially to ensure that these treatments are used effectively and safely – for the right patient at the right time. Reflecting on the logistics, we may hope that effective treatments reduce disability and long-term symptoms – and turn chronic conditions into acute ones. In so doing neurologists should become available to take on the new work.

While the expanding landscape of pharmacological and other treatments holds great promise, the art of Neurology will always require balancing scientific advancements in Pharmacology and other fields with patient-centred care. By combining innovative treatments with clarity as to each patient's unique journey, neurologists continue to make a meaningful difference even in the absence of a definitive cure.

REFERENCES

1. Steinmetz JD, Seeher KM, Schiess N, Nichols E, Cao B, Servili C, et al. Global, regional, and national burden of disorders affecting the nervous system, 1990–2021: A systematic analysis for the global burden of disease study 2021. Lancet Neurol. 2024;23(4):344–81.
2. The top 10 causes of death [Internet]. 2024. [cited 2025 Jan 28]. Available from: https://www.who.int/news-room/fact-sheets/detail/the-top-10-causes-of-death
3. Kingwell E, Marriott JJ, Jetté N, Pringsheim T, Makhani N, Morrow SA et al. Incidence and prevalence of multiple sclerosis in Europe: A systematic review. BMC Neurol. 2013;13:128.
4. Stroke – Causes and Risk Factors | NHLBI, NIH [Internet]. 2023 [cited 2025 May 11]. Available from: https://www.nhlbi.nih.gov/health/stroke/causes
5. Filippi M, Bar-Or A, Piehl F, Preziosa P, Solari A, Vukusic S et al. Multiple sclerosis. Nat Rev Dis Primer. 2018;4(1):1–27.
6. van den Berg B, Walgaard C, Drenthen J, Fokke C, Jacobs BC, van Doorn PA. Guillain-barré syndrome: Pathogenesis, diagnosis, treatment and prognosis. Nat Rev Neurol. 2014;10(8):469–82.
7. Armstrong MJ, Okun MS. Diagnosis and treatment of parkinson disease: A review. JAMA. 2020;323(6):548–60.
8. van Es MA, Hardiman O, Chio A, Al-Chalabi A, Pasterkamp RJ, Veldink JH et al. Amyotrophic lateral sclerosis. Lancet Lond Engl. 2017;390(10107):2084–98.
9. Whitley RJ. Herpes simplex encephalitis: Adolescents and adults. Antiviral Res. 2006;71(2–3):141–8.
10. Marais S, Thwaites G, Schoeman JF, Török ME, Misra UK, Prasad K et al. Tuberculous meningitis: A uniform case definition for use in clinical research. Lancet Infect Dis. 2010;10(11):803–12.
11. Price M, Ballard C, Benedetti J, Neff C, Cioffi G, Waite KA et al. CBTRUS statistical report: Primary brain and other Central nervous system tumors diagnosed in the United States in 2017–2021. Neuro-Oncol. 2024;26(Supplement_6):vi1–85.
12. Hazell AS, Butterworth RF. Hepatic encephalopathy: An update of pathophysiologic mechanisms. Proc Soc Exp Biol Med. 1999;222(2):99–112.
13. Wijnia JW. A Clinician's view of wernicke-korsakoff syndrome. J Clin Med. 2022;11(22):6755.
14. Stone J, Carson A. Functional neurologic disorders. Contin Minneap Minn. 2015;21(3 Behavioral Neurology and Neuropsychiatry):818–37.

15. Walker FO. Huntington's disease. Lancet Lond Engl. 2007;369(9557):218–28.
16. Giess D, Erdos J, Wild C. An updated systematic review on spinal muscular atrophy patients treated with nusinersen, onasemnogene abeparvovec (at least 24 months), risdiplam (at least 12 months) or combination therapies. Eur J Paediatr Neurol. 2024;51:84–92.
17. (PDF) Proposed definition and classification of cerebral palsy, April 2005 - Introduction. ResearchGate [Internet]. [cited 2025 May 15]; Available from: https://www.researchgate.net/publication/231747224_Proposed_definition_and_classification_of_cerebral_palsy_April_2005_-_Introduction

2 Physiological Barriers of Targeting the Diseased Brain

Hakan Sahin

2.1 INTRODUCTION

The central nervous system (CNS) relies on specialized protective barriers that regulate molecular exchange and preserve the delicate environment required for neuronal function. In the adult brain, different types of interfaces have been identified, each contributing in distinct but complementary ways to homeostasis [1, 2]. Recognition that such barriers exist dates to the late 19th and early 20th centuries. Paul Ehrlich (1854–1915) observed that systemically injected dyes stained most organs but not the brain, while Edwin E. Goldmann (1862–1913) later showed that a dye placed directly into cerebrospinal fluid (CSF) readily labeled brain tissue, reasoning that a vascular barrier, rather than the tissue itself, accounted for this selective exclusion [3–5]. Among the various protective restrictions of the CNS, the blood-brain barrier (BBB) is the most prominent, while additional barriers such as the blood-cerebrospinal fluid barrier (BCSFB) [6], the meningeal barrier [7], the specialized interfaces of the circumventricular organs (CVOs) [8], and the glia limitans [9] are essential for maintaining neural homeostasis. Acting together, these structures form a multilayered defense that shields the brain from harmful influences, regulates communication between blood and neural tissue, and preserves the precise microenvironment needed for CNS function.

In this chapter, the principal types of brain barriers will be introduced, with emphasis on how they restrict molecular flux at the cellular level. The developmental formation of these barriers and their transition into the adult configuration will also be described. Finally, an overview of barrier alterations under different conditions will be provided as a foundation for understanding their role in disease.

2.2 BLOOD-BRAIN BARRIER

The BBB is a highly specialized, semi-permeable interface that protects the CNS by regulating the exchange of molecules between the bloodstream and brain tissue. It is primarily formed by brain microvascular endothelial cells (BMECs), which line the cerebral capillaries and are sealed together by extensive tight junctions. These cells, together with pericytes, astrocytic end-feet, basement membrane components, neurons, and extracellular matrix, constitute the neurovascular unit (NVU). This multicellular system is critical for maintaining CNS homeostasis by shielding neural tissue from harmful substances while simultaneously supplying nutrients essential for neuronal function [6, 10]. The concept of the BBB has evolved significantly since its initial discovery. In the late 19th century, Paul Ehrlich demonstrated that systemically injected dyes did not penetrate the brain, although he did not attribute this phenomenon to specialized vascular properties. Subsequent observations by Lewandowsky [11], Biedl & Krause [12], and others [13] strengthened the hypothesis that cerebral vessels restrict the passage of many compounds. Later, Goldmann's studies with trypan blue highlighted the importance of the choroid plexus and CSF in regulating access to the brain [3]. The term "blood–brain barrier" was first introduced in 1921 by Stern and Gautier, who described it as a protective system that both excluded harmful molecules and selectively permitted the entry of compounds necessary for metabolism [14]. The advent of electron microscopy in the 1960s finally localized the BBB to the endothelial lining of cerebral capillaries, with Reese and

DOI: 10.1201/9781003482406-2

Karnovsky demonstrating that tight junctions between endothelial cells form the structural basis of the barrier [2, 15].

Physiologically, the BBB functions through several complementary mechanisms. Its physical barrier is maintained by tight and adherens junction proteins, which restrict paracellular flux, while a low rate of vesicular transport minimizes uncontrolled transcellular passage. In addition, its metabolic barrier relies on a repertoire of enzymes capable of degrading or inactivating xenobiotics and therapeutic molecules before they can reach neural tissue. Selectivity is further refined by a system of influx and efflux transporters, which regulate the movement of nutrients such as glucose, amino acids, and nucleosides into the brain, while simultaneously extruding toxic metabolites and many pharmacological agents back into the circulation [2, 10].

The structural and functional integrity of the BBB is essential for preserving the stable microenvironment required by neurons and glia. By maintaining ionic homeostasis, the barrier ensures proper synaptic transmission and prevents excessive crosstalk between neurotransmitter pools in the central and peripheral compartments. At the same time, it provides immune surveillance while minimizing inflammatory responses that could otherwise compromise neural function. However, this protective selectivity also presents major challenges for therapy. More than 98% of small-molecule drugs and virtually all macromolecular therapeutics fail to cross the BBB under physiological conditions. Only small lipophilic molecules, generally with molecular weights under 400–600 Da, can diffuse passively, and even then with significant limitations [16].

Endothelial cells are widely regarded as the central anatomical and functional component of the BBB. They line the cerebral microvessels and interact with other cellular partners of the NVA to establish and maintain barrier integrity. Compared with peripheral endothelial cells, BMECs display distinct morphological and functional specializations that account for the unique selectivity of the barrier. Their barrier-forming characteristics are not innate but result from continuous communication with surrounding astrocytes, pericytes, and neurons, which provide essential cues for their differentiation and maintenance. Morphologically, BMECs are joined by extensive junctional complexes that tightly seal the intercellular cleft. Tight junctions and adherens junctions together create a continuous barrier that restricts the passage of water, ions, and solutes while preserving the polarity of endothelial cells. These structures also generate high transendothelial electrical resistance, measured at approximately 1800 $\Omega \cdot cm^2$, which is far greater than that of peripheral capillaries. Endothelial cells in the BBB lack fenestrations, display a low rate of vesicular transport, and are enriched in mitochondria, adaptations that support active transport processes and reduce nonspecific permeability [17, 18].

Tight junctions are the principal structural and functional components responsible for limiting paracellular diffusion and maintaining BBB homeostasis. Also known as occluding junctions, they function as gates and fences by sealing the interendothelial cleft and preventing lateral diffusion of membrane proteins and lipids. These junctions are composed of multiple transmembrane proteins, the most notable of which are claudins, occludin, and junctional adhesion molecules, supported by cytoplasmic proteins such as zonula occludens (ZO)-1, ZO-2, and ZO-3 that link them to the cortical actin cytoskeleton. Claudin-5 is the most abundant isoform in BMECs and is indispensable for barrier selectivity; deletion of claudin-5 increases permeability to small molecules. Occludin also plays a regulatory role, as changes in its phosphorylation state modulate BBB permeability, and its loss has been associated with barrier disruption in conditions such as sepsis, diabetes, and multiple sclerosis. Junctional adhesion molecules further contribute to endothelial cohesion and leukocyte transmigration, while accessory proteins, including cingulin, atrial fibrillation (AF)-6, and kinase-like molecules, regulate junctional dynamics. Together, these proteins form a tightly organized complex that is highly sensitive to pathological insults. Reductions in their expression or dissociation from the junctional complex are consistently linked with BBB breakdown in conditions such as hypoxia, trauma, seizures, and neurodegenerative diseases [10, 18–21].

Adherens junctions complement tight junctions by providing structural stability and ensuring proper assembly of junctional complexes. Composed primarily of vascular endothelial cadherin and cytoplasmic partners such as catenins, p120, and plakoproteins, these complexes connect to the actin cytoskeleton and form adhesive bridges between adjacent cells. Vascular endothelial cadherin, in particular, is essential for maintaining cell–cell adhesion, while catenins stabilize this interaction and coordinate with tight junction proteins through ZO-1. Adherens junctions are also involved in modulating cell tension, migration, and angiogenesis, thereby reinforcing the overall stability of the endothelial barrier. Although their precise role in BBB physiology is still being clarified, disruption of adherens junction proteins has been consistently associated with barrier failure and increased vascular permeability in pathological states [18, 22, 23].

Functionally, BMECs are highly specialized. Their luminal surface carries a net negative charge that impedes the entry of negatively charged compounds, and they express very low levels of leukocyte adhesion molecules, reducing immune cell infiltration under normal conditions. They also harbor an array of transport systems, including solute carriers and ATP-binding cassette transporters, that regulate the influx of essential nutrients such as glucose and amino acids while actively excluding xenobiotics and many pharmacological agents. These adaptations, supported by tight and adherens junctions, generate the exceptional selectivity of the BBB and maintain CNS homeostasis [24–26].

The phenotype and performance of endothelial cells depend strongly on their interactions with other elements of the NVA. Astrocytic signals enhance junctional complexity and increase electrical resistance, while pericyte association further strengthens junctional organization and protects endothelial cells against apoptosis. Conversely, loss of these cellular partners or downregulation of junctional proteins compromises barrier function, leading to increased permeability and pathological infiltration of circulating molecules. In this way, the endothelial cell layer of the BBB exemplifies a highly dynamic but tightly regulated interface, capable of maintaining brain protection under normal physiology but prone to dysfunction when its molecular or cellular supports are disturbed [27, 28].

Astrocytes, also referred to as astroglia, are the most abundant glial cells in the CNS, exhibiting highly polarized and heterogeneous morphologies across different brain regions. Traditionally, they are classified as protoplasmic astrocytes, which predominate in the vascularized gray matter, and fibrous astrocytes, which are more common in the less vascularized white matter. Their processes extend to the vasculature, where astrocytic endfeet form direct associations with the basement membrane through proteins such as aquaporin-4 and the dystrophin–dystroglycan complex, anchored by the proteoglycan agrin. This anatomical positioning establishes astrocytes as key modulators of the NVA. They clear metabolic waste, regulate blood flow, control vascular tone, maintain ion homeostasis, and balance neuroimmune responses, thereby linking neuronal activity to vascular function [18, 29–31].

Although their role in BBB development has been debated, it is widely accepted that astrocytes strongly influence endothelial phenotype and barrier stability. Some studies suggest that endothelial cells initially establish the barrier independently, while others argue that astrocytes induce barrier properties in developing cerebral vasculature. In either case, the BBB operates as a coordinated system, and astrocytes serve as an interface between neurons, BMECs, and pericytes. Interestingly, in invertebrates that lack a vascularized circulatory system, astrocyte-like glia act as the main cellular barrier separating circulating fluids from neural tissue, underscoring their fundamental role in CNS protection [18, 32].

Astrocytic endfeet create a perivascular sheath that regulates barrier permeability in both directions. They release soluble factors that strengthen endothelial junctions and control transport processes, while also providing metabolic support to neurons by recycling neurotransmitters, stabilizing extracellular ion concentrations, and maintaining water balance. These functions are critical during normal brain activity and become even more important during pathological conditions, when astrocytes transition from a resting to a reactive state. Reactive astrocytes proliferate, form glial scars, and alter their signaling capacity, changes that can either protect or exacerbate BBB dysfunction depending on context. For example, impaired astrocytic regulation of potassium and hydrogen ions elevates extracellular concentrations, disrupts homeostasis, and contributes to neuronal hyperexcitability [21].

Astrocytes secrete a wide range of molecules that directly influence barrier permeability. Among these are vascular endothelial growth factor (VEGF), nitric oxide, matrix metalloproteinases, and endothelin, all of which increase vascular permeability when dysregulated. Conversely, astrocyte-derived factors such as sonic hedgehog, angiopoietin-1, glial-derived neurotrophic factor, and apolipoprotein E contribute to the reinforcement and repair of the barrier. Integrin signaling further links astrocytes to extracellular matrix components like laminin and fibronectin, ensuring endothelial stability. Loss of integrin function reduces astrocyte coverage around vessels and destabilizes barrier properties, while abnormal activation can trigger inflammatory cascades. Astrocytes also influence BBB regulation through cross-talk with endothelial signaling pathways. Growth factor signaling, including VEGF/VEGF-receptor and transforming growth factor β pathways, modulates angiogenesis and permeability but becomes pathogenic when overactivated, leading to tight junction degradation and barrier leakage. Other mechanisms, such as Wnt and Notch signaling, intersect with astrocyte-derived cues to shape endothelial phenotype. In addition, astrocytes regulate vascular function through calcium-dependent signaling, translating neuronal activity into vascular responses and thereby coupling neural demand with blood supply [21, 33–37].

Pericytes are mural cells strategically positioned along the walls of cerebral capillaries, embedded within the basement membrane and abluminal to endothelial cells. In the CNS, their coverage of microvessels is particularly extensive, approaching complete association with endothelial surfaces. Their location within the NVA places them in constant communication with BMECs and astrocytes, making them central to the regulation of BBB integrity. The platelet-derived growth factor β (PDGF-β) pathway exemplifies this cross-talk: endothelial cells release PDGF-β, which binds PDGF-ß receptor on pericytes to recruit them to developing vessels, while pericytes, in turn, secrete factors that shape endothelial junctional density and influence the polarization of astrocytic endfeet. A reduction in pericyte number diminishes tight junctions and compromises barrier selectivity, highlighting their essential role in maintaining BBB function [18, 38, 39].

Historically described as Rouget's cells in 1873 and later named pericytes, these cells cover approximately one-third of the capillary wall and form direct structural and functional associations with endothelial cells [40, 41]. Their stellate processes, distributed along tight junctions, wrap around the vascular basement membrane and regulate capillary permeability. Pericytes possess contractile capacity reminiscent of smooth muscle cells, enabling them to adjust cerebral blood flow and contribute to angiogenesis. They also synthesize key components of the basement membrane, including type IV collagen, laminin, and polysaccharides, which reinforce vascular stability. The extent of pericyte coverage is inversely correlated with vascular permeability, and experimental depletion results in marked increases in capillary leakage, further underscoring their importance for barrier maintenance [21, 42].

Pericytes interact with endothelial cells through gap junctions, focal adhesions, and N-cadherin-mediated contacts, facilitating ionic communication, paracrine signaling, and adhesion. These interactions enhance endothelial survival, promote the expression of tight junction proteins such as claudin-5 and occludin, and increase transendothelial resistance. In co-culture systems, endothelial cells associated with pericytes display reduced intercellular gaps, enhanced resistance to apoptosis, and greater junctional integrity than those cultured alone. Conversely, pericyte loss not only elevates permeability but also alters endothelial gene expression, increasing molecules such as angiopoietin-2, plasmalemma vesicle-associated protein, and leukocyte adhesion molecules, thereby permitting the entry of macromolecules into the brain parenchyma [21].

Beyond barrier regulation, pericytes perform phagocytic functions, clearing cellular debris and participating in microglial proliferation during brain injury [43]. They can also transdifferentiate into multipotent vascular stem cells under pathological conditions, adopting microglia-like phenotypes with immune and phagocytic capacity. Through their ability to direct astrocytic endfeet toward endothelial surfaces and modulate basement membrane composition, pericytes orchestrate angiogenesis, hemodynamics, and BBB stability [44, 45].

These restrictive properties have profound implications for drug development in neurological diseases. Because nearly every neuron lies within 25 μm of a brain capillary, the vascular route is the most direct path for delivering therapeutics to the CNS. Yet, the very features that protect the brain from toxic exposure also obstruct effective pharmacological intervention in disorders such as brain tumors and neurodegenerative diseases. To address this paradox, research has increasingly focused on strategies to modulate BBB permeability and to design delivery systems capable of exploiting receptor-mediated transport, carrier-mediated uptake, or transient disruption of barrier integrity [18, 46, 47].

2.3 BLOOD–CEREBROSPINAL FLUID BARRIER

CSF provides both mechanical protection and metabolic support to the CNS, acting as a cushion for the brain and spinal cord while distributing nutrients and clearing waste. It is continuously produced within the lateral, third, and fourth ventricles, predominantly by the choroid plexus, although cerebral capillaries also contribute. CSF is similar in composition to interstitial fluid and exchanges freely across pial surfaces, allowing dynamic communication between the two compartments. After being secreted in the ventricles, CSF flows through the cisterna magna and over the convexities of the cerebral hemispheres before being reabsorbed into the venous circulation at arachnoid granulations [1, 48].

The primary interface regulating CSF composition is the BCSFB, located at the choroid plexus. Morphologically, the choroid plexus consists of a vascularized stroma with fenestrated, highly permeable capillaries enveloped by a single layer of epithelial cells derived from specialized ependyma. While the underlying capillaries do not exhibit BBB properties, barrier function is provided by the epithelial layer, whose apical tight junctions limit the paracellular movement of solutes. These cuboidal epithelial cells display polarity, with their apical membrane enriched in transporters and microvilli that amplify surface area for exchange. The BCSFB thereby establishes a controlled environment within the ventricles, selectively permitting entry of ions and nutrients while restricting the passage of macromolecules and toxins [49, 50].

The molecular architecture of the BCSFB resembles but is distinct from that of the BBB. Tight junction proteins such as claudin-1, claudin-2, claudin-11, occludin, and ZO form the core of the epithelial seal. Adherens junctions, mediated by E-cadherin and associated catenins, provide structural integrity and polarity, while gap junctions composed of connexins enable intercellular communication and coordination of barrier function [51, 52]. This arrangement not only maintains electrical resistance across the epithelial lining but also ensures that barrier permeability can adapt to physiological and pathological cues [53, 54]. Therefore, the BCSFB fulfills critical roles in neuroimmune and metabolic regulation [48, 55]. Epithelial cells of the choroid plexus express an array of ATP-binding cassette and solute carrier transporters that mediate both efflux and influx of compounds, including metabolites, xenobiotics, and therapeutic drugs. For example, multidrug resistance proteins such as multidrug resistance protein 1, expressed basolaterally, export waste products from CSF into the blood, while P-glycoprotein modulates selective entry of compounds into the ventricular space. The barrier also participates in immune surveillance: macrophages, dendritic cells, and epiplexus (Kolmer's) cells reside within the stroma or on the epithelial surface, while cytokine and chemokine secretion by epithelial cells helps regulate T-cell trafficking [56, 57].

Taken together, the BCSFB complements the BBB in safeguarding the CNS. By producing CSF, maintaining its composition, and controlling bidirectional exchange with the bloodstream, the choroid plexus epithelium ensures homeostatic stability. Its strategic location, complex molecular machinery, and susceptibility to pathological disruption make the BCSFB both a crucial regulator of brain physiology and a promising therapeutic target in neurological disease.

2.4 MENINGEAL BARRIER

The meninges of the CNS are organized into two major coverings: the dura mater, or pachymeninx, and the leptomeninges, composed of the arachnoid and pia mater. Both the dura and arachnoid encase the brain and spinal cord, while the arachnoid specifically provides the outer barrier that confines CSF within the subarachnoid space (SAS). The pia mater adheres more closely to the CNS surface, yet it is separated from the parenchyma by a thin subpial compartment that contains blood vessels and varying amounts of connective tissue. Traditionally viewed as passive protective layers, the meninges and particularly the leptomeninges are now understood to serve dynamic roles in regulating fluid flow, solute exchange, and cellular trafficking at the brain's surface and within the parenchyma. The SAS itself functions as a distribution pathway, supporting vascular branching patterns that differ from those in most other organs. Rather than entering at a single point, arteries divide into smaller branches within the SAS before penetrating neural tissue, forming a dense vascular network. Increasingly, the outer coverings of the brain are also recognized as morphogenetic signaling centers, active transport sites, stem cell niches, and regulators of immune entry into the CNS [7, 58].

Among the leptomeninges, the arachnoid barrier is the most prominent structural interface. It surrounds the outer surface of the CNS and, owing to its impermeability, ensures that CSF remains contained within the SAS while also preventing exchange with the fenestrated vasculature of the dura [59, 60]. This barrier is composed of epithelial-like leptomeningeal cells interconnected by tight and adherens junctions, which form a selective blockade that limits molecular passage and restricts the invasion of immune cells from systemic circulation [61, 62]. Leptomeningeal cells further reinforce barrier function by lining collagen fibers within the SAS, creating a smooth interface between CSF and underlying connective tissue. While thin leptomeningeal layers, such as the pia, can be penetrated by inflammatory cells, the thicker arachnoid layer provides more effective resistance. Importantly, although veins in the SAS are ensheathed by leptomeningeal cells, the primary control point for immune-cell entry into CSF remains the vascular endothelium of these veins [7, 59].

Beyond containment, leptomeningeal cells actively facilitate CSF circulation and clearance. They line arachnoid granulations that mediate CSF absorption into venous blood, as well as lymphatic outflow pathways linking CSF to nasal lymphatics. They also extend along the outer surfaces of arteries, guiding CSF entry into the CNS parenchyma. Functionally, the arachnoid barrier contributes not only to CSF compartmentalization but also to immune regulation, preventing activated leukocytes from infiltrating the SAS under physiological conditions, while also providing attachment surfaces for immune cells, tumor cells, and pathogens during disease [7].

Taken together, the meninges form far more than a mechanical covering. They constitute active barrier systems that integrate structural, transport, and immunological functions, contributing to CNS homeostasis while playing key roles in fluid circulation and defense. Their protective and facilitative nature places them alongside the BBB and BCSFB as critical regulators of the brain's internal environment.

2.5 CIRCUMVENTRICULAR ORGANS BARRIER

CVOs are small, midline specializations positioned along the walls of the third and fourth ventricles that couple the brain to the composition of blood and CSF. They include the organum vasculosum of the lamina terminalis (OVLT), subfornical organ (SFO), median eminence (ME), neurohypophysis, subcommissural organ, pineal gland, and the area postrema. Although grouped together, these structures are functionally diverse: the OVLT, SFO, and area postrema act primarily as sensory interfaces that detect circulating signals such as ions, osmolality, pH, amino acids, and peptides, and relay this information to regulatory circuits controlling fluid balance, thermoregulation, cardiovascular tone, and aversive responses such as nausea and vomiting. The ME, neurohypophysis, and pineal gland are predominantly secretory, releasing hypothalamic hormones into the portal vasculature, neurohypophyseal peptides into the systemic circulation (neurohypophysis), and melatonin (pineal) [8, 63–65].

Anatomically, CVOs are distinguished by a dense plexus of fenestrated capillaries that lack the typical endothelial tight junctions of the BBB. As a result, low-molecular-weight solutes readily equilibrate between blood and CVO parenchyma. However, this does not equate to unrestricted access. High-molecular-weight proteins are largely excluded from penetrating beyond the inner basement membrane, and diffusion out of CVO cores into neighboring brain is constrained by tanycytes and astrocytes that form tight, highly polarized cellular walls [66–68]. Tanycytes, which are specialized ependymal cells whose processes extend deep into adjacent hypothalamic regions, form a secondary interface at CVO margins. They express junctional proteins, create honeycomb-like seals at their ventricular surfaces, and function as bidirectional conduits through receptor-mediated and caveolar transcytosis, moving selected molecules between CSF, CVOs, and surrounding parenchyma. These features collectively make the CVOs "windows of the brain", permitting carefully filtered blood-to-brain communication without collapsing the broader BBB architecture [63, 66, 67, 69, 70].

2.6 GLIA-LIMITANS

One of the most conserved glial roles across evolution is the construction of cellular barriers that fence off the brain parenchyma. In primates and humans, the glia limitans consists of astrocytic endfeet anchored to a basement membrane and appears in two principal configurations: along the CNS surface as the glia limitans superficialis (or externa), and around the vasculature as the glia limitans perivascularis. A third variant forms at sites of injury, where a rim of reactive astrocytes organizes a glial border which is called glia limitans perilesionis that abuts the fibrotic scar. The basement membrane underlying these astrocytic endfeet is composed of extracellular matrix proteins and is positioned beneath the pia and around all penetrating vessels. Endfeet adhesion is stabilized by the dystrophin-associated glycoprotein complex, which links the astrocyte membrane to its cytoskeleton [9].

Region-specific variants underscore the diversity of this barrier system. Across most of the brain and spinal cord, endfeet forming the glia limitans arise from marginal (surface) and perivascular astrocytes; in the cerebellum, Bergmann glia dominate the perimeningeal barrier. At perivascular sites, astrocytes contribute matrix components to the basement membrane, whereas at the surface, the meningeal compartment adds fibrillar collagens (types I, III, VI), laminins, nidogen, fibronectin, collagen IV, and heparan sulfate proteoglycans. Species differences are notable: the classical, continuous endfoot sheet at the pial surface is a prominent feature of primate brains, while in rodents, densely packed astrocyte somata, together with endfeet, help constitute the peri-meningeal barrier [71, 72]. In portions of the posterior prefrontal and amygdaloid cortices, a specialized population of surface-associated astrocytes organizes the superficial barrier differently: their somata lie at the cortical surface and extend one process downward into layer I and another to ensheath pial vessels. These cells form an overlapping, highly coupled network via abundant gap junctions, creating a continuous sheet that is functionally equivalent to other endfoot-based glia limitans structures. Perivascular variants also incorporate juxtavascular microglial processes into the barrier architecture [9, 73].

2.7 FETAL CSF-BRAIN BARRIER

A long tradition of ultrastructural work, combining thin-section electron microscopy with tracer studies using intravenously administered horseradish peroxidase, has sought to define when the principal brain interfaces acquire competence to exclude proteins. Early reports, especially from chick and rat embryos, often characterized fetal barriers as immature, extending conclusions from protein penetration data to broad claims about barrier function. Reinterpretation of these results, together with freeze-fracture imaging and rigorously controlled tracer paradigms, indicates otherwise. The molecular components that seal endothelial and epithelial junctions appear to be present from the earliest stages at which these cells can be recognized [74].

In the meninges, classical thin-section and freeze-fracture studies in adult laboratory animals identified the meningeal diffusion block within a compact arachnoid barrier layer that adjoins the dura, where continuous tight junctions form a cellular seal between dural vessels and the SAS. Although embryonic development is less completely documented, freeze-fracture imaging in early chick embryos has revealed a lattice of tight junctions within the immature pia-arachnoid, suggesting that meningeal barrier assembly occurs early and subsequently strengthens as the SAS differentiates [74, 75].

Comparative embryonic timelines reinforce this principle of early competence. In chicks, endothelial tight junctions are already evident by day 4 to 6 of incubation. Some investigators noted less complex junctional networks and easier tracer penetration before day 10, yet even in those cases, many endothelial and choroid plexus epithelial contacts were already established before that time. Freeze-fracture analyses later confirmed complete apical junctional belts in choroid plexus epithelium at the earliest stages studied, around day 8, with no qualitative differences between embryos, hatchlings, and juveniles despite large shifts in CSF protein concentration. In rats, compatible ultrastructural and tracer data demonstrate that cerebral vessels are impermeable to albumin-sized molecules by embryonic day 14 to 15, and choroid plexus tight junctions that prevent horseradish peroxidase passage are in place by embryonic day 14, the earliest stage examined. Comparable evidence from nonhuman primates shows that developing intraocular and retinal vessels possess continuous endothelium with zonular tight junctions that exclude circulating tracer molecules from the outset [62, 74–79].

Limited but valuable human embryonic material fits a similar pattern. Freeze-fracture replicas and serial sections from 7 to 8 weeks of gestation reveal well-formed tight junctions in cerebral endothelium, choroid plexus epithelium, and pia-arachnoid layers. Moreover, the junctional network of the choroid plexus appears robust as soon as the plexus anlage is recognizable. These findings undermine the long-standing assertion that the human fetal BBB is intrinsically leaky. Once endothelial and epithelial lineages differentiate, they display mature-looking sealing strands. Early human neuroependyma also exhibits distinctive lateral junctions known as strap junctions, which spiral along adjacent cells near the ventricular surface. These contacts are thought to provide a temporary CSF to the brain barrier during early gestation, diminishing as the epithelium matures into ependyma with prominent gap junctions around 20 to 22 weeks [74].

Embryologically, two CSF compartments exist initially: an inner fluid filling the closing neural tube and primitive ventricles, and an outer compartment that becomes the SAS. Before the foramina of Magendi and Luschka open, these compartments remain separate. Arachnoid granulations, major clearance routes in the adult, do not appear until late gestation and enlarge postnatally. Neonatal drainage is therefore likely to rely heavily on lymphatic pathways, particularly across the cribriform plate into nasal lymphatics. In parallel, vascularization proceeds from a primary meninx that envelops the neural tube and later differentiates into dura and leptomeninges. A primitive perineural vascular plexus forms at the surface of the neural tissue and remodels into the dense leptomeningeal vascular plexus that supplies the mature SAS [7, 80].

Barrier maturation at the surface of the brain follows a distinct timetable from that of the BBB and the BCSFB. In humans, claudin 11, a defining tight junction protein of the arachnoid barrier, is first detected around 7 weeks post-conception, appearing in the posterior fossa and then spreading rostrally. By 12 weeks, the SAS is fully delineated. In rats, claudin 11 immunoreactivity emerges by embryonic day 15, and a continuous arachnoid barrier is evident by day 18. Physiological discontinuities, such as over the cribriform plate, median eminence, and pineal gland, remain, likely reflecting specialized sites for fluid outflow and neuroendocrine signaling. Concurrently, pial microvessels within the SAS express claudin 5 from the earliest stages, providing endothelial barrier properties consistent with plasma protein exclusion. Aquaporin channels and transporters such as excitatory amino-acid transporter 1 appear in meningeal and astroglial domains, further regulating fluid and solute movement. Claudin 11-positive leptomeningeal cells can even be traced along arachnoid trabeculae toward the tela choroidea, suggesting a continuum of barrier-forming elements that extend into the choroid plexus stroma [7, 81].

Taken together, cross-species developmental evidence supports a unified view: once endothelial and epithelial lineages are specified, tight junction-based barriers at cerebral vessels and choroid plexus are already functional; a transient neuroependymal junctional system restricts protein access during peak CSF protein states; and the arachnoid barrier matures progressively to separate outer CSF from fenestrated dural vasculature. Rather than representing an immature system, the embryonic brain displays early competence, with barrier structures that evolve to meet developmental needs for growth, signaling, and protection.

2.8 ALTERATIONS OF THE BRAIN BARRIERS

The integrity of the BBB is essential for neuronal homeostasis, and alterations in its structure or function are closely linked with disease. Many disorders of the CNS show evidence of barrier disruption, including epilepsy, ischemic stroke, multiple sclerosis, traumatic brain injury, and Alzheimer's disease. Systemic illnesses can also accelerate barrier failure with neurological manifestations, as seen in sepsis and in poorly controlled hypertension [82]. In this context, barrier dysfunction is best understood as any gain or loss of function that disturbs the normal balance of molecular exchange or signaling at the vascular interface. Broadly, this includes processes that increase paracellular movement of solutes or cells, processes that alter transcellular transport, and disturbances in the bidirectional communication between BMECs and supporting cells that together shape the local microenvironment. Increased paracellular transport permits entry of circulating proteins, hormones, and leukocytes, while changes in the glycocalyx or basement membranes can further modulate flux. Downstream consequences include oxidative stress, neuroinflammation, and neuronal injury [83].

Hemodynamic and mechanical forces are constant influences on barrier physiology. BMECs experience wall shear stress from blood flow, as well as circumferential and axial stresses that vary with pressure and vessel geometry. Within the autoregulatory range of cerebral perfusion, blood flow remains relatively stable, but rapid or extreme deviations can strain cell-to-cell junctions and increase permeability. Chronic elevation of arterial pressure promotes vascular remodeling, heightened oxidative stress, and region-specific permeability changes, while acute surges can produce transient barrier leaks that resolve when pressure normalizes. Exercise can transiently raise systolic pressure to high levels, yet in healthy individuals this does not necessarily cause barrier dysfunction, underscoring the importance of context and dose [83, 84].

Externally applied energy can modulate the barrier in useful or harmful ways. High-intensity focused ultrasound can ablate tissue and disrupt junctional integrity, whereas low-energy pulsed ultrasound with microbubbles is used to produce a controlled, reversible increase in permeability that facilitates drug delivery. The extent and duration of opening depend on acoustic frequency, pressure, pulse characteristics, and microbubble properties, and may be accompanied by inflammatory signaling [85, 86]. Thermal stress provides another example. Systemic hyperthermia sufficient to raise core and brain temperatures can increase permeability, produce microhemorrhage, and trigger inflammatory cascades, with vulnerability influenced by age and exposure time. Local heating strategies for tumor therapy, such as laser interstitial thermal therapy, intentionally create a limited zone of injury and a surrounding window of increased permeability that can be leveraged for adjuvant treatment. By contrast, controlled hypothermia can enhance barrier function and reduce secondary injury in select acute conditions [83, 87, 88].

Traumatic brain injury illustrates the interplay between primary mechanical forces and secondary biological responses. Rapid acceleration, direct impact, or blast exposure imposes shear, tensile, and compressive loads that damage microvessels and parenchyma [89]. Barrier disruption often follows a biphasic course, with an early phase within hours and a delayed phase over days [90]. Endothelial transcytosis can rise early, tight junction proteins may later decline, and neovascularization with remodeling proceeds over subsequent days [83].

Ionizing radiation can impair barrier function through endothelial injury, inflammatory activation, and microvascular rarefaction. At a cellular level, radiation can induce senescence, reduce angiogenic repair, and promote cytokine-driven changes in junctional organization [91].

Hypoxia is a common barrier stressor in stroke and at high altitude [92]. During focal ischemia, oxygen and glucose deprivation initiate a cascade that includes rapid metabolic failure, production of reactive oxygen species, changes in endothelial vesicular trafficking, and later degradation or mislocalization of tight junction proteins [93–95]. Astrocytes and pericytes respond with swelling, altered end foot contact, migration, and paracrine signaling that further modulates permeability [96].

Exogenous chemical agents can open or damage the barrier. Intra-arterial administration of hyperosmotic solutions shrinks endothelial cells, increases tensile stress at junctions, and produces a transient, focal increase in paracellular pathways that has been used to enhance the delivery of therapeutics [97]. Vasoactive compounds, membrane-active peptides, and other pharmacologic strategies have been explored with variable reproducibility [98]. Recreational substances can also impair barrier function. Chronic alcohol exposure is associated with oxidative damage to the endothelium, loss of tight junctions, and basement membrane changes, while certain stimulants raise body temperature and blood pressure, intensify oxidative injury, and increase leukocyte trafficking, together producing marked leakage [99, 100]. Anesthetic agents can transiently increase permeability, in part through inflammatory signaling and osmotic-like effects, with greater susceptibility in aged brains [101].

Endogenous influences shape barrier health across the lifespan. The gut-brain axis modulates endothelial phenotype through microbial metabolites, immune signaling, and intestinal permeability. Diet, obesity, and micronutrient status affect transporter expression, inflammatory tone, and redox balance [102]. Aging brings decreased microvascular density, thickened basement membranes, pericyte degeneration, astrocyte end foot changes, reduced expression of key transporters, and modest increases in permeability that may remain clinically silent until combined with a second hit such as systemic inflammation [103, 104].

Common molecular pathways recur across these scenarios. Oxidative stress alters junctional proteins, stimulates protease production that degrades the basement membrane, and disrupts signaling within the neurovascular unit (NVA). Inflammation recruits leukocytes, increases cytokine and chemokine release, and reorganizes paracellular and transcellular routes. Angiogenesis, whether reactive or therapeutic, transiently lowers barrier selectivity during sprout formation and maturation until pericyte coverage and basement membrane are restored. Finally, diverse pathogens reach the brain through multiple routes, including paracellular passage, transcellular pathways, or carriage within immune cells, and they often cause barrier breakdown through direct cytopathic effects and robust inflammatory responses [83].

The BCSFB undergoes structural and functional changes across the lifespan. With aging, epithelial cells shrink, the basement membrane thickens, and deposits such as Biondi bodies accumulate, leading to diminished CSF secretion [105]. These changes are exacerbated in neurodegenerative disorders, particularly Alzheimer's disease, where decreased CSF turnover, impaired clearance of amyloid-β, and dysregulated immune signaling have been documented [106]. Altered transporter expression, downregulation of tight junction proteins, and increased inflammatory activity further contribute to compromised barrier function. Such dysfunction not only reduces the barrier's protective capacity but also accelerates disease progression by permitting toxic accumulation and heightened neuroinflammation [107].

REFERENCES

1. M. J. Cipolla, *Chapter 6, Barriers of the CNS*, in *The Cerebral Circulation, Morgan & Claypool Life Sciences* (2009).
2. R. Pandit, L. Chen, and J. Götz, The blood-brain barrier: Physiology and strategies for drug delivery, Adv Drug Deliv Rev 165–166, 1 (2020).
3. E. E. Goldmann, Die äussere und innere Sekretion des gesunden und. Beiträge zur klinischen Chirurgie: mittheilungen aus der chirurgischen Klinik zu Tübingen, 64, 192 (1909).
4. P. Ehrlich, Farben-therapeutische Versuche bei Trypanosomerkankung, Klin Wochenschr 14, 362 (1904).
5. M. Bentivoglio and K. Kristensson, Tryps and trips: Cell trafficking across the 100-year-old blood–brain barrier, Trends Neurosci 37, 325 (2014).
6. A. Dotiwala, C. McCausland, and N. Samra, *Anatomy, Head and Neck: Blood Brain Barrier*, in StatPearls. StatPearls Publishing (2023).
7. R. O. Weller, M. M. Sharp, M. Christodoulides, R. O. Carare, and K. Møllgård, The meninges as barriers and facilitators for the movement of fluid, cells and pathogens related to the rodent and human CNS, Acta Neuropathol 135, 363 (2018).
8. L. Descarries, M. Riad, and M. Parent, *Ultrastructure of the Serotonin Innervation in the Mammalian Central Nervous System*, in C. P. Müller & K. A. Cunningham (Eds.), *Handbook of Behavioral Neuroscience: Serotonin* (Vol. 31). (Elsevier, 2020), pp. 49–90.
9. A. Verkhratsky and A. M. Butt, *Astroglial Physiology*, in *Neuroglia* (Elsevier, 2023), pp. 89–197.
10. J. J. Lochhead, J. Yang, P. T. Ronaldson, and T. P. Davis, Structure, function, and regulation of the blood-brain barrier tight junction in central nervous system disorders, Front Physiol 11, (2020).
11. M. Lewandowsky, Zur lehre der cerebrospinalflussigkeit, Z Klin Med 40, 480 (1909).
12. A. Biedl and R. Krause, Über eine bisher unbekannte toxische Wirkung der Gallensäuren auf das Centralnervensystem, Centralblatt Inn Med 1185 (1898).
13. E. Roux and A. Borrel, Tétanos cérébral et immunité contre le tétanos., Ann Inst Pasteur 4 225 (1898).
14. L. Stern and R. Gautier, Recherches sur Le liquide céphalo-rachidien: I.–Les rapports entre Le liquide céphalo-rachidien et la circulation sanguine, Arch Int Physiol 17, 138 (1921).
15. T. S. Reese and M. J. Karnovsky, Fine structural localization of a blood-brain barrier to exogenous peroxidase, J Cell Biol 34, 207 (1967).
16. E. A. Neuwelt et al., Engaging neuroscience to advance translational research in brain barrier biology, Nat Rev Neurosci 12, 169 (2011).
17. A. M. Butt, H. C. Jones, and N. J. Abbott, Electrical resistance across the blood-brain barrier in anaesthetized rats: a developmental study, J Physiol 429, 47 (1990).
18. D. Wu, Q. Chen, X. Chen, F. Han, Z. Chen, and Y. Wang, The blood–brain barrier: structure, regulation, and drug delivery, Signal Transduct Target Ther 8, (2023).
19. S. Tsukita, M. Furuse, and M. Itoh, Multifunctional strands in tight junctions, Nat Rev Mol Cell Biol 2, 285 (2001).
20. A.-C. Luissint, C. Artus, F. Glacial, K. Ganeshamoorthy, and P.-O. Couraud, Tight junctions at the blood brain barrier: physiological architecture and disease-associated dysregulation, Fluids Barriers CNS 9, 23 (2012).
21. Y. Gong, M. Wu, Y. Huang, X. He, J. Yuan, and B. Dang, Research developments in the neurovascular unit and the blood-brain barrier (Review), Biomed Rep 22, 1 (2025).
22. T. S. Malinova and S. Huveneers, Sensing of cytoskeletal forces by asymmetric adherens junctions, Trends Cell Biol 28, 328 (2018).
23. M. Kwak et al., Adherens junctions organize size-selective proteolytic hotspots critical for Notch signalling, Nat Cell Biol 24, 1739 (2022).
24. Y. Zhao, Y. Liu, Y. Xu, K. Li, L. Zhou, H. Qiao, Q. Xu, and J. Zhao, The role of ferroptosis in blood–brain barrier injury, Cell Mol Neurobiol 43, 223 (2023).
25. K. Rössler, C. Neuchrist, K. Kitz, O. Scheiner, D. Kraft, and H. Lassmann, Expression of leucocyte adhesion molecules at the human blood-brain barrier (BBB), J Neurosci Res 31, 365 (1992).
26. A. R. Santa-Maria et al., Flow induces barrier and glycocalyx-related genes and negative surface charge in a lab-on-a-chip human blood-brain barrier model, Journal of Cerebral Blood Flow & Metabolism 41, 2201 (2021).
27. G. W. Goldstein, Endothelial cell-astrocyte interactions, Ann N Y Acad Sci 529, 31 (1988).

28. G. Li, J. Gao, P. Ding, and Y. Gao, The role of endothelial cell–pericyte interactions in vascularization and diseases, J Adv Res 67, 269 (2025).
29. D. R. Manu, M. Slevin, L. Barcutean, T. Forro, T. Boghitoiu, and R. Balasa, Astrocyte involvement in blood–brain barrier function: A critical update highlighting novel, complex, neurovascular interactions, Int J Mol Sci 24, 17146 (2023).
30. M. V. Sofroniew and H. V. Vinters, Astrocytes: biology and pathology, Acta Neuropathol 119, 7 (2010).
31. D.C. Wei and E. H. Morrison, *Histology, Astrocytes*, in StatPearls. StatPearls Publishing (2023).
32. Z. Zhao, A. R. Nelson, C. Betsholtz, and B. V. Zlokovic, Establishment and dysfunction of the blood-brain barrier, Cell 163, 1064 (2015).
33. A. I. Rodriguez-Perez, A. Borrajo, C. Diaz-Ruiz, P. Garrido-Gil, and J. L. Labandeira-Garcia, Crosstalk between insulin-like growth factor-1 and angiotensin-II in dopaminergic neurons and glial cells: role in neuroinflammation and aging, Oncotarget 7, 30049 (2016).
34. A. Zacharek, J. Chen, A. Li, X. Cui, Y. Li, C. Roberts, Y. Feng, Q. Gao, and M. Chopp, Angiopoietin1/TIE2 and VEGF/FLK1 induced by MSC treatment amplifies angiogenesis and vascular stabilization after stroke, J Cereb Blood Flow Metab 27, 1684 (2007).
35. Y. Li, Y. Xia, Y. Wang, L. Mao, Y. Gao, Q. He, M. Huang, S. Chen, and B. Hu, Sonic hedgehog (SHH) regulates the expression of angiogenic growth factors in oxygen–glucose-deprived astrocytes by mediating the nuclear receptor NR2F2, Mol Neurobiol 47, 967 (2013).
36. M. R. Mizee, P. G. Nijland, S. M. A. van der Pol, J. A. R. Drexhage, B. van het Hof, R. Mebius, P. van der Valk, J. van Horssen, A. Reijerkerk, and H. E. de Vries, Astrocyte-derived retinoic acid: A novel regulator of blood–brain barrier function in multiple sclerosis, Acta Neuropathol 128, 691 (2014).
37. I. Kim, S.-O. Moon, S. K. Park, S. W. Chae, and G. Y. Koh, Angiopoietin-1 reduces VEGF-stimulated leukocyte adhesion to endothelial cells by reducing ICAM-1, VCAM-1, and E-selectin expression, Circ Res 89, 477 (2001).
38. R. J. H. A. Tillie, T. L. Theelen, K. van Kuijk, L. Temmerman, J. de Bruijn, M. Gijbels, C. Betsholtz, E. A. L. Biessen, and J. C. Sluimer, A switch from cell-associated to soluble PDGF-B protects against atherosclerosis, despite driving extramedullary hematopoiesis, Cells 10, 1746 (2021).
39. P. Lindahl, B. R. Johansson, P. Levéen, and C. Betsholtz, Pericyte loss and microaneurysm formation in PDGF-B-deficient mice, Science (1979) 277, 242 (1997).
40. D. Ribatti, The discovery of the pericytes: A historical note, Clinical Anatomy 38, 606 (2025).
41. K. Kisler et al., Pericyte degeneration leads to neurovascular uncoupling and limits oxygen supply to brain, Nat Neurosci 20, 406 (2017).
42. M.-P. Dehouck et al., Quantitative targeted absolute proteomics for better characterization of an in vitro human blood–brain barrier model derived from hematopoietic stem cells, Cells 11, 3963 (2022).
43. E. A. Winkler, A. P. Sagare, and B. V. Zlokovic, The pericyte: A forgotten cell type with important implications for Alzheimer's disease?, Brain Pathol 24, 371 (2014).
44. H. Thurgur and E. Pinteaux, Microglia in the neurovascular unit: Blood–brain barrier–microglia interactions after central nervous system disorders, Neuroscience 405, 55 (2019).
45. Y. Huang, S. Chen, Y. Luo, and Z. Han, Crosstalk between inflammation and the BBB in stroke, Curr Neuropharmacol 18, 1227 (2020).
46. Y. Song, C. Hu, Y. Fu, and H. Gao, Modulating the blood–brain tumor barrier for improving drug delivery efficiency and efficacy, VIEW 3, (2022).
47. C. Greene and M. Campbell, Tight junction modulation of the blood brain barrier: CNS delivery of small molecules, Tissue Barriers 4, e1138017 (2016).
48. R. Mineiro, M. Rodrigues Cardoso, A. Catarina Duarte, C. Santos, J. Cipolla-Neto, F. Gaspar do Amaral, D. Costa, and T. Quintela, Melatonin and brain barriers: The protection conferred by melatonin to the blood-brain barrier and blood-cerebrospinal fluid barrier, Front Neuroendocrinol 75, 101158 (2024).
49. N. Strazielle and J.-F. Ghersi-Egea, Choroid Plexus in the Central Nervous System: Biology and Physiopathology, J Neuropathol Exp Neurol 59, 561 (2000).
50. Z. Redzic and M. Segal, The structure of the choroid plexus and the physiology of the choroid plexus epithelium, Adv Drug Deliv Rev 56, 1695 (2004).
51. G. G. Kovacs, *Cellular Reactions of the Central Nervous System*, in (2018), pp. 13–23.
52. H. Wolburg, K. Wolburg-Buchholz, A. Reichenbach, and A. F. Mack, *Ependymal Cells* ☆, in *Reference Module in Biomedical Sciences* (Elsevier, 2015).

53. Y. Saito and E. M. Wright, Regulation of bicarbonate transport across the brush border membrane of the bull-frog choroid plexus, J Physiol 350, 327 (1984).
54. Z. Redzic, Molecular biology of the blood-brain and the blood-cerebrospinal fluid barriers: similarities and differences, Fluids Barriers CNS 8, 3 (2011).
55. E.-M. Krämer-Albers, Extracellular vesicles at CNS barriers: Mode of action, Curr Opin Neurobiol 75, 102569 (2022).
56. P. Solár, A. Zamani, L. Kubíčková, P. Dubový, and M. Joukal, Choroid plexus and the blood–cerebrospinal fluid barrier in disease, Fluids Barriers CNS 17, 35 (2020).
57. J.-F. Ghersi-Egea, N. Strazielle, M. Catala, V. Silva-Vargas, F. Doetsch, and B. Engelhardt, Molecular anatomy and functions of the choroidal blood-cerebrospinal fluid barrier in health and disease, Acta Neuropathol 135, 337 (2018).
58. J. Derk et al., Formation and function of the meningeal arachnoid barrier around the developing mouse brain, Dev Cell 58, 635 (2023).
59. B. Engelhardt, P. Vajkoczy, and R. O. Weller, The movers and shapers in immune privilege of the CNS, Nat Immunol 18, 123 (2017).
60. B. Engelhardt, R. O. Carare, I. Bechmann, A. Flügel, J. D. Laman, and R. O. Weller, Vascular, glial, and lymphatic immune gateways of the central nervous system, Acta Neuropathol 132, 317 (2016).
61. Y. Uchida et al., Involvement of Claudin-11 in disruption of blood-brain, -spinal cord, and -arachnoid barriers in multiple sclerosis, Mol Neurobiol 56, 2039 (2019).
62. S. Nabeshima, T. S. Reese, D. M. D. Landis, and M. W. Brightman, Junctions in the meninges and marginal glia, Journal of Comparative Neurology 164, 127 (1975).
63. S. Miyata, Glial functions in the blood-brain communication at the circumventricular organs, Front Neurosci 16, (2022).
64. J. Ruberte, M. Navarro, A. Carretero, H. E. König, and L. Puelles, *Nervous System*, in *Morphological Mouse Phenotyping* (Elsevier, 2017), pp. 377–474.
65. J. C. Rios, M. W. Galper, and T. P. Naidich, *Ventricles and Intracranial Subarachnoid Spaces*, in *Imaging of the Brain* (Elsevier, 2013), pp. 245–271.
66. S. Morita, E. Furube, T. Mannari, H. Okuda, K. Tatsumi, A. Wanaka, and S. Miyata, Heterogeneous vascular permeability and alternative diffusion barrier in sensory circumventricular organs of adult mouse brain, Cell Tissue Res 363, 497 (2016).
67. S. Miyata, New aspects in fenestrated capillary and tissue dynamics in the sensory circumventricular organs of adult brains, Front Neurosci 9, (2015). https://www.frontiersin.org/journals/neuroscience/articles/10.3389/fnins.2015.00390/full
68. S. Morita and S. Miyata, Different vascular permeability between the sensory and secretory circumventricular organs of adult mouse brain, Cell Tissue Res 349, 589 (2012).
69. F. Langlet, A. Mullier, S. G. Bouret, V. Prevot, and B. Dehouck, Tanycyte-like cells form a blood–cerebrospinal fluid barrier in the circumventricular organs of the mouse brain, J Comp Neurol 521, 3389 (2013).
70. A. Mullier, S. G. Bouret, V. Prevot, and B. Dehouck, Differential distribution of tight junction proteins suggests a role for tanycytes in blood-hypothalamus barrier regulation in the adult mouse brain, J Comp Neurol 518, 943 (2010).
71. X. Liu, Z. Zhang, W. Guo, G. Burnstock, C. He, and Z. Xiang, The superficial glia limitans of mouse and monkey brain and spinal cord, Anat Rec 296, 995 (2013).
72. J. Sievers, F. W. Pehlemann, S. Gude, and M. Berry, Meningeal cells organize the superficial glia limitans of the cerebellum and produce components of both the interstitial matrix and the basement membrane, J Neurocytol 23, 135 (1994).
73. A. Verkhratsky and A. Pivoriūnas, Astroglia support, regulate and reinforce brain barriers, Neurobiol Dis 179, 106054 (2023).
74. Møllgård K and Saunders N.R., The development of the human blood-brain and blood-CSF barriers, Neuropathol Appl Neurobiol 12, 337 (1986).
75. J. J. Van Rybroek and F. N. Low, Intercellular junctions in the developing arachnoid membrane in the chick, J Comp Neurol 204, 32 (1982).
76. S. Wakai and N. Hirokawa, Development of blood-cerebrospinal fluid barrier to horseradish peroxidase in the avian choroidal epithelium, Cell Tissue Res 214, (1981).
77. R. Dermietzel, K. Meller, W. Tetzlaff, and M. Waelsch, In vivo and in vitro formation of the junctional complex in choroid epithelium, Cell Tissue Res 181, (1977).
78. K. Meller, Ultrastructural aspects of the choroid plexus epithelium as revealed by the rapid-freezing and deep-etching techniques, Cell Tissue Res 239, 189 (1985).

79. S. Roy, A. Hirano, J. A. Kochen, and H. M. Zimmerman, The fine structure of cerebral blood vessels in chick embryo, Acta Neuropathol 30, 277 (1974).
80. D. G. Gómez, A. T. DiBenedetto, A. M. Pavese, A. Firpo, D. B. Hershan, and D. G. Potts, Development of arachnoid villi and granulations in man, Cells Tissues Organs 111, 247 (1981).
81. K. Møllgård, K. M. Dziegielewska, C. B. Holst, M. D. Habgood, and N. R. Saunders, Brain barriers and functional interfaces with sequential appearance of ABC efflux transporters during human development, Sci Rep 7, 11603 (2017).
82. M. Kaya and B. Ahishali, Basic physiology of the blood-brain barrier in health and disease: A brief overview, Tissue Barriers 9, 1840913 (2021).
83. N. Zhao, T. D. Chung, Z. Guo, J. J. Jamieson, L. Liang, R. M. Linville, A. F. Pessell, L. Wang, and P. C. Searson, The influence of physiological and pathological perturbations on blood-brain barrier function, Front Neurosci 17, (2023).
84. P. Campinho, A. Vilfan, and J. Vermot, Blood flow forces in shaping the vascular system: a focus on endothelial cell behavior, Front Physiol 11, (2020).
85. D. McMahon, C. Poon, and K. Hynynen, *Increasing BBB Permeability via Focused Ultrasound: Current Methods in Preclinical Research*, in (2019), pp. 267–297.
86. A. H. Mesiwala, L. Farrell, H. J. Wenzel, D. L. Silbergeld, L. A. Crum, H. R. Winn, and P. D. Mourad, High-intensity focused ultrasound selectively disrupts the blood-brain barrier in vivo, Ultrasound Med Biol 28, 389 (2002).
87. E. J. Yerxa, An introduction to occupational science, a foundation for occupational therapy in the 21st century, Occup Ther Health Care 6, 1 (1990).
88. H. Sharma and P. Dey, Probable involvement of 5-hydroxytryptamine in increased permeability of blood-brain barrier under heat stress in young rats, Neuropharmacology 25, 161 (1986).
89. N. H. Guley et al., A novel closed-head model of mild traumatic brain injury using focal primary overpressure blast to the cranium in mice, J Neurotrauma 33, 403 (2016).
90. Y. Xiong, A. Mahmood, and M. Chopp, Animal models of traumatic brain injury, Nat Rev Neurosci 14, 128 (2013).
91. K. Lumniczky, T. Szatmári, and G. Sáfrány, Ionizing radiation-induced immune and inflammatory reactions in the brain, Front Immunol 8, (2017).
92. E. Ortiz-Prado, J. F. Dunn, J. Vasconez, D. Castillo, and G. Viscor, Partial pressure of oxygen in the human body: A general review., Am J Blood Res 9, 1 (2019).
93. M. Krueger, I. Bechmann, K. Immig, A. Reichenbach, W. Härtig, and D. Michalski, Blood—brain barrier breakdown involves four distinct stages of vascular damage in various models of experimental focal cerebral ischemia, J Cereb Blood Flow Metab 35, 292 (2015).
94. A. Durukan and T. Tatlisumak, Acute ischemic stroke: Overview of major experimental rodent models, pathophysiology, and therapy of focal cerebral ischemia, Pharmacol Biochem Behav 87, 179 (2007).
95. R. Prakash and S. T. Carmichael, Blood–brain barrier breakdown and neovascularization processes after stroke and traumatic brain injury, Curr Opin Neurol 28, 556 (2015).
96. S. Liu, D. Agalliu, C. Yu, and M. Fisher, The role of pericytes in blood-brain barrier function and stroke, Curr Pharm Des 18, 3653 (2012).
97. R. M. Linville, J. G. DeStefano, M. B. Sklar, C. Chu, P. Walczak, and P. C. Searson, Modeling hyperosmotic blood–brain barrier opening within human tissue-engineered in vitro brain microvessels, J Cereb Blood Flow Metab 40, 1517 (2020).
98. C. V. Borlongan and D. F. Emerich, Facilitation of drug entry into the CNS via transient permeation of blood brain barrier: Laboratory and preliminary clinical evidence from bradykinin receptor agonist, Cereport, Brain Res Bull 60, 297 (2003).
99. A. Rubio-Araiz, F. Porcu, M. Pérez-Hernández, M. S. García-Gutiérrez, M. A. Aracil-Fernández, M. D. Gutierrez-López, C. Guerri, J. Manzanares, E. O'Shea, and M. I. Colado, Disruption of blood–brain barrier integrity in postmortem alcoholic brain: Preclinical evidence of TLR4 involvement from a binge-like drinking model, Addiction Biology 22, 1103 (2017).
100. J. Wei, L. Qin, Y. Fu, Y. Dai, Y. Wen, and S. Xu, Long-term consumption of alcohol exacerbates neural lesions by destroying the functional integrity of the blood–brain barrier, Drug Chem Toxicol 45, 231 (2022).
101. S. Yang et al., Anesthesia and surgery impair blood–brain barrier and cognitive function in mice, Front Immunol 8, (2017).
102. S. E. Kanoski, Y. Zhang, W. Zheng, and T. L. Davidson, The effects of a high-energy diet on hippocampal function and blood-brain barrier integrity in the rat, Journal of Alzheimer's Disease 21, 207 (2010).

103. E. F. Goodall, C. Wang, J. E. Simpson, D. J. Baker, D. R. Drew, P. R. Heath, M. J. Saffrey, I. A. Romero, and S. B. Wharton, Age-associated changes in the blood-brain barrier: comparative studies in human and mouse, Neuropathol Appl Neurobiol 44, 328 (2018).
104. A. C. Yang et al., Physiological blood–brain transport is impaired with age by a shift in transcytosis, Nature 583, 425 (2020).
105. I. C. M. Verheggen, W. M. Freeze, J. J. A. de Jong, J. F. A. Jansen, A. A. Postma, M. P. J. van Boxtel, F. R. J. Verhey, and W. H. Backes, Application of contrast-enhanced magnetic resonance imaging in the assessment of blood-cerebrospinal fluid barrier integrity, Neurosci Biobehav Rev 127, 171 (2021).
106. D. G. Gadhave, V. V. Sugandhi, S. K. Jha, S. N. Nangare, G. Gupta, S. K. Singh, K. Dua, H. Cho, P. M. Hansbro, and K. R. Paudel, Neurodegenerative disorders: Mechanisms of degeneration and therapeutic approaches with their clinical relevance, Ageing Res Rev 99, 102357 (2024).
107. K. Mathias, R. S. Machado, T. Cardoso, A. dal B. Tiscoski, N. Piacentini, J. S. Prophiro, J. S. Generoso, T. Barichello, and F. Petronilho, The blood-cerebrospinal fluid barrier dysfunction in brain disorders and stroke: Why, how, what for?, Neuromolecular Med 26, 38 (2024).

3 Poly (ADP-Ribose) Polymerase (PARP) as a Novel Therapeutic Target for Repurposing Cancer Drugs for Treating Neurological and Neurodegenerative Conditions

Saheed O. Benson and Amos A. Fatokun

3.1 INTRODUCTION

Neurological and neurodegenerative conditions such as stroke, Alzheimer's disease and Parkinson's disease are associated with significant morbidity and mortality globally, and their incidence and prevalence are expected to rise exponentially in the next few decades (Gaskin, Gomes, and Darshan, 2017; Gooch, Pracht, and Borenstein, 2017). While major advances have been made in understanding the cellular and molecular mechanisms underpinning these pathologies and developing novel therapeutic agents for managing them, current drugs are still largely symptomatic and, therefore, inadequate. There is, therefore, a critical unmet need to develop disease-modifying neurotherapeutics that will target and modify the course of disease and translate to significantly improved health outcomes for patients. Unfortunately, drug discovery is prohibitively expensive and takes many years. One of the ways to address this notable challenge is to repurpose existing drugs for novel indications (Jourdan et al., 2020). This chapter illustrates an excellent example of such repurposing, in the context of the molecular target poly (ADP-ribose) polymerase (PARP), which is known to contribute to the DNA damage response (DNA repair). PARP inhibitors (PARPi) have been years in development, on the basis of the genetic concept of synthetic lethality (Hartwell et al., 1997), and a number of PARPi are now in clinical use as targeted therapy against *BRCA*-mutated cancers.

Interestingly, PARP has also now been sufficiently shown to contribute to neuronal death, engendering parthanatos, or PARP-mediated cell (neuronal) death, and PARPi have been demonstrated to be neuroprotective, thus heralding the prospect of using PARPi as disease-modifying neurotherapeutics, although there are still drawbacks (Fatokun, Dawson, and Dawson, 2014; Berger et al., 2018).

This chapter briefly explores PARP, especially PARP1, the most abundant and most studied of all PARPs, as an established therapeutic target in targeted cancer therapy, and then as a potential therapeutic target in neurological and neurodegenerative conditions. It discusses opportunities and challenges in the therapeutic domain of applying PARPi as disease-modifying neurotherapeutics.

3.2 THE PARP SUPERFAMILY

PARPs represent a family of post-translational modification enzymes that regulate a variety of biological processes such as DNA repair, DNA replication, chromatin remodelling, transcriptional regulation, antiviral response, programmed cell death, and mitosis (Wei and Yu, 2016; Guo et al.,

DOI: 10.1201/9781003482406-3

2020; Wu et al., 2021; Padella et al., 2022). In humans, the family comprises 17 members with a conserved C-terminal catalytic domain and a very diverse N-terminal domain (Figure 3.1), and they are encoded by different genes (Hottiger et al., 2010). PARPs catalyse the transfer of ADP-ribose from nicotinamide adenine dinucleotide (NAD^+) onto acceptor proteins such as histones, topoisomerases, p53, and PARP itself, a process known as ADP-ribosylation, during which PARPs covalently attach their ADP-ribose units onto the carboxyl group of arginine, aspartic acid, glutamic acid, cysteine, lysine, serine, and tyrosine residues of target proteins via an ester bond (Wei and Yu, 2016; Chen et al., 2018; Wang et al., 2019; Li et al., 2020).

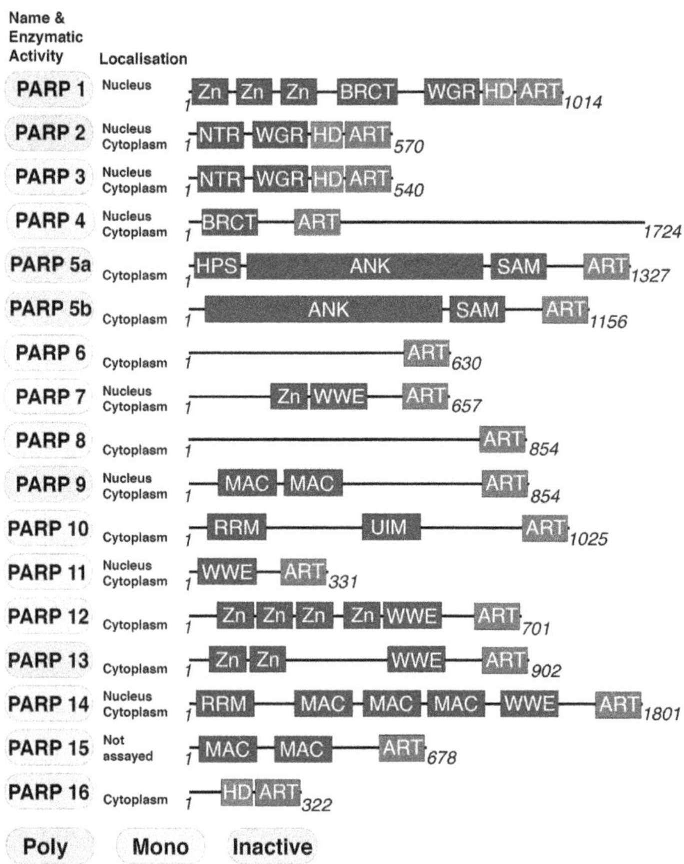

FIGURE 3.1 Domain architecture, protein size, ribosylation pattern, and subcellular localisation of the PARP superfamily. The 17 members share a conserved C-terminal catalytic domain (CAT), shown in pink/red, and a very diverse N-terminal domain, denoted in blue. The CAT domain comprises the helical subdomain (HD) and the ADP-ribosyl transferase (ART) subdomain. PARP1, PARP2, and PARP3 are DNA-dependent PARPs and share a common tryptophan, glycine, and arginine (WGR) motif that interacts with DNA. Some PARPs have zinc fingers (Zn), which are involved in sensing and binding DNA strand breaks. The BRCA1 carboxy-terminal motifs (BRCT) in PARP1 are important for protein–protein interaction. PARP5a and PARP5b are known as the tankyrases, with their unique sterile alpha motif and ankyrin domain repeats. The ankyrin domain is a 33-residue motif comprising two α-helices separated by loops. Tankyrase 1 (PARP5a) has a histidine, proline, serine region of unknown function. Macro domains (MAC) are found in PARP9, PARP14, and PARP15, and they have high-affinity ADP-ribose-binding ability (Langelier et al., 2018; Eisemann and Pascal, 2020; van Beek et al., 2021). NTR means N-terminal region; WWE means tryptophan-tryptophan-glutamate motif; RRM means RNA recognition motifs; and UIM means ubiquitin-interacting motif. Figure created by authors.

Although they are referred to as PARPs, only 4 members (PARP1, PARP-2, PARP-5a, and PARP-5b—the last two known as the tankyrases) are responsible for poly ADP-ribosylation (PARylation) that causes the synthesis of linear or branched poly (ADP-ribose) (PAR) polymers. The other members, with the exception of PARP-9 and PARP13, are only capable of mono ADP-ribosylation (MARylation), which means they are only able to transfer a single ADP-ribose unit (i.e., mono (ADP-ribose) (MAR) unit) onto their target proteins (Wei and Yu, 2016; Langelier et al., 2018). While all catalytically active members of the PARP family are able to 'transfer' the ADP-ribose moiety from NAD^+ onto various acceptors, not all engage in the polymerisation of ADP-ribose units (ADPrs); thus, all PARPs are 'transferases' but not all are 'polymerases'. Due to this transferase activity, PARP enzymes are sometimes referred to as ADP-ribosyltransferases. Lüscher et al. (2021) recommended a change in nomenclature that better reflects the biochemical and structural diversity of this superfamily of enzymes which are involved in mono- and poly (ADP-ribose)-mediated cellular processes. In addition, the term 'polymerase' is often associated with enzymes involved in DNA or RNA synthesis, not proteins involved in protein–protein modification (Lüscher et al., 2021). Based on evolutionary similarity in the structures of bacterial ADP-ribosyltransferases and mammalian PARPs, the term 'ADP-ribosyltransferase diphtheria toxin-like' (ARTD) has been suggested to be more accurate than 'PARP' (Kuny and Sullivan, 2016). However, currently, most publications tend to use both ARTD and PARP nomenclature simultaneously (e.g., PARP1 described as ARTD1/PARP1). Apart from the irksomeness of writing two names for the same enzyme at the same time, there is a potential for confusion, as the new ARTD isoform numbering does not always correspond with the PARP nomenclature isoform numbering (e.g., ARTD14 is PARP7, while ARTD7 is PARP15). Consequently, for ease of reading and for consistency, this chapter will adopt the 'PARP' nomenclature and classification.

During PARylation, multiple ADPrs are transferred to acceptor proteins and are joined through 1'-2' or 2'-1' glycosidic ribose-ribose bonds, which lead to a long, linear poly (ADP-ribose) (PAR) polymer structure or a complex, branched structure, respectively (Figure 3.2) (Kalisch et al., 2012;

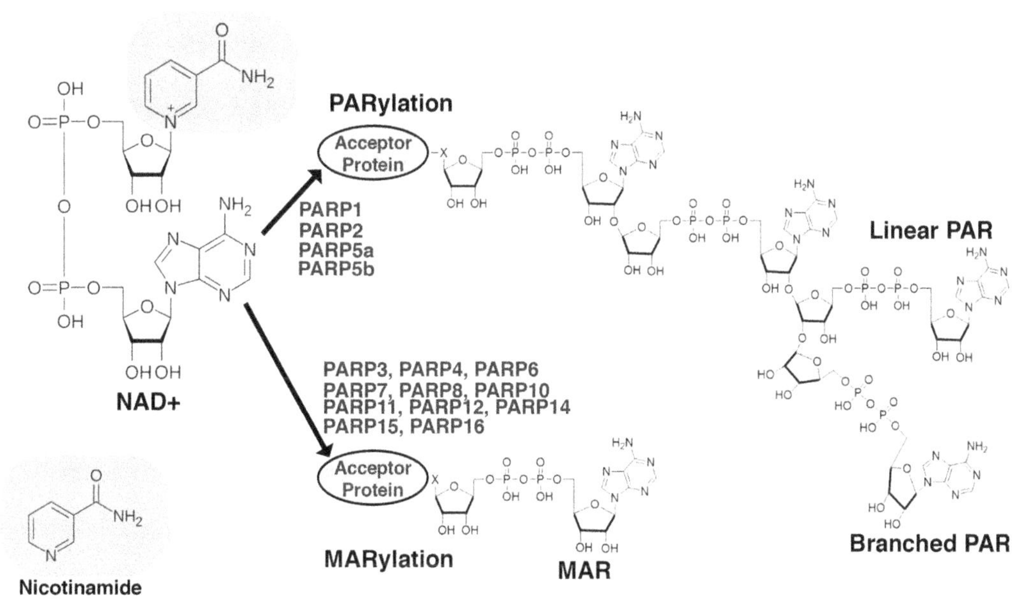

FIGURE 3.2 Poly ADP-ribosylation (PARylation) and mono ADP-ribosylation (MARylation) by PARPs. PARPs can produce mono (ADP-ribose) (MAR) and poly (ADP-ribose) (PAR) from NAD^+, releasing nicotinamide as a by-product. PARP1, the most prolific member of the PARP family, can produce PAR chains of up to 200 ADP-ribose units, with branching occurring every 20–50 units. This causes a huge variation in the PAR chain's length and branching shape (Kamaletdinova, Fanaei-Kahrani and Wang, 2019). Figure created by authors.

Bian et al., 2019). PARylation is a transient and reversible post-translational modification of proteins that plays an important role in various cellular processes, such as maintaining genomic stability, regulating transcription, telomere length homeostasis, chromatin structure stabilisation, DNA methylation, and cell signalling. At the molecular level, PARylation modifies the biochemical and functional properties of its target proteins, which in turn dictate the protein–protein or protein-nucleic acid (both DNA and RNA) interactions, thus promoting the formation of scaffolds that direct DDR proteins to a precise site and regulating other protein modifications such as ubiquitylation (Krietsch et al., 2013).

On the other hand, little is known about the biological importance of MARylation, even though most members of the PARP superfamily are involved only in this type of enzymatic reaction (Vyas et al., 2013). Recent studies have revealed that most of these PARP 'monoenzymes' are localised in the cytosol and regulate cellular processes such as RNA metabolism, cellular transport, focal adhesion, and stress responses (Challa, Stokes, and Kraus, 2021). Of the at least 17 members of PARP now identified in eukaryotic cells, PARP1 is the most abundant and extensively studied (Fatokun, Dawson, and Dawson, 2014; Di Girolamo and Fabrizio, 2019; Wang et al., 2019).

3.3 POLY (ADP-ribose) polymerase-1

PARP1 (EC 2.4.2.30) accounts for 85-90% of total PARP enzyme activity and the production of over 95% of total PAR polymers in most cells (Cayuela et al., 2001). PARP1 is a 116-KDa nuclear protein consisting of 1014 amino acid residues and is involved primarily in DNA damage signalling by detecting and rapidly binding to the site of DNA breaks, especially single-strand breaks. Apart from the initiation of DNA damage response, PARP1 has also been implicated in the surveillance of genomic integrity and other cellular homoeostatic activities such as cell division, apoptosis, and regulation of gene transcription and amplification (Kumar et al., 2021). Upon sensing DNA damage, PARP1 activity can be stimulated up to 500-fold over basal levels, allowing it to transfer 50 to 200 ADPrs to numerous nuclear proteins, including PARP1 itself (Alvarez-Gonzalez and Althaus, 1989; Yu et al., 2002; Jannetti et al., 2020).

PARP1 structure is divided into three main domains: an N-terminal DNA-binding domain, a central auto modification domain, and a C-terminal catalytic domain (Figure 3.3). The N-terminal DNA-binding domain has two homologous zinc fingers (Zn1 and Zn2) and a nuclear localisation sequence (NLS) used for DNA binding. Zn1 and Zn2 recognise and bind DNA breaks, which

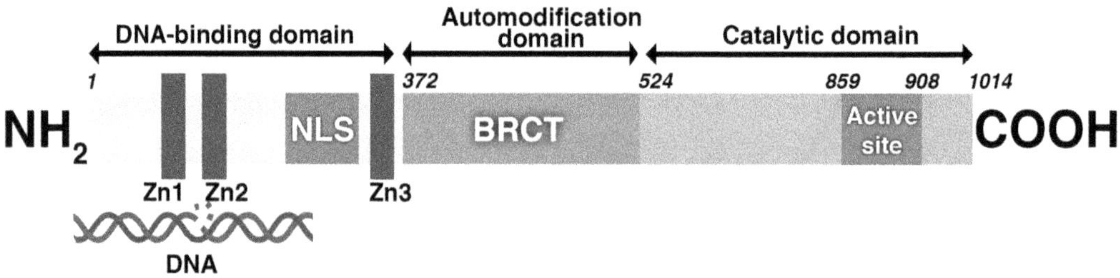

FIGURE 3.3 The primary structure of PARP1, with three main domains. (1) an N-terminal DNA-binding domain containing two zinc-finger motifs (Zn1 and Zn2), which are responsible for detection of DNA breaks, an NLS containing a caspase-cleavage site, and a third zinc-finger (Zn3) motif for protein–protein interactions; (2) a Central automodification domain with a breast cancer-associated Gene 1 BRCT motif for protein–protein interaction and which plays an important role in regulating PARP1 activity; and (3) a C-terminal catalytic domain containing the nicotinamide adenine dinucleotide (NAD)-binding site and the poly (ADP-ribose) polymer (PAR)-synthesising domain (Fatokun, Dawson and Dawson, 2014). Figure created by authors.

increases PARP1 activity up to 500-fold. There is another zinc finger in this domain, Zn3, which is structurally and functionally distinct from the previous two. Zn3 is uniquely found in PARP1 compared to the other PARPs. These Zn fingers play different roles in the enzymatic function of PARP1. Zn1 is involved in the sensing and binding of DSBs, interacting with the catalytic site and activating PARP1. Zn2 plays a crucial role in the recognition of Single-Strand Breaks (SSBs), while Zn3 is responsible for protein–protein interactions during autoPARylation by initiating hydrolysis of the NAD^+ substrate (Wang et al., 2017; Jannetti et al., 2020).

The central auto modification domain serves as an acceptor of ADP-ribose units, thereby allowing the enzyme to transfer poly (ADP-ribose) to itself (automodification or autoPARylation). AutoPARylation occurs after DNA repair and causes the PARP1 enzyme to dissociate from the DNA due to the accumulated high negative charge of the PAR chain. This dissociation allows the other DDR proteins recruited by PARP1 to have access to the DNA damage site (Murai et al., 2012; Mitra et al., 2019). This domain also contains a BRCT repeat motif, which plays an important role in protein targeting in the DNA damage and repair response.

The WGR is the tryptophan–glycine–arginine-rich domain that participates in binding DNA near the 5′-terminus and mediates domain-domain contact essential for DNA-dependent activity. Several studies have indicated that the WGR domain is critical for PARP1 activation. The WGR domain has been observed to move closer to form a tighter-binding complex after binding to a DNA break. This is accompanied by the transmission of binding information to the active site to yield the catalytic activation of PARP1 (Li et al., 2013; Rudolph, Mahadevan, and Luger, 2020).

PARP1, the most abundant and extensively studied PARP, has been reported to be about 62% identical in its catalytic domain to PARP2, its closest homologue, and PARP2 is responsible for 10%–15% of human cellular PARP activity (Hoch and Polo, 2019). A closer study of the PARP1 catalytic domain showed that it has two sub-pockets, which are occupied by the substrate NAD^+. One sub-pocket (NI site) binds to the nicotinamide-ribose part of NAD^+, while the other (AD site) binds to the adenine-ribose moiety. The NI site allows NAD^+-mimicking PARP1 inhibitors to bind through hydrogen bonds with residues Ser904 and Gly863, while residue Tyr907 provides supplementary π–π stacking interaction. On the other hand, the AD site is large enough to tolerate the binding of structurally diverse PARPi. Several potent PARPi, such as olaparib, have been reported to exploit the AD sub-pocket to increase their potency (Zhong et al., 2019). PARP1's catalytic domain is involved in three different stages of PAR synthesis, namely: initiation – this is the attachment of the first ADP-ribose unit to the amino acid residue in an acceptor protein; elongation – the formation of a (2′-1″) ribose-ribose glycosidic bond; and branching – the formation of a (2″-1‴) ribose-ribose bond between ADP-ribose units (Alemasova and Lavrik, 2019). A better understanding of the enzymatic structure of PARP1, especially the catalytic site, will help in the development of more potent PARPi that will have less undesirable effects on other NAD^+-consuming enzymes such as sirtuins or HDACs (histone deacetylases), CD38 (Cluster of Differentiation 38), CD157, CD73, SARM1 (sterile alpha and TIR motif containing 1) and NNMT (nicotinamide N-methyltransferase) (Wei and Yu, 2016).

3.3.1 Degradation of Poly (ADP-ribose) Polymer (PAR) by Poly (ADP-ribose) Glycohydrolase

It is important to note that while PARPs are responsible for the formation of MAR or PAR, the degradation of the PAR chains is mainly catalysed by poly (ADP-ribose) glycohydrolase (PARG). PARG is the only enzyme that can hydrolyse the ribose-ribose bonds of PAR on PARylated proteins (Slade, 2020). PARG counteracts the action of PARP enzymes, thereby preventing the build-up of PAR which could have detrimental consequences for the cell. Studies have demonstrated that a deficiency or pharmacological inhibition of PARG leads to increased sensitivity to PARPi and chemotherapeutic agents (Steffen et al., 2011). Similarly, ADP-ribosylhydrolase 3 (ARH3) has been shown to catalyse the exoglycosidic hydrolysis of PAR and MAR. Both PAR-hydrolysing enzymes, but predominantly PARG, play important roles in regulating PARP enzyme activities through their dePARylation of modified proteins (O'Sullivan et al., 2019; Harrision et al., 2020).

3.3.2 PARP1 Expression and Activity

The PARP1 protein is encoded by the *PARP1* gene located on chromosome 1 (1q42.12) in human cells (Auer et al., 1989). It operates by sensing SSBs and utilising nicotinamide adenine dinucleotide (NAD^+) as a substrate to catalyse the formation of poly(ADP)-ribose (PAR) chains, which recruit other DNA repair proteins and nucleases to the site of the DNA damage (Yu et al., 2002).

The nuclear concentrations of PARP1 range from 2×10^5 to 1×10^6 enzymes/nucleus in eukaryotic cells. Within thirty seconds of detecting DNA damage, PARP1 autoPARylates itself, which leads to about a 500-fold increase in its activity over its basal level (Alvarez-Gonzalez and Althaus, 1989; Jannetti et al., 2020). The increase in PARP1 activity correlates to higher PAR levels in the cells. In cancerous cells, where the PARP1 enzyme activity is found to be higher than in healthy cells, remarkably increased PAR formation has been reported compared to the levels in adjacent non-cancerous cells (Nosho et al., 2006).

Interestingly, PARP1 enzymatic activity and functions are regulated through various post-translational modifications. PARP1 can be modified through ADP-ribosylation by itself (automodification) and by other members of the PARP family, such as PARP2 and PARP3, which enhances its activity. Likewise, other ADP-ribosylating enzymes such as sirtuin 6 (SIRT6) can act on PARP1, promoting double-strand DNA repair in oxidative stress. Phosphorylation of PARP1 by the kinases ERK1/2, JNK1, and activated calcium-dependent protein kinase (CaMKII) can modulate its catalytic activity. Other post-translational modifications are acetylation and ubiquitylation/SUMOylation by protein acetyltransferases (p300 and CBP) and E3 ligases, respectively (Luo and Kraus, 2012).

The binding of PARP1 with DNA, nucleosomes and other nuclear proteins such as KIF4, histone variant mH2A1.1, Ku and condensin I/XRCC1 repair complex can activate its enzymatic activity. NMNAT-1, a cellular NAD^+ synthase, catalyses the final step of NAD^+ (the substrate for PARP1) biosynthesis. NMNAT-1 binds to PAR on activated PARP1, thereby stimulating poly ADP-ribosylation (Berger, Lau, and Ziegler, 2007).

PARP1 enzymatic activation consumes high amounts of NAD^+. As a result, its end product, PAR, exhibits mild regulation of PARP1 by conferring on it a strong negative charge, which repels the enzyme from the negatively charged DNA, thereby stopping ADP-ribosylation. This action limits the consumption of NAD^+. With a similar goal of converting ATP, which requires NAD^+ for its production, PARP1 activity is regulated during apoptosis by its cleavage. Caspase 3, a key enzyme in the apoptotic pathway, breaks PARP1 into two fragments between the second and third zinc-binding domains, one larger 89 kDa, and a shorter 24 kDa. This alters its DNA-binding ability, consequently affecting its activity. PARP1 cleavage has also been reported in autophagic and necrotic cell death pathways by the actions of various cathepsins (Ko and Ren, 2012).

3.3.3 PARP1 and Synthetic Lethality

PARP1 is a nuclear protein that plays a vital role in sensing DNA damage and facilitating DNA base excision repair (BER), which is key in SSB repair. Research has demonstrated that knocking down PARP1 expression with siRNA or inhibiting PARP1 activity with small molecules significantly reduced the cell's ability to repair its SSBs. This implies that PARP1 must be required for this DDR pathway (Mateo et al., 2019; Amin, Chaabouni, and George, 2020). Tumour cells carrying germline or somatic *BRCA1* and/or *BRCA2* gene mutations, such as in some breast, prostate and ovarian cancers, utilise NHEJ for the repair of DSBs due to their deficiency in the HR pathway. The error-prone NHEJ is not effective and cannot cope with a large-scale DSB repair.

Blocking the action of PARP1 in these cancer cells will cause excessive SSBs in them, which will eventually lead to DSBs. The sheer level of DSBs will then overwhelm the NHEJ pathway, causing serious genomic instability and ultimately resulting in cell death via apoptosis (Shah et al., 2014). Thus, PARP1 inhibition in HR-deficient cells is synthetically lethal. The genetic concept of synthetic lethality was first described in 1946 by Dobzhansky as a situation where a mutation in either one of two genes alone has no effect but combining mutations in both genes leads to cell death. However, it was not until 1997 that this concept was introduced into cancer therapeutics (Hartwell et al., 1997).

The synthetically lethal effect of PARP1 inhibition is not limited to cancer cells with *BRCA1/2* mutations alone but also extends to cells that show other key gene dysfunctions in the HR pathway. For example, it has been demonstrated that 30%–50% of high-grade serous tumours may be sensitive to PARPi due to mutations in other HR repair genes (Luo et al., 2020b). Tumours with loss of HR proteins, including RAD51, RPA1, NBS1, ATR, ATM, CHK1, CHK2, FANCD2, FANCA, and FANCC, have similar biological and clinical behaviours to those with *BRCA1/2* mutations. This HR pathway deficiency confers a unique vulnerability to PARP inhibition and is described as 'BRCAness'. Studies have demonstrated that BRCAness can be induced. For example, Luo et al. (2020a) created double BRCAness in NSCLC cell models (which lack *BRCA*1/2 mutation) by knocking down the *BRCA1* and *RAD50* genes. The clonogenic survival assay revealed that the knockdown cells were significantly sensitive to the effect of fluzoparib (a PARP1 inhibitor) (Luo et al., 2020a). Recently, genome-wide profiling of synthetic genetic lethality revealed cyclin-dependent kinase 12 as one of the additional genes conferring sensitivity to PARPi (Keung et al., 2020). Similarly, the therapeutic benefit of this BRCAness has been highlighted in triple-negative breast cancers and such benefits include increased sensitivity to PARPi and overcoming chemoresistance (Underhill, Toulmonde, and Bonnefoi, 2011; Lu et al., 2018; Boussios et al., 2019; Parkes, McCabe, and Kennedy, 2019).

There are two distinct mechanisms that underlie PARPi-induced death of cancer cells (i.e., the efficacy of PARPi): enzymatic inhibition, which is the original mechanism that has been known for a while and which first underpinned the development of PARPi; and PARP trapping, which is a relatively recent understanding of the effects of PARP inhibition.

3.3.3.1 Enzymatic Inhibition

PARP1 recognises and binds to the sites of SSBs through the zinc-finger DNA-binding domains. It then catalyses the transfer of ADPrs from NAD^+ to acceptor proteins that repair SSBs. The resulting poly (ADP-ribose) polymer (PAR) has a negative charge that helps recruit other positively charged DDR proteins necessary for the repair of SSBs. This enzymatic activity plays a major role in the BER pathway. However, the enzymatic activity of PARP1 is targeted by PARPi, which compete with NAD^+ at the catalytic site, thus preventing PARylation. Subsequently, PARPi disrupt the repair of SSBs, leading to the accumulation of double-strand DNA breaks (DSBs) through subsequent DNA replications. In normal cells, such DSBs are repaired mainly by homologous recombination repair (HR) using sister chromatids as a template, but in cancer cells with defective HR, DSB repair is attempted by an alternative pathway, non-homologous end-joining (NHEJ). NHEJ is error-prone and not effective in dealing with the sheer volume of DSBs that results from unrepaired SSBs. The accumulation of these DSBs leads to increased genomic instability, eventually resulting in cell death (Figure 3.4). This is a key mechanism by which PARPi exert cytotoxic activity (D'Andrea, 2018).

3.3.3.2 PARP Trapping

A turning point in the understanding of the mechanisms of action of PARPi was reached when Murai et al. (2012) first demonstrated that PARPi can trap the PARP1 and PARP2 enzymes at damaged DNA. They also showed that trapped PARP-DNA complexes were more cytotoxic than the accumulation of unrepaired SSBs caused by the pharmacological inhibition or genetic depletion of

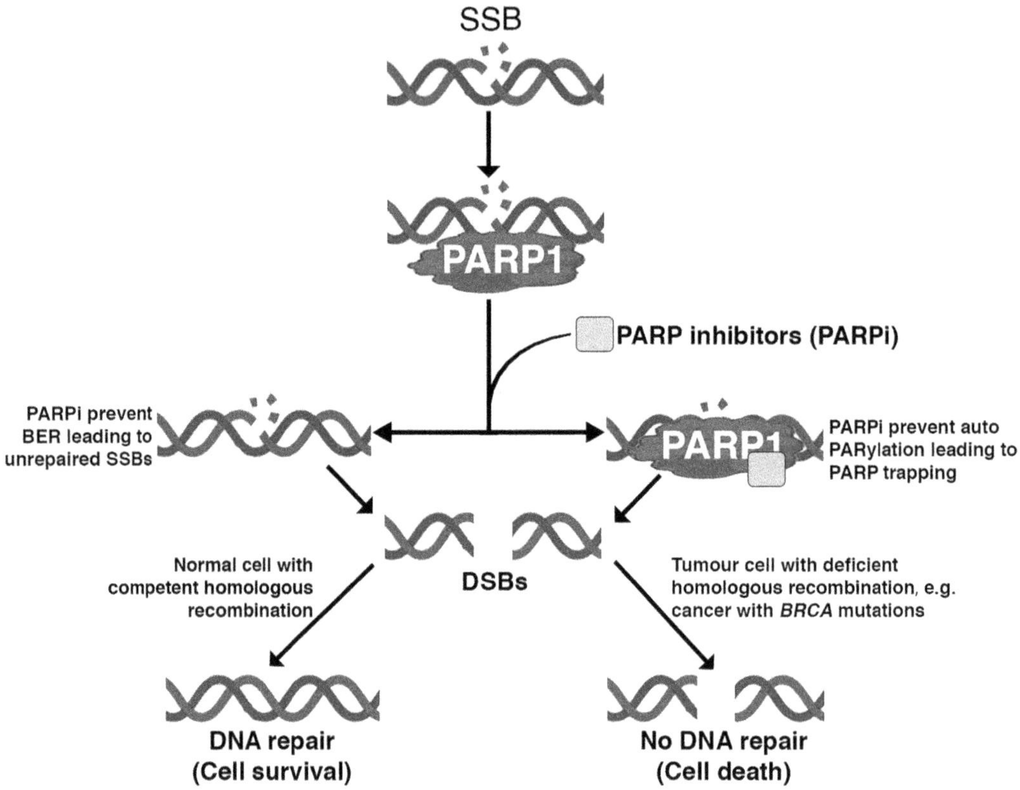

FIGURE 3.4 Mechanisms of PARP1 inhibition and trapping on the damaged DNA, leading to cell death. A PARPi will prevent the repair by PARP of single-strand DNA breaks (enzymatic inhibition of PARP) but could also trap PARP on the DNA damage site to form PARP-DNA complexes (PARP trapping), ultimately leading to double-strand breaks. In HR-deficient tumours, the resulting DSBs cannot be repaired, leading to cell death. Figure created by authors.

PARP. Although the precise mechanisms of locking or trapping PARP1 on damaged DNA are yet to be understood, there are speculations that PARP1 inhibitors prevent the release of PARP1 from DNA by inhibiting autoPARylation or causing allosteric changes in PARP1 structure that increase its stickiness to the DNA. Either way, the trapped PARP1 will cause stalling and the collapse of replication forks (Figure 3.4).

Further work has shown that the length of time for which PARP is trapped on DNA varies significantly, based on the doses of the different PARPi. This implies that there is a link between their inhibitory potencies and their trapping potential. It has been demonstrated that weak PARP1 trappers, such as veliparib, do not give the same level of synthetic lethality in preclinical models when compared with effective trappers such as olaparib, rucaparib, and niraparib. Consequently, veliparib, because of its lower trapping potency, has not demonstrated significantly increased anti-tumour activity when combined with neoadjuvant platinum-based chemotherapy. Current studies seem to show that new inhibitors with better PARP trapping are more potent PARPi, suggesting that PARP trapping is important in improving the efficacy of PARPi in cancer treatment. Trapped PARP-DNA complexes were more cytotoxic than unrepaired, single-strand breaks caused by PARP inactivation (Murai et al., 2012). However, this high trapping potential might prove to be a liability for other promising non-oncological applications of PARPi, such as for neuroprotection (Boussios et al., 2019; Hoch and Polo, 2019; Mateo et al., 2019).

3.3.4 Other Non-DNA Repair Roles of PARP1

Apart from its DNA repair roles, PARP1 is involved in various biological activities. Therefore, PARPi have been suggested or shown to be potentially therapeutic in several different conditions (see Berger et al., 2018). For example, PARP1's pro-inflammatory roles have been reported in different cancers. PARP1 regulates the inflammatory process by acting as transcriptional coactivator/corepressor of nuclear factor-κB, a family of proteins that are involved in acute and chronic inflammatory processes. For instance, the acetylation of PARP1 by acetyltransferases P300/CBP leads to the activation of NF-κB. Moreover, the activation of the NF-κB pathway is involved in proliferative signalling, tumour invasion and metastasis, and tumour-promoting inflammation. Various studies have indicated the potential benefits of PARPi in chronic inflammatory and immune-mediated diseases. For example, the PARPi minocycline reduces apoptosis and alleviates inflammatory response after myocardial infarction through inhibiting PARP1 (Zhao, Zhang, and Hong, 2018). Furthermore, PARP1 activates extracellular regulated kinase (ERK) signalling, which promotes metastasis and angiogenesis. Studies have demonstrated that PARP1 is also involved in chromatin compaction, de-condensation and remodelling. This is achieved either by direct poly ADP-ribosylation of histones and other chromatin-associated proteins, or indirect control of the recruitment of chromatin remodelling factors. Additionally, PARP1 can regulate transcriptional processes by directly binding to promoters or other transcription factors. For example, in healthy melanocytes, inactive PARP1 binds to the CXC ligand 1 (CXCL1) promoter, thus preventing the binding of NF-κB. However, in melanoma, the activation of PARP and PAR modification releases the CXCL1 to bind NF-κB. This enhances the expression of CXCL1 and the production of CXCL1 chemokine, a key effector of tumour growth (Bi et al., 2019).

PARP1 inhibition has also been reported to improve microvascular function in type 2 diabetes mellitus, prevent hypertension-induced cardiac remodelling, block established asthma, mitigate elastase-induced lung inflammation/emphysema, and promote sleep (Choi et al., 2012; Ghonim et al., 2015; Wang et al., 2018; Zada et al., 2021). More recently, Curtin et al. (2020) reviewed the possibility of repurposing PARPi to combat the then global pandemic (COVID-19) caused by SARS-CoV-2. One of the serious complications of COVID is acute lung injury in which upregulation of PARP1 gene expression is involved in a cascade of events with other receptors. Murine models and preclinical trials have shown that PARPi can decrease key interleukins such as IL-1, IL-6, and TNF-α in SARS-CoV-2-induced cytokine storm and thus alleviate lung fibrosis after infection (Curtin et al., 2020; Ummarino, Hausman, and Di Ruscio, 2021).

As described later in this chapter, PARP has been shown to be involved in the pathophysiology of neurological and neurodegenerative diseases, and thus a current and major application of the concept of repurposing is in developing PARPi as potentially disease-modifying neurotherapeutics.

3.4 PARPi AND THEIR MECHANISMS OF ACTION

The activity of PARP enzymes was first reported in 1963 (Chambon, Weill, and Mandel, 1963). It was revealed that they were dependent on DNA and NAD^+ as a substrate. Mechanistically, PARPI work as targeted anti-cancer agents on the basis of synthetic lethality. Although initially used only as chemosensitisers in chemotherapy and radiotherapy, PARPi were later identified as synthetically lethal partners with cancers arising from a *BRCA* gene mutation, and thus represent the first class of targeted cancer therapeutics to explore synthetic lethality.

In 1980, Purnell and Whish reported the first synthetic PARPi, 3-aminobenzamide (3-AB), a structural analogue of NAD^+, following their observation that nicotinamide and 5-methylnicotinamide competed with NAD^+ as a PARP1 substrate. 3-AB inhibits PARP to 96% but is required in millimolar concentrations, and its lack of specificity also inhibits de novo purine synthesis (Purnell and Whish, 1980; Milam and Cleaver, 1984). However, it was not until 1996 that the

PARP-inhibitory activity of 3-aminobenzamide was noted as a chemosensitiser of cisplatin in the treatment of ovarian cancer (Bernges and Zeller, 1996). This led to the rise of the first generation of PARPi, which were merely substituted nicotinamide or benzamide. A major advancement in the use of PARPi as chemosensitisers in cancers occurred in 2005 when it was first demonstrated in a proof-of-concept trial that PARPi were synthetically lethal in *BRCA1−/−* and *BRCA2−/−* cells, but not in *BRCA⁺/−* cells (Bryant et al., 2005; Farmer et al., 2005). Following this discovery, several analogues of benzamide with greater potency were developed as second-generation PARPi. We now have third-generation inhibitors with greatly improved properties belonging to a diverse family of chemical structures, such as derivatives of imidazopyridine, imidazoquinolinone, and isoquinolindione (Fatokun, Dawson, and Dawson, 2014; Moreno-Lama et al., 2020).

Currently, four PARPi have been approved by the US Food and Drug Administration (FDA) and the European Medicines Agency and are in clinical use. These are (in the order of their approvals from 2014 to 2018): olaparib, rucaparib, niraparib, and talazoparib (Figure 3.5). In order to identify patients who are eligible for and will benefit the most from PARPi, several companion diagnostic tests (CDx) have been developed to detect somatic or germline *BRCA1/2* mutations. The companion diagnostics not only identify *BRCA* mutations but also alterations in the other HR pathway genes, which are expressed as HRD status. Patients with *BRCA* mutations and/or HRD-positive status have shown significantly higher progression survival benefits from PARPi compared to patients with wild-type *BRCA*/HRD-negative or HR-proficient tumours (Mateo et al., 2019; Kyo et al., 2022; Paulet et al., 2022).

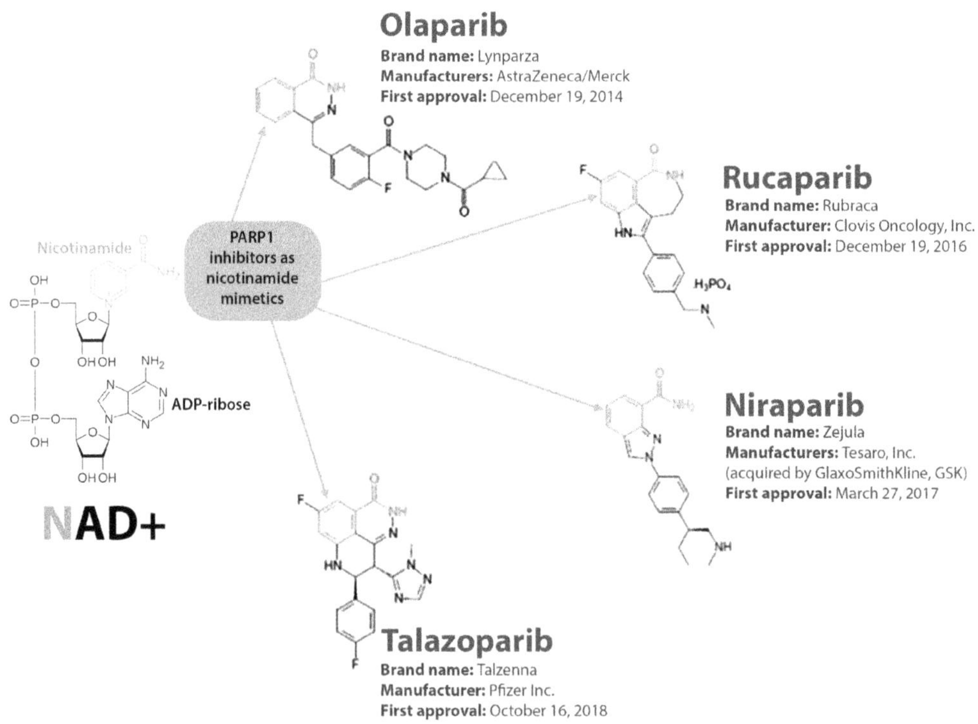

FIGURE 3.5 Chemical structures and manufacturers of the four approved and clinically used PARPi. Most PARPi are designed to mimic the nicotinamide moiety of NAD⁺ (in red), enabling them to compete for the NAD⁺-binding site on PARP1. In niraparib, this mimicry is achieved using a rotationally constrained primary benzamide, while in olaparib, talazoparib, and rucaparib, the benzamide motif is embedded in a ring. Some PARPi possess a fluorine atom in the meta-position of their NAD⁺-mimicking benzamide moiety, which enhances their ability to bind to their target. All the approved PARPi are PARP1/2 inhibitors, while olaparib and rucaparib also inhibit PARP3. In addition, rucaparib inhibits PARP5a and PARP5b (tankyrases 1 and 2). The brand names, the pharmaceutical manufacturers, and the dates of first approval are highlighted.

3.5 PARP IN NEURONAL DEATH (PARTHANATOS)

While PARP has a distinct and important physiological role, the understanding of which has been successfully harnessed for targeted cancer therapy, as already explained, it has also been known to contribute to pathological scenarios. The switch from a beneficial physiological role to a detrimental pathological role is underpinned by the degree of PARP activation, which consequently determines the amount and complexity of the polymer (PAR) that it generates. PARP activation that generates mild-to-moderate levels of PAR is consistent with its physiological function, while PARP over-activation that produces a surplus amount of PAR is known to induce cell death. The latter is particularly relevant to neuronal death. In this context, the death that is caused by over-activation of PARP1 has been termed 'parthanatos', a portmanteau word that comprises 'thanatos', the personification of death in Greek mythology, and 'par', the polymer generated by PARP, thus portraying 'parthanatos' as 'PAR-induced cell death', with PAR being generated by PARP, meaning that parthanatos could also be defined as 'PARP-mediated cell death'. It should be noted that while PARP does not leave the nucleus where it is localised, PAR as its messenger can exit the nucleus to act as a death signal (Andrabi et al., 2006).

Galluzzi et al. (2018) defined parthanatos as a form of regulated cell death driven by the hyperactivation of PARP1. Parthanatos is known to be induced by exposure to DNA-alkylating agents, such as MNNG, or exposure to cellular insults that can trigger DNA damage such as oxidative stress, nitrosative stress, inflammation, and glutamate-induced excitotoxicity, especially in neurones (Galluzzi et al., 2018).

Parthanatos has been widely implicated in many neurological conditions and neurodegenerative diseases, such as Parkinson's disease, Alzheimer's disease, stroke, Amyotrophic lateral sclerosis, Huntington's disease and ischaemia-reperfusion injury (Yang et al., 2020; Park et al., 2022). When PARP1 is over-activated, it produces a prodigious amount of PAR of varying complexities. PAR, which has been demonstrated to be a death signal, causes nuclear translocation of the mitochondrially-localised apoptosis-inducing factor (AIF), resulting in large-scale DNA fragmentation and nuclear condensation, and ultimately cell death (Andrabi et al., 2006; Fatokun, Dawson, and Dawson, 2014) (Figure 3.6). It has now been further clarified that AIF causes the recruitment of the nuclease macrophage migration inhibitory factor (MIF) to the nucleus, where MIF cleaves genomic DNA into large fragments (Wang et al., 2016). MIF has thus been recognised as a PARP1-dependent, AIF-associated nuclease (PAAN) (Wang et al., 2016). The entire process as described, from the insult-induced overactivation of PARP to the eventual parthanatic death of neurones, is described as the parthanatos cascade, which offers opportunities for pharmacological or similar interventions at various levels of the cascade in order to achieve neuroprotection.

It should be mentioned that, previously, the death of neurones as a result of PARP overactivation was thought to be due to the excessive consumption and thus depletion of NAD^+ and ATP, which caused the neurones to die in a characteristically necrotic manner (as apoptosis could not proceed in the absence of energy) (the suicide hypothesis) (Ha and Snyder, 1999), but it is now known that, rather than energy depletion, it is the PARP overactivation-mediated, PAR-induced triggering of AIF translocation and the consequent raft of further downstream signalling events as earlier described that is principally responsible for the cell death (Yu et al., 2002). Interestingly, parthanatos is a caspase-independent cell death and remarkably different from apoptosis. Although it shares some cytological and morphological features of apoptosis and necrosis, the co-involvement of PARP1, PAR and AIF characterises parthanatos (Fatokun, Dawson, and Dawson, 2014) (Table 3.1).

Recent studies have exploited the unique features of parthanatos, with a view to using protection against it as a surrogate for anti-PARP activity or activity against the PARP signalling cascade. A PARPi in this context would then be neuroprotective. One of these studies prospecting for neuroprotective PARPi revealed that two natural or naturally derived compounds, 4'-methoxyflavone and 3',4'-dimethoxyflavone, protected against parthanatos and were thus characterised as neuroprotective PARPi (Fatokun et al., 2013).

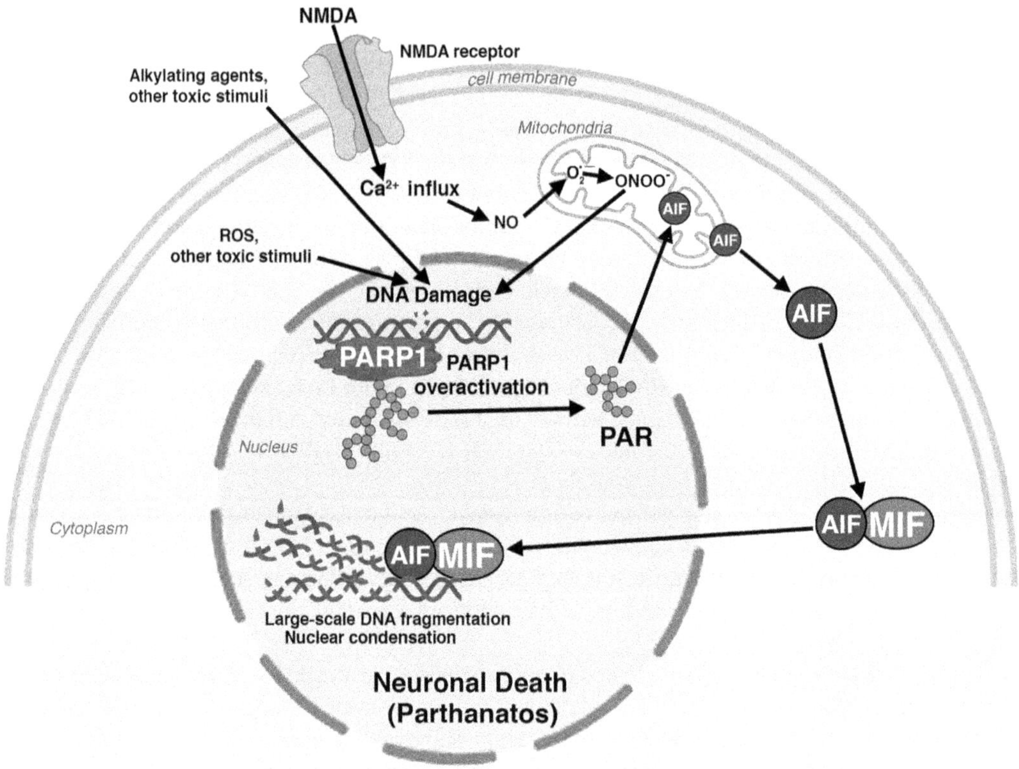

FIGURE 3.6 Schematic representation of PARP1-mediated cell (neuronal) death (parthanatos). Overactivation of PARP1 caused by insults such as oxidative stress induced by reactive oxygen species (ROS) leads to excessive production of poly (ADP-ribose) (PAR). PAR triggers a cascade that involves AIF release from the mitochondria and the translocation of the AIF/MIF complex to the nucleus, where it ultimately causes cell death because of strong endonuclease activity. Figure created by authors.

While the use of PARPi to prevent or treat disease conditions involving parthanatos, especially conditions involving neuronal death, seems promising, certain aspects of the molecular mechanisms of parthanatos are yet to be clarified.

3.6 REPURPOSING PARPi FOR THE TREATMENT OF NEUROLOGICAL AND NEURODEGENERATIVE CONDITIONS: OPPORTUNITIES AND CHALLENGES

Drug repurposing aims to develop an existing drug (or drug candidate) for use in treating diseases other than the one(s) for which it was originally developed. This strategy is usually more systematic and considerably less expensive and time-consuming, compared to conventional drug discovery programmes. Drug repurposing explores the pharmacological data of approved, preclinical and even failed therapeutics to discover new medical uses (Sahaboglu et al., 2016). Many *in vitro* and *in vivo* studies have confirmed the involvement of PARP in other pathologies apart from cancers. Therefore, although approved and preclinical PARPi are currently focused mostly on cancer therapy, there is now mounting evidence suggesting that PARPi could be repurposed for other diseases, especially in the context of achieving neuroprotection in neurological and neurodegenerative conditions such as stroke, Parkinson's disease, amyotrophic lateral sclerosis, Huntington's disease, and Alzheimer's disease (Martire, Mosca, and d'Erme, 2015; Berger et al., 2018; Thapa et al., 2021; Zhao et al., 2021). Such potential is described in the subsequent parts of this chapter.

TABLE 3.1
Distinctive Features of Four Different Molecular Mechanisms of Cell Death. While Several Cell Death Modalities Have Now Been Established (Galluzzi et al., 2018), the Table Compares Parthanatos with Three Major Morphologically Distinct Types of Cell Death, Namely Apoptosis, Necroptosis and Autophagic Cell Death. Adapted from Fatokun, Dawson and Dawson (2014) and Park et al. (2020).

	Parthanatos	Apoptosis	Necroptosis	Autophagic Cell Death
Examples of triggers	Reactive oxygen species, reactive nitrogen species, excitotoxicity, ischaemia, genotoxic stress, DNA damage	Death receptor signalling, trophic factor withdrawal, DNA damage	Activation of death receptors, Inhibition of caspase-8	Amino acid starvation, serum starvation, protein aggregates, rapamycin
Morphological and biochemical changes	Loss of membrane integrity, Phosphatidyl serine externalisation, Chromatin condensation, Large-scale DNA fragmentation, Mitochondrial depolarisation, Cell shrinkage	Cell membrane blebbing, Caspase activation, Apoptotic body formation, Phosphatidyl serine externalisation, Chromatin condensation, Small DNA fragmentation, Mitochondrial depolarisation, Cell shrinkage	Cellular swelling, Membrane rupture, Mitochondrial swelling	Cell membrane blebbing, Autophagic vacuoles, Increased lysosomal activity
Main mediators/ regulators	PARP-1, PAR, AIF, MIF	Caspases, Bax, Bak, APAF1, cytochrome *c*	RIP1/3, MLKL, calpains	Atg proteins, Beclin-1, VPS34, AMBRA-1
Energy (ATP) requirement?	No	Yes	No	Yes
Inflammatory response component?	No	No	Yes	No
Inhibitors	PARP inhibitors, PARG, MIF nuclease inhibitors (putative), Iduna	Caspase inhibitors, Bcl-2	RIP1 inhibitors (e.g., necrostatin-1), calpain inhibitors	VPS34 inhibitors, e.g., 3-methyladenine and wortmannin
Biochemical/ Cellular markers	PARP1 activation, PAR accumulation, AIF and MIF translocation to nucleus	Caspase activation	RIPK3 activation	Autophagosome formation, Lysosomal degradation

Cerebral ischaemia or ischaemic stroke is a global health issue that affects more than 3% of the adult population, resulting in lifelong disability or death. Up until now, the recombinant tissue plasminogen activator (tPA) is the only medicine approved by the FDA for the treatment of ischaemic stroke. However, the safety and effectiveness of tPA is limited to 3 to 4.5 hours after ischaemic stroke onset, and a vast majority of patients do not receive tPA administration due to its narrow therapeutic time window (Venkat et al., 2017). In view of this dire need for novel therapies for stroke, reports of PARP1 inhibition significantly reducing post-ischaemia/reperfusion(I/R)-induced brain damage and thus suggesting a promising opportunity to reposition PARPi for treating stroke are reassuring. In 2016, Xu et al. demonstrated that PARP inhibitors DPQ, veliparib, rucaparib, olaparib and talazoparib each attenuated neuronal cell death in an ischaemic, cell-based model derived by oxygen and glucose deprivation (Xu et al., 2016). Hamby et al. (2007) reported that PJ34, a PARP inhibitor, reduced neuronal cell death by over 80%, even up to 8 hours after transient ischaemia–reperfusion in mouse models. This was corroborated more recently by Jiao and Li (2021) when they demonstrated that inhibition of PARP1 by PJ34 mitigated I/R-induced brain damage via targeting over-activation of inflammatory response. Similarly, the PARPi JPI-289 was able to effect up to 53% decrease in infarct volume as late as 10 hours after stroke onset in a transient middle cerebral artery occlusion (tMCAO) stroke model in monkeys. Also, other PARP inhibitors, such as MP-124, nicotinamide, 3-aminobenzamide, 5-aminoisoquinolinone, 4-amino-1,8-napthalimide, and 1,5 isoquinolinediol have demonstrated protection against cell death induced by cerebral I/R injury and neurotoxins (Hamby et al., 2007; Han et al., 2020; Alhosaini et al., 2021; Jiao and Li, 2021; Yilmaz, Alkan, and Ballar Kirmizibayrak, 2021).

Another neuropathology for which PARPi repurposing is promising is Alzheimer's disease (AD). This neurodegenerative disease is the most common type of dementia. AD is characterised by accumulation of extracellular amyloid-β (Aβ) peptide and hyperphosphorylated tau protein, which eventually trigger excessive production of intracellular free radicals. The increased oxidative/nitrosative stress leads to DNA damage and subsequently overactivation of PARP, which thus implicates PAR-mediated neuronal cell death (parthanatos) (Feng et al., 2019). The PARPi PJ34 has been reported to suppress Aβ-induced microglial activation and neurotoxicity in a mouse model (Kauppinen et al., 2011). Recently, Maggiore et al. (2022) suggested two different mechanisms of neuroprotection by PARP1 inhibition in *Drosophila* models of AD. They demonstrated that the PARPi olaparib and MC2050 antagonised the aggregation of toxic Aβ oligomers while also suppressing the AD-induced transposable element activation by epigenetically regulating chromatin structure and function (Maggiore et al., 2022), a finding that reveals other mechanisms that might be involved in neuroprotection by PARPi in AD, apart from the PARPi preventing the accumulation of PAR.

Parthanatos has also been implicated in Parkinson's disease (PD), another common neurodegenerative condition and the most common movement disorder; therefore, PARPi can play a role in treating PD. In PD mouse models simulated by the injection of 6-hydroxydopamine (6-OHDA), genetic deletion of PARP1 prevented PD-related symptoms by suppressing AMP protein kinase and the nuclear translocation of AIF (Park et al., 2020). Kam et al. (2018) showed that the inhibition or deletion of PARP1 significantly reduced neurotoxicity induced by the transmission of pathologic α-synuclein in a PD murine model (Kam et al., 2018).

There are other CNS indications where the use of PARPi has been shown to be promising, including traumatic brain injury and amyotrophic lateral sclerosis, but a more extensive treatment of these scenarios could be found elsewhere (e.g., see Fatokun, Dawson, and Dawson, 2014; Berger et al., 2018; Guttman et al., 2025).

Despite the promise of repurposing PARPi for use in CNS conditions, there are a number of aspects that need to be considered, and which warrant further studies, including the potential risk of genotoxicity, PARP1 inhibition versus effects on other PARP isoforms, sex differences in the protective effects of PARPi (it was found in some, but not all, studies, that PARPi might be more efficacious in males than in females), and subsets of patients in whom PARPi would be most efficacious. In addressing these and related concerns, consideration should be given to whether alternative therapies exist for the indication and how effective they are (e.g., disease-modifying vs. symptomatic), and what the duration of PARPi treatment will be (acute vs. chronic). Furthermore, in repurposing PARPi for CNS conditions, factors such as preclinical efficacy data and the risk-benefit ratio of treating patients with a drug that inhibits an enzyme that has physiological roles in the regulation of DNA repair have to be considered (Schmid et al., 2020). On the basis of these factors, indications that have been recommended include those where the severity of the disease is high, available therapeutic options are limited, the duration of administration of the PARPi could be short, or, in cases of chronic devastating diseases, where alternative therapeutic options cannot halt disease development. Such CNS indications include acute ischaemic stroke, acute severe traumatic brain injury, Parkinson's disease, and progressive multiple sclerosis (Berger et al., 2018).

3.7 CONCLUSION

This review explores the physiological and pathological roles of PARP, first in cancer, where PARPi are currently used clinically as targeted therapy, on the basis of synthetic lethality, as well as PARP's roles in a wide range of other pathological conditions, focusing on neurological and neurodegenerative diseases, where PARPi have been suggested or shown to be neurotherapeutic and disease-modifying. However, drawbacks to the potential use of PARPi in treating CNS conditions remain, which could be overcome through further mechanistic studies and extensive and rigorous clinical trials before PARPi can progress to become part of the armamentarium for treating such CNS diseases, which will be a major advancement in achieving disease-modifying outcomes in this therapeutic domain.

ACKNOWLEDGMENT

A significant part of this work has been drawn from Saheed O. Benson's PhD thesis.

REFERENCES

Alemasova, E. E. and Lavrik, O. I. (2019) 'Poly(ADP-ribosyl)ation by PARP1: Reaction mechanism and regulatory proteins', *Nucleic Acids Research*, 47(8), pp. 3811–3827.

Alhosaini, K., Ansari, M. A., Nadeem, A., Bakheet, S. A., Attia, S. M., Alhazzani, K., Albekairi, T. H., Al-Mazroua, H. A., Mahmood, H. M. and Ahmad, S. F. (2021) '5-aminoisoquinolinone, a PARP-1 inhibitor, ameliorates immune abnormalities through upregulation of anti-inflammatory and downregulation of inflammatory parameters in t cells of BTBR mouse model of autism', *Brain Science*, 11(2).

Alvarez-Gonzalez, R. and Althaus, F. R. (1989) 'Poly(ADP-ribose) catabolism in mammalian cells exposed to DNA-damaging agents', *Mutation Research/DNA Repair*, 218(2), pp. 67–74.

Amin, N., Chaabouni, N. and George, A. (2020) 'Genetic testing for epithelial ovarian cancer', *Best Practice & Research: Clinical Obstetrics & Gynaecology*, 65, pp. 125–138.

Andrabi, S., Kim, N., Yu, S.-W., Wang, H., Koh, D., Sasaki, M., Klaus, J., Otsuka, T., Zhang, Z., Koehler, R., Hurn, P., Poirier, G., Dawson, V. and Dawson, T. (2006) 'Poly(ADP-ribose) (PAR) polymer is a death signal', *Proceedings of the National Academy of Sciences of the United States of America*, 103, pp. 18308–13.

Auer, B., Nagl, U., Herzog, H., Schneider, R. and Schweiger, M. (1989) 'Human nuclear NAD+ ADP-ribosylt ransferase(polymerizing): Organization of the gene', *DNA*, 8(8), pp. 575–580.

Berger, F., Lau, C. and Ziegler, M. (2007) 'Regulation of poly(ADP-ribose) polymerase 1 activity by the phosphorylation state of the nuclear NAD biosynthetic enzyme NMN adenylyl transferase 1', *Proceedings of the National Academy of Sciences*, 104(10), pp. 3765–3770.

Berger, N. A., Besson, V. C., Boulares, A. H., Bürkle, A., Chiarugi, A., Clark, R. S., Curtin, N. J., Cuzzocrea, S., Dawson, T. M., Dawson, V. L., Haskó, G., Liaudet, L., Moroni, F., Pacher, P., Radermacher, P., Salzman, A. L., Snyder, S. H., Soriano, F. G., Strosznajder, R. P. and Szabo, C. (2018) 'Opportunities for the repurposing of PARP inhibitors for the therapy of non-oncological diseases', *British Journal of Pharmacology*, 175(2), pp. 192–222.

Bernges, F. and Zeller, W. J. (1996) 'Combination effects of poly(ADP-ribose) polymerase inhibitors and DNA-damaging agents in ovarian tumor cell lines —With special reference to cisplatin', *Journal of Cancer Research and Clinical Oncology*, 122, pp. 665–67.

Bi, B., Tang, L., Liang, R., Wu, H., Huang, Q. and Pan, L. (2026) 'CXCL1-overexpressing cancer-associated fibroblasts stimulate hepatocellular carcinoma progression via neutrophil recruitment, NET formation, and immune suppression', *Cancer Immunology, Immunotherapy*, 75(4), p. 100. DOI: 10.1007/s00262-026-04304-3.

Bian, C., Zhang, C., Luo, T., Vyas, A., Chen, S.-H., Liu, C., Kassab, M. A., Yang, Y., Kong, M. and Yu, X. (2019) 'NADP+ is an endogenous PARP inhibitor in DNA damage response and tumor suppression', *Nature Communications*, 10(1), p. 693.

Boussios, S., Karihtala, P., Moschetta, M., Karathanasi, A., Sadauskaite, A., Rassy, E. and Pavlidis, N. (2019) 'Combined strategies with poly (ADP-ribose) polymerase (PARP) inhibitors for the treatment of ovarian cancer: A literature review', *Diagnostics*, 9(3).

Bryant, H. E., Schultz, N., Thomas, H. D., Parker, K. M., Flower, D., Lopez, E., Kyle, S., Meuth, M., Curtin, N. J. and Helleday, T. (2005) 'Specific killing of BRCA2-deficient tumours with inhibitors of poly(ADP-ribose) polymerase', *Nature*, 434(7035), pp. 913–917.

Cayuela, M. L., Carrillo, A., Ramirez, P., Parrilla, P. and Yelamos, J. (2001) 'Genomic instability in a PARP-1(-/-) cell line expressing PARP-1 DNA-binding domain', *Biochemical and Biophysical Research Communications*, 285(2), pp. 289–94.

Challa, S., Stokes, M. S. and Kraus, W. L. (2021) 'MARTs and MARylation in the cytosol: Biological functions, mechanisms of action, and therapeutic potential', *Cells*, 10(2). DOI: 10.3390/cells10020313

Chambon, P., Weill, J. D. and Mandel, P. (1963) 'Nicotinamide mononucleotide activation of new DNA-dependent polyadenylic acid synthesizing nuclear enzyme', *Biochemical and Biophysical Research Communications*, 11, pp. 39–43.

Chen, X., Huan, X., Liu, Q., Wang, Y., He, Q., Tan, C., Chen, Y., Ding, J., Xu, Y., Miao, Z. and Yang, C. (2018) 'Design and synthesis of 2-(4,5,6,7-tetrahydrothienopyridin-2-yl)-benzoimidazole carboxamides as novel orally efficacious poly(ADP-ribose)polymerase (PARP) inhibitors', *The European Journal of Medicinal Chemistry*, 145, pp. 389–403.

Choi, S.-K., Galán, M., Kassan, M., Partyka, M., Trebak, M. and Matrougui, K. (2012) 'Poly(ADP-ribose) polymerase 1 inhibition improves coronary arteriole function in type 2 diabetes mellitus', *Hypertension*, 59(5), pp. 1060–1068.

Curtin, N., Bányai, K., Thaventhiran, J., Le Quesne, J., Helyes, Z. and Bai, P. (2020) 'Repositioning PARP inhibitors for SARS-CoV-2 infection(COVID-19); a new multi-pronged therapy for acute respiratory distress syndrome?, *British Journal of Pharmacology*, 177(16), pp. 3635–3645.

D'Andrea, A. D. (2018) 'Mechanisms of PARP inhibitor sensitivity and resistance', *DNA Repair*, 71, pp. 172–176.

Di Girolamo, M. and Fabrizio, G. (2019) 'Overview of the mammalian ADP-ribosyl-transferases clostridia toxin-like (ARTCs) family', *Biochemical Pharmacology*, 167, pp. 86–96.

Eisemann, T. and Pascal, J. (2020) 'Poly(ADP-ribose) polymerase enzymes and the maintenance of genome integrity', *Cellular and Molecular Life Sciences*, 77.

Farmer, H., McCabe, N., Lord, C. J., Tutt, A. N. J., Johnson, D. A., Richardson, T. B., Santarosa, M., Dillon, K. J., Hickson, I., Knights, C., Martin, N. M. B., Jackson, S. P., Smith, G. C. M. and Ashworth, A. (2005) 'Targeting the DNA repair defect in BRCA mutant cells as a therapeutic strategy', *Nature*, 434(7035), pp. 917–921.

Fatokun, A. A., Dawson, V. L. and Dawson, T. M. (2014) 'Parthanatos: mitochondrial-linked mechanisms and therapeutic opportunities', *British Journal of Pharmacology*, 171(8), pp. 2000–2016.

Fatokun, A. A., Liu, J. O., Dawson, V. L. and Dawson, T. M. (2013) 'Identification through high-throughput screening of 4'-methoxyflavone and 3',4'-dimethoxyflavone as novel neuroprotective inhibitors of parthanatos', *British Journal of Pharmacology*, 169(6), pp. 1263–1278.

Feng, S.-T., Wang, Z.-Z., Yuan, Y.-H., Sun, H.-M., Chen, N.-H. and Zhang, Y. (2019) 'Mangiferin: A multipotent natural product preventing neurodegeneration in Alzheimer's and Parkinson's disease models', *Pharmacological Research*, 146, p. 104336.

Galluzzi, L., Vitale, I., Aaronson, S. A., Abrams, J. M., Adam, D., Agostinis, P., Alnemri, E. S., Altucci, L., Amelio, I., Andrews, D. W., Annicchiarico-Petruzzelli, M., Antonov, A. V., Arama, E., Baehrecke, E. H., Barlev, N. A., Bazan, N. G., Bernassola, F., Bertrand, M. J. M., Bianchi, K. and Kroemer, G. (2018) 'Molecular mechanisms of cell death: recommendations of the Nomenclature Committee on Cell Death 2018', *Cell Death & Differentiation*, 25(3), pp. 486–541.

Gaskin, J., Gomes, J. and Darshan, S. et al (2017) 'Burden of neurological conditions in Canada', *Neurotoxicology*, 61, pp. 2–10. DOI: 10.1016/j.neuro.2016.05.001.

Ghonim, M. A., Pyakurel, K., Ibba, S. V., Al-Khami, A. A., Wang, J., Rodriguez, P., Rady, H. F., El-Bahrawy, A. H., Lammi, M. R., Mansy, M. S., Al-Ghareeb, K., Ramsay, A., Ochoa, A., Naura, A. S. and Boulares, A. H. (2015) 'PARP inhibition by olaparib or gene knockout blocks asthma-like manifestation in mice by modulating CD4+ T cell function', *Journal of Translational Medicine*, 13(1), p. 225.

Gooch, C. L., Pracht, E. and Borenstein, A. R. (2017) 'The burden of neurological disease in the United States: a summary report and call to action', *Annals of Neurology*, 81(4), pp. 479–484. DOI: 10.1002/ana.24897.

Guo, C., Zhang, F., Wu, X., Yu, X., Wu, X., Shi, D. and Wang, L. (2020) 'BTH-8, a novel poly (ADP-ribose) polymerase-1 (PARP-1) inhibitor, causes DNA double-strand breaks and exhibits anticancer activities in vitro and in vivo', *International Journal of Biological Macromolecules*, 150, pp. 238–245.

Guttman, L. C., Yang, L., Liu, M., Dawson, V. L. and Dawson, T. M. (2025) 'Targeting PAAN/MIF nuclease activity in parthanatos-associated brain diseases', *Advances in Pharmacology*, 102, pp. 1–26. Academic Press.

Ha, H. C. and Snyder, S. H. (1999) 'Poly(ADP-ribose) polymerase is a mediator of necrotic cell death by ATP depletion', *Proceedings of the National Academy of Sciences*, 96(24), pp. 13978–13982.

Hamby, A. M., Suh, S. W., Kauppinen, T. M. and Swanson, R. A. (2007) 'Use of a poly(SDP-ribose) polymerase inhibitor to suppress inflammation and neuronal death after cerebral ischemia-reperfusion', *Stroke*, 38(2), pp. 632–636.

Han, S., Kim, Y. H., Choi, H. Y., Soh, D.-J., Kim, J., Nam, J., Kim, J.-W., Bae, K.-S. and Lim, H.-S. (2020) 'First-in-human evaluation of the safety, tolerability, and pharmacokinetics of a neuroprotective poly (adp-ribose) polymerase-1 inhibitor, jpi-289, in healthy volunteers', *Drug Design, Development and Therapy*, 14, pp. 3189–3199.

Harrision, D., Gravells, P., Thompson, R. and Bryant, H. E. (2020) 'Poly(ADP-ribose) glycohydrolase (PARG) vs. poly(ADP-Ribose) polymerase (PARP) – function in genome maintenance and relevance of inhibitors for anti-cancer therapy', *Fronters in Molecular Biosciences*, 28(7), p. 191.

Hartwell, L. H., Szankasi, P., Roberts, C. J., Murray, A. W. and Friend, S. H. (1997) 'Integrating genetic approaches into the discovery of anticancer drugs', *Science*, 278(5340), pp. 1064–8. DOI: 10.1126/science.278.5340.1064.

Hoch, N. C. and Polo, L. M. (2019) 'ADP-ribosylation: From molecular mechanisms to human disease', *Genetics and Molecular Biology*, 43(1 suppl 1), p. e20190075.

Hottiger, M. O., Hassa, P. O., Lüscher, B., Schüler, H. and Koch-Nolte, F. (2010) 'Toward a unified nomenclature for mammalian ADP-ribosyltransferases', *Trends in Biochemical Sciences*, 35(4), pp. 208–219.

Jannetti, S. A., Zeglis, B. M., Zalutsky, M. R. and Reiner, T. (2020) 'Poly(ADP-ribose)polymerase (PARP) inhibitors and radiation therapy', *Frontiers in Pharmacology*, 3(11), p. 170.

Jiao, Y. and Li, G. (2021) 'PARP inhibitor PJ34 ameliorates cognitive impairments induced by transient cerebral ischemia/reperfusion through its anti-inflammatory effects in a rat model', *Neuroscience Letters*, 764, p. 136202.

Jourdan, J. P., Bureau, R., Rochais, C. and Dallemagne, P. (2020) 'Drug repositioning: A brief overview', *Journal of Pharmacy and Pharmacology*, 72, pp. 1145–51.

Kalisch, T., Amé, J.-C., Dantzer, F. and Schreiber, V. (2012) 'New readers and interpretations of poly(ADP-ribosyl)ation', *Trends in Biochemical Sciences*, 37(9), pp. 381–390.

Kam, T.-I., Mao, X., Park, H., Chou, S.-C., Karuppagounder, S. S., Umanah, G. E., Yun, S. P., Brahmachari, S., Panicker, N., Chen, R., Andrabi, S. A., Qi, C., Poirier, G. G., Pletnikova, O., Troncoso, J. C., Bekris, L. M., Leverenz, J. B., Pantelyat, A., Ko, H. S. and Dawson, V. L. (2018) 'Poly(ADP-ribose) drives pathologic α-synuclein neurodegeneration in Parkinson's disease', *Science*, 362(6414), p. eaat8407.

Kamaletdinova, T., Fanaei-Kahrani, Z. and Wang, Z.-Q. (2019) 'The enigmatic function of PARP1: From PARylation activity to PAR readers', *Cells*, 8(12). DOI: 10.3390/cells8121625.

Kauppinen, T. M., Suh, S. W., Higashi, Y., Berman, A. E., Escartin, C., Won, S. J., Wang, C., Cho, S.-H., Gan, L. and Swanson, R. A. (2011) 'Poly(ADP-ribose)polymerase-1 modulates microglial responses to amyloid β', *Journal of Neuroinflammation*, 8(1), p. 152.

Keung, M. Y., Wu, Y., Badar, F. and Vadgama, J. V. (2020) 'Response of breast cancer cells to PARP inhibitors is independent of BRCA status', *Journal of Clinical Medicine*, 9(4), p. 940.

Ko, H. L. and Ren, E. C. (2012) 'Functional Aspects of PARP1 in DNA Repair and Transcription', *Biomolecules*, 2(4), pp. 524–548. DOI: 10.3390/biom2040524.

Krietsch, J., Rouleau, M., Pic, É, Ethier, C., Dawson, T. M., Dawson, V. L., Masson, J.-Y., Poirier, G. G. and Gagné, J.-P. (2013) 'Reprogramming cellular events by poly(ADP-ribose)-binding proteins', *Molecular Aspects of Medicine*, 34(6), pp. 1066–1087.

Kumar, V., Kumar, A., Mir, K. U. I., Yadav, V. and Chauhan, S. S. (2021) 'Pleiotropic role of PARP1: An overview', *3 Biotech*, 12(1), p. 3.

Kuny, C. V. and Sullivan, C. S. (2016) 'Virus–host interactions and the ARTD/PARP family of enzymes', *PLOS Pathogens*, 12(3), p. e1005453.

Kyo, S., Kanno, K., Takakura, M., Yamashita, H., Ishikawa, M., Ishibashi, T., Sato, S. and Nakayama, K. (2022) 'Clinical landscape of PARP inhibitors in ovarian cancer: molecular mechanisms and clues to overcome resistance', *Cancers*, 14(10).

Langelier, M. F., Eisemann, T., Riccio, A. A. and Pascal, J. M. (2018) 'PARP family enzymes: regulation and catalysis of the poly(ADP-ribose) posttranslational modification', *Current Opinion in Structural Biolo gy*, 53, pp. 187–198.

Li, S., Li, X. Y., Zhang, T. J., Zhu, J., Xue, W. H., Qian, X. H. and Meng, F. H. (2020) 'Design, synthesis and biological evaluation of erythrina derivatives bearing a 1,2,3-triazole moiety as PARP-1 inhibitors', *Bioorganic Chemistry*, 96, p. 103575.

Li, X., Erden, O., Li, L., Ye, Q., Wilson, A. and Du, W. (2013) 'Binding to WGR domain by salidroside activates PARP1 and protects hematopoietic stem cells from oxidative stress', *Antioxidants & Redox Signaling*, 20(12), pp. 1853–1865.

Lu, Y., Liu, Y., Pang, Y., Pacak, K. and Yang, C. (2018) 'Double-barreled gun: Combination of PARP inhibitor with conventional chemotherapy', *Pharmacology & Therapeutics*, 188, pp. 168–175.

Luo, J., Dai, X., Hu, H., Chen, J., Zhao, L., Yang, C., Sun, J., Zhang, L., Wang, Q., Xu, S., Xu, Y., Liu, N., Ying, G. and Wang, P. (2020a) 'Fluzoparib increases radiation sensitivity of non-small cell lung cancer (NSCLC) cells without BRCA1/2 mutation, a novel PARP1 inhibitor undergoing clinical trials', *Journal of Cancer Research and Clinical Oncology*, 146(3), pp. 721–737.

Luo, T., Yu, Q., Zou, H., Zhao, H., Gu, J., Yuan, Y., Zhu, J., Bian, J. and Liu, Z. (2020b) 'Role of poly (ADP-ribose) polymerase-1 in cadmium-induced cellular DNA damage and cell cycle arrest in rat renal tubular epithelial cell line NRK-52E', *Environmental Pollution*, 261, p. 114149.

Luo, X. and Kraus, W. L. (2012) 'On PAR with PARP: cellular stress signaling through poly(ADP-ribose) and PARP-1', *Genes & Development*, 26(5), pp. 417–432.

Lüscher, B., Ahel, I., Altmeyer, M., Ashworth, A., Bai, P., Chang, P., Cohen, M., Corda, D., Dantzer, F., Daugherty, M. D., Dawson, T. M., Dawson, V. L., Deindl, S., Fehr, A. R., Feijs, K. L. H., Filippov, D. V., Gagné, J.-P., Grimaldi, G., Guettler, S. and Ziegler, M. (2021) 'ADP-ribosyltransferases, an update on function and nomenclature', *The FEBS Journal*, 289(23), pp. 7399–7410.

Maggiore, A., Casale, A. M., Toscanelli, W., Cappucci, U., Rotili, D., Grieco, M., Gagné, J.-P., Poirier, G. G., d'Erme, M. and Piacentini, L. (2022) 'Neuroprotective effects of PARP inhibitors in drosophila models of Alzheimer's disease', *Cells*, 11(8). DOI: 10.3390/cells11081284.

Martire, S., Mosca, L. and d'Erme, M. (2015) 'PARP-1 involvement in neurodegeneration: a focus on Alzheimer's and Parkinson's diseases', *Mechanisms of Ageing and Development*, 146, pp. 53–64.

Mateo, J., Lord, C. J., Serra, V., Tutt, A., Balmaña, J., Castroviejo-Bermejo, M., Cruz, C., Oaknin, A., Kaye, S. B. and de Bono, J. S. (2019) 'A decade of clinical development of PARP inhibitors in perspective', *Annals of Oncology*, 30(9), pp. 1437–1447.

Milam, K. M. and Cleaver, J. E. (1984) 'Inhibitors of poly(adenosine diphosphate-ribose) synthesis: Effect on other metabolic processes', *Science*, 223(4636), pp. 589–591.

Mitra, A., Biswas, R., Bagchi, A. and Ghosh, R. (2019) 'Insight into the binding of a synthetic nitro-flavone derivative with human poly (ADP-ribose) polymerase 1', *International Journal of Biological Macromolecules*, 141, pp. 444–459.

Moreno-Lama, L., Galindo-Campos, M. A., Martinez, C., Comerma, L., Vazquez, I., Vernet-Tomas, M., Ampurdanes, C., Lutfi, N., Martin-Caballero, J., Dantzer, F., Quintela-Fandino, M., Ali, S. O., Jimeno, J. and Yelamos, J. (2020) 'Coordinated signals from PARP-1 and PARP-2 are required to establish a proper T cell immune response to breast tumors in mice', *Oncogene*, 39(13), pp. 2835–2843.

Murai, J., Huang, S.-y. N., Das, B. B., Renaud, A., Zhang, Y., Doroshow, J. H., Ji, J., Takeda, S. and Pommier, Y. (2012) 'Trapping of PARP1 and PARP2 by Clinical PARP Inhibitors', *Cancer Research*, 72(21), pp. 5588–5599.

Nosho, K., Yamamoto, H., Mikami, M., Taniguchi, H., Takahashi, T., Adachi, Y., Imamura, A., Imai, K. and Shinomura, Y. (2006) 'Overexpression of poly(ADP-ribose) polymerase-1 (PARP-1) in the early stage of colorectal carcinogenesis', *European Journal of Cancer*, 42(14), pp. 2374–2381.

O'Sullivan, J., Tedim Ferreira, M., Gagné, J.-P., Sharma, A. K., Hendzel, M. J., Masson, J.-Y. and Poirier, G. G. (2019) 'Emerging roles of eraser enzymes in the dynamic control of protein ADP-ribosylation', *Nature Communications*, 10(1), pp. 1182.

Padella, A., Ghelli Luserna Di Rorà, A., Marconi, G., Ghetti, M., Martinelli, G. and Simonetti, G. (2022) 'Targeting PARP proteins in acute leukemia: DNA damage response inhibition and therapeutic strategies', *Journal of Hematology & Oncology*, 15(1), p. 10.

Park, H., Kam, T.-I., Dawson, T. M. and Dawson, V. L. (2020) 'Chapter one - poly (ADP-ribose) (PAR)-dependent cell death in neurodegenerative diseases', in Spetz, J.K.E. and Galluzzi, L. (eds.) *International Review of Cell and Molecular Biology*: Academic Press, pp. 1–29.

Park, H., Kam, T.-I., Peng, H., Chou, S.-C., Mehrabani-Tabari, A. A., Song, J.-J., Yin, X., Karuppagounder, S. S., Umanah, G. K., Rao, A. V. S., Choi, Y., Aggarwal, A., Chang, S., Kim, H., Byun, J., Liu, J. O., Dawson, T. M. and Dawson, V. L. (2022) 'PAAN/MIF nuclease inhibition prevents neurodegeneration in Parkinson's disease', *Cell*, 185(11), pp. 1943–1959.e21.

Parkes, E., McCabe, N. and Kennedy, R. (2019) 'Development of PARP Inhibitors for BRCA-Deficient Epithelial Ovarian Cancer', in Dammacco, F., Silvestris, F. (eds.) Oncogenomics: From Basic Research to Precision Medicine. Academic Press, pp. 521–532. https://doi.org/10.1016/B978-0-12-811785-9.00036-3.

Paulet, L., Trecourt, A., Leary, A., Peron, J., Descotes, F., Devouassoux-Shisheboran, M., Leroy, K., You, B. and Lopez, J. (2022) 'Cracking the homologous recombination deficiency code: how to identify responders to PARP inhibitors', *European Journal of Cancer*, 166, pp. 87–99.
Purnell, M. R. and Whish, W. J. (1980) 'Novel inhibitors of poly(ADP-ribose) synthetase', *Biochemical Journal*, 185(3), pp. 775–777.
Rudolph, J., Mahadevan, J. and Luger, K. (2020) 'Probing the conformational changes associated with DNA binding to PARP1', *Biochemistry*, 59(21), pp. 2003–2011.
Schmid, S., Omlin, A., Higano, C., Sweeney, C., Martinez Chanza, N., Mehra, N., Kuppen, M. C. P., Beltran, H., Conteduca, V., Vargas Pivato de Almeida, D., Cotait Maluf, F., Oh, W. K., Tsao, C. K., Sartor, O., Ledet, E., Di Lorenzo, G., Yip, S. M., Chi, K. N., Bianchini, D., De Giorgi, U., Hansen, A. R., Beer, T. M., Lavaud, P., Morales-Barrera, R., Tucci, M., Castro, E., Karalis, K., Bergman, A. M., Le, M. L., Zürrer-Härdi, U., Pezaro, C., Suzuki, H., Zivi, A., Klingbiel, D., Schär, S. and Gillessen, S. (2020) 'Activity of platinum-based chemotherapy in patients with advanced prostate cancer with and without DNA repair gene aberrations', *JAMA Network Open*, 3(10), p. e2021692. DOI: 10.1001/jamanetworkopen.2020.21692.
Sahaboglu, A., Barth, M., Secer, E., Amo, E. M., Urtti, A., Arsenijevic, Y., Zrenner, E. and Paquet-Durand, F. (2016) 'Olaparib significantly delays photoreceptor loss in a model for hereditary retinal degeneration', *Scientific Reports*, 6(1), p. 39537.
Shah, M. M., Dobbin, Z. C., Nowsheen, S., Wielgos, M., Katre, A. A., Alvarez, R. D., Konstantinopoulos, P. A., Yang, E. S. and Landen, C. N. (2014) 'An ex vivo assay of XRT-induced Rad51 foci formation predicts response to PARP-inhibition in ovarian cancer', *Gynecologic Oncology*, 134(2), pp. 331–7.
Slade, D. (2020) 'PARP and PARG inhibitors in cancer treatment', *Genes & Development*, 34(5–6), pp. 360–394.
Steffen, J. D., Coyle, D. L., Damodaran, K., Beroza, P. and Jacobson, M. K. (2011) 'Discovery and structure–activity relationships of modified salicylanilides as cell permeable inhibitors of poly(ADP-ribose) glycohydrolase (PARG)', *Journal of Medicinal Chemistry*, 54(15), pp. 5403–5413.
Thapa, K., Khan, H., Sharma, U., Grewal, A. K. and Singh, T. G. (2021) 'Poly (ADP-ribose) polymerase-1 as a promising drug target for neurodegenerative diseases', *Life Sciences*, 267, pp. 118975.
Ummarino, S., Hausman, C. and Di Ruscio, A. (2021) 'The PARP way to epigenetic changes', *Genes*, 12(3). DOI: 10.3390/genes12030446.
Underhill, C., Toulmonde, M. and Bonnefoi, H. (2011) 'A review of PARP inhibitors: from bench to bedside', *Annals of Oncology*, 22(2), pp. 268–279.
van Beek, L., McClay, É, Patel, S., Schimpl, M., Spagnolo, L. and Maia de Oliveira, T. (2021) 'PARP power: a structural perspective on PARP1, PARP2, and PARP3 in DNA damage repair and nucleosome remodelling', *International Journal of Molecular Sciences*, 22(10). DOI: 10.3390/ijms22105112.
Venkat, P., Shen, Y., Chopp, M. and Chen, J. (2017) 'Cell-based and pharmacological neurorestorative therapies for ischemic stroke', *Neuropharmacology*, 134(Pt B), pp. 310–322.
Vyas, S., Chesarone-Cataldo, M., Todorova, T., Huang, Y.-H. and Chang, P. (2013) 'A systematic analysis of the PARP protein family identifies new functions critical for cell physiology', *Nature Communications*, 4(1), p. 2240.
Wang, H., Yang, X., Yang, Q., Gong, L., Xu, H. and Wu, Z. (2018) 'PARP-1 inhibition attenuates cardiac fibrosis induced by myocardial infarction through regulating autophagy', *Biochemical and Biophysical Research Communications*, 503(3), pp. 1625–1632.
Wang, L., Liang, C., Li, F., Guan, D., Wu, X., Fu, X., Lu, A. and Zhang, G. (2017) 'PARP1 in carcinomas and PARP1 inhibitors as antineoplastic drugs', *International Journal of Molecular Sciences*, 18(10). DOI: 10.3390/ijms18102111.
Wang, L., Yang, C., Xie, C., Jiang, J., Gao, M., Fu, L., Li, Y., Bao, X., Fu, H. and Lou, L. (2019) 'Pharmacologic characterization of fluzoparib, a novel poly(ADP-ribose) polymerase inhibitor undergoing clinical trials', *Cancer Science*, 110(3), pp. 1064–1075.
Wang, Y., An, R., Umanah, G. K., Park, H., Nambiar, K., Eacker, S. M., Kim, B., Bao, L., Harraz, M. M., Chang, C., Chen, R., Wang, J. E., Kam, T.-I., Jeong, J. S., Xie, Z., Neifert, S., Qian, J., Andrabi, S. A., Blackshaw, S. and Dawson, T. M. (2016) 'A nuclease that mediates cell death induced by DNA damage and poly(ADP-ribose) polymerase-1', *Science*, 354(6308), p. aad6872.
Wei, H. and Yu, X. (2016) 'Functions of PARylation in DNA damage repair pathways', *Genomics, Proteomics & Bioinformatics*, 14(3), pp. 131–139.
Wu, W., Zhao, J., Xiao, J., Wu, W., Xie, L., Xie, X., Yang, C., Yin, D. and Hu, K. (2021) 'CHFR-mediated degradation of RNF126 confers sensitivity to PARP inhibitors in triple-negative breast cancer cells', *Biochemical and Biophysical Research Communications*, 573, pp. 62–68.

Xu, J.-C., Fan, J., Wang, X., Eacker, S. M., Kam, T.-I., Chen, L., Yin, X., Zhu, J., Chi, Z., Jiang, H., Chen, R., Dawson, T. M. and Dawson, V. L. (2016) 'Cultured networks of excitatory projection neurons and inhibitory interneurons for studying human cortical neurotoxicity', *Science Translational Medicine*, 8(333), pp. 333ra48–333ra48.

Yang, D., Shu, T., Zhao, H., Sun, Y., Xu, W. and Tu, G. (2020) 'Knockdown of macrophage migration inhibitory factor (MIF), a novel target to protect neurons from parthanatos induced by simulated post-spinal cord injury oxidative stress', *Biochemical and Biophysical Research Communications*, 523(3), pp. 719–725.

Yilmaz, S., Alkan, T. and Ballar Kirmizibayrak, P. (2021) 'A new underlying mechanism for the neuroprotective effect of bosutinib: Reverting toxicity-induced PARylation in SIN1-mediated neurotoxicity', *Journal of Biochemical and Molecular Toxicology*, 35(12), p. e22915.

Yu, S.-W., Wang, H., Poitras Marc, F., Coombs, C., Bowers William, J., Federoff Howard, J., Poirier Guy, G., Dawson Ted, M. and Dawson Valina, L. (2002) 'Mediation of poly(ADP-ribose) polymerase-1-dependent cell death by apoptosis-inducing factor', *Science*, 297(5579), pp. 259–263.

Zada, D., Sela, Y., Matosevich, N., Monsonego, A., Lerer-Goldshtein, T., Nir, Y. and Appelbaum, L. (2021) 'Parp1 promotes sleep, which enhances DNA repair in neurons', *Molecular Cell*, 81(24), pp. 4979–4993.e7.

Zhao, H., Tang, J., Chen, H., Gu, W., Geng, H., Wang, L. and Wang, Y. (2021) '14,15-EET reduced brain injury from cerebral ischemia and reperfusion via suppressing neuronal parthanatos', *Int J Mol Sci*, 22(18).

Zhao, H., Zhang, J. and Hong, G. (2018) 'Minocycline improves cardiac function after myocardial infarction in rats by inhibiting activation of PARP-1', *Biomedicine & Pharmacotherapy*, 97, pp. 1119–1124.

Zhong, Y., Meng, Y., Xu, X., Zhao, L., Li, Z., You, Q. and Bian, J. (2019) 'Design, synthesis and evaluation of phthalazinone thiohydantoin-based derivative as potent PARP-1 inhibitors', *Bioorganic Chemistry*, 91, p. 103181.

4 Targeting Microglia for the Treatment of the Diseased Brain

Touraj Ehtezazi, Luca Peruzzotti-Jametti, Stefano Pluchino, and Darren Sexton

4.1 INTRODUCTION

Central nervous system (CNS) inflammation plays a key role in human non-communicable neurological disorders such as traumatic brain injury (TBI), (1) multiple sclerosis (MS), (2) treatment with chemotherapeutic drugs (3), Alzheimer's disease (AD) (4) [due to Aβ deposition or presence of tau oligomers in the brain] (5,6), chronic colitis (7), CDKL5 deficiency (8), subarachnoid haemorrhage (9), high fat diet (10), systemic injection of lipopolysaccharide (11), abdominal surgery (12), germinal matrix haemorrhage (GMH) (13), and intracerebral haemorrhage (14,15). CNS inflammation is linked to the activation of brain-resident immune cells (microglia) through mechanisms such as the suppression of suppressor MEK1 (16) and the infiltration of circulating leukocytes into the brain and spinal cord. Astrocytes are also activated in neuroinflammation, which is associated with neurodegenerative disorders (17), secreting IL-1β, which contributes to the breakdown of the blood-brain barrier (BBB) (18). There are also other mechanisms that astrocytes facilitate BBB disruption in MS (19). Activated astrocytes and microglia both exhibit increased gene expression (type I interferon response and MHC class I antigen presentation) in severe TBI (20), suggesting a coordinated immune response to the insult. In addition, neuroinflammation was associated with overexpression of heat shock protein family member 8 in the spinal cord ischemia–reperfusion injury, leading to the breakdown of the blood-spinal cord barrier (21). There are neuroinflammatory responses that are positive, which hand in hand with the intrinsic repair system recover brain injuries such as TBI (22). However, when the neuroinflammation becomes a persistent feature of the CNS, then this leads to significant accumulation of proinflammatory cytokines.

Several compounds have been studied to prevent activation of microglia. These include: α–linolenic acid (at 11 μg/mL *in vitro* cell culture concentration, typical serum level for ALA (α–linolenic acid) is 1.44 μg/mL (4, 23), phloretin (at the dose of 50 mg/kg administered intraperitoneally) (24), glycolysis inhibitors (2-deoxy-D-glucose at the dose of 1000 μM), (25) luteolin (in CDKL5 deficiency disorder at the dose of 10 mg/kg *in vivo*) (8), GW9508 (administered intranasally in rat model of GMH) (13), siponimod (25 μg/mL *in vitro*) (26), rutin (at the dose of 100 mg/kg in male Tau-P301S AD mouse model) (6), fibrin-derived $\gamma^{377\text{-}395}$ peptide (YSMKKTTMKIIPFNRLTIG at 200 μM *in vitro* for fibrinogen activated microglia) (27), and glatiramer acetate (28) in patients with MS (at 20-mg dose subcutenous injection daily) (29).

In this chapter, we will explore the critical role of microglia in neuroinflammation and its implications for brain diseases. We will review the complex process of microglial activation, with a particular focus on the cellular metabolic changes that occur during this process. This chapter also provides a comprehensive review of current clinical and pre-clinical therapies aimed at modulating microglial activation. We will critically evaluate these interventions, discussing their efficacy, limitations, and potential for future development in treating neuroinflammatory conditions.

DOI: 10.1201/9781003482406-4

4.2 MYELOID CELLS IN THE HEALTHY AND DISEASED CNS

4.2.1 Myeloid Cells in the Healthy Brain

Myeloid cells in the CNS are referred to as microglia, CNS-associated macrophages, and monocytes. Under the healthy conditions, microglia present in the brain parenchyma, and CNS-associated macrophages are located within the boundaries of the CNS parenchyma (perivascular space, subarachnoid space, and the choroid plexus) (30). Investigating wild type mouse and human CNS revealed a diverse population of immune cells in the healthy state (30). Microglia formed 80% of leukocytes (30) in the brain of healthy mice with homogeneous three-dimensional distribution in the brain parenchyma, a typical cell-to-cell distance of 50-60 μm, and a volume density of 6.5×10^3 cells/mm^3 (31). Under normal conditions, microglia have several long arms (expansions) with irregular shapes and numerous branches and spines (Figure 4.1A) (32). However, the microglia cell body swells under Pathological conditions (Figure 4.1B–F) (32). Border-associated macrophages (BAMs) comprised about 10% of leukocyte populations in the healthy mouse brain (30). These are blood-borne macrophages located in perivascular spaces (the area between brain endothelial cells and basement membrane) (34), choroid plexus (within the stroma), leptomeninges (pia mater or pial leptomeninges), and dura mater (within the connective tissue itself) (35).

Although blood-borne macrophages invade the perivascular space, they remain there and do not enter into the brain parenchyma under healthy conditions (34). These observations suggest that microglia are the only resident macrophages of the brain parenchyma in the healthy state. There are other immune cells in the CNS with less than 3% each (total 10%): neutrophils, monocytes, dendritic cells (DCs, classic and plasmacytoid), monocyte-derived cells, B cells, T cells, natural killer cells, innate lymphoid cells, eosinophils, and mast cells. These are located within the pia mater, perivascular space, choroid plexus, and dura mater (30). T cells form the majority of immune cells in

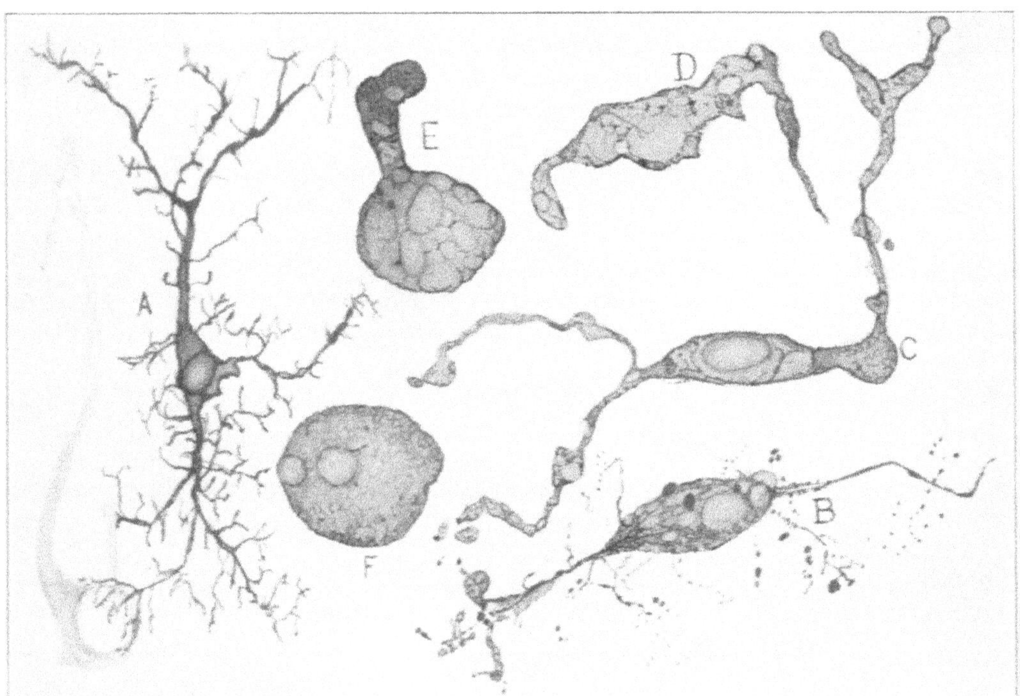

FIGURE 4.1 Microscopic photograph of human brain microglia at rest (A), (B)–(F) are activated microglia from areas adjacent to (A) brain tumour. Reproduced with permission from reference (33).

the cerebrospinal fluid (CSF) of healthy individuals (36), and border compartments (choroid plexus, pia mater, dura mater) may serve as the entry ports to the CNS. It was found that only 5% of brain leukocytes were blood-borne in the healthy murine CNS (30). DCs were located only in the choroid plexus, pia mater and dura mater (30). The population of myeloid cells was recently published for the human CNS with general similarities to mice, with exceptions such as the presence of microglia-like Kolmer cells in the choroid plexus (37).

Microglia are called to be at the resting state in a normal brain. Against the name, they are highly active at this state. They continuously survey the brain microenvironment through their processes and extensions; and the BBB disruptions immediately activate microglia (31). Microglia prevent the formation of extra connections between neurons and maintain suitable neuronal synaptic connectivity by removing unnecessary neuronal connections in the healthy adult brain (38). In addition, microglia phagocytose apoptotic cells (Figure 4.2) in the subgranular zone of the dentate gyrus and the subventricular zone during neurogenesis in the healthy adult brain (39).

Ionised calcium-binding adapter molecule 1 (IBA-1) is a marker expressed by microglia (40) in different morphological phenotypes (41), but not all the time, which was considered abnormal (42). IBA-1 expressing microglia form four different clusters in healthy human brains: SPP1^{-}CTSD^{-}IBA-1^{+}, SPP1^{-}CTSD^{+}IBA-1^{+}, SPP1^{+}CTSD^{-}IBA-1^{+}, and SPP1^{+}CTSD^{+}IBA-1^{+}, with SPP1^{-}CTSD^{-}IBA-1^{+} gene-expressing microglia forming 98%–99% of the microglia population. Examining human grey and white matter microglia revealed that both expressed IBA-1. However, white matter microglia significantly expressed more HLA-DR and CD68 than grey matter (43). It should be noted that microglia cell lines such as BV2 or RAW264.7 did not present the signature gene expression of mouse adult microglia. In addition, these cell lines did not strongly express human adult microglia genes (*P2RY12, GPR34, MERTK, C1QA, PROS1, GAS6*) (44). Both human and mouse microglia share the same gene and miRNA expression (44). TGF-β1 was required for the *in vivo* development and presence of microglia in the brain (44).

4.2.2 Microglia in Aged and Diseased Brain

There was heterogeneity among the activated microglia, which were grouped into 10 clusters in the brains of patients with MS (45). The gene expression patterns for the clusters of microglia completely changed with SPP1^{-}CTSD^{-}IBA-1^{+} gene-expressing microglia forming only 26%–53% of the population. While SPP1^{+}CTSD^{+}IBA-1^{+}cluster population increased to the range of 15%–51% (45). Therefore, there are subpopulations of microglia in the brain; however, the signature changes under pathological conditions. Disease-associated microglia (DAM) have been identified in gliomas (GAMs) with 14 distinguishable clusters. Certain clusters expressed *CD163* gene (anti-inflammatory role), while others expressed *Ccl2* (pro-inflammatory role) (43). GAMs had different gene expressions compared to homeostatic microglia with downregulation of homeostatic genes (such as *CX3CR1*) but upregulation of inflammatory genes (such as *TREM2*) (43). A Higher proportion of microglia present SPP1 in the AD patients compared to control subjects, with increased phagocytic activity is SPP1-positive microglia (46). SPP1 protein was released from perivascular macrophages of an AD mouse model, which activated microglia to engulf synapses in the hippocampus region (47). From the drug delivery perspective, this may be translated to targeting the perivascular macrophages and inhibition of SPP1 release to avoid neuronal damages by microglia. There were significant depositions of C1q and C3 (components of classical complement activation) on synapses of hippocampal and frontal cortex regions of AD mouse models (effects from oligomeric Aβ), which mediated engulfment of these synapses by microglia (48). Microglia contributed to the transmission of tau aggregates from one neuron to another by phagocytosis and exocytosis of the aggregates through exosomes (49). The activated microglia metabolism switches to anaerobic glycolysis, leading to the impairment of Aβ phagocytosis (50,51).

The CNS immune cell distribution and number significantly change in the brains with ageing and neurodegenerative diseases such as AD, MS, TBI, and glioblastoma (30, 37). The number

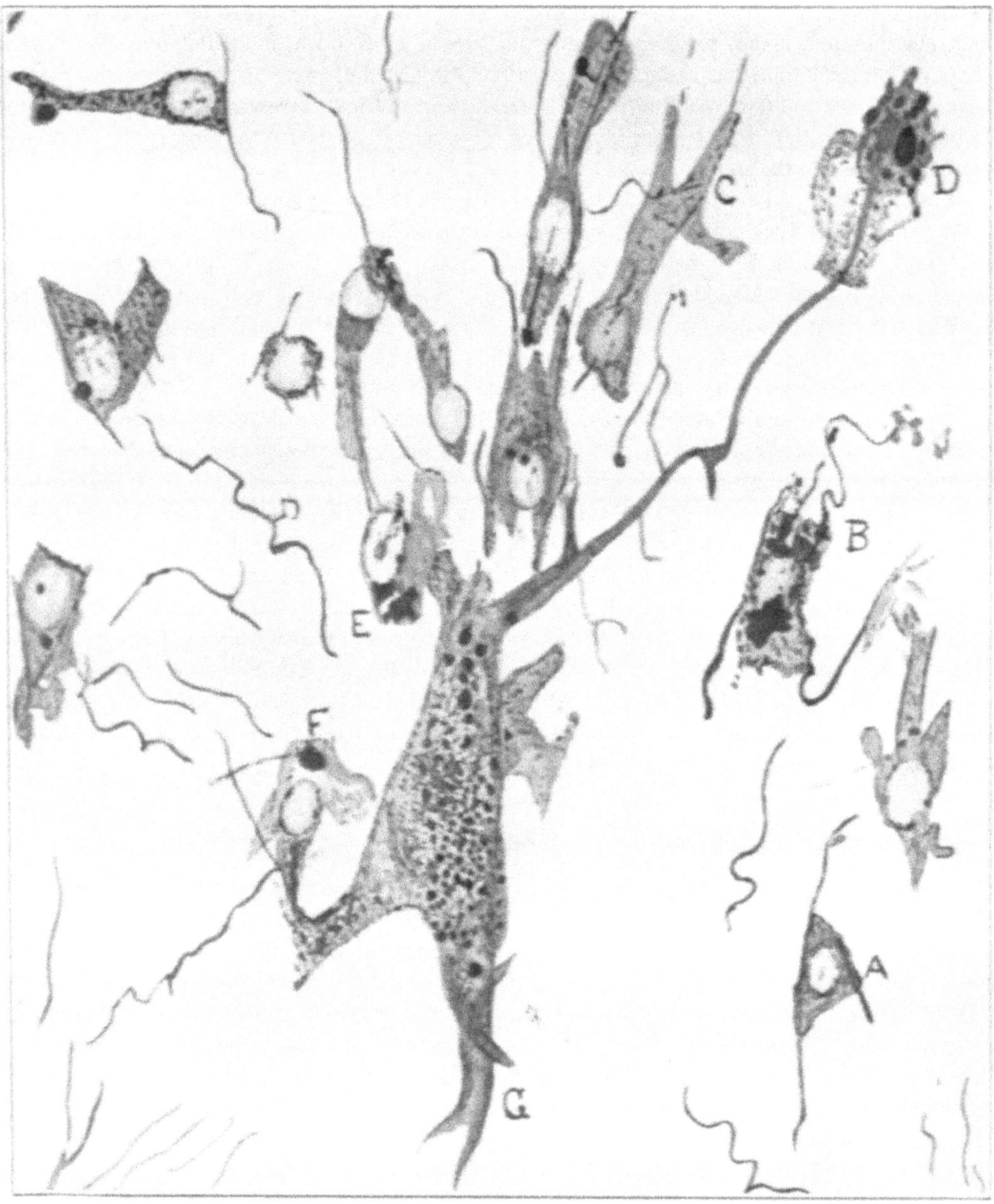

FIGURE 4.2 As part of microglia function, these cells phagocytose dead neurons in the human brain. Microglia ingesting neuron fibre fragments A, B, and C. Microglia are in the process of ingesting the remaining branches of a neuron D, E, and F. There are parts left such as branch G and cell body that perhaps waiting/ not attracted for phagocytosis. Reproduced with permission from reference (33).

of microglia increased by ageing in the corpus callosum and cerebellum of wild-type mice (52). The number and lengths of branches reduced in microglia with ageing both in mice (53) and humans (54). The ramifications are also reduced in aged microglia. (53) Furthermore, aged microglia become less ramified and dynamic in response to injury (53). Extracellular myelin fragments increase with ageing, which are cleared by microglia. However, this leads to intracellular myelin storage and age-associated immune activation (over expression of MHC-II) (55). With ageing, microglia become more vigilant to sense infectious microbial ligands, and at the same time they become more neuroprotective through upregulation of Stat 3 and Neuregulin-1 pathways (56). As a result, the microglia morphology change to more of hyper-ramified, round and amoeboid, indicating higher activation as well as increased cytokine release including TNF-α, TGFβ, IL-1β, and IL-6. While this activation, apparently protecting the neurons, suppresses neurogenesis through reduction of neural stem cell proliferation (57). Interestingly, a unique subset of microglia has been identified in both AD patients and mouse model of AD (known as DAMs) having typical markers of CD9, Itgax (CD11c), Clec7a, and CD63 (58). DAMs were localised near AD plaques and expressed high levels of phagocytic and lipid metabolism pathways, which are beneficial to AD. Therefore, inducing DAM could be a therapeutic target. Aβ plaques are decorated with Gas6 protein in human subjects with AD, which promotes engulfment of Aβ plaques by microglia through Ax1 and Mer receptors (59). The leakage of the BBB was clearly demonstrated in the hippocampus regions of the brains of patients with mild cognitive impairment (applying advanced dynamic contrast-enhanced MRI with using MultiHance as the contrast agent having a molar mass of 667.72) (60), and AD (an exclusive image was published in this review paper) (61). Fibrinogen (with molecular weight of approximately 340k Da) leaked into only certain regions of the brains of AD transgenic mice, which activated CD11b receptor of microglia and eliminated neuronal spines (62). These observations suggest that although the BBB becomes leaky in AD, only in certain regions large molecules and nanoparticles (NPs) may cross the BBB. Therefore, a transferrin receptor antibody was conjugated to triggering receptor expressed on myeloid cells 2 (TREM2) agonist antibody to produce a bispecific antibody. This improved the brain entry of the antibody 10-fold (compared to unconjugated TREM2 antibody) following intravenous injection and targeting microglia (63). The TREM2 agonist antibody strongly activated (100-fold) TREM2. Strong TREM2 activation promoted microglia phagocytosis. The expression of TREM2 on microglia contributed to anti-inflammatory responses and phagocytic activities, and stimulation by lipopolysaccharide (LPS) downregulated *TREM2* expression (64). The restoration of microglia phagocytic capacity by N,N′-diacetyl-p-phenylenediamine improved cognitive function in AD transgenic mice (65). Furthermore, the expression of proinflammatory cytokine genes was also reduced in microglia *(TNFα, IL-1β, IL-6, iNOS)* (65), suggesting that downregulation of these pro-inflammatory genes by alternative compounds may promote phagocytic capacity of microglia. This approach might be useful for the treatment of Parkinson's disease (PD) and MS.

The population of T cells significantly increased (almost 10 times) in aged mice compared to adult mice (30). Meanwhile the number decreased for plasmacytoid DCs and natural killer cells in the aged mice (30). In contrast to the healthy brain, T cells were identified in the brain parenchyma surrounding Aβ plaques as well as DAM in the mouse model of AD (30, 58). As reported previously, the number of infiltrating immune cells was increased in the brains of experimental autoimmune encephalomyelitis (EAE) mice. As a result, the percentage of microglial population decreased (30). In addition, macrophages were identified in the spinal cord parenchyma of EAE mice (66), and in the cortical demyelinating lesions of patients with MS (67). Although macrophages presented anti-inflammatory characteristics after ingestion of brain myelin (68), the number of T cells increased in the CSF of EAE rats. They migrated from leptomeninges through the CNS parenchyma into the CSF (69). Immune cells were identified in anatomically dissected glioblastoma from patients. These included DCs and macrophages (tumour-associated) drawn into the tumour due to the hypoxia within the regions that contained parenchyma and perivascular space (37).

When there is destruction of cerebral tissue, activated microglia migrate to the site of the lesion (33). In addition, immune cells infiltrate through damaged meningeal vascular network to the site of the injury following TBI. In particular, wound-healing macrophages localise to the lesion perimeter, and classical monocytes scavenge dead cells in the meninges (70). Furthermore, monocyte-macrophages engulf and digest fibrin at the site of injury to promote angiogenesis (70). It should be noted that there are re-injury and secondary injury post-TBI. Re-injury means that there is a physical trauma to the brain and most likely to the same primary damaged site. This additional damage will rapidly recruit vascular neutrophils to the damaged site (70). The secondary injury is a physiological response to the initial injury and the immune reactions form part of the physiological response. Investigating the CSF of TBI patients showed alarming levels of TNF-α cytokine, up to 757 pg/ml, which could promote observed neurodegeneration in TBI patients (71). TNF-α levels were 22.3 ± 9.5 pg/ml in the CSF of control subjects (72). Therefore, the immune reaction following TBI plays crucial roles in the repair of the injury. However, the response appears to be dysregulated, which leads to neuroinflammation and consequent neurodegeneration.

Activated microglia were noticed in the brains of patients with PD (73), Huntington's disease (74), MS (75), and dementia with Lewy bodies (76) using the $[^{11}C](R)$PK11195 PET imaging technique. Recently, it was discovered that ^{11}C-PBR28 bound to the microglia marker 18 kDa translocator protein. This is a better candidate due to having a higher signal-to-noise ratio compared to $[^{11}C](R)$PK11195 (77). Activated microglia have been implicated in the disease progression of neurodegenerative diseases such as MS (75). Microglia morphology changes upon activation. Morphologically, there are four types of microglia: (1) resting (ramified), (2) amoeboid, (3) activated, and (4) dystrophic (41, 78), as shown in Figure 4.3.

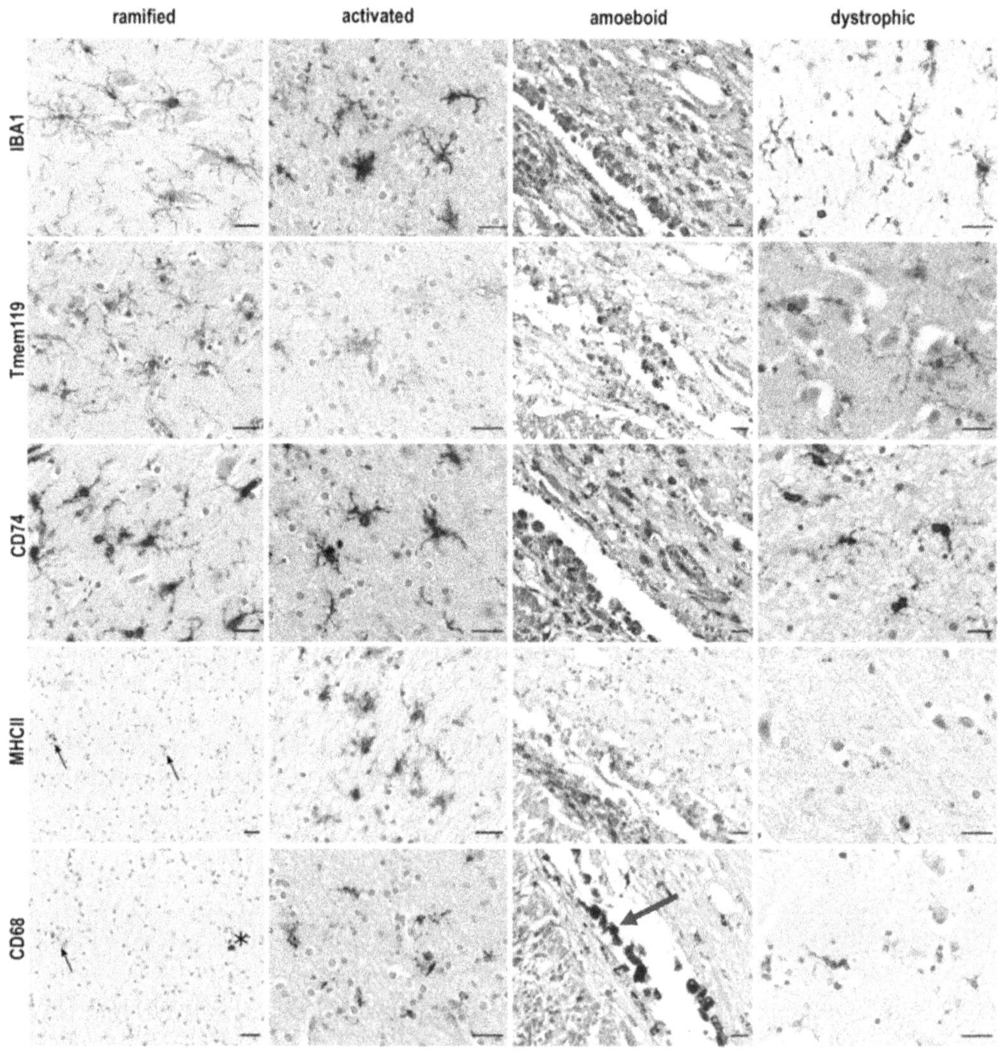

FIGURE 4.3 Different microglia morphological phenotypes and expression of markers in the human tissue from the frontal cortex and hippocampus. Reproduced with permission from reference (41).

4.3 THE TCA CYCLE IN IMMUNE CELLS

The immunometabolism is introduced for both macrophages and microglia in this section, as in certain brain disorders activated microglia and macrophages are involved in the brain. The tricarboxylic acid (TCA) cycle plays an important role in the phenotype of macrophages and microglia (79). The TCA cycle is intact in inactivated microglia and macrophages, which drives oxidative phosphorylation to produce ATP (Figure 4.4).

In oxidative phosphorylation, electrons are transferred from NADH or $FADH_2$ to O_2, and as a result ADP is converted to ATP. The oxidative phosphorylation rate is increased in M2 macrophages (80) and microglia (81). Acetyl-CoA cannot freely cross the mitochondria membrane, therefore it is transported to the cytosol in the form of citrate, or acetylcarnitine by carnitine acetyltransferase (CAT). Acetylcarnitine is translocated to the nucleus, where it is converted back to acetyl-CoA by CAT and becomes the source of acetyl for histone acetylation, (82,83) such as histone H3 at lysine 9 (H3K9) in microglia (84). On the other hand, the TCA cycle is broken at the stage of isocitrate conversion to α-ketoglutarate in M1 macrophages due to the downregulation of isocitrate dehydrogenase (IDH) enzyme (85), leading to the accumulation of citrate and potentially isocitrate in M1 macrophages and LPS activated microglia (Figure 4.5) (85,86). LPS activation of microglia led to reduction of acetate levels in microglia and therefore, hypoacetylation of H3K9, followed by production of increased pro-inflammatory cytokines (84). Acetate supplementation of microglia reversed LPS-induced hypoacetylation of H3K9 and LPS-induced expression of the pro-inflammatory cytokine proteins such TNF-α (84).

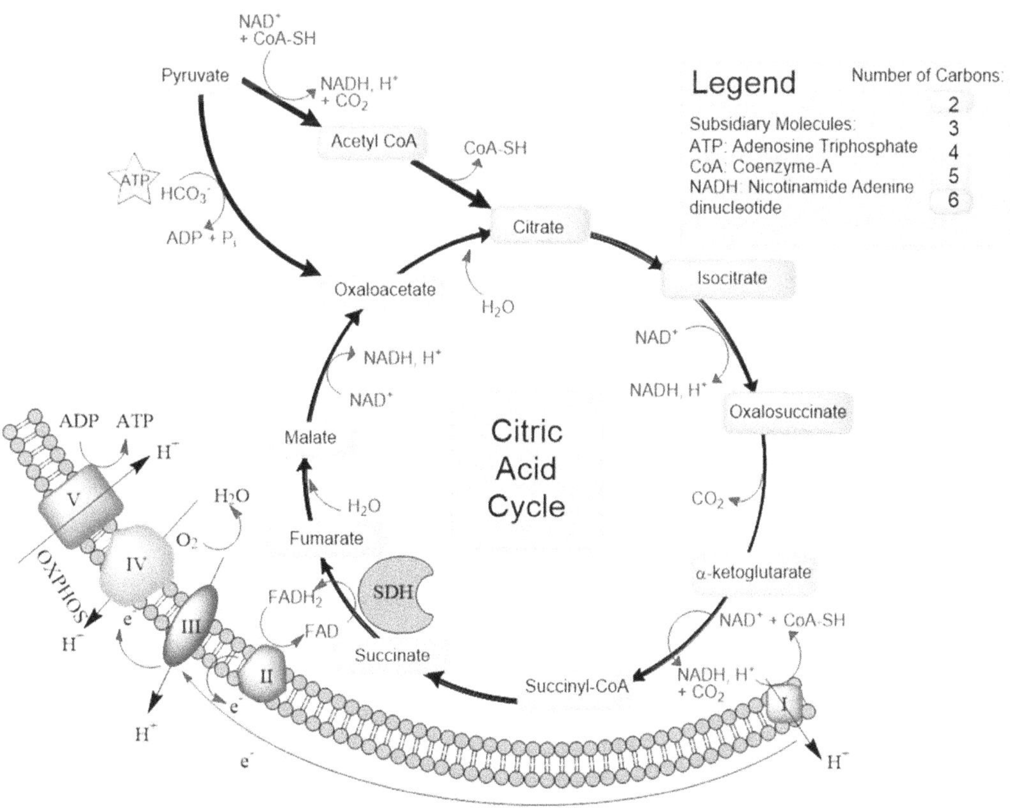

FIGURE 4.4 The TCA cycle in M2 microglia. The ATP is produced through oxidative phosphorylation.

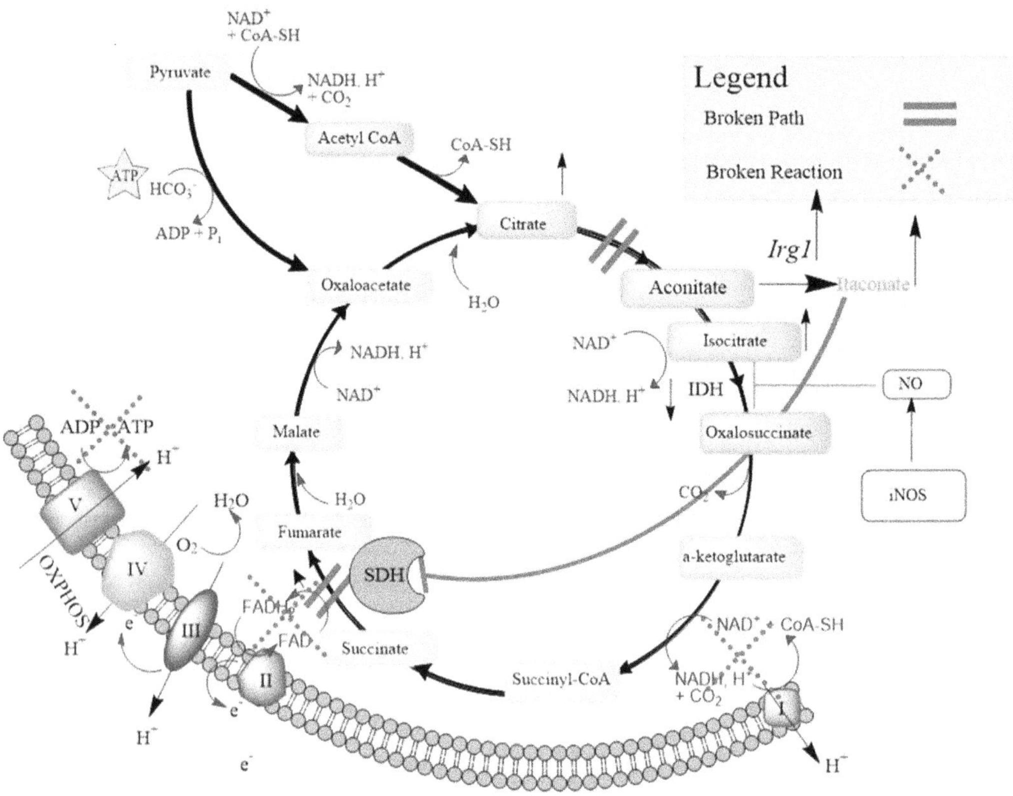

FIGURE 4.5 TCA cycle is broken at the stage of isocitrate conversion to α-ketoglutarate in M1 macrophages.

Furthermore, the expression of immune responsive gene 1 (Irg1) increases in activated macrophages (87) and LPS-activated microglia (88). Irg1 converts cis-aconitate to itaconate. As itaconate is cell impermeable, then dicarboxylate, citrate, and oxoglutarate carriers transport itaconate across the mitochondrial inner membrane (89). Accumulated itaconate forms itaconic acid in the cell, and activated macrophages by INF-γ and LPS produced itaconic acid, which was identified in the cells and cell culture supernatant (90). The accumulated itaconate in the mitochondria supresses succinate dehydrogenase enzyme (SDH) resulting in the accumulation of succinate in activated microglia (86) and macrophages (91, 92). SDH activity was inhibited by itaconate, due to an active aspartate-argininosuccinate shunt (85). The accumulated succinate can be transported to the cytosol by dicarboxylic acid transporter (93), and enhances the production of IL-1β and TNF-α in LPS activated macrophages. (91). The levels of fumarate and malate slightly increased in LPS activated macrophages (91), and microglia (86), compared to control levels. The accumulated succinate was also released to the extracellular space in EAE mice (animal model of MS), which led to significant increased CSF succinate levels compared to healthy controls and infiltration of macrophages into the CNS (94). In addition, the expression of citrate carrier increased, which transports citrate from mitochondria into the cytosol. Consequently, the citrate levels increased in the macrophages (95).

4.3.1 Glycolysis and Oxidative Phosphorylation

LPS stimulation switched macrophages (91) and microglia (96) to glycolysis. Pro-inflammatory cytokines such as IL-1β and INFγ switched microglia to glycolysis (97). Under the glycolysis condition, glucose is converted to pyruvate in the cytoplasm, (98) which then is transported to mitochondria by MPC-1 transporter and converted to acetyl-CoA by pyruvate dehydrogenase (99). Then citrate is formed from acetyl-CoA by citrate synthase, and *cis*-aconitate is formed from citrate by aconitase (100). The TCA cycle is also affected by the elevated nitric oxide (NO) in LPS+INFγ-stimulated macrophages. NO modulates the TCA cycle metabolites through the modulation of IDH (101).

Cells generate ATP for energy by employing both glycolysis and oxidative phosphorylation (OXPHOS). However, OXPHOS is more efficient as a source of energy compared to glycolysis. In fact, glycolysis produces 2 ATP per cycle, and OXPHOS produces 36 ATP per cycle. Glycolysis and OXPHOS work together to supply ATP needed by the cells (102). When a macrophage is activated, for example by LPS, the macrophage switches to glycolysis (103); and the glycolytic activities of the macrophage increase independent of the activation method. Then glucose becomes the main source of acetyl-CoA (82), as well as nuclear glycogen (82). The TCA cycle breaks leading to generating alarming metabolites and pro-inflammatory cytokines, while the cell requires ATP, if not more than before, to fulfil its mission. As a result, the number of glucose receptors increases to uptake more glucose molecules as needed to produce required ATP and achieve the biological targets. As pro-inflammatory immune cells rely on glucose to perform the inflammatory functions, then reducing the uptake of glucose may limit their ability to accomplish their mission (104).

4.3.2 Activation and Immune Response of CNS Myeloid Cells

Macrophages can be activated by three main modes: innate, classic, and alternative pathways (103). These three different methods lead to the expression of specific genes and the release of pro- or anti-inflammatory cytokines. LPS and INF-γ induce classic macrophage (101) and microglia (105) activation. Macrophage innate activation is mediated through Toll-like receptor (TLR) activation (using LPS), and alternative activation is achieved by using cytokines like IL-4/IL-13, or IL-10. More precisely, macrophages can be activated through LPS/INF-γ or Toll-like receptor 2 (TLR2), TLR3, TLR4, and TLR9 (103). The activation of TLR4, TLR2, and TLR9 by LPS caused increased expression of glucose transporter 1 and switched DCs to glycolysis through the phosphatidylinositol 3'-kinase (PI3K)/AKT pathway (106). Microglia also express TLR4 (107), which mediates activation by LPS (108). This shows that immune cells communicate with stimulants at least through TLRs. The manipulation of TLR-associated metabolic pathways could have therapeutic applications and inhibit excessive immune responses (106). For example, IL-10 inhibited the metabolic switch to glycolysis by TLR agonists in DCs (106). TLR-activated DCs rely on glucose as the source of nutrients and glucose limitation made them more sensitive to death (106). Migration of DCs into the CNS plays a crucial role in the pathogenesis of MS (109). DCs may be alarmed by the products of dying or damaged cells (110). Succinate and TLR agonists were required to induce the production of TNFα from DCs (110). The aerobic glycolysis increased in TLR-activated DCs but the levels of intracellular citrate and isocitrate decreased (111).

Both macrophages (112) and microglia (105) immune responses are also controlled by the circadian clock protein BMAL1, with the deficiency of BMAL1 increasing the rate of glycolysis and IL-1β expression in LPS- activated macrophages, compared to control LPS-activated macrophages (112). This indicates that BMAL1 plays an anti-inflammatory role in macrophages. Similar role was also observed for BMAL1 in microglia. The expression of BMAL1 was higher in the hippocampal microglia at the beginning of the day light and then gradually decreased until middle of the night for adult male Sprague–Dawley rats (105). As a result, LPS-activated hippocampal microglia produced minimal TNF-α at the beginning of the day light but reached maximum at the middle of the day, when BMAL1 expression was low in the microglia. Although the levels of BMAL1 were minimal in the microglia at the middle of the night, the expression of TNF-α was also minimal at that time. Therefore, microglia appeared to be more reactive during the middle of the day compared to the night (105). These observations would suggest that anti-inflammatory drugs may have a better efficacy at a certain time of the day.

4.4 TARGETING IMMUNOMETABOLISM TO INHIBIT MYELOID CELL ACTIVATION

An anti-inflammatory role has been suggested for itaconate (89). Pre-treatment of LPS-activated mouse bone marrow derived macrophages (BMDMs) with 7.5 mM itaconic acid [macrophages were pre-treated for 12 h then activated for 4 h] significantly reduced pro-IL-1β levels while significantly increased IFN-β levels in the cells, but did not change levels of TNF (113). Furthermore, endogenous itaconic acid (produced naturally in the cell) was able to reduce production of inflammatory cytokines like IL-6, IL-1β, and TNF-α in RAW264.7 cells (monocyte/macrophage) activated with LPS compared to cells in which *Irg1* was knocked down, however, the levels still were significantly higher than unstimulated macrophages (114). Itaconate was generated in large quantities following the activation of macrophages with LPS+IFNγ (90, 92) through the induction of Irg1 expression (92), and secreted into extracellular space. However, the quantities do not seem high enough to trigger anti-inflammatory responses (92). Interestingly, exogenous itaconate increased IL-1β production from BMDMs at 2.5 mM concentration, but then it decreased at 7.5 mM itaconate concentration [Extended Data Fig. 7C of ref (113)]. Furthermore, Atpenin A5 (a SDH inhibitor) significantly reduced the secretion of TNFα at 20 μM concentration from human activated T cells. However, addition of 1 mM itaconic acid decreased TNFα secretion from these cells but not significantly (115). Therefore, it is likely that the low levels of itaconate are either stimulatory in the cytosol rather than inhibitory for IL-1β production or not significantly effective in reduction of TNF-α expression. On the other hand, miR93 promoted M2 polarisation by reducing the expression of Irg1 and itaconate production (116,117). These observations could not lead to clear conclusions about the use of itaconic acid as an immunomodulatory agent.

As exogenous itaconate was not taken up intracellularly, then derivatives of itaconic acid, such as 4-octyl itaconate (OI) and dimethyl itaconate (DMI), were employed with the aim that these compounds would be hydrolysed to itaconate in the activated macrophages leading to enhance its immunomodulatory effects (89, 92). OI reduced serum IFN-β and IL-1β levels in mice injected with LPS, as well as IL-1β and TNF-α levels in LPS activated human peripheral blood mononuclear cells (89). In addition, serum IL-1β levels were reduced in mice pre-treated with OI and then stimulated with LPS compared to LPS-stimulated only mice (89). Initial treatment of bone marrow-derived DCs with OI at 320 μM (77 μg/mL) and then stimulation with R837 (TLR7 ligand) significantly reduced secretion of IL-1β compared to the OI-untreated cells (118). Pre-treatment with OI at 25 μM also significantly reduced secretion of IL-1β, IL-6, and TNF-α from post LPS activated peripheral blood mononuclear cells of patients with systemic lupus erythematosus (119). Similarly, DMI showed anti-inflammatory roles (significantly reduced pro-IL-1β levels in LPS activated macrophages) at a much lower concentration (0.25 mM) compared to itaconate itself (7.5 mM) (113) in mouse BMDMs. However, the anti-inflammatory role of DMI was not through the release of itaconate but via a strong electrophilic stress response (113), as DMI could not be metabolised to itaconate in mouse BMDMs (120). In addition, DMI reduced secretion of IL-6, but not TNFα, in LPS stimulated mouse BMDM at 40 μg/mL (250 μM) concentration leading to immuno-regulatory effects (121). The immunomodulatory effects of DMI were through the induction of Nrf2 response by reacting with cellular glutathione (121). It should be noted that DMI exerted cytotoxic effects against microglia at concentrations above 200 μM (32 μg/mL), but DMI significantly reduced the release of IL-1β from activated (LPS +ATP) microglia at 50 μM concentration (122). Furthermore, DMI was injected IP in mice with the amounts of 20 mg/500 μl (equivalent to 1000 mg/kg) for the treatment of IMQ-induced psoriasis, one day before the IMQ application and then daily for seven days. Mouse treated with DMI showed no detectable skin changes (121). Similarly, DMI was not effective in ameliorating disease severity in EAE mice at 300 mg/kg, but therapeutic effects were observed at 400 mg/kg doses, and DMI reduced the number of activated microglia in the brain and spinal cord (123). To summarise, itaconic acid or its derivatives suppressed production of pro-inflammatory cytokines when they were added in high concentrations extracellularly.

4.5 TARGETING MICROGLIA FOR NEURODEGENERATIVE DISEASE TREATMENT

As activated microglia play important roles in the development of neuroinflammation, then interventions to target microglia have become promising therapeutic strategies. (124,125). Table 4.1 lists the outcomes of clinical trials with effects on activated microglia. These are re-purposed agents such as metformin (used in the treatment of type 2 diabetes), which also reduced the release of

TABLE 4.1
Clinical Trials Conducted on the Reduction of Microglia Activation

API	Mechanism of Action	Disease	Main Finding	Ref
TB006	Monoclonal antibody targeting galectin 3, which its inhibition ↓ TNFα from α-synuclein activated microglia (128)	AD	↓ CDR-SB	(129)
Minocycline	↓TNFα from activated microglia (130)	Multiple-System-Atrophy	[^{11}C](*R*)-PK11195-PET imaging demonstrated ↓ microglia activation	(131)
Nilotinib	↓secretion of TNF-α, IL-1β from activated microglia (132)	AD	↓CSF Aβ40, Aβ42	(133)
NLY01	GLP-1R agonist, ↓TNF-α release from activated microglia (134)	PD	Motor benefit in younger participants	(135)
Exenatide	↑M2-microglia, ↓TNF-α (136)	PD	↑ Motor scores	(137)
Verdiperstat[a]	MPO inhibitor, ↓CD11b in microglia (138)	PD	↓Volume distribution of activated microglia across the nigrostriatal regions	(77)
Fluoxetine	↓TNFα and IL-6 from LPS activated microglia (139)	AD	↑Memory and cognition (preliminary study)	(140)
Ibudilast	↓TNFα, IL-1β, and IL-6 from LPS activated microglia (141)	MS	↑Median time to first relapse, ↓brain atrophy, and sustained disability progression	(142)
Masitinib	↓CSF-1R kinase activity in M-CSF treated microglia, ↓ M-CSF-induced microglia proliferation, ↓TNFα from microglia isolated from aged symptomatic SOD1^{G93A} rats (143)	ALS AD MS	↑ALSFRS-R ↑ADAS-cog ↓EDSS	(144) (145) (146)
RNS60 (Revalesio)	↓iNOS mRNA in LPS activated microglia (147) ↑ IL-4 in microglia from C57BL/6-SOD1^{G93A} mice (148)	ALS AIS	↓ALSAQ40 ↓Infarct growth	(149) (150)
Metformin	↓IL-1β, ↑IL-10 in LPS activated microglia (151)	Type 2 diabetes	↓Dementia incidence	(152)
Pepinemab	Antibody to block Sema4D, which promotes ramification of microglia (153), Pepinemab (VX15/MAB 2503) prevented microglia activation (154)	HD	↓OTS	(155)
IONIS-HTT$_{RX}$ (synthetic oligomer, given intrathecally)	↓mHTT, which promotes pro-inflammatory activity of microglia (128)	HD	↓Mutant HTT in Patients' CSF	(156)

[a] Also known as: 1-(2-Isopropoxyethyl)-2-thioxo-1, 2,3,5-tetrahydro-pyrrolo[3,2-d] pyrimidin-4-one and AZD3241

IL-1β from LPS activated microglia. AL002 is humanised monoclonal IgG1 antibody that activates the TREM2 signalling pathway in microglia with satisfactory outcomes in phase 1 trial (tested on healthy volunteers) (126). A phase 2 study is underway to evaluate the efficacy and safety of AL002 in participants with early AD (127). Table 4.2 presents pharmacological APIs to target microglia at the pre-clinical stages. In addition, there are NP-based drug delivery systems that have been developed to specifically target microglia. Table 4.3 summarises nano-formulations developed to target microglia in the diseased brain, and two examples are explained below with more details in relation to the design of NPs.

4.6 DESIGNING NPs TO TARGET MICROGLIA IN DISEASED BRAIN

NPs have been designed specifically to target the microglia for the treatment of PD. Poly(vinylpyrrolidone)-quercetin-copper-selenium (CSPQ)-based ultrasmall NPs were prepared with the size of 3 nm. The NPs were surface coated with MES23.5 cell membrane (CSPQ@CM), which contained VCAM-1. It was found that VCAM-1 interacted with α4β1-integrin expressed on the surface of microglia cells, promoting uptake of CSPQ@CM NPs by microglia. In fact, *in vitro* studies confirmed that CSPQ@CM NPs were taken up significantly more than CSPQ-NPs by microglia. CSPQ@CM NPs had a size of 25 nm. PD mice received 200 μL of 0.6 mg/mL of CSPQ@CM NPs intravenously for three days and showed significantly improved memory function and less IBA-1 expression in the brain. These NPs did not have brain-targeting ligand; and, therefore, focused ultrasound was used to facilitate NP crossing the BBB. It was suggested that CSPQ@CM NPs scavenged ROS in the brain of PD mice and, as a result, promoted M2 microglia phenotype, which was confirmed by increased expression of CD206 marker in the brain of PD mice (179).

In another approach, self-assembled peptide-based NPs were designed to deliver siRNA to microglia. The amphiphilic peptide contained poly-L-lysine (30 residues) to form a complex with siRNA. Poly-L-lysine was conjugated to mannose at the N-terminus to specifically targeting activated microglia (180). Mannose has been employed for targeting NPs to microglia. (181) Mannose binds to mannose receptor 1, which is also known as CD206, overexpressed in M2 type microglia (182), but considerably less expressed in M1 type microglia (183,184). In addition, it contained poly-leucine (30 residues), which formed the hydrophobic part of the amphiphilic peptide; and it was linked to poly L-lysine by an ROS-responsive thioketal (2-mercaptoacetic acid derivative). This amphiphilic peptide was denoted as LSL. Then, siRNA was mixed with LSL (N/P =5) to form core NPs (LSL@siRNA), with poly-L-Leucine at the core and siRNA with mannose facing towards the surface. Finally, the LSL-siRNA complexes were coated with another peptide (GMA). GMA contained Angiopep-2 (TFFYGGSRGKRNNFKTEEY) at the C terminus, to target the BBB, with an MMP-cleavable sequence (GPLGIAGQ) at the middle, and poly gamma benzyl glutamate (residues not specified) at the N terminus to cover siRNA and provide protection (180). These LSLGMA NPs had a diameter of 238 nm and a zeta potential of 1.25 mV, with LSL@siRNA to GMA ratio of 1. The LSLGMA NPs reduced NF-κB expression both at protein and mRNA levels in LPS-activated microglia when the NPs were loaded with siNF-κB (siRNA). Additionally, the expressions of TNF-α and IL-6 were reduced, which could be due to the ROS scavenging properties of the NPs and knockdown of NF-κB. LSLGMA NPs were intravenously administered at the dose of (1 mg/kg of siRNA) every three days for 10 administrations. The NPs improved behavioural experiments of MPTP induced PD mice, such as increased total travelled distance. Furthermore, the expressions of α-syn, NF-kB, IL-1β, TNF-α and IBA-1 were reduced in the brains of the PD mouse models.

4.7 CONCLUSIONS

Microglia play key roles in the maintenance of CNS functions both in health and disease. Microglia activation is required to remove cell debris when brain injuries happen such as TBI. However, chronic microglia activation leads to neurodegeneration, which requires switching activated

TABLE 4.2
Microglia Targeting APIs at the Pre-Clinical Stages

API	Mechanism of Action/*In Vitro* Outcomes	Disease/Route/Dose/Duration	Main *In Vivo* Outcomes	Ref
N, N'-Diacetyl-p-phenylenediamine	Restores microglial phagocytosis	AD/IP/2mg/kg/day for two months in transgenic mice	↑Cognitive function	(65)
TGF-β1	↓ *IL-6, TNF-α, Ccl2* gene expression in thrombin stimulated microglia	ICH/injection into the brain site of ICH (striatum)/10 ng/10 min prior to ICH	↑Behavioural outcomes	(157)
Dapansutrile	NLRP3 inflammasome inhibitor/↓ TNF-α, IL-6, IL-1β from WT primary microglia cells were cultured and stimulated with 1 µg/mL LPS	AD/oral/7.5 g/kg/for 3 months in APP/PS1 mice	↓ IBA-1 expression, TNF-α, IL-1β, IL-6, Aβ plaques in the brain, ↑cognitive function, number of dendritic spines	(158)
Desloratadine	Antagonist of $5HT_{2A}$ receptor/upregulating transcriptions of phagocytic receptors TLR2 and TLR4	AD/ oral gavage /20 mg/kg/day for three months	↓Aβ plaque in the brain of APP/PS1 mice	(159)
DW14006	A direct *AMPKα1* activator, ↑ microglia Aβ phagocytosis, ↓IL-1β, TNF-α mRNA and molecules activated by $Aβ_{25-35}$.	AD/ 30 mg/kg/day/IP/APP/PS1 mice/ for 100 days	↓ Aβ plaque, $iNOS^+$ cells, TNF-α, IL-1β in the brain, ↑cognitive function, IBA-1 expressing microglia around Aβ plaques, No. of dendritic spines	(160)
X15856: A hexacyclic hybrid of two peptide units and four peptoid[1] units	↓Phosphorylation of p38 MAPK, p42/44 MAPK, ERK 1/2, PKC, NF-κB, ↓IL-6, TNF-α, CCL2, CCL3, CXCL2, CXCL3, CXCL10 in LPS activated BV2 microglia	NI/25µM/pre-treated with X15856	*In vitro* study	(161)
Biphasic tetravalent TREM2 antibody (TREM2 agonist monoclonal Ab (Ab18))	The antibody targets transferrin receptor, enhances brain entry of the antibody as well as agonist antibody (Ab18) to ↑TREM2 activity receptor on microglia	AD/IP/20 mg/kg/day for 14 weeks	↓Aβ plaque in the brain of 5XFAD mice	(63)
Saikosaponin B2	↓TLR4 expression in LPS activated microglia	CUMS/ intragastric/10 mg/kg for 21 days	↓TLR4, IL-6, and IL-1β, ↑IL-10 in the hippocampus of stressed-induced mice	(162)
Sodium rutin	↑ Phagocytosis receptor in microglia + anaerobic glycolysis leading mitochondrial OXPHOS	AD/oral/18-25 mg/kg/day in drinking water for 7 months	↓Aβ deposition in hippocampal DG regions of 5XFAD mice	(50)

(Continued)

TABLE 4.2 *(Continued)*
Microglia Targeting APIs at the Pre-Clinical Stages

API	Mechanism of Action/*In Vitro* Outcomes	Disease/Route/Dose/Duration	Main *In Vivo* Outcomes	Ref
INF-γ	↑Glycolytic metabolism of microglia, ↑microglia immune response,	AD/IP/1x10^5 U/once a week/for 12 weeks	↓Brain Aβ load of 5XFAD mice	(51)
STF-083010	IRE1α inhibitor, ↓IBA-1, ↓IL-1β in activated microglia of HIE brain	HIE/Intra nasal/45μg/Rat pup/over 2 min/ single treatment	↓Infarct volume, IL-1β, IL-8, NLRP1 in HIE rat model	(163)
ROS-responsive polymeric micelle system, micelles had targeting peptide of KLVFFAED, which helped crossing the BBB in the hippocampus	↓Brain ROS,	AD/IV/5 mg/kg/once a week/ for 12 weeks	↓Microglia activation, ↓Aβ burden, ↓brain TNF-α in APP/PS1 mice.	(164)
MCC950	Inhibitor of NLRP3 inflammasome, ↓IL-1β from LPS+ATP or LPS+Aβ activated microglia, no effects on TNFα	AD/IP/10 mg/kg/every second day for 3 months	↓Microglia activation in APP/PS1 mice	(165)
		CUMS/IP/10 mg/kg/ once a day for 4 weeks	↓Microglia hyper-ramification (length, area, volume, number of branches, number of terminal points), ↓IL-1β, TNF-α, IL-6 in the cortex of stressed-induced mice	(166)

[1] Peptoids are compounds that mimic peptides but with improved stability against proteases.

TABLE 4.3
Nanoparticle Formulations to Target Activated Microglia at the Pre-Clinical Stage

Nanocarrier	API/Disease	Size (nm)/Charge (mv)	Delivery Route/Dose/ Duration	Brain Targeting Strategy/ Ligand	Main Finding	Ref
PLGA	miRNA-129-5p/NI	468 nm (for NP loaded with 250 µg of API+PEI)/-25 mV	In vitro/250 µg/mL/24 h with LPS then 72 h with NP	-	↓ IL-6, IL-1β, TNFα, NO, ↑ IL-10, VEGF, IL-13 in LPS activated microglia	(167)
Gold NPs	*Ephedra sinica Stapf*/NI	35 nm/-24 mV	*In vitro*/40 µg/mL/microglia Pre-treated with NPs for 1 h then 24 h LPS	-	↓IL-1β, TNFα from LPS activated microglia	(168)
PHSA@PF/pTREM2 Gene delivery system	↑TREM2 in microglia/AD	83 nm/19 mV	AD/IV/5 µg/mouse/ every 4 days for 5 times	CGRILARGEINFK peptide used to target both BBB and microglia, fluorinated PEI 1.8 kDa used for endosomal escape strategy	↓memory deficit in APP/PS1 mice	(169)
Amphiphilic macromolecules of PEG	NPs bind to CD36 receptors of microglia/PD	227 nm/-38 mV for most effective NP (T12P5)	*In Vitro*/ pre-treatment with 75 µM of αsyn- activated microglia for 24 h, then with 10 µM of αsyn for 24 h	-	↓TNFα and NO from αsyn-activated microglia	(170)
PLGA	PINK1 siRNA/ Ischemic stroke	211 nm/-34 mV	IT/20 µL of 10 µg/µL/once/ NPs were administered prior to stroke	-	↑M2 microglia, ↓Infarct Volume, NPs were taken up by microglia, ↓PINK1 in the brain	(171)
Polydopamine (PDA)	PDA/optic nerve injury	215 nm/-31 mV	Intravitreally injection/4µg/ once, sample collection 7 days post injection	-	↓Microglia density, IBA-1 in the retina	(172)
Plant based exosomes	*Panax notoginseng* saponins/ ischemic stroke	151 nm/-8mV	IV/single 3 mg/kg dose 2 h post reperfusion	Intrinsically cross the BBB	↓Infarct volume, ↑behaviour outcomes, internalised by microglia, ↓IL-6, IL-10,	(172)
Lipid NPs	siRNA (PU.1)/NI	76 nm/-13 mV	Intracisternal/70 µg/kg/every third day for 14 days (5 injections)	-	↓IBA-1 expression in transgenic CK-p25 mice	(173)

(Continued)

TABLE 4.3 *(Continued)*
Nanoparticle Formulations to Target Activated Microglia at the Pre-Clinical Stage

Nanocarrier	API/Disease	Size (nm)/Charge (mv)	Delivery Route/Dose/Duration	Brain Targeting Strategy/Ligand	Main Finding	Ref
Liposomes	IL-10/ICH	153 nm/-23 mV	IN/3 μg IL-10/mouse, 2 h post ICH for 7 days/ once a day	-	↓Injury volume, IL-1β, IL-6, ↑microglia soma size, $CD36^{+}IBA\text{-}1^{+}$ cell number	(174)
Dual targeting liposomes	Danshensu/ischemic stroke	101 nm/-17 mV	IV/2.5 mg/kg of Danshensu (5.7 mg/kg of liposomes)/ once a day for 3 days after reperfusion	TF for targeting BBB and PS for uptake by activated microglia	↓TNF-α, IL-6, IL-1β, IBA-1, infarct volume	(175)
Mn-porphyrin-BSA NPs coated with microglia membrane	Mn^{2+} + porphyrin+aPD -1/GB	184 nm/-15 mV	IV/5 mg/kg dose of Mn-porphyrin (23 mg/kg of NPs)	Use of microglia membrane	NPs were not designed to target microglia but stimulated them, ↑TNFα, survival rate	(176)
18β-glycyrrhetinic acid-conjugated-Diethylaminoethylen-dextran NPs	18β-glycyrrhetinic acid/ Ischemic Stroke	105 nm/30 mV	ICV/ 2μL of 175 μg/mL of NPs/once post stroke	-	↓Infarct volume/ ↓microglia M1-type/ ↑Microglia M2-type	(177)

TREM2 is not exclusively expressed on microglia, it is also expressed on myeloid cells (178).

microglia to the resting state. This knowledge has led to the repurposing approved medicines or development of new therapeutic agents, some of which are in clinical trials with results to be published in the near future. Novel pharmacological approaches and NP-formulations form part of therapeutic agents intended for targeting microglia to treat neuroinflammation. In particular, novel multifunctional NP-formulations have been developed that not only cross the BBB but also specifically target microglia. In the light of promising preclinical outcomes, the way forward would be testing these new approaches in humans.

REFERENCES

1. Mondal K, Takahashi H, Cole J II, Del Mar NA, Li C, Stephenson DJ, et al. Systemic elevation of n-3 polyunsaturated fatty acids (n-3-PUFA) is associated with protection against visual, motor, and emotional deficits in mice following closed-head mild traumatic brain injury. Molecular Neurobiology. 2021;58(11):5564–80.
2. Peruzzotti-Jametti L, Pluchino S. Targeting mitochondrial metabolism in neuroinflammation: Towards a therapy for progressive multiple sclerosis. Trends Mol Med. 2018;24(10):838–55.
3. Umfress A, Speed HE, Tan C, Ramezani S, Birnbaum S, Brekken RA et al. Neuropathological effects of chemotherapeutic drugs. ACS Chem Neurosci. 2021;12(16):3038–48.
4. Desale SE, Chinnathambi S. α- Linolenic acid modulates phagocytosis and endosomal pathways of extracellular Tau in microglia. Cell Adh Migr. 2021;15(1):84–100.
5. Flores-Aguilar L, Iulita MF, Orciani C, Tanna N, Yang J, Bennett DA et al. Cognitive and brain cytokine profile of non-demented individuals with cerebral amyloid-beta deposition. J Neuroinflammation. 2021;18(1):147.
6. Sun X-y, Li L-j, Dong Q-X, Zhu J, Huang Y-r, Hou S-j et al. Rutin prevents tau pathology and neuroinflammation in a mouse model of Alzheimer's disease. J Neuroinflammation. 2021;18(1):131.
7. He X-f, Li L-l, Xian W-b, Li M-y, Zhang L-y, Xu J-h et al. Chronic colitis exacerbates NLRP3-dependent neuroinflammation and cognitive impairment in middle-aged brain. J Neuroinflammation. 2021;18(1):153.
8. Galvani G, Mottolese N, Gennaccaro L, Loi M, Medici G, Tassinari M et al. Inhibition of microglia overactivation restores neuronal survival in a mouse model of CDKL5 deficiency disorder. J Neuroinflammation. 2021;18(1):155.
9. Dodd WS, Noda I, Martinez M, Hosaka K, Hoh BL. NLRP3 inhibition attenuates early brain injury and delayed cerebral vasospasm after subarachnoid hemorrhage. J Neuroinflammation. 2021;18(1):163.
10. Kacířová M, Železná B, Blažková M, Holubová M, Popelová A, Kuneš J et al. Aging and high-fat diet feeding lead to peripheral insulin resistance and sex-dependent changes in brain of mouse model of tau pathology THY-Tau22. J Neuroinflammation. 2021;18(1):141.
11. Millischer V, Heinzl M, Faka A, Resl M, Trepci A, Klammer C et al. Intravenous administration of LPS activates the kynurenine pathway in healthy male human subjects: A prospective placebo-controlled cross-over trial. J Neuroinflammation. 2021;18(1):158.
12. Oberman K, Hovens I, de Haan J, Falcao-Salles J, van Leeuwen B, Schoemaker R. Acute pre-operative ibuprofen improves cognition in a rat model for postoperative cognitive dysfunction. J Neuroinflammation. 2021;18(1):156.
13. Xiao J, Cai T, Fang Y, Liu R, Flores JJ, Wang W et al. Activation of GPR40 attenuates neuroinflammation and improves neurological function via PAK4/CREB/KDM6B pathway in an experimental GMH rat model. J Neuroinflammation. 2021;18(1):160.
14. Xue M, Yong VW. Neuroinflammation in intracerebral haemorrhage: Immunotherapies with potential for translation. Lancet Neurol. 2020;19(12):1023–32.
15. Zhou Y, Wang Y, Wang J, Anne Stetler R, Yang QW. Inflammation in intracerebral hemorrhage: From mechanisms to clinical translation. Prog Neurobiol. 2014;115:25–44.
16. Duan R-N, Yang C-L, Du T, Liu A, Wang A-R, Sun W-J et al. Smek1 deficiency exacerbates experimental autoimmune encephalomyelitis by activating proinflammatory microglia and suppressing the IDO1-AhR pathway. J Neuroinflammation. 2021;18(1):145.
17. Ding ZB, Song LJ, Wang Q, Kumar G, Yan YQ, Ma CG. Astrocytes: A double-edged sword in neurodegenerative diseases. Neural Regen Res. 2021;16(9):1702–10.
18. Didier N, Romero IA, Créminon C, Wijkhuisen A, Grassi J, Mabondzo A. Secretion of interleukin-1beta by astrocytes mediates endothelin-1 and tumour necrosis factor-alpha effects on human brain microvascular endothelial cell permeability. J Neurochem. 2003;86(1):246–54.

19. Yi W, Schlüter D, Wang X. Astrocytes in multiple sclerosis and experimental autoimmune encephalomyelitis: Star-shaped cells illuminating the darkness of CNS autoimmunity. Brain Behav Immun. 2019;80:10–24.
20. Todd BP, Chimenti MS, Luo Z, Ferguson PJ, Bassuk AG, Newell EA. Traumatic brain injury results in unique microglial and astrocyte transcriptomes enriched for type i interferon response. J Neuroinflammation. 2021;18(1):151.
21. Mi J, Yang Y, Yao H, Huan Z, Xu C, Ren Z et al. Inhibition of heat shock protein family a member 8 attenuates spinal cord ischemia–reperfusion injury via astrocyte NF-κB/NLRP3 inflammasome pathway. J Neuroinflammation. 2021;18(1):170.
22. DiSabato DJ, Quan N, Godbout JP. Neuroinflammation: The devil is in The details. J Neurochem. 2016;139(Suppl 2):136–53.
23. Greupner T, Kutzner L, Nolte F, Strangmann A, Kohrs H, Hahn A et al. Effects of a 12-week high-α-linolenic acid intervention on EPA and DHA concentrations in red blood cells and plasma oxylipin pattern in subjects with a low EPA and DHA status. Food Funct. 2018;9(3):1587–600.
24. Dierckx T, Haidar M, Grajchen E, Wouters E, Vanherle S, Loix M et al. Phloretin suppresses neuroinflammation by autophagy-mediated Nrf2 activation in macrophages. J Neuroinflammation. 2021;18(1):148.
25. Cheng J, Zhang R, Xu Z, Ke Y, Sun R, Yang H et al. Early glycolytic reprogramming controls microglial inflammatory activation. J Neuroinflammation. 2021;18(1):129.
26. Gruchot J, Lein F, Lewen I, Reiche L, Weyers V, Petzsch P et al. Siponimod modulates the reaction of microglial cells to pro-inflammatory stimulation. Int J Mol Sci. 2022;23(21).
27. Adams RA, Bauer J, Flick MJ, Sikorski SL, Nuriel T, Lassmann H et al. The fibrin-derived gamma377-395 peptide inhibits microglia activation and suppresses relapsing paralysis in central nervous system autoimmune disease. J Exp Med. 2007;204(3):571–82.
28. Kasindi A, Fuchs DT, Koronyo Y, Rentsendorj A, Black KL, Koronyo-Hamaoui M. Glatiramer acetate immunomodulation: Evidence of neuroprotection and cognitive preservation. Cells. 2022;11(9).
29. Ratchford JN, Endres CJ, Hammoud DA, Pomper MG, Shiee N, McGready J et al. Decreased microglial activation in MS patients treated with glatiramer acetate. J Neurol. 2012;259(6):1199–205.
30. Mrdjen D, Pavlovic A, Hartmann FJ, Schreiner B, Utz SG, Leung BP et al. High-dimensional single-cell mapping of Central nervous system immune cells reveals distinct myeloid subsets in health, aging, and disease. Immunity. 2018;48(2):380–95.e6.
31. Nimmerjahn A, Kirchhoff F, Helmchen F. Resting microglial cells are highly dynamic surveillants of brain parenchyma in vivo. Science. 2005;308(5726):1314–8.
32. Sierra A, de Castro F, Del Río-Hortega J, Rafael Iglesias-Rozas J, Garrosa M, Kettenmann H. The "Big-Bang" for modern glial biology: Translation and comments on Pío del Río-Hortega 1919 series of papers on microglia. Glia. 2016;64(11):1801–40.
33. Penfield W. Microglia and the process of phagocytosis in gliomas. Am J Pathol. 1925;1(1):77–90.15.
34. Bechmann I, Kwidzinski E, Kovac AD, Simbürger E, Horvath T, Gimsa U et al. Turnover of rat brain perivascular cells. Exp Neurol. 2001;168(2):242–9.
35. McMenamin PG. Distribution and phenotype of dendritic cells and resident tissue macrophages in the dura mater, leptomeninges, and choroid plexus of the rat brain as demonstrated in wholemount preparations. J Comp Neurol. 1999;405(4):553–62.
36. Svenningsson A, Andersen O, Edsbagge M, Stemme S. Lymphocyte phenotype and subset distribution in normal cerebrospinal fluid. J Neuroimmunol. 1995;63(1):39–46.
37. Sankowski R, Süß P, Benkendorff A, Böttcher C, Fernandez-Zapata C, Chhatbar C et al. Multiomic spatial landscape of innate immune cells at human central nervous system borders. Nat Med. 2024;30(1):186–98.
38. Liu YJ, Spangenberg EE, Tang B, Holmes TC, Green KN, Xu X. Microglia elimination increases neural circuit connectivity and activity in adult mouse cortex. J Neurosci. 2021;41(6):1274–87.
39. Fourgeaud L, Través PG, Tufail Y, Leal-Bailey H, Lew ED, Burrola PG et al. TAM receptors regulate multiple features of microglial physiology. Nature. 2016;532(7598):240–4.
40. Ito D, Imai Y, Ohsawa K, Nakajima K, Fukuuchi Y, Kohsaka S. Microglia-specific localisation of a novel calcium binding protein, Iba1. Brain Res Mol Brain Res. 1998;57(1):1–9.
41. Lier J, Streit WJ, Bechmann I. Beyond activation: Characterizing microglial functional phenotypes. Cells. 2021;10(9).
42. Lier J, Winter K, Bleher J, Grammig J, Mueller WC, Streit W et al. Loss of IBA1-Expression in brains from individuals with obesity and hepatic dysfunction. Brain Res. 2019;1710:220–9.

43. Sankowski R, Böttcher C, Masuda T, Geirsdottir L, Sagar, Sindram E et al. Mapping microglia states in the human brain through the integration of high-dimensional techniques. Nat Neurosci. 2019;22(12):2098–110.
44. Butovsky O, Jedrychowski MP, Moore CS, Cialic R, Lanser AJ, Gabriely G et al. Identification of a unique TGF-β–dependent molecular and functional signature in microglia. Nat Neurosci. 2014;17(1):131–43.
45. Masuda T, Sankowski R, Staszewski O, Böttcher C, Amann L, Sagar, et al. Spatial and temporal heterogeneity of mouse and human microglia at single-cell resolution. Nature. 2019;566(7744):388–92.
46. Dräger NM, Sattler SM, Huang CT, Teter OM, Leng K, Hashemi SH et al. A CRISPRi/a platform in human iPSC-derived microglia uncovers regulators of disease states. Nat Neurosci. 2022;25(9):1149–62.
47. De Schepper S, Ge JZ, Crowley G, Ferreira LSS, Garceau D, Toomey CE et al. Perivascular cells induce microglial phagocytic states and synaptic engulfment via SPP1 in mouse models of Alzheimer's disease. Nat Neurosci. 2023;26(3):406–15.
48. Hong S, Beja-Glasser VF, Nfonoyim BM, Frouin A, Li S, Ramakrishnan S et al. Complement and microglia mediate early synapse loss in Alzheimer mouse models. Science. 2016;352(6286):712–6.
49. Asai H, Ikezu S, Tsunoda S, Medalla M, Luebke J, Haydar T et al. Depletion of microglia and inhibition of exosome synthesis halt tau propagation. Nat Neurosci. 2015;18(11):1584–93.
50. Pan RY, Ma J, Kong XX, Wang XF, Li SS, Qi XL et al. Sodium rutin ameliorates Alzheimer's disease-like pathology by enhancing microglial amyloid-β clearance. Sci Adv. 2019;5(2):eaau6328.
51. Baik SH, Kang S, Lee W, Choi H, Chung S, Kim JI et al. A breakdown in metabolic reprogramming causes microglia dysfunction in Alzheimer's disease. Cell Metab. 2019;30(3):493–507.e6.
52. Poliani PL, Wang Y, Fontana E, Robinette ML, Yamanishi Y, Gilfillan S et al. TREM2 sustains microglial expansion during aging and response to demyelination. J Clin Invest. 2015;125(5):2161–70.
53. Damani MR, Zhao L, Fontainhas AM, Amaral J, Fariss RN, Wong WT. Age-related alterations in the dynamic behavior of microglia. Aging Cell. 2011;10(2):263–76.
54. Davies DS, Ma J, Jegathees T, Goldsbury C. Microglia show altered morphology and reduced arborization in human brain during aging and Alzheimer's disease. Brain Pathol (Zurich, Switzerland). 2017;27(6):795–808.
55. Safaiyan S, Kannaiyan N, Snaidero N, Brioschi S, Biber K, Yona S et al. Age-related myelin degradation burdens the clearance function of microglia during aging. Nat Neurosci. 2016;19(8):995–8.
56. Hickman SE, Kingery ND, Ohsumi TK, Borowsky ML, Wang LC, Means TK et al. The microglial sensome revealed by direct RNA sequencing. Nat Neurosci. 2013;16(12):1896–905.
57. Solano Fonseca R, Mahesula S, Apple DM, Raghunathan R, Dugan A, Cardona A et al. Neurogenic niche microglia undergo positional remodeling and progressive activation contributing to age-associated reductions in neurogenesis. Stem Cells Dev. 2016;25(7):542–55.
58. Keren-Shaul H, Spinrad A, Weiner A, Matcovitch-Natan O, Dvir-Szternfeld R, Ulland TK et al. A unique microglia type associated with restricting development of Alzheimer's disease. Cell. 2017;169(7):1276–90.e17.
59. Huang Y, Happonen KE, Burrola PG, O'Connor C, Hah N, Huang L et al. Microglia use TAM receptors to detect and engulf amyloid β plaques. Nature Immunology. 2021;22(5):586–94.
60. Montagne A, Barnes Samuel R, Sweeney Melanie D, Halliday Matthew R, Sagare Abhay P, Zhao Z et al. Blood-brain barrier breakdown in the aging human hippocampus. Neuron. 2015;85(2):296–302.
61. Montagne A, Nation DA, Pa J, Sweeney MD, Toga AW, Zlokovic BV. Brain imaging of neurovascular dysfunction in Alzheimer's disease. Acta Neuropathol. 2016;131(5):687–707.
62. Merlini M, Rafalski VA, Rios Coronado PE, Gill TM, Ellisman M, Muthukumar G et al. Fibrinogen induces microglia-mediated spine elimination and cognitive impairment in an Alzheimer's disease model. Neuron. 2019;101(6):1099–108.e6.
63. Zhao P, Xu Y, Jiang L, Fan X, Li L, Li X et al. A tetravalent TREM2 agonistic antibody reduced amyloid pathology in a mouse model of Alzheimer's disease. Sci Transl Med. 2022;14(661):eabq0095.
64. Liu W, Taso O, Wang R, Bayram S, Graham AC, Garcia-Reitboeck P et al. Trem2 promotes anti-inflammatory responses in microglia and is suppressed under pro-inflammatory conditions. Hum Mol Genet. 2020;29(19):3224–48.
65. Park MH, Lee M, Nam G, Kim M, Kang J, Choi BJ et al. N,N'-Diacetyl-p-phenylenediamine restores microglial phagocytosis and improves cognitive defects in Alzheimer's disease transgenic mice. Proc Natl Acad Sci U S A. 2019;116(47):23426–36.
66. Tran EH, Hoekstra K, van Rooijen N, Dijkstra CD, Owens T. Immune invasion of the central nervous system parenchyma and experimental allergic encephalomyelitis, but not leukocyte extravasation from blood, are prevented in macrophage-depleted mice1. J Immunol. 1998;161(7):3767–75.

67. Lucchinetti CF, Popescu BFG, Bunyan RF, Moll NM, Roemer SF, Lassmann H et al. Inflammatory cortical demyelination in early multiple sclerosis. New Engl J Med. 2011;365(23):2188–97.
68. Boven LA, Van Meurs M, Van Zwam M, Wierenga-Wolf A, Hintzen RQ, Boot RG et al. Myelin-laden macrophages are anti-inflammatory, consistent with foam cells in multiple sclerosis. Brain. 2005;129(2):517–26.
69. Schläger C, Körner H, Krueger M, Vidoli S, Haberl M, Mielke D et al. Effector T-cell trafficking between the leptomeninges and the cerebrospinal fluid. Nature. 2016;530(7590):349–53.
70. Russo MV, Latour LL, McGavern DB. Distinct myeloid cell subsets promote meningeal remodeling and vascular repair after mild traumatic brain injury. Nature Immunology. 2018;19(5):442–52.
71. Morganti-Kossman MC, Lenzlinger PM, Hans V, Stahel P, Csuka E, Ammann E et al. Production of cytokines following brain injury: Beneficial and deleterious for the damaged tissue. Mol Psychiatry. 1997;2(2):133–6.
72. Mogi M, Harada M, Riederer P, Narabayashi H, Fujita K, Nagatsu T. Tumor necrosis factor-α (TNF-α) increases both in the brain and in the cerebrospinal fluid from parkinsonian patients. Neuroscience Letters. 1994;165(1):208–10.
73. Gerhard A, Pavese N, Hotton G, Turkheimer F, Es M, Hammers A et al. In vivo imaging of microglial activation with [11C](R)-PK11195 PET in idiopathic Parkinson's disease. Neurobiol Dis. 2006;21(2):404–12.
74. Tai YF, Pavese N, Gerhard A, Tabrizi SJ, Barker RA, Brooks DJ et al. Microglial activation in presymptomatic Huntington's disease gene carriers. Brain. 2007;130(Pt 7):1759–66.
75. Debruyne JC, Versijpt J, Van Laere KJ, De Vos F, Keppens J, Strijckmans K et al. PET visualization of microglia in multiple sclerosis patients using [11C]PK11195. Eur J Neurol. 2003;10(3):257–64.
76. Surendranathan A, Su L, Mak E, Passamonti L, Hong YT, Arnold R et al. Early microglial activation and peripheral inflammation in dementia with Lewy bodies. Brain. 2018;141(12):3415–27.
77. Jucaite A, Svenningsson P, Rinne JO, Cselényi Z, Varnäs K, Johnström P et al. Effect of the myeloperoxidase inhibitor AZD3241 on microglia: A PET study in Parkinson's disease. Brain. 2015;138(Pt 9):2687–700.
78. Gopinath A, Collins A, Khoshbouei H, Streit WJ. Microglia and other myeloid cells in central nervous system health and disease. J Pharmacol Exp Ther. 2020;375(1):154–60.
79. Lepiarz-Raba I, Gbadamosi I, Florea R, Paolicelli RC, Jawaid A. Metabolic regulation of microglial phagocytosis: Implications for Alzheimer's disease therapeutics. Transl Neurodegener. 2023;12(1):48.
80. Mills EL, O'Neill LA. Reprogramming mitochondrial metabolism in macrophages as an anti-inflammatory signal. Eur J immunol. 2016;46(1):13–21.
81. Song S, Yu L, Hasan MN, Paruchuri SS, Mullett SJ, Sullivan MLG et al. Elevated microglial oxidative phosphorylation and phagocytosis stimulate post-stroke brain remodeling and cognitive function recovery in mice. Commun Biol. 2022;5(1):35.
82. Feron O. The many metabolic sources of acetyl-CoA to support histone acetylation and influence cancer progression. Ann Transl Med. 2019;7(Suppl 8):S277–S.
83. Izzo LT, Trefely S, Demetriadou C, Drummond JM, Mizukami T, Kuprasertkul N et al. Acetylcarnitine shuttling links mitochondrial metabolism to histone acetylation and lipogenesis. Sci Adv. 2023; 9(18):eadf0115.
84. Soliman ML, Puig KL, Combs CK, Rosenberger TA. Acetate reduces microglia inflammatory signaling in vitro. J Neurochem. 2012;123(4):555–67.
85. Jha Abhishek K, Huang Stanley C-C, Sergushichev A, Lampropoulou V, Ivanova Y, Loginicheva E et al. Network integration of parallel metabolic and transcriptional data reveals metabolic modules that regulate macrophage polarization. Immunity. 2015;42(3):419–30.
86. Benito A, Hajji N, O'Neill K, Keun HC, Syed N. β-hydroxybutyrate oxidation promotes the accumulation of immunometabolites in activated microglia cells. Metabolites. 2020;10(9).
87. Degrandi D, Hoffmann R, Beuter-Gunia C, Pfeffer K. The proinflammatory cytokine-induced IRG1 protein associates with mitochondria. J Interferon Cytokine Res. 2009;29(1):55–67.
88. Ni L, Xiao J, Zhang D, Shao Z, Huang C, Wang S et al. Immune-responsive gene 1/itaconate activates nuclear factor erythroid 2-related factor 2 in microglia to protect against spinal cord injury in mice. Cell Death Dis. 2022;13(2):140.
89. Mills EL, Ryan DG, Prag HA, Dikovskaya D, Menon D, Zaslona Z et al. Itaconate is an anti-inflammatory metabolite that activates Nrf2 via alkylation of KEAP1. Nature. 2018;556(7699):113–7.
90. Strelko CL, Lu W, Dufort FJ, Seyfried TN, Chiles TC, Rabinowitz JD et al. Itaconic acid is a mammalian metabolite induced during macrophage activation. J Am Chem Soc. 2011;133(41):16386–9.

91. Tannahill GM, Curtis AM, Adamik J, Palsson-McDermott EM, McGettrick AF, Goel G et al. Succinate is an inflammatory signal that induces IL-1β through HIF-1α. Nature. 2013;496(7444): 238–42.
92. Lampropoulou V, Sergushichev A, Bambouskova M, Nair S, Vincent EE, Loginicheva E et al. Itaconate links inhibition of succinate dehydrogenase with macrophage metabolic remodeling and regulation of inflammation. Cell Metab. 2016;24(1):158–66.
93. Selak MA, Armour SM, MacKenzie ED, Boulahbel H, Watson DG, Mansfield KD et al. Succinate links TCA cycle dysfunction to oncogenesis by inhibiting HIF-α prolyl hydroxylase. Cancer Cell. 2005;7(1):77–85.
94. Peruzzotti-Jametti L, Bernstock JD, Vicario N, Costa ASH, Kwok CK, Leonardi T et al. Macrophage-derived extracellular succinate licenses neural stem cells to suppress chronic neuroinflammation. Cell Stem Cell. 2018;22(3):355–68.e13.
95. Infantino V, Iacobazzi V, Menga A, Avantaggiati ML, Palmieri F. A key role of the mitochondrial citrate carrier (SLC25A1) in TNFα- and IFNγ-triggered inflammation. Biochim Biophys Acta. 2014;1839(11):1217–25.
96. Nair S, Sobotka KS, Joshi P, Gressens P, Fleiss B, Thornton C et al. Lipopolysaccharide-induced alteration of mitochondrial morphology induces a metabolic shift in microglia modulating the inflammatory response in vitro and in vivo. Glia. 2019;67(6):1047–61.
97. Geric I, Schoors S, Claes C, Gressens P, Verderio C, Verfaillie CM et al. Metabolic reprogramming during microglia activation. Immunometabolism. 2019;1(1).
98. Lewis MT, Kasper JD, Bazil JN, Frisbee JC, Wiseman RW. Quantification of mitochondrial oxidative phosphorylation in metabolic disease: Application to type 2 diabetes. Int J Mol Sci. 2019;20(21):5271.
99. 99, Loussouarn C, Pers Y-M, Bony C, Jorgensen C, Noël D. Mesenchymal stromal cell-derived extracellular vesicles regulate the mitochondrial metabolism via transfer of miRNAs. Front Immunol. 2021;12(171).
100. Veech RL Tricarboxylic acid cycle. In: Lennarz WJ, Lane MD, editors. Encyclopedia of Biological Chemistry (Second Edition). Waltham: Academic Press; 2013. p. 436–40.
101. Bailey JD, Diotallevi M, Nicol T, McNeill E, Shaw A, Chuaiphichai S et al. Nitric oxide modulates metabolic remodeling in inflammatory macrophages through TCA cycle regulation and itaconate accumulation. Cell Rep. 2019;28(1):218–30.e7.
102. Zheng J. Energy metabolism of cancer: Glycolysis versus oxidative phosphorylation (Review). Oncol Lett. 2012;4(6):1151–7.
103. Rodríguez-Prados J-C, Través PG, Cuenca J, Rico D, Aragonés J, Martín-Sanz P et al. Substrate fate in activated macrophages: A comparison between innate, classic, and alternative activation. J Immunol. 2010;185(1):605–14.
104. Pålsson-McDermott EM, O'Neill LAJ. Targeting immunometabolism as an anti-inflammatory strategy. Cell Res. 2020;30(4):300–14.
105. Fonken LK, Frank MG, Kitt MM, Barrientos RM, Watkins LR, Maier SF. Microglia inflammatory responses are controlled by an intrinsic circadian clock. Brain Behav Immun. 2015;45:171–9.
106. Krawczyk CM, Holowka T, Sun J, Blagih J, Amiel E, DeBerardinis RJ et al. Toll-like receptor-induced changes in glycolytic metabolism regulate dendritic cell activation. Blood. 2010;115(23):4742–9.
107. Yao L, Kan EM, Lu J, Hao A, Dheen ST, Kaur C et al. Toll-like receptor 4 mediates microglial activation and production of inflammatory mediators in neonatal rat brain following hypoxia: Role of TLR4 in hypoxic microglia. J Neuroinflammation. 2013;10(1):785.
108. Hines DJ, Choi HB, Hines RM, Phillips AG, MacVicar BA. Prevention of LPS-induced microglia activation, cytokine production and sickness behavior with TLR4 receptor interfering peptides. PLoS One. 2013;8(3):e60388.
109. De Laere M, Berneman ZN, Cools N. To the brain and back: migratory paths of dendritic cells in multiple sclerosis. J Neuropathol Exp Neurol. 2018;77(3):178–92.
110. Rubic T, Lametschwandtner G, Jost S, Hinteregger S, Kund J, Carballido-Perrig N et al. Triggering the succinate receptor GPR91 on dendritic cells enhances immunity. Nat Immunol. 2008;9(11):1261–9.
111. Everts B, Amiel E, Huang SC, Smith AM, Chang CH, Lam WY et al. TLR-driven early glycolytic reprogramming via the kinases TBK1-IKKε supports the anabolic demands of dendritic cell activation. Nat Immunol. 2014;15(4):323–32.
112. Timmons GA, Carroll RG, O'Siorain JR, Cervantes-Silva MP, Fagan LE, Cox SL et al. The circadian clock protein BMAL1 acts as a metabolic sensor in macrophages to control the production of Pro IL-1β. Front Immunol. 2021;12:700431.

113. Swain A, Bambouskova M, Kim H, Andhey PS, Duncan D, Auclair K et al. Comparative evaluation of itaconate and its derivatives reveals divergent inflammasome and type I interferon regulation in macrophages. Nat Metab. 2020;2(7):594–602.
114. Zhu X, Guo Y, Liu Z, Yang J, Tang H, Wang Y. Itaconic acid exerts anti-inflammatory and antibacterial effects via promoting pentose phosphate pathway to produce ROS. Sci Rep. 2021;11(1):18173.
115. Nastasi C, Willerlev-Olsen A, Dalhoff K, Ford SL, Gadsbøll A, Buus TB et al. Inhibition of succinate dehydrogenase activity impairs human T cell activation and function. Sci Rep. 2021;11(1):1458.
116. Yang C, Liu T, Shi G-P. Therapeutic potential of tricarboxylic acid cycle metabolite itaconate in cardiovascular diseases. EBioMedicine. 2020;59:102938.
117. Ganta VC, Choi MH, Kutateladze A, Fox TE, Farber CR, Annex BH. A MicroRNA93-interferon regulatory factor-9-immunoresponsive gene-1-itaconic acid pathway modulates M2-like macrophage polarization to revascularize ischemic muscle. Circulation. 2017;135(24):2403–25.
118. Muri J, Wolleb H, Broz P, Carreira EM, Kopf M. Electrophilic Nrf2 activators and itaconate inhibit inflammation at low dose and promote IL-1β production and inflammatory apoptosis at high dose. Redox biology. 2020;36:101647.
119. Tang C, Wang X, Xie Y, Cai X, Yu N, Hu Y et al. 4-octyl itaconate activates Nrf2 signaling to inhibit pro-inflammatory cytokine production in peripheral blood mononuclear cells of systemic lupus erythematosus patients. Cell Physiol Biochem. 2018;51(2):979–90.
120. ElAzzouny M, Tom CT, Evans CR, Olson LL, Tanga MJ, Gallagher KA et al. Dimethyl itaconate is not metabolized into itaconate intracellularly. J Biol Chem. 2017;292(12):4766–9.
121. Bambouskova M, Gorvel L, Lampropoulou V, Sergushichev A, Loginicheva E, Johnson K et al. Electrophilic properties of itaconate and derivatives regulate the IκBζ–ATF3 inflammatory axis. Nature. 2018;556(7702):501–4.
122. Yang S, Zhang X, Zhang H, Lin X, Chen X, Zhang Y et al. Dimethyl itaconate inhibits LPS-induced microglia inflammation and inflammasome-mediated pyroptosis via inducing autophagy and regulating the Nrf-2/HO-1 signaling pathway. Mol Med Rep. 2021;24(3).
123. Kuo P-C, Weng W-T, Scofield BA, Paraiso HC, Brown DA, Wang P-Y et al. Dimethyl itaconate, an itaconate derivative, exhibits immunomodulatory effects on neuroinflammation in experimental autoimmune encephalomyelitis. J Neuroinflammation. 2020;17(1):138.
124. Liu CY, Wang X, Liu C, Zhang HL. Pharmacological targeting of microglial activation: new therapeutic approach. Front Cell Neurosci. 2019;13:514.
125. Gao C, Jiang J, Tan Y, Chen S. Microglia in neurodegenerative diseases: mechanism and potential therapeutic targets. Signal Transduct Target Ther. 2023;8(1):359.
126. Ward M, Long H, Schwabe T, Rhinn H, Tassi I, Salazar SV et al. A phase 1 study of AL002 in healthy volunteers. Alzheimers Dementia. 2021;17(S9):e054669.
127. Paul R, Jackson S, Ward M, Joshi A, Castro A, Yeh FL et al. INVOKE-2: A phase 2 randomized, double-blind, placebo-controlled study to evaluate the efficacy and safety of AL002 in participants with early Alzheimer's disease. Alzheimers Dementia. 2021;17(S9):e054615.
128. Crotti A, Benner C, Kerman BE, Gosselin D, Lagier-Tourenne C, Zuccato C et al. Mutant Huntingtin promotes autonomous microglia activation via myeloid lineage-determining factors. Nat Neurosci. 2014;17(4):513–21.
129. Sun, D., Haig, G., Rasool, S.. Results from a clinical study of an anti-galectin-3 monoclonal antibody in patients with moderate to severe Alzheimer's disease. 15th Conference Clinical Trials Alzheimer's Disease, November 29- December 2, 2022, San Francisco, CA, USA: Symposia - Oral Communications - Late Breaking Abstracts (Clinical Trial Alzheimer's Disease). J Prev Alzheimers Dis. 2022;9(1):8–50.
130. Giuliani F, Hader W, Yong VW. Minocycline attenuates T cell and microglia activity to impair cytokine production in T cell-microglia interaction. J Leukoc Biol. 2005;78(1):135–43.
131. Dodel R, Spottke A, Gerhard A, Reuss A, Reinecker S, Schimke N et al. Minocycline 1-year therapy in multiple-system-atrophy: Effect on clinical symptoms and [11C] (R)-PK11195 PET (MEMSA-trial). Mov Disord. 2010;25(1):97–107.
132. Wu J, Xu X, Zheng L, Mo J, Jin X, Bao Y. Nilotinib inhibits microglia-mediated neuroinflammation to protect against dopaminergic neuronal death in Parkinson's disease models. Int Immunopharmacol. 2021;99:108025.
133. Turner RS, Hebron ML, Lawler A, Mundel EE, Yusuf N, Starr JN et al. Nilotinib effects on safety, tolerability, and biomarkers in Alzheimer's disease. Ann Neurol. 2020;88(1):183–94.
134. Yun SP, Kam TI, Panicker N, Kim S, Oh Y, Park JS et al. Block of A1 astrocyte conversion by microglia is neuroprotective in models of Parkinson's disease. Nat Med. 2018;24(7):931–8.

135. McGarry A, Rosanbalm S, Leinonen M, Olanow CW, To D, Bell A et al. Safety, tolerability, and efficacy of NLY01 in early untreated Parkinson's disease: A randomised, double-blind, placebo-controlled trial. Lancet Neurol. 2024;23(1):37–45.
136. Lin M-H, Cheng P-C, Hsiao P-J, Chen S-C, Hung C-H, Kuo C-H et al. The GLP-1 receptor agonist exenatide ameliorates neuroinflammation, locomotor activity, and anxiety-like behavior in mice with diet-induced obesity through the modulation of microglial M2 polarization and downregulation of SR-A4. Int Immunopharmacol. 2023;115:109653.
137. Athauda D, Maclagan K, Skene SS, Bajwa-Joseph M, Letchford D, Chowdhury K et al. Exenatide once weekly versus placebo in Parkinson's disease: A randomised, double-blind, placebo-controlled trial. Lancet. 2017;390(10103):1664–75.
138. Stefanova N, Georgievska B, Eriksson H, Poewe W, Wenning GK. Myeloperoxidase inhibition ameliorates multiple system atrophy-like degeneration in a transgenic mouse model. Neurotox Res. 2012;21(4):393–404.
139. Liu D, Wang Z, Liu S, Wang F, Zhao S, Hao A. Anti-inflammatory effects of fluoxetine in lipopolysaccharide(LPS)-stimulated microglial cells. Neuropharmacology. 2011;61(4):592–9.
140. Mowla A, Mosavinasab M, Pani A. Does fluoxetine have any effect on the cognition of patients with mild cognitive impairment? A double-blind, placebo-controlled, clinical trial. J Clin Psychopharmacol. 2007;27(1):67–70.
141. Mizuno T, Kurotani T, Komatsu Y, Kawanokuchi J, Kato H, Mitsuma N et al. Neuroprotective role of phosphodiesterase inhibitor ibudilast on neuronal cell death induced by activated microglia. Neuropharmacology. 2004;46(3):404–11.
142. Oskarsson B, Maragakis N, Bedlack RS, Goyal N, Meyer JA, Genge A et al. MN-166 (ibudilast) in amyotrophic lateral sclerosis in a Phase IIb/III study: COMBAT-ALS study design. Neurodegener Dis Manag. 2021;11(6):431–43.
143. Trias E, Ibarburu S, Barreto-Núñez R, Babdor J, Maciel TT, Guillo M et al. Post-paralysis tyrosine kinase inhibition with masitinib abrogates neuroinflammation and slows disease progression in inherited amyotrophic lateral sclerosis. J Neuroinflammation. 2016;13(1):177.
144. Mora JS, Genge A, Chio A, Estol CJ, Chaverri D, Hernández M et al. Masitinib as an add-on therapy to riluzole in patients with amyotrophic lateral sclerosis: a randomized clinical trial. Amyotroph Lateral Scler Frontotemporal Degener. 2020;21(1–2):5–14.
145. Dubois B, López-Arrieta J, Lipschitz S, Doskas T, Spiru L, Moroz S et al. Masitinib for mild-to-moderate Alzheimer's disease: Results from a randomized, placebo-controlled, phase 3, clinical trial. Alzheimers Res Ther. 2023;15(1):39.
146. Vermersch P, Brieva-Ruiz L, Fox RJ, Paul F, Ramio-Torrenta L, Schwab M et al. Efficacy and safety of masitinib in progressive forms of multiple sclerosis: A randomized, phase 3, clinical trial. Neurol Neuroimmunol Neuroinflamm. 2022;9(3).
147. Khasnavis S, Jana A, Roy A, Mazumder M, Bhushan B, Wood T et al. Suppression of nuclear factor-κB activation and inflammation in microglia by physically modified saline. J Biol Chem. 2012;287(35):29529–42.
148. Vallarola A, Sironi F, Tortarolo M, Gatto N, De Gioia R, Pasetto L et al. RNS60 exerts therapeutic effects in the SOD1 ALS mouse model through protective glia and peripheral nerve rescue. J Neuroinflammation. 2018;15(1):65.
149. Beghi E, Pupillo E, Bianchi E, Bonetto V, Luotti S, Pasetto L et al. Effect of RNS60 in amyotrophic lateral sclerosis: a phase II multicentre, randomized, double-blind, placebo-controlled trial. Eur J Neurol. 2023;30(1):69–86.
150. Ghosh S, Dubow J, Sutherland J, Kalmes A, Baird G, Mock J, et al., editors. RESCUE: A randomized, blinded, placebo-controlled, parallel group study design to determine the safety of RNS60 in large vessel occlusion stroke patients undergoing endovascular thrombectomy (P10-5.029). Neurology. 2024;(7_supplement_1):6499.
151. Wu Y-q, Xiong J, He Z-l, Yuan Y, Wang B-n, Xu J-y et al. Metformin promotes microglial cells to facilitate myelin debris clearance and accelerate nerve repairment after spinal cord injury. Acta Pharmacol Sin. 2022;43(6):1360–71.
152. Zimmerman SC, Ferguson EL, Choudhary V, Ranatunga DK, Oni-Orisan A, Hayes-Larson E et al. Metformin cessation and dementia incidence. JAMA Network Open. 2023;6(10):e2339723-e.
153. Toguchi M, Gonzalez D, Furukawa S, Inagaki S. Involvement of Sema4D in the control of microglia activation. Neurochem Int. 2009;55(7):573–80.
154. Smith ES, Jonason A, Reilly C, Veeraraghavan J, Fisher T, Doherty M et al. SEMA4D compromises blood–brain barrier, activates microglia, and inhibits remyelination in neurodegenerative disease. Neurobiol Dis. 2015;73:254–68.

155. Feigin A, Evans EE, Fisher TL, Leonard JE, Smith ES, Reader A et al. Pepinemab antibody blockade of SEMA4D in early Huntington's disease: A randomized, placebo-controlled, phase 2 trial. Nat Med. 2022;28(10):2183–93.
156. Tabrizi SJ, Leavitt BR, Landwehrmeyer GB, Wild EJ, Saft C, Barker RA et al. Targeting Huntingtin expression in patients with Huntington's disease. New Engl J Med. 2019;380(24):2307–16.
157. Taylor RA, Chang CF, Goods BA, Hammond MD, Mac Grory B, Ai Y et al. TGF-β1 modulates microglial phenotype and promotes recovery after intracerebral hemorrhage. J Clin Invest. 2017;127(1):280–92.
158. Lonnemann N, Hosseini S, Marchetti C, Skouras DB, Stefanoni D, D'Alessandro A et al. The NLRP3 inflammasome inhibitor OLT1177 rescues cognitive impairment in a mouse model of Alzheimer's disease. Proc Natl Acad Sci U S A. 2020;117(50):32145–54.
159. Lu J, Zhang C, Lv J, Zhu X, Jiang X, Lu W et al. Antiallergic drug desloratadine as a selective antagonist of 5HT(2A) receptor ameliorates pathology of Alzheimer's disease model mice by improving microglial dysfunction. Aging Cell. 2021;20(1):e13286.
160. Lv J, Wang W, Zhu X, Xu X, Yan Q, Lu J et al. DW14006 as a direct AMPKα1 activator improves pathology of AD model mice by regulating microglial phagocytosis and neuroinflammation. Brain Behav Immun. 2020;90:55–69.
161. Sun L, Wilke Saliba S, Apweiler M, Akmermer K, Herlan C, Grathwol C et al. Anti-neuroinflammatory effects of a macrocyclic peptide-peptoid hybrid in lipopolysaccharide-stimulated BV2 microglial cells. Int J Mol Sci. 2024;25(8).
162. Wang X, Li S, Yu J, Wang W, Du Z, Gao S et al. Saikosaponin B2 ameliorates depression-induced microglia activation by inhibiting ferroptosis-mediated neuroinflammation and ER stress. J Ethnopharmacol. 2023;316:116729.
163. Huang J, Lu W, Doycheva DM, Gamdzyk M, Hu X, Liu R et al. IRE1α inhibition attenuates neuronal pyroptosis via miR-125/NLRP1 pathway in a neonatal hypoxic-ischemic encephalopathy rat model. J Neuroinflammation. 2020;17(1):152.
164. Lu Y, Guo Z, Zhang Y, Li C, Zhang Y, Guo Q et al. Microenvironment remodeling micelles for Alzheimer's disease therapy by early modulation of activated microglia. Adv Sci (Weinh). 2019;6(4):1801586.
165. Dempsey C, Rubio Araiz A, Bryson KJ, Finucane O, Larkin C, Mills EL et al. Inhibiting the NLRP3 inflammasome with MCC950 promotes non-phlogistic clearance of amyloid-β and cognitive function in APP/PS1 mice. Brain Behav Immun. 2017;61:306–16.
166. Liu Q, Zhang MM, Guo MX, Zhang QP, Li NZ, Cheng J et al. Inhibition of microglial NLRP3 with MCC950 attenuates microglial morphology and NLRP3/Caspase-1/IL-1β signaling in stress-induced mice. J Neuroimmune Pharmacol. 2022;17(3–4):503–14.
167. Kalashnikova I, Cambell H, Kolpek D, Park J. Optimization and characterization of miRNA-129-5p-encapsulated poly (lactic-co-glycolic acid) nanoparticles to reprogram activated microglia. Nanoscale Adv. 2023;5(13):3439–52.
168. Park SY, Yi EH, Kim Y, Park G. Anti-neuroinflammatory effects of Ephedra sinica Stapf extract-capped gold nanoparticles in microglia. Int J Nanomed. 2019;14:2861–77.
169. Wang P, Yang P, Qian K, Li Y, Xu S, Meng R et al. Precise gene delivery systems with detachable albumin shell remodeling dysfunctional microglia by TREM2 for treatment of Alzheimer's disease. Biomaterials. 2022;281:121360.
170. Zhao N, Francis NL, Song S, Kholodovych V, Calvelli HR, Hoop CL et al. CD36-binding amphiphilic nanoparticles for attenuation of α-synuclein-induced microglial activation. Adv NanoBiomed Res. 2022;2(6):2100120.
171. Choi SG, Shin J, Lee KY, Park H, Kim SI, Yi YY et al. PINK1 siRNA-loaded poly(lactic-co-glycolic acid) nanoparticles provide neuroprotection in a mouse model of photothrombosis-induced ischemic stroke. Glia. 2023;71(5):1294–310.
172. Li S, Zhang R, Wang A, Li Y, Zhang M, Kim J et al. Panax notoginseng: Derived exosome-like nanoparticles attenuate ischemia reperfusion injury via altering microglia polarization. J Nanobiotechnol. 2023;21(1):416.
173. Ralvenius WT, Andresen JL, Huston MM, Penney J, Bonner JM, Fenton OS et al. Nanoparticle-mediated delivery of Anti-PU.1 siRNA via localized intracisternal administration reduces neuroinflammation. Adv Mater. 2024;36(8):2309225.
174. Han R, Lan X, Han Z, Ren H, Aafreen S, Wang W et al. Improving outcomes in intracerebral hemorrhage through microglia/macrophage-targeted IL-10 delivery with phosphatidylserine liposomes. Biomaterials. 2023;301:122277.

175. Bai M, Cui N, Liao Y, Guo C, Li L, Yin Y et al. Astrocytes and microglia-targeted Danshensu liposomes enhance the therapeutic effects on cerebral ischemia-reperfusion injury. J Control Release. 2023;364:473–89.
176. Fan Q, Kuang L, Wang B, Yin Y, Dong Z, Tian N et al. Multiple synergistic effects of the microglia membrane-bionic nanoplatform on mediate tumor microenvironment remodeling to amplify glioblastoma immunotherapy. ACS nano. 2024;18(22):14469–86.
177. Jin L, Zhu Z, Hong L, Qian Z, Wang F, Mao Z. ROS-responsive 18β-glycyrrhetic acid-conjugated polymeric nanoparticles mediate neuroprotection in ischemic stroke through HMGB1 inhibition and microglia polarization regulation. Bioact Mater. 2023;19:38–49.
178. Colonna M. The biology of TREM receptors. Nat Rev Immunol. 2023;23(9):580–94.
179. Liu H, Han Y, Wang T, Zhang H, Xu Q, Yuan J et al. Targeting microglia for therapy of Parkinson's disease by using biomimetic ultrasmall nanoparticles. Journal of the Am Chem Soc. 2020;142(52):21730–42.
180. Wu Y, Wang W, Qiu X, Lu Z, Ji W, Shen J et al. A STIR nucleic acid drug delivery system for stirring phenotypic switch of microglia in Parkinson's disease treatments. Nano Res. 2023;16(5):7216–26.
181. Sharma A, Porterfield JE, Smith E, Sharma R, Kannan S, Kannan RM. Effect of mannose targeting of hydroxyl PAMAM dendrimers on cellular and organ biodistribution in a neonatal brain injury model. J Control Release. 2018;283:175–89.
182. von Ehr A, Attaai A, Neidert N, Potru PS, Ruß T, Zöller T et al. Inhibition of microglial TGFβ signaling increases expression of Mrc1. Front Cell Neurosci. 2020;14.
183. Yuan J, Ge H, Liu W, Zhu H, Chen Y, Zhang X et al. M2 microglia promotes neurogenesis and oligodendrogenesis from neural stem/progenitor cells via the PPARγ signaling pathway. Oncotarget. 2017;8(12):19855–65.
184. Marzolo MP, von Bernhardi R, Inestrosa NC. Mannose receptor is present in a functional state in rat microglial cells. J Neurosci Res. 1999;58(3):387–95.

5 Novel Nanoparticles as Drug Delivery Systems to the Brain

Touraj Ehtezazi and Changiz Taghibiglou

5.1 INTRODUCTION

According to the World Health Organization (WHO) (https://www.who.int/news/item/14-03-2024-over-1-in-3-people-affected-by-neurological-conditions-the-leading-cause-of-illness-and-disability-worldwide) and a recent report published in *the Lancet Neurology*, more than 3 billion people worldwide were living with a neurological condition. More than 3.4 billion people, or 43.1% of the world population, were affected by a neurological condition in 2021 (1). A wide range of diseases and disorders affects central nervous system (CNS) and spinal cord including vascular disorders (such as stroke), infections (such as meningitis, and encephalitis), structural disorders (such as brain and spinal cord tumours), seizure disorders (such as epilepsy), neuro-degenerative diseases (such as Parkinson's, Alzheimer's and Huntington's diseases (HD), and Amyotrophic lateral sclerosis), Autoimmune or inflammatory disorders [multiple sclerosis (MS), Guillain-Barré syndrome, etc.], and mental health disorders (such as mood disorders, depression, and schizophrenia).

The delivery of safe and sufficient pharmacological agents to CNS has always been a challenging issue, mostly due to the existence of a fixed structure called the blood brain barrier (BBB). The BBB multicellular structure has been composed of variety of cells including pericytes, neurons, astrocytes, microglia and a supportive basal membrane. The main physiological function of the BBB is to protect CNS from the outside environment including chemical, biological moieties, and among them therapeutic agents. The integrity of the BBB is important for well-being and normal healthy brain function as its breakdown has been linked to a wide range of diseases, such as neurodegenerative disorders, schizophrenia and ischemic stroke (IS) (reviewed in (2–8)). The anatomical structure and function of the BBB have been comprehensively discussed elsewhere (9–11). Several mechanisms are involved in the regulated transport through the BBB including passive membrane diffusion, paracellular diffusion, transport-mediated uptake, and transcytosis. The complex biological interface and structural integrity of the BBB is a major obstacle in effective transport of most drugs from the vasculature into the parenchyma of the nervous system. Cellular structure of BBB allows part of small molecular weight drugs (MW $<$ 400 Da) to pass through via diffusion process (12); however, most drugs are larger molecules not being able to cross the BBB. To overcome this issue, several strategies have been proposed and tested, including cerebrospinal fluid (CSF) delivery, intracranial delivery, intranasal (IN) delivery systems, ultrasound-based technologies, membrane transporters, receptor-mediated transcytosis, and nanotechnology-based therapeutic systems (reviewed in (13–15), with novel nanotechnology discussed in the later part of this chapter. These approaches have their advantages and disadvantages which have been extensively reviewed by the others (16, 17). In this chapter, we will discuss recent novel advances in the BBB targeting cell penetrating peptides (CPPs) and associated nanotechnology.

 DOI: 10.1201/9781003482406-5

5.2 CELL PENETRATING PEPTIDES FOR THE BRAIN DELIVERY OF BIOACTIVE PEPTIDES AND PROTEINS

CPPs have emerged as a promising tool for delivering peptide-based drugs to the CNS, overcoming the significant challenge posed by the blood-brain barrier (BBB). Recent advancements, highlight their potential in enhancing CNS drug delivery, particularly for treating neurodegenerative diseases, brain tumors, and other CNS disorders. CPPs, typically short sequences of 5–30 amino acids, are valued for their ability to translocate across cellular membranes, including the BBB, with low toxicity and high efficiency. They can carry various cargos—such as peptides, proteins, nucleic acids, and nanoparticles (NPs)—into the CNS via energy-dependent (e.g., endocytosis) or energy-independent mechanisms. Notable CPPs like *trans*-activator of transcription (for human immunodeficiency virus type 1, HIV-1), penetratin (from Drosophila), and polyarginine have been widely studied, with newer variants like cyclic CPPs and glycosylated CPPs showing improved stability and penetration.

Historically, a breakthrough in peptide-mediated drug delivery happened in 1988, when two independent groups reported their first *in vitro* data on the effective cellular internalization of the *trans*-activator of transcription (TAT) protein of HIV-1 (18, 19). Later studies resulted in identifying key amnio acid (aa) sequences of TAT essential in its cell membrane penetrating function and introduction of the short truncated 9 aa TAT (20, 21). Later, other CPPs capable of entering cells were introduced including the homeodomain of Antennapedia, a homeoprotein of *Drosophila melanogaster (22)*, penetratin (mimicking the third helix of the Antennapedia homeodomain) (23). Historical and development aspects of CPPs have been comprehensively reviewed by the others (24, 25). More recently, designing novel CPPs has been advanced by using artificial intelligence and machine learning, leading to an explosion of novel CPP candidates with optimized stability and kinetics (26–41).

Classical CPPs, are categorised based on their physicochemical properties as either cationic, hydrophobic (reviewed in (42)), or amphipathic water-soluble short peptides (5-42 aa) composed of mostly positively charged amino acids arginine and lysine residues (43). They can also be categorized based on their origin into natural, synthetic, or chimeric peptides. CPPs are usually bind to their cargos by covalent conjugation and non-covalent complexation strategies (reviewed in (44–46)). Although their exact mechanisms of transporting cargo across biological membranes are still obscure, some pathways and factors have been proposed for CPPs' functions including electrostatic interactions between positively charged amino acids of CPPs and negatively charged membrane glycosaminoglycans, particularly heparan sulfate proteoglycans, and also interactions with lipid components of the plasma membrane (PM) (reviewed in (47, 48)).

Among different CPPs, TAT, penetratin, polyarginines, and PEP-1 are widely used in cargo delivery of bioactive peptides to the brain in both pre-clinical and clinical studies. We therefore discuss these CPPs in more detail. Several recently published comprehensive review articles have covered physicochemical and structural characteristics of a wide range of CPP molecules (12, 49–52).

5.2.1 *Trans* Activator of Transcription Protein of HIV-1

TAT is one of the commonly used CPPs for bioactive peptide delivery to the CNS, with several TAT-fused therapeutic peptides currently in different phases of clinical trials. A few of them have even successfully finalised to Phase 3 clinical trials. Here we discuss those TAT-fused bioactive peptides in more details.

5.2.1.1 TAT-NR2B9c (NA-1, Nerinetide)

Stroke is the leading cause of adult disability in developed countries, as well as the third leading cause of death in the world including the North America. Every 10 minutes a Canadian suffers a stroke. The cost of stroke in Canada is more than 3 billion dollars annually (– equating to ~$65,000 per person in the first six months). In Canada, ~2.7% of the adult population have survived a stroke. The majority (~ 87%) of all strokes are ischemic one causing neuronal cell death through major cellular and molecular mechanisms including neuroinflammation, oxidative stress, and overexcitation of glutamate receptors (glutamate excitotoxicity). In the glutamate excitotoxicity, massive neuronal Ca^{2+} influx leads to the activation of Ca^{2+} dependent pro-death signalling pathways (calcium toxicity) and brain tissue damage. The N-methyl-D-aspartate (NMDA) glutamate receptors (NMDARs) particularly those containing GluN2B subunit are considered the major culprit in the excitotoxicity-induced neuronal death in stroke and neurodegenerative diseases. The role of glutamate excitotoxicity and NMDARs in the pathogenesis of stroke and neurodegenerative diseases has been comprehensively reviewed by others (53, 54).

There are two main strategies for the stroke treatment: (1) restoring brain blood supply by the thrombolytic recombinant tissue plasminogen activator (rt-PA) and/or endovascular therapy for large vessel occlusions; and (2) to prevent and/or rescue damaged neurons and prevent further neuronal loss in the area surrounding the core region (neurons immediately die) of stroke area called penumbra. If not rescued, neurons in the penumbra area die within a few days via programmed cell death. The penumbra area provides sufficient time for therapeutic intervention to save the neurons. Rescuing penumbra neurons has been the target of extensive research by academic laboratories and pharmaceutical companies.

NMDAR-dependent glutamate excitotoxicity is one of the major mechanisms in neuronal apoptosis following stroke and neurodegenerative diseases. One of its main pathological manifestations is the GluN2B subunit containing NMDAR-dependent massive Ca^{2+} influx and activation of calcium-dependent pro-death signaling pathways (reviewed in (55)). Michael Tymianski's group discovered that following excitotoxicity, Glu2B dependent Ca^{2+} influx reaches enough quanta in the microenvironment within the close vicinity of the C-tail of GluN2B attached to the scaffolding protein post-synaptic density protein 95 (PSD-95). In the affected neurons, this would activate nearby toxic neuronal nitric oxide synthase (nNOS), which triggers cell death. The PSD-95 is a member of the Discs large (DLG) family of scaffolding proteins, and a class of a membrane-associated guanylate kinases (MAGUKs) (reviewed in (56–58)). PSD-95 interacts with the GluN2B subunit of NMDARs and nNOS through specific protein–protein interaction domains, known as PSD-95/Disc large/Zonula occludens-1 (PDZ) domains. PSD-95 and nNOS contain three and one PDZ domain(s), respectively (reviewed in (59)). Michael Tymianski's group demonstrated that PDZ2 domains of PSD-95 bind to the 9 amino acid residues of the C-tail of GluN2B ((Lys-Leu-Ser-Ser-Ile-Glu-Ser-Asp-Val; called NR2B9c) and in fact, PDZ protein interactions are underlying GluN2B-mediated excitotoxicity and neuroprotection by PSD-95 inhibitors (60). They synthesised this short 9 amino acid peptide and rendered it cell permeant by fusing it to the 11 amino acids of the cell-membrane transduction domain of the human immunodeficiency virus type 1 (HIV-1) TAT protein (Tyr-Gly-Arg-Lys-Lys-Arg-Arg-Gln-Arg-Arg-Arg) and used it as an interfering peptide to separate the complex of PSD-95 and nNOS from GluN2B c-terminal (as depicted in Figure 5.1).

They called the therapeutic peptide TAT-NR2Bc, NA-1, and, most recently, Nerinetide (reviewed in (61)). They first discovered the neuroprotective effects of TAT-NR2Bc peptide in a cellular model of stroke and later in a rat model of stroke (middle cerebral artery occlusion, MCAO) (62). The middle cerebral artery occlusion is considered as a gold standard *in vivo* model in preclinical stroke research (reviewed in (63, 64)). The therapeutic effectiveness of TAT-NR2Bc peptide has also been reported after a single intravenous (i.v.) injection 1 or 3 hours after stroke in permanent and transient focal ischemia in the rat (65, 66). Another group administered NR2B9c NPs intranasally in a rat model of ischemic brain injury and observed effective neuroprotection against excitotoxicity, reduced ischemic brain injury and improved neurological function deficits (67). Tymianski's team later in a pre-clinical translation study demonstrated the stroke neuroprotective effects of Tat-NR2B9c in gyrencephalic nonhuman primates (68, 69).

Following the recommendations and guidelines for stroke drug development established by the Stroke Academic Industry Roundtable (STAIR) committee, Tymianski's team developed

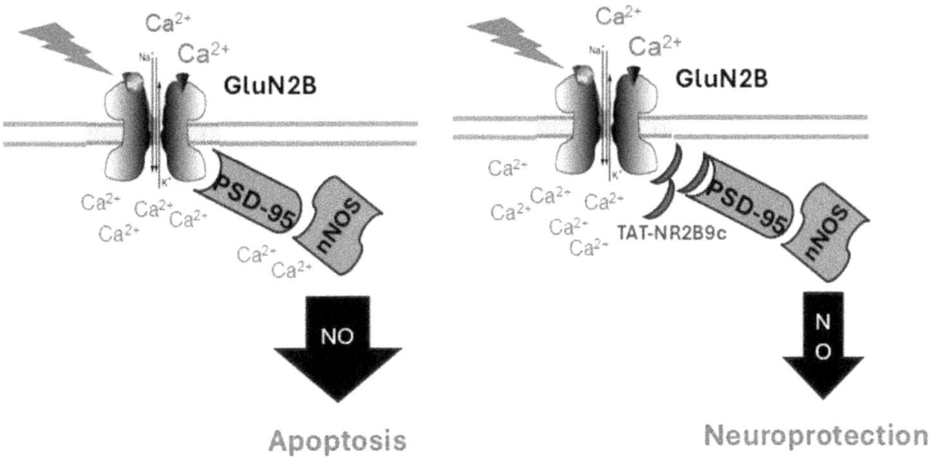

FIGURE 5.1 Excitotoxicity-induced neuronal apoptosis through function of tertiary complex of GluN2B subunit of NMDAR-PSD95-nNOS and formation of toxic NO. TAT-NR2B9c (NA-1 or nerinetide) by disrupting this complex formation and protects neurons.

and conducted a clinical trial called ENACT (the neuroprotectant tested in the Evaluating Neuroprotection in Aneurysm Coiling Therapy, NA-1 (also known as Tat-NR2B9c) (70, 71). ENACT provided a direct step to testing the neuroprotective effects of TAT-NR2Bc (NA-1) in human subjects. In a double-blind, randomised, controlled study, they recruited patients aged 18 years or older from 14 hospitals in Canada (three sites) and the USA (11 sites) who had a ruptured or unruptured intracranial aneurysm amenable to endovascular repair (phase 2 clinical study) (72). The ENACT study provided supportive evidence for the potential neuroprotective effects of NA-1 peptide in human ischaemic stroke (IS). In another large multicentre, double-blind, randomised, placebo-controlled study (ESCAPE-NA1) done in 48 acute care hospitals in eight countries, they enrolled patients with acute ischaemic stroke due to large vessel occlusion within a 12-hour treatment window (73). This trial was registered with ClinicalTrials.gov, NCT02930018. Patients were randomly assigned to receive i.v. TAT-NR2Bc peptide (nerinetide) in a single dose of 2.6 mg/kg, up to a maximum dose of 270 mg or normal saline as a placebo group. Surprisingly, initial results indicated that TAT-NR2Bc peptide (nerinetide) did not improve the proportion of patients achieving good clinical outcomes after endovascular thrombectomy compared with patients receiving placebo. After more detailed and careful analysis of data, the research team realized that the therapeutic peptide treatment was not effective in those patients who received an i.v. alteplase (recombinant tissue plasminogen activator (tPA) protease) prior to the nerinetide injection. The lack of effectiveness in patients who received alteplase was attributed to the protease activity of this medication, resulting in the degradation of the TAT-fused peptide (nerinetide) in circulation before reaching to its target in the CNS. To overcome this issue and to prevent drug–drug interactions between alteplase and nerinetide, Tymianski's team designed protease-resistant agents such as d-Tat-l-2B9c (74). They later demonstrated that stroke patients in the nerinetide group who were not treated with concurrent alteplase showed improved clinical outcomes and the treatment effect was larger among patients with favorable Alberta Stroke Program Early CT Score (ASPECTS) profiles (75). Overall, in acute IS patients with large vessel occlusion undergoing endovascular treatment without concurrent i.v. alteplase, the ESCAPE-NA1 trial has demonstrated that intravenous nerinetide not only improved clinical outcomes but also was beneficial from health-economic standpoints (76). More extended reports from these clinical trials demonstrated beneficial effects of nerinetide peptide on infarct volume (77, 78), clinical aspects(79–81), patient safety (75, 82, 83), functional and cognitive outcomes

(84, 85). Following successful outcome of the ESCAPE-NA1 clinical phase 3 trial, nerinetide is currently completing two more phase 3 clinical trials (ESCAPE-NEXT and FRONTIER) for treating acute IS with promising results.

Since neuronal GluN2B-dependent overexcitation of NMDARs and massive Ca^{2+} influx are attributed in pathogenesis of a wide range of neurological and CNS diseases. Several preclinical studies have been conducted to study potential therapeutic effects of nerinetide peptide in animal models of neonatal hypoxic ischemic brain injury (86), type-2 diabetic neuropathic pain (87), spinal nociceptive plasticity and neuropathic pain (88), oxaliplatin-induced neuropathic pain (89), brain injury with intracerebral hemorrhage (90), status epilepticus (91, 92), traumatic brain injury (TBI) (93), retinal ischemia-reperfusion injury/glaucoma (94), subarachnoid hemorrhage (95), and cocaine-seeking behaviour attenuation (96). Nerinetide peptide is a promising TAT-fused peptide which most likely opens new avenues in peptide neuropharmacology.

5.2.1.2 TAT-K13 Peptide

Phosphatase and tensin homolog deleted on chromosome 10 (PTEN) is an important tumour suppressor with lipid and protein phosphatase activities (reviewed in (97)). A physical association between GluN2B subunit of NMDAR and PTEN had been previously reported (98). This association was involved in neuronal apoptosis following excitotoxicity and its downregulation found to be neuroprotective (98). Following two groundbreaking Nature reports in 2005 and 2006 by Pandolfi and colleagues on identifying tumour suppressing effects of PTEN and its connection with p53-dependent cellular senescence, in addition to inhibiting PISK/Akt pro-survival signaling pathway (99, 100), they later published two very important back-to-back reports (101, 102) and identified NEDD4-1 as a proto-oncogenic ubiquitin ligase for PTEN which was responsible for ubiquitination of PTEN including mono-ubiquitination of two key lysine amino acids (K13 and K289) and subsequent PTEN nuclear import and tumor suppression. Their reports prompted us to study whether PTEN mono-ubiquitination nuclear translocation is involved in neuronal apoptosis following excitotoxicity and stroke. We observed nuclear translocation of PTEN in the brain penumbra following stroke as well as neurons exposed to glutamate/NMDA excitotoxicity. We then designed two short TAT-fused decoy peptides each containing either K13 or K289 and some flanking amino acids from each side of these lysine amino acids mimicking mono- ubiquitination domains of PTEN (Tat-$PTEN_{K13}$: $KEIVSRNK_{13}RRYQED$-Tat, and Tat-$PTEN_{K289}$: Tat-$PEETSEK_{289}VENGS$). In a series of *in vitro* and *in vivo* experiments, we identified only K13 amino acid responsible for the nuclear translocation of PTEN and subsequent neuronal apoptosis in stroke and TAT-fused K13 peptide provided effective neuroprotection (as schematically depicted in Figure 5.2) (103). PTEN nuclear translocation enhances neuronal injury after hypoxia-ischemia via modulation of the nuclear factor-κB signalling pathway (104). PTEN has therefore appeared to be a viable therapeutic target for the IS (105). Recently, Wang et al. reported an LC-MS/MS assay for determining TAT-K13 in human plasma usable for pharmacokinetics studies (106). Later, the same group conducted the first-in-human safety, tolerability, and pharmacokinetics of TAT-K13 (SY-007) in healthy Chinese subjects (107). They observed a Nonlinear pharmacokinetics with this TAT-fused peptide with a safe dose up to 60 mg. We and the others also reported PTEN activation and nuclear accumulation of PTEN in other neurological disorders including epilepsy (108), and neurodegenerative diseases (109, 110), which may open more potential therapeutic application for TAT-K13 in future.

5.2.1.3 TAT-Indip

Sterol regulatory element binding proteins (SREBPs) are lipid transcription factors that are mainly involved in the regulation of cellular cholesterol and lipid homeostasis. There are three isoforms of SREBPs (SREBP-1a, 1c, and 2), which regulate both fatty acid and cholesterol synthesis (111). Under physiological conditions, immature SREBP-1 is retained inactive in the endoplasmic reticulum (ER) membrane as part of a complex with SREBP cleavage-activating protein (SCAP), which is attached to ER membrane proteins: insulin-induced gene 1 (Insig1). Decreased ER cholesterol

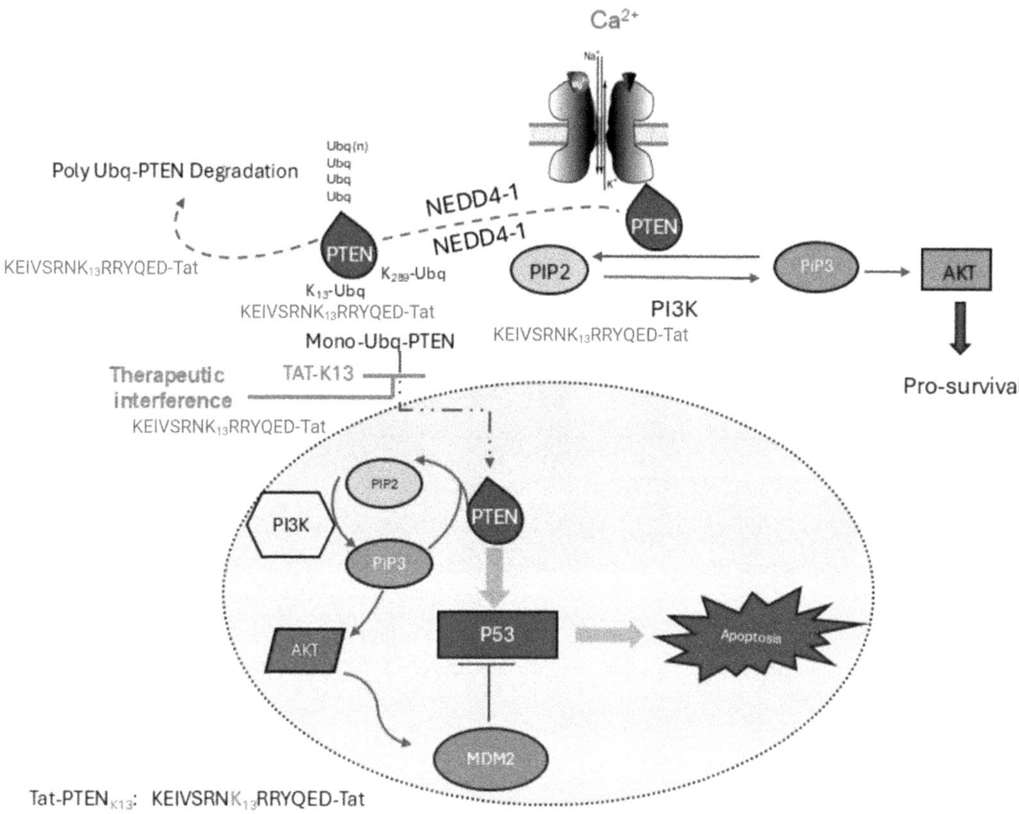

FIGURE 5.2 GluN2B-dependent glutamate excitotoxicity and cellular calcium influx leads to mono-ubiquitination of cytosolic PTEN and subsequent nuclear translocation and P53-dependent cell death. The decoy peptide TAT-K13 exerts neuroprotection by inhibiting PTEN nuclear translocation.

content, anoxia/hypoxia (112), and ER stress trigger degradation of Insig1 resulting in Golgi translocation of immature SREBPs and their subsequent proteolytic activation by S1P and S2P enzymes.

Sterol regulatory element-binding proteins are involved in numerous biological processes, such as ER stress, inflammation, autophagy, and apoptosis. SREBPs are also connected to a variety of diseases, including obesity, dyslipidemia, diabetes, non-alcoholic fatty liver disease, non-alcoholic steatohepatitis, chronic kidney disease, neurodegenerative diseases, and cancers (113, 114). We have previously identified the activation of SREBP-1 as a key step in the pathogenesis of stroke (115), amyotrophic lateral sclerosis (116), Alzheimer's disease (AD) (117), Parkinson's disease (unpublished data), and epilepsy (118).

We had previously identified that the activation of the SREBP-1 transcription factor in neurons plays an important role in GluN2B subunit of NMDAR-mediated excitotoxic neuronal death (115). One of the requirements of activation and maturation of SREBP-1 is poly ubiquitination and subsequent degradation of ER membrane localized Insig1 protein. To prevent this degradation (and later SREBP-1 activation), we designed a TAT-fused short peptide called INSIG-1-derived interference peptide (Indip) containing two key lysine amino acids to mimic poly ubiquitination domain of Insig1 (YGRKKRRQRRRG-EPHKFKREW). Indip acted as decoy peptide and by stabilizing Insig1, inhibited SREBP-1 activation in *in vitro* neuronal challenges with simvastatin (cholesterol reduction) and oxygen and glucose deprivation. We demonstrated that the interference peptide Indip, by blocking the nuclear translocation of SREBP-1 (N-SREBP1) exerted a neuroprotective effect both *in vitro* and *in vivo* (as depicted in Figure 5.3). We also reported the neuroprotective effect of Indip in

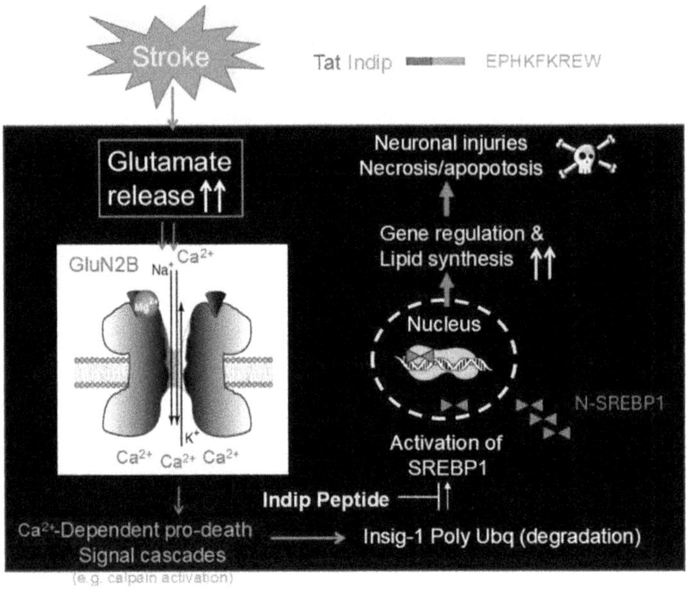

FIGURE 5.3 GluN2B-dependent glutamate excitotoxicity and cellular calcium influx cause insig-1 protein poly-ubiquitination and subsequent proteasomal degradation which facilitates immature SREBP1 transportation to the Golgi apparatus and proteolytic activation. TAT-Indip peptide by stabilizing insig-1 protein, inhibits SREBP1 activation and exerts neuroprotection.

an *in vitro* model of ALS (116). Further studies are needed for the stability, pharmacokinetics, and metabolism of Indip peptide. There are efforts underway to find more potential application for the TAT-Indip peptides.

5.2.1.4 Tat-NR2B$_{CT1292-1304}$ and TAT p53DM

Death-associated protein kinase 1 (DAPK1) is a tumour-suppressor gene that is also involved in a variety of cellular processes, including apoptosis, autophagy, and the cell cycle (119). The dysregulation of DAPK1, a ubiquitously expressed in the CNS, contributes, to various neurological diseases including AD (120), IS, epileptic seizure (121), depression (122) and neurodegenerative diseases (reviewed in (123)).

In 2010, Tu et al. discovered an interaction between DAPK1 and c-terminal of GluN2B in excitotoxicity and a mouse model of stroke. They identified amino acid 1292-1304 in C-tail of GluN2B as interacting domain with DAPK1 which resulted in phosphorylation of Ser-1303 and subsequent enhancement of NMDA channel conductance (124). They synthesised a TAT-fused interfering peptide (TAT-^{1292}KKNRNKLRRQHSY^{1304}) called Tat-NR2B$_{CT1292-1304}$. Later another group predicted an interaction between DAPK1 and GluN2B-CT$_{1290-1310}$ by computational simulation methods including protein and peptide docking and molecular dynamics (MD) simulation (125). Although the original TAT-fused peptide (Tat-NR2B$_{CT1292-1304}$) proved to be effective and neuroprotective, the therapeutic effects of the longer TAT-fused peptide (GluN2B-CT$_{1290-1310}$) require further studies. Schmidt et al demonstrated pathological role of DAPK1 in the YAC128 mouse model of HD by enhancing extrasynaptic GluN2B S1303 phosphorylation (126). It would be thus plausible to suggest potential therapeutic effects of TAT-GluN2B-CT peptides in HD. To our knowledge, neither of these TAT-GluN2B-CT peptides has not been studies in clinical trials yet.

There is a direct interaction between DAPK1 and DNA-binding motif (DM) of the tumour suppressor protein p53 (p53DM), which is known as a transcriptional regulator of pro-death genes (recently reviewed in (127)). An interaction between activated DAPK1 and P53 resulted in serine-23 and nuclear translocated of activation of P53, which subsequently triggers expression of pro-death genes (128, 129). Moreover, an interaction between $P53^{pserine-23}$ and cyclophilin D (CypD) in the mitochondrial matrix mediates necrosis. It was previously reported that the DNA binding domain of P53 interacts with CypD, however, a recent report identified the N-terminal (aa 1–70 and further refined to aa 20–70) of P53 as the main interacting domain with CypD (130). Wang X. et al. designed and synthesised a TAT-fused decoy peptide resembled interacting domain/DNA binding domain of P53 (aa 271-282) with DAPK1 (TAT- RVCACPGRDRRT) to separate P53 from DAPK1 (131). They tested their TAT-fused peptide in a mouse MCAO model of stroke and demonstrated its neuroprotective effects and improved behavioural deficits when injected 6 h post stroke. Interestingly, this peptide blocked both nuclear and mitochondrial pro-death pathways (131). Another TAT-fused peptide called $p53\text{-}DBD^{271-282}$, was designed by Yan et al. to separate interaction between DNA binding domain of P53 (aa 271-282) and maternally expressed gene 3 (MEG3) which functions as a long non-coding RNA (lncRNA). They demonstrated that an interaction between P53 and MEG3 mediated ischemic neuronal death, and the above interfering peptide exerted neuroprotection in both *in vitro* and *in vivo* models of cerebral ischemia (132). There are other TAT-fused P53 based peptides in non-CNS fields and mostly designed for pre-clinical cancer treatment which are out of scope of this review (e.g., (133–138)).

5.2.1.5 TAT-Panx$_{308}$

The pannexin (PANX) family is a family of ATP-permeable channels and exists in three members of PANX1, 2 and 3 in human genome. PANX1 is the most abundant and widely expressed member plays important roles in both physiological processes (e.g. regulation of blood pressure, glucose uptake, inflammation and cell death) and pathological conditions (e.g. stroke, epilepsy, migraine, chronic pain, diabetes, glaucoma, cancers, hypertension, AD) (recently reviewed in (139–142)). This makes PANX1 a promising therapeutic target for many diseases. In the CNS, PANX1 is also related to neural stem cell maintenance, neuronal maturation, and synaptic plasticity. Recent single-particle cryo-electron microscopy (cryo-EM) identified PANX1 as a heptamer channel (143). Under normal cellular conditions, C-terminal tail (CTT) of PANX1 functions as its auto-inhibitor. However, membrane depolarization, intracellular calcium release, extracellular potassium, tyrosine phosphorylation or mechanical stretching cause PANX1 channel activation. Moreover, during excitotoxicity and neuronal apoptosis, activation of caspases 3 and 7 causes cleavage of the CTT (removal of autoinhibition), subsequent channel activation, and ATP release. Thomson's team first demonstrated that anoxia-induced NMDA receptor activation opens pannexin channels via Src family kinases (SFK) (144). They developed a TAT-conjugated Panx1 Y308 interfering peptide that targeted the SFK consensus-like sequence of PANX1 (sequence YGRKKRRQRRR-LKVYEILPTFDVLH) and attenuated the anoxic depolarization without affecting SFK activation. In stroke and excitotoxicity, NMDA glutamate receptors, Src kinase and pannexin 1 form a pro-death signalling complex. Weilinger et al. used their TAT-Panx308 interfering peptide to disrupt the NMDAR-Src-Panx1 signalling complex which resulted in neuroprotection in both *in vitro* (oxygen and glucose deprivation challenge) and *in vivo* (5 mg/kg, 2 hours after 60-min transient MCAO) models of stroke (145). Figure 5.4 depicts NMDAR-Src kinase dependent activation of pannexin1 and mechanism of action of neuroprotective TAT-Panx308 peptide.

5.2.1.6 TAT-GluR2$_{3Y}$ Peptide

More than 25 years ago, Yu Tian Wang's team first reported the functionally distinct constitutive and regulated clathrin-dependent endocytosis of the alpha-amino-3-hydroxy-5-methylisoxazole-4-propionic acid (AMPA) subtype of glutamate receptors in hippocampal CA1 neurons(146). They demonstrated that this endocytosis was GluR2 (GluA2) subunit of AMPAR dependent leading to the long-term depression (LTD) form of synaptic plasticity and accelerated by insulin and low-frequency stimulation. They later identified 3 tyrosine (Y) amino acids (Y869, Y873, and Y876) at the c-terminal of GluA2 necessary for the regulated AMPAR clathrin-dependent endocytosis and

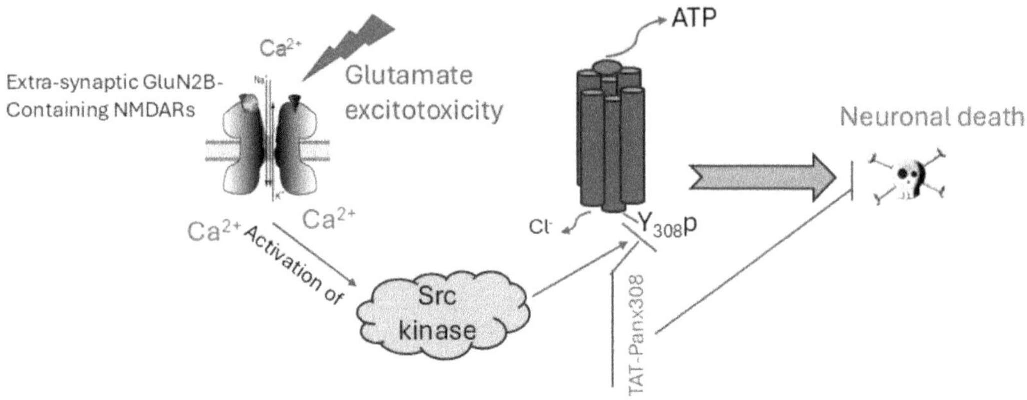

FIGURE 5.4 NMDAR-Src kinase-dependent activation of pannexin1 and mechanism of action of neuroprotective TAT-Panx308 peptide.

expression of certain forms of synaptic plasticity plastic (147). They designed a TAT-fused interfering peptide containing amnio acids 869-877 of C-tail of GluA2 and called it TAT-GluR2$_{3Y}$ peptide (depicted in Figure 5.5).

This bioactive peptide interferes with clathrin-mediated endocytosis of AMPAR and subsequently LTD (148). Interestingly, GluN2B subunit of NMDAR regulates AMPAR endocytosis and LTD and this is an essential step for NMDAR-mediated excitotoxic neuronal death. An inhibition of clathrin-mediated AMPAR endocytosis with the TAT-GluR2$_{3Y}$ peptide prevents NMDAR-mediated neuronal apoptosis without affecting NMDAR activity which makes it a potential neuroprotective peptide. Several preclinical studies have demonstrated potential therapeutic implications of

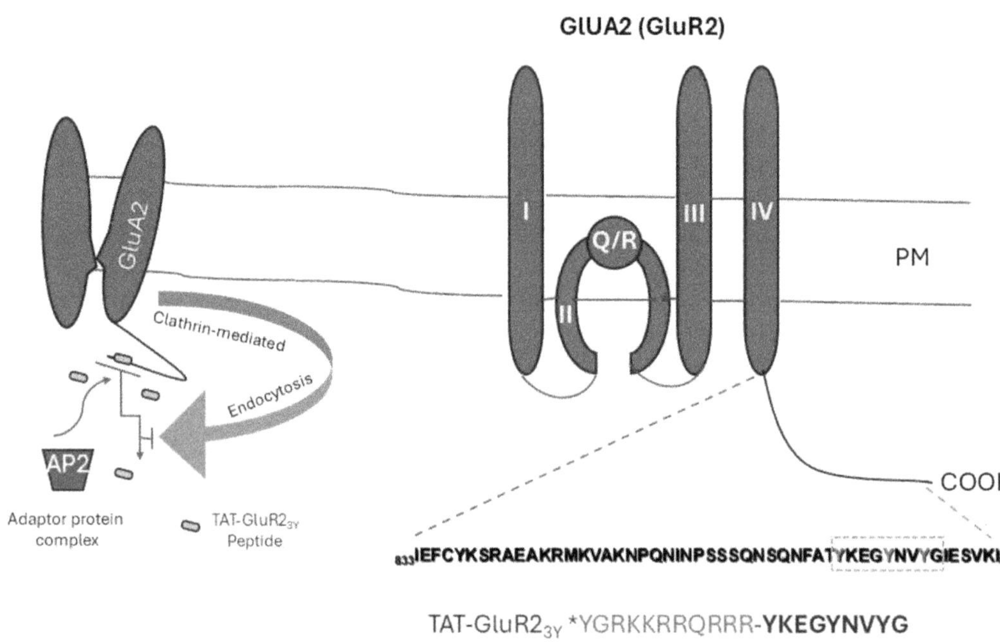

FIGURE 5.5 The TAT-GluR2$_{3Y}$ peptide by acting as a decoy peptide, interacts with the adaptor protein complex 2 (AP2) components of the clathrin-mediated endocytic machinery and suppresses AMPA glutamate receptor endocytosis.

TAT-GluR2$_{3Y}$ peptide in rat models of morphine addiction (149, 150) and drug-seeking behaviour in a mouse model of methamphetamine addiction (151). Zhang et al. recently used Tat-GluR2$_{3Y}$ peptide in a rat model of migraine and demonstrated an alleviating migraine-like pain sensitization (152). Ashourpour et al. observed amelioration of cognitive dysfunction after chronic administration of Tat-GluR2$_{3Y}$ peptide (3µmol/kg/2 weeks) in rats with amyloid beta neurotoxicity (153). They later reported more evidence that AMPAR endocytosis inhibition by Tat-GluR2$_{3Y}$ peptide facilitated memory consolidation and retrieval amyloid beta-induced memory-impaired rats (154, 155). Other researchers also reported rescue of memory retrieval deficits after infusion of Tat-GluR2$_{3Y}$ peptide (156, 157). On the contrary, Howland's team reported that infusion of this peptide before the memory retrieval phase resulted in object recognition memory disruption (158). Similarly, Ferrara et al. reported that the endocytosis of GluR2-containing AMPARs in rat amygdala regulates retrieval-induced strengthening of memories for traumatic events (159). Role of AMPAR including effects of Tat-GluR2$_{3Y}$ peptide in the memory retrieval has been reviewed elsewhere (157, 160). Babaei et al. have recently used Tat-GluR2$_{3Y}$ peptide with D-serine in a streptozotocin-induced spatial memory impaired postmenopausal rats and observed synergistically improvement in their cognitive functions (161). The same TAT-fused peptide was used for preventing the natural forgetting of long-term memories in rats (162). Ashby et al. also used Tat-GluR2$_{3Y}$ peptide to block LTD in rat's central nucleus of the amygdala which may have some implication in latent inhibition (LI) adaptive learning which is impaired in certain forms of mental illness (163). Liu et al. demonstrated therapeutic potential of Tat-GluR2$_{3Y}$ peptide in alleviating neuropathic pain by inhibiting GluA2 internalization (164). Tat-GluR2$_{3Y}$ peptide has been used to block AMPAR endocytosis in more pre-clinical studies in different neurological contexts by several other groups (165–169).

5.2.1.7 Other TAT-fused CNS Peptides

There is a long list of TAT-fused CNS peptide which is out of scop or limit of this chapter. For example, most recently Yi et al (170) amino acids 408-429 of presenilin-1 (PS1) as the critical domain for binding to ELK1 (E-twenty-six (ETS)-like protein 1) protein driving AD pathogenesis. They designed a TAT-fused peptide with the interacting domain (TAT-PS1$_{408\text{-}420}$) to selectively block this interaction (Tat-PS1408–429: YGRKKRRQRRR-IACFVAILIGLCLTLLLLAIFK). Blocking this interaction by the TAT-fused peptide resulted in PS1 degradation. In an *in vivo* study in a APP23/PS45 double transgenic mouse model of AD, daily intraperitoneal administrations of (10 mg/kg body weight) TAT-PS1$_{408\text{-}420}$ resulted in Aβ reduction and rescued memory impairment without any toxicity, precisely inhibiting the pathogenic interaction. In all excitable cells, ion channels initiate electrical signalling, and modulatory proteins customised their functions. The voltage-gated sodium (Na(V)) channel is one of these ion channels which triggers an action potential necessary for normal physiological functions of the heart and brain. Pathological dysfunction of this ion channel resulted in amplification of the late/persistent sodium current (I(NaL)), observed in various human diseases, including cardiac arrhythmias and epilepsy (171–173). Most recently Mahling et al., using a computational protein design platform, engineered a TAT-fused *de novo* peptide modulator, engineered late-current inhibitor X by inactivation-gate release (ELIXIR), with a very high binding affinity to Na(V) channels (174). This TAT-fused engineered peptide called ELIXIR (YGRKKRRQRRR**GSGMSPRREALYRGFRASYDVLRH** MW:4826.48 g/mol) effectively reversed gain-of-function defects and dysfunction of Na(V) linked to both cardiac arrhythmias and epilepsy in cellular and murine models. Their groundbreaking study set the stage for the rational design of future peptide-based therapeutic approaches.

Autophagy or the lysosomal degradation pathway is not only essential for the healthy cellular homeostasis but also has a crucial role in many pathological conditions including aging, stroke, cancers, and neurodegenerative disease (175–177). Beclin 1 is a protein necessary for initiating the cellular process of autophagy. Shoji-Kawata et al. developed an autophagy-inducing peptide TAT-beclin 1 (YGRKKRRQRRR-**GG-TNVFNATFEIWHDGEFGT**) with potential efficacy in the treatment of human diseases (178). This peptide was derived from a region of the autophagy protein, beclin 1. They demonstrated that the Tat-beclin 1 decreased the accumulation of polyglutamine expansion protein aggregates and the replication of several pathogens (including HIV-1) *in vitro*, and reduced mortality in mice infected with chikungunya or West Nile virus. Mishra and Thakur (179) reported that in scopolamine-induced amnesic male mice, TAT-beclin 1 peptide improved contextual and recognition memory as well as mitochondrial ultrastructure by normalizing mitochondrial length and area. Similarly, De Risi et al. (180), in a mouse model of mild cognitive impairment (MCI), demonstrated that sub-chronic treatment with autophagy agonists spermidine as well as Tat-beclin 1 rescued memory capacity. In addition, intrahippocampal infusion of TAT-Beclin-1 improved long-term memory in mice (181).

Mike Salter's team recently reported that in male mice (not in female mice) intrathecal delivery of this TAT-fused peptide reversed inflammation- and nerve injury-induced mechanical hypersensitivity and prevented mechanical hypersensitivity pain (182). Mathew et al. in an *in vitro* R28 neuronal cell line model of oxygen glucose deprivation (OGD) as well as an *in vivo* rat model of retinal ischemic injury demonstrated that TAT-Beclin 1 by stimulating autophagy rescued retinal neurons from cell death (183). The stroke-prone spontaneously hypertensive (SHRSP) rat is a suitable animal model for the stroke associated with high blood pressure. Forte et al. (184) used Tat-beclin 1 peptide to selectively reactivate autophagy/mitophagy which resulted in reduced stroke occurrence, restored autophagy/mitophagy and improved mitochondrial function.

TAT-Pep5 (H- YGRKKRRQRRR-CFFRGGFFNHNPRYC-OH, Cyclic, MW: 3404.9 g/mol) is a cell-permeable p75^{NTR} signalling inhibitor that controls the biological activity of p75^{NTR}. This TAT-fused peptide is primarily used for neuroscience applications. Neuroprotective effects of TAT-Pep5 have been investigated by several research teams in different pathological conditions(185–190). Buhusi et al reported that intra-hippocampal infusions of TAT-Pep5, significantly improved learning and memory in aged mice (191). Similarly, Sun et al. reported that intra-hippocampal CA1 injection TAT-Pep5 effectively mitigated the contextual memory deficits in a rodent model (192). TAT-Pep5 has been also used for non-CNS related pathological conditions.

Several studies indicate that sphingosine kinase 2 (SPK2) is implicated in stroke, preconditioning and neuronal autophagy (193–196). Liu et al. had identified through sequence analysis that SPK2 contains a 9-amino acid motif of 219-LNGLLDRPD-227, which was similar to that present in BH3-only proteins, a pro-apoptotic sequence (197). Song et al. designed TAT-SPK2 peptide by fusing 11 amino acids of TAT with a GG linker to 18 amino acids (aa 210-227) from the SPK2 containing 9 amino acids (aa 219-227) of BH3 domain (198). They observed that TAT-SPK2 peptide activated autophagy and protected neurons against OGD injury (an *in vitro* model of stroke). Here is their peptide structure: YGRKKRRQRRR-G-G-SGDGLLYEVLNGLLDRPD. The same researchers later demonstrated potential neuroprotective effects of TAT-SPK2 peptide in animal models of ischemia-reperfusion injury (199) and neonatal hypoxia brain damage (200).

Currently monoclonal antibodies (mAbs) are major protein-based therapeutics in the biopharmaceutical market, as more than half of 200 approved biologicals are mAbs. The CNS delivery of protein-based therapeutics including mAbs remains a challenging matter.

There are many other cell penetrating peptides assisted CNS drug delivery including TAT-fused peptides with potential therapeutic applications have been extensively discussed by the others (28, 30, 45, 47, 51, 201–211).

5.2.2 Penetratin

Penetratin is an amphipathic CPP derived from *drosophila antennapedia* homeodomain (RQIKIWFQNRRMKWKK). Several studies have reported penetretin as a CPP for delivering peptides and protein-based therapeutic agents to CNS in preclinical animal models. Bolton et al. first demonstrated suitability of a dose of 1 microgram dose of penetratin peptide for directing agents to small and discrete areas of the brain, by direct injecting it into the rat striatum or lateral ventricles (212). Most recently, Spiegel et al. reported an enhanced antibody uptake across the olfactory epithelium (IN application) into the brain and spinal cord when co-administered with penetratin (213). In a transgenic AD mouse model, Gothwal et al. demonstrated IN delivery of pApoE2 to the brain via penetratin-mannose multi-functionalised chitosan polymeric micelles, which resulted in a reduction of total tau and phosphorylated tau burden (214). Khafagy et al. studied effective delivery of leptin to the brain through IN co-administration with L-penetratin in rat (215). They demonstrated that a single dose administration in normal rats resulted in efficient leptin absorption into the systemic circulation and accumulated in the anterior part of the brain. In addition, chronic delivery of leptin through repeated IN coadministrations with L-penetratin significantly suppressed the appetite and the body weight increase of the rats and lowered their plasma triglyceride levels. Delivering antibodies against amyloid beta is considered a therapeutic measure against AD. Kamei et al. reported that in an aged amyloid precursor protein (APP) knock-in mouse model of AD (App KI mice), IN co-administration of human immunoglobulin G (IgG) with L-penetratin more efficiently delivered human IgG (as a model molecule for Anti-Aß), to the brain compared with i.v. injection and multiple IN treatment of L-penetratin with Anti-Aß resulted in significant improvement in learning and memory in APP KI mice (216–218). The same group had previously reported intranasal co-administration of insulin with L- or D-penetratin significantly enhanced insulin brain delivery as it reached to the cerebral cortex, cerebellum, and brain stem (219). They also demonstrated positive effects of efficient IN brain delivery of insulin combined with penetratin on dementia during the mild and progressive stages of cognitive dysfunction in a mild and progressive memory loss mouse model (220). Most recently, Wang Y et al, conjugated penetratin with a potent amyloid-beta protein (Aß) inhibiting protein "transthyretin" and demonstrated a significant beneficial effect in a *C. elegans* model of AD when compared with transthyretin alone (221). When the cellular prion protein (PrPc) is pathologically misfolded and converted to the abnormal, infectious form of the prion protein (PrPsc) can cause fatal neurodegenerative diseases like Creutzfeldt-Jakob disease in humans and bovine spongiform encephalopathy in cattle (recently reviewed in (222, 223). Skrlj et al. designed a single-chain antibody fragment (scFvs) against PrPsc and used penetratin as a linker between the two variable domains of the scFv. They demonstrated that intravenously injection of scFv-penetratin was efficiently delivered to the infected mouse brain cells (224). Spencer et al. also used penetratin to enhance transport of scFVs against alpha-synuclein across the BBB (225). Fibronectin is one of the components of the extracellular matrix (ECM) which contains a functional sequence (GRGDSPASSPISC) beneficial for improving the angiogenesis both in *in vivo* and *in vitro* models (226, 227). Yang et al. fused this functional sequence with penetratin to improve the penetration of peptides through the BBB into the ischemic hemisphere in the focal cerebral ischemic reperfusion injury rat models (132). They demonstrated significant reduction in the cerebral ischemic volume and the improvement of neurologic score in the stroke animal model.

In an interesting study, Harreither et al. identified structural similarity between penetratin and the 16 amino acids of third helix of the human octamer-binding transcription factor 4 (OCT-4) homeodomain called Oct4 protein transduction domain (Oct4-PTD: DVVRVWFCNRRQKGKR) (228). Oct-4 *is* a transcription factor protein that is encoded by the *POU5F1* gene and is part of the POU (Pit-Oct-Unc) family. This transcription factor is vital for maintaining pluripotency and the self-renewal of stem cells (recently reviewed in (229)). Harreither et al. demonstrated translocation of full length Oct4 into human and mouse cell lines without the addition of any kind of cationic fusion CPPs (228). They suggested that some transcription factors such as Oct-4 might be part of an alternative paracrine signalling pathway, by transferring to neighbouring cells thereby actively changing the behaviour of the recipient cell. Other natural peptides with intrinsic CPP properties were also identified. For example, peptides derived from the prion protein N terminus composed of a hydrophobic sequence followed by a basic segment (mouse mPrP(1-28) or bovine bPrP(1-30)) are known to have cell-penetrating ability. Lofgren et al. also reported their potent anti-prion effects (230).

5.2.3 Polyarginine

It was long known that in rodents intracarotid infusion of polycations including poly-L-arginine alters the BBB permeability (231). Since then, polyargenine peptides have been used for the CNS peptide delivery in pre-clinical studies. A chain of 6 to 11 arginine residues (R6-R11) has been successfully used to deliver different cargoes to the brain (reviewed in (232)). A single 44-residue peptide called PLR (P21-LK15-8R) containing a polyarginine domain (8R) stimulated uptake and enhanced insulin and oxytocin transcytosis over a 6 h period by 7.8-fold and 5-fold, respectively (233). Cheng et al. designed a peptide by combining 8 arginine amino acids with a segment derived from the core region of Aβ ($A\beta^{25\text{-}35}$). They studied the efficiency of IN application of R8-$A\beta^{25\text{-}35}$ peptide on amyloid reduction and cognitive improvement in APP/PS1 double transgenic mice and observed a significant reduction in Aβ accumulation and amelioration of the memory deficits in the transgenic mice (234). In an established retinal *in vivo* NMDA toxicity model, hepta-arginine (R7) linked to the pro-survival PSD-95/PDZ-binding cyclic peptide (CN2097), efficiently delivered its cargo to mitochondria and exerted neuroprotection (235). More polyarginine-based CPP molecules have been also designed for cellular cargo delivery. Komin et al. reported the design of a CPP molecule, designated the CL peptide, that increased the delivery of small-molecule cargoes across model epithelium approximately 10-fold. The CL peptide contains a helix-like motif and a polyarginine tail (R7) (236). Gotanda et al. studied kinetics and the efficiency of 11R-mediated peptide delivery in the normal and ischemic brain in mice (237). In control animals, 11R signal was observed in the microvessels and neurons throughout the brain 1 h after systemic administration, but it was faint after 2 h; whereas in a transient MCAO mouse model, 11R signal was enhanced and remained detectable in the cells on the ipsilateral side for 8 h post administration (237). Pham et al. also studied a myristoylated polyarginine backbone for the brain delivery of imaging probes in mouse (238). A contrast agent, MN-MPAP-Cy5.5 conjugated to myristoylated polyarginine peptides was also proposed for magnetic resonance imaging (MRI) (239). Most recently, an arginine-rich nuclear localization signal (ArgiNLS) strategy has been proposed for streamlined image segmentation of single cells (240). Liu et al. had previously introduced a tandem peptide by combination of R8 and c(RGDfK) to overcome the non-selective property of CPPs and achieved a successful BBB transporting and brain glioma targeting effect (241). Later the same team, enhanced the glioma tissue penetrating ability; by designing a dual receptor recognizing cell penetrating peptide with a sequence of RRRRRRRRdGR (R8-dGR, lower case letter represents D-amino acid residue) (242). This peptide was very target selective by recognizing both integrin $\alpha_v\beta_3$ and NRP-1 receptors. Matsushita et al. demonstrated that 11 arginines (11R) were highly efficient in delivering proteins into the neurons of brain slices. They fused 11R and nuclear localization signal (NLS) with PKA inhibitory peptide which resulted in exclusive delivery to the nuclear compartment of neurons in brain slices (243). The nuclear factor of activated T cells (NFAT) transcription factor is overexpressed in many tumours including the most common primary brain tumour, gliomas (reviewed in (244)). Moreover, Kamei et al. using microvascular endothelial model bEnd.3 cells studied CPPs including octa-arginine (R8) for non-covalent brain delivery of biopharmaceuticals. They identified R8 as the fastest CPPs for the insulin delivery to bEnd.3 cells. An *in vivo* study in rats showed an efficient accumulation of insulin in hippocampus after the i.v. coadministration of insulin with D-R8 without altering the insulin disposition in plasma (245). Recently, Akita et al. successfully used octa-arginine for the nose-to-brain delivery of a glucagon-like peptide-2 derivative. They demonstrated that the R8-Cargo migrated from the trigeminal nerve to the CNS via the principal sensory trigeminal nucleus and then through the trigeminal lemniscus (246). Similarly, Sasaki-Hamada et al. used octa-arginine for the IN delivery of a neuropeptide Neuromedin U (NMU) to the brain in a mouse model of inflammation-mediated amnesia (247). They observed significant accumulation of the cargo in the hippocampus and paraventricular hypothalamic nucleus 30 min after its IN administration and subsequent improvement in inflammation-mediated amnesia.

Despite all successful CNS delivery of peptide cargoes by polyarginine CPPs, there are a few reports on failure of peptide cargo delivery by polyarginine. For instance, Ellert-Miklaszewska et al. used a 11R fused VIVIT peptide (11R-VIVIT, RRRRRRRRRRR-GGG-MAGPHPVIVITGPHEE) to interfere in the calcinurin binding to NFAT. However, 11R-VIVIT failed to inhibit NFAT signaling in glioma cells. They concluded that 11R was not a suitable CPP for the VIVIT delivery (248).

Besides being a CPP molecule, a potent neuroprotective effect of cationic arginine-rich and poly-arginine peptides (up to 18Rs) has been also investigated in preclinical models of stroke and traumatic brain injuries by several groups (249–262). Meloni et al. also compared neuroprotective effects of different CPPS including TAT, penetratin, arg-9, and Prp-1 in primary cortical neurons exposed to glutamic acid and kainic acid (263). They reported that Arg-9, penetratin, and TAT-D displayed consistent and high-level neuroprotective activity in both the glutamic acid (IC50: 0.78, 3.4, 13.9 muM) and kainic acid (IC50: 0.81, 2.0, 6.2 muM) injury models, while Pep-1 was ineffective. Further studies are needed to clarify the neuroprotective function of CPPs.

5.2.4 Pep-1

Morris et al. (264) designed a 21-residue peptide carrier, called Pep-1 (KETWWETWWTEWSQP KKKRKV). Pep-1 was composed of three distinct domains: (1) a hydrophobic tryptophan-rich motif containing five tryptophan residues (KETW WETWWTEW), for efficient targeting to the cell membrane and for forming hydrophobic interactions with proteins; (2) a hydrophilic lysine-rich domain (KKKRKV) derived from the NLS of simian virus 40 (SV-40) large T antigen, to improve intracellular delivery and solubility; and (3) a spacer domain (SQP), containing a proline residue, to improve the flexibility of both the hydrophobic and the hydrophilic domains (264). Because of high stability, a lack of toxicity, and a lack of sensitivity to serum, Pep-1, a short amphipathic peptide carrier, has been extensively used for the CNS delivery of therapeutic peptides in various pre-clinical studies. Here, we only mention some of those studies.

PIN1 is an important enzyme in a variety of cellular processes, including intracellular signalling and apoptosis (reviewed in (265)). Park et al. recently constructed a PEP-1-PIN1 fusion protein and investigated PIN1's function in HT-22 hippocampal cells as well as in a brain ischemic injury gerbil model (266). PEP-1-PIN1, inhibited cell death in the H_2O_2-treated HT-22 hippocampal cells and protected hippocampal neurons in an ischemic injury gerbil model. Choi et al. investigated potential therapeutic effects of Pep-1 fused to the cellular antioxidant enzyme glutaredoxin-1 (GLRX1) in protecting SH-SY5Y dopaminergic neurons against 1-methyl-4-phenylpyridinium (MPP(+)) and 1-methyl-4-phenyl-1,2,3,6-tetrahydropyridine (MPTP) toxicity as well as in a MPTP-induced Parkinson's (PD) disease animal model (267). They observed a significant reduction in cell death in both *in vitro* and *in vivo* experiments. The same PeP-1 peptide was also demonstrated neuroprotection against ischemic injury (268). Similarly, Sohn et al. used PEP-1-metallothionein-III protein to ameliorate the oxidative stress-induced neuronal cell death and brain ischemic insults (269). Copper, zinc-superoxide dismutase 1 (SOD1) is one of the major cellular defensive mechanisms against the deleterious effects of reactive oxygen species (ROS) after ischemia. Zhang et al. reported that i.v. injected PEP-1-SOD1 fusion proteins protected brain from ischemic injury in an experimental asphyxial cardiac arrest rat model (270). PON1 is a potent protective enzyme against oxidative stress-induced inflammation. Kim et al. designed a PRP-1 conjugated PON1 proteins and demonstrated that exogenous PEP-1-PON1 proteins inhibited oxidative stress-induced inflammation in microglia and suppress mitochondrial dysfunction and neurotoxicity in neurons (271). PEP-1 has been used as a CPP for delivering a wide range of bioactive peptide cargoes including PEP-1-catalase (PEP-1-CAT) (272, 273), glutathione S-transferase alpha 2 (PEP-1-GSTA2) (274), glial cell line-derived neurotrophic factor (PEP-1-GDNF) (275), heat shock protein 27 (PEP-1-HSP27) (276, 277), heme-oxygenase-1 (PEP-1-HO-1) (278, 279), sulfiredoxin 1 (PEP-1-Srxn1) (280), Cu, Zn-superoxide dismutase (PEP-1-SOD1) (281–288), carbonyl reductase 1 (PEP-1-CBR1) (289), glutathione S-transferase pi (PEP-1-GSTpi) (290), frataxin (PEP-1-frataxin fusion protein)

(291), ribosomal protein S3 (PEP-1-rpS3) (292–294), pyridoxal-5'-phosphate phosphatase (PEP-1-PLPP fusion protein) (295), peroxiredoxin-2 (PEP-1-PRX2 fusion protein) (282, 296), phosphoprotein enriched in astrocytes 15 (PEP-1-PEA-15) (297), phosphatidylethanolamine-binding protein 1 (PEP-1-PEBP1 fusion protein) (298, 299), interleukin-1 receptor antagonist (IL-1RA) (IL-1RA-PEP) (298–300), phosphoglycerate kinase 1 (PEP-1-PGK1) (301), CREKA peptide(Cys-Arg-Glu-Lys-Ala) a targeting ligand for the fibrin–fibronectin complexes in the tumour ECM (CREKA and PEP-1 peptides) (302), frataxin (PEP-1- frataxin) (303), phosphoglycerate mutase 5 (PEP-1-PGAM5) (304), copper chaperone for SOD (PEP-1-CCS) (305), and Sirtuin 2 (PEP-1-SIRT2 fusion protein) (306). In addition to bioactive peptide brain delivery by PEP-1, Chang et al. recently studied PEP-1 for IN delivery of mitochondria for treatment of PD in a 6-hydroxydopamine lesioned rat model which resulted in significant recovery and improvement (307).

5.3 NOVEL APPROACHES FOR DRUG DELIVERY TO THE BRAIN

5.3.1 The Use of Small Molecules and Monoclonal Antibodies for the Treatment of the CNS Disorders

There are small molecules such as donepezil that are used for the treatment of AD or several antibodies such as natalizumab for the treatment of MS. Small molecules such as donepezil would cross the BBB via diffusion, but may also through carrier-mediated transport (308). Under healthy conditions, the choroid plexus permeability decreases with increasing the molecular weight of hydrophilic molecules such as dextran from 400 Da to 500 kDa (309) with facilitated diffusion for large molecules with 800 kDa molecular weight (310). However, this permeability decreases in AD (311) and depression (312). On the other hand, the permeability of the choroid plexus increases for antibodies such as IgG under severe neuroinflammation (313). This could explain the reduction of brain Aβ load in patients with AD in response to aducanumab administration (314), although the therapeutic outcomes took a long time (about a year) to be detected in the brain following one-year monthly i.v. infusions of the antibody. The CSF/serum albumin quotient (Qalb) is a marker of blood-cerebrospinal fluid barrier (BCSFB) permeability, which increased with the choroid plexus volume in patients with amyotrophic lateral sclerosis. Typical Qalb is 6.8 for individuals under 45 years of age and increases to 10.2 for individuals over 45 years of age (315), but Qalb increased to over 15 in patients with ALS. Therefore, antibodies cannot cross the BBB on their own, but they may penetrate the brain through the choroid plexus depending on the disease and age.

5.3.2 The Use of Invasive Methods for Drug Delivery to the Brain

Due to physiological barriers in accessing the CNS, invasive methods have been developed and used in clinical trials such as convection enhanced delivery (CED) (316), intrathecal administration (317, 318), lumbar puncture (318), and intracerebral implantation of human neural stem cells (319). Low-intensity focused ultrasound (FUS) guided by MRI is a non-invasive procedure that has been employed to reversibly open the BBB in patients with AD for the delivery of aducanumab into the brain (320). These methods have shown effects in the clinic such as the reduction of the Aβ load in the regions of the brain where FUS was applied in patients treated with aducanumab compared to the non-FUS-treated regions (320). This technique requires an MRI-guided FUS device with a hemispherical phased-array transducer with more than 1000 ultrasound sources that converged on a focal point in the brain with real-time guidance by MRI. Due to the technical requirements, this clinical trial managed to recruit only three patients with a treatment protocol over six months with monthly FUS application (320), as opposed to the 139 million patients that would be affected by AD by 2050. Lumbar puncture is typically associated with headaches and rare occasions of meningitis (321). Limb weakness and headache were also detected during the use of CED (322); and sepsis was reported with intracerebral implantation of human neural stem cells (319).

5.3.3 Use of Nasal Route to Target the Brain

As stated in other chapters, the BBB and the BCSFB impose significant barriers for most drugs reaching the brain. IN administration has become an attractive method of drug delivery to the brain, due to its unique set of anatomical characteristics, when daily drug administration is required (323). Nasal delivery provides rapid onset of action (324). A nasal atomiser may be used for drug delivery, administered twice daily (BID) with doses in the range of 30–150 mg per nostril (323). The olfactory region of the nasal cavity has an area of around 5 cm,[2] in adult humans, which contains olfactory filaments. These are formed by a cluster of hundreds to several thousand olfactory receptor axons bundled together (325). Olfactory filaments are part of what constitutes cranial nerve I. These filaments transfer smell information to the brain via neurons descending from the brain. Olfactory filaments are spread over three regions of the nasal cavity: (1) superior turbinate, (2) middle turbinate, and (3) nasal septum. Studies using MRI tractography and endoscopic techniques revealed that olfactory filaments cover 177, 77, and 96 mm^2 surface area in the nasal septum, middle turbinate and superior turbinate, respectively (326, 327), while the cerebral cortex vascular surface is estimated to be 16,350 mm^2 in an adult human (328). The neurons in the olfactory filaments provide direct access to the brain. The drug molecules or NPs need to enter these neurons, travel across the cells, and exit from the other end. In addition to challenges in cellular uptake of the drug molecules or NPs, these therapeutic agents may cause necrosis/apoptosis in the neurons in the olfactory region (329) or become metabolised by the cells (330). Since these neurons have not been adapted to transport compounds to the brain. The turbinate region of the nasal cavity has a limited volume in the range of 3.94 ± 1.03 cm^3 (females) to 4.02 ± 1.41 cm^3 (males) in humans (331), suggesting that the nasal cavity can accommodate only a small volume of a formulation. Furthermore, mucociliary clearance, the presence of a mucus layer, and local enzymes are other factors that can hamper drug delivery to the brain through the IN route (332). These challenges have made systemic delivery of NPs to the brain more popular than the IN route. This trend was confirmed by 270 article citations for the nasal route versus 627 for i.v. administration for drug delivery to the brain when searching only on PubMed; and a similar trend was observed with 40,100 article citations for the nasal route versus 138,000 for i.v. administration, when searching Google Scholar in June 2025.

5.3.4 Use of NPs for the Treatment of the CNS Disorders

NPs such as liposomes have been investigated for the treatment of the CNS disorders, dating back to 1989, and employing implantable pumps was found to be more effective for the treatment of brain cancers (333). The NP formulations have evolved since 1996, when poly(butylcyanoacrylate) NPs were loaded with dalargin by surface adsorption over 3 h (334). The drug-loaded NPs were surface coated with polysorbate 80 just before i.v. administration to mice to evaluate the analgesic effects of dalargin. Interestingly, dalargin bound to NPs and coated with polysorbate 80 showed statistically significant enhancement of hot plate latency at a dose of 5.0 mg/kg compared to the control group that received only empty NPs (334). More recently, bacterial outer membrane vesicles were developed that were loaded with si*Cd47* small-interfering RNA (siRNA) and doxorubicin-intercalated aptamer (335). The vesicles were surface decorated with aniopep-2 peptide to facilitate crossing the BBB following i.v. administration to mice bearing a glioblastoma (GBM) orthotopic tumour model. The role of si*Cd47* was to reduce the expression of CD47 levels in GBM cells, thereby promoting phagocytosis of GBMs by macrophages and microglia (335). NP formulations have now entered clinical trials for both systemic administration (Clinical Trial IDs: NCT04899908 and NCT02820454) and intrathecal administration via CED (Clinical Trial ID: NCT03566199) for the treatment of brain cancers.

In the following sections of the chapter, the latest and novel examples of NP formulations will be presented for common non-communicable CNS disorders. The emphasis will be on the formulations that have shown efficacy at least *in vivo* for the treatment of MS, AD, PD, TBI, IS, and GBM. Despite the significant developments in the formulation of NPs for the treatment of brain disorders, only those for brain cancer (even a limited number of these formulations) and Parkinson's disease are in clinical trials.

5.4 NOVEL NPs FOR THE TREATMENT OF MS

MS is a chronic inflammatory demyelinating disease of the CNS that affects approximately 2.8 million people worldwide. The disease manifests in several forms, with progressive multiple sclerosis (PMS) representing the most challenging variant, characterised by continuous neurological deterioration without distinct relapses or remissions. PMS encompasses both primary progressive MS (PPMS) and secondary progressive MS (SPMS), both of which lead to irreversible disability accumulation and significantly reduced quality of life for patients. The pathophysiology of PMS involves complex mechanisms including chronic neuroinflammation, microglial activation, oligodendrocyte dysfunction, and progressive neurodegeneration. One treatment approach is based on antigen-specific tolerization for the treatment of autoimmune disease, which may selectively blunt autoimmunity without compromising normal immune function or putting the patients at risk of developing serious infections and other diseases such as cancer. Therefore, a noninflammatory mRNA vaccine has been developed for the treatment of MS (336). In this approach, a liposome formulation was employed for the delivery of mRNA encoding MOG_{35-55} peptide (MEVGWYRSPFSRVVHLYRNGK), the epitope of myelin oligodendrocyte glycoprotein. The formulation was administered i.v. to MOG_{35-55}-induced experimental autoimmune encephalomyelitis (EAE) mice, a clinically relevant mouse model of MS. The number of MOG_{35-55}-specific $CD4^+$ T cells was significantly reduced in the spinal cord of the EAE mice compared to control treatments (336). In another vaccination approach, mesoporous polydopamine (mPDA) NPs were loaded with MOG_{35-55} peptide and were intravenously injected into EAE mice (337). MOG_{35-55}-loaded mPDA inhibited the development of the disease upon vaccination (337). MOG_{35-55}-loaded self-assembled rapamycin NPs also showed efficacy in EAE mice through immunosuppressive effects of rapamycin and immunotolerance by MOG_{35-55} peptide (338). Poly(lactide-co-glycolide) (PLGA) NPs loaded with MOG_{35-55} and LABL peptide (ITDGEATDSG, blocking the ICAM-1 signalling process) (339) and PLGA NPs loaded with proteolipid protein $PLP_{139-151}$ peptide [HSLGKWLGHPDKF (340, 341)] (342) were other immune-tolerogenic approaches that showed efficacy in EAE mice.

The resolution of neuroinflammation was achieved by promoting transdifferentiation of proinflammatory T-helper type 17 (Th17) cells into immunosuppressive regulatory T (Treg) cells (343). This approach was based on the overwhelming number of CNS-infiltrating Th17 cells in the CNS lesions of MS, as *post-mortem* examination of CNS lesions from a patient with aggressive relapsing-remitting MS revealed accumulation of Th17 cells in these regions (344). Since Th17 cells have the ability to cross the BBB and reach the CNS lesions, Th17 cells were coupled with nanocapsules containing (Aminooxy)-acetic acid (Th17-AOA-*n*C). Upon exposure of nanocapsules to ROS in the CNS lesions, (Aminooxy)-acetic acid was released and promoted transdifferentiation of Th17 cells into Treg cells. The IL-17 levels were also significantly decreased in the spinal cord of EAE mice following treatment with Th17-AOA-*n*C cells. IL-17 represents a pro-inflammatory cytokine, which is released by Th17 cells (345). One of the interesting novelties of this work is targeting inflamed MS niches (specific regions of the brain lesions such as the inflamed ring around demyelinated areas (346)) in the brain.

A clinical trial was conducted to study the effect of crocin-selenium NPs on the cognition and oxidative stress markers of MS patients (347). The patients received oral capsules of crocin-selenium, and the aim was to increase antioxidant levels in the blood to improve cognitive function. Indeed, the patients demonstrated significant improvements in cognitive function. The importance and novelty of this study are the use of NPs in MS clinical trials.

5.5 NOVEL NPs FOR THE TREATMENT OF AD

AD is a major public health threat with the number of affected individuals increasing each year. AD is characterised by the gradual degeneration of neurons, and impairment of neuronal connectivity, leading to a decline in cognitive function and memory loss. The onset and progression of AD are associated with excessive up-regulation of receptors for advanced glycation end products (RAGE) on the BBB. Therefore, enhanced RAGE degradation could be a therapeutic avenue. PLGA-PEG based NPs were developed with surface decoration of RAP peptide (ELKVLMEKEL) to facilitate targeting of RAGE receptors on the BBB (348). The NPs were also loaded with simvastatin to upregulate the brain endothelial low-density lipoprotein receptor-related protein 1 (LRP1)-involved in the clearance of Aβ peptides across the BBB- on the abluminal side of the BBB. The NPs were designed to interact with RAGE receptors to promote internalization of the receptors into lysosomes, followed by degradation of the proteins in these vesicles. These NPs improved memory function in AD mouse models (348). To target multifaceted pathological conditions associated with AD, engineered exosomes were developed (349). The core of these NPs was formed by dopamine polymerization, forming polydopamine (PDA) polymer, which is considered to be biodegradable and biocompatible. PDA NPs were also able to convert near-infrared light irradiation to heat. PDA NPs were coated with membranes of exosomes derived from brain endothelial cells to facilitate the BBB crossing following i.v. administration. Furthermore, PDA NPs were coated with membranes of RAW264.7-derived exosomes to facilitate targeting of microglia. Fully exosome-membrane-coated-PDA NPs were surface-decorated with cholesterol-labelled Aβ nucleic acid aptamers to capture Aβ oligomers and Aβ filaments. The rationale was that the surface-captured Aβ filament and Aβ oligomer NPs would be phagocytised by microglia, which would be mediated through the surface coating with RAW264.7 exosome membranes.

Finally, PDA NPs were also surface-coated with exosome membranes but core-loaded with resveratrol, which was released upon irradiation with near-infrared beams. *In vitro* studies demonstrated the release of resveratrol upon irradiation with NIR808 (laser with 808 nm wavelength) and degradation of Aβ filaments as well as phagocytosis of Aβ-surface-adsorbed NPs by microglia cell lines (BV2). Both *in vitro* BBB models and *in vivo* experiments confirmed crossing of the PDA NPs through the barrier and targeting of the brain. The PDA NPs improved memory function in AD mouse models as well as reduced Aβ-load in the hippocampus and cortex regions of the brain. In another approach, sialic acid-coated PDA NPs were loaded with methylene blue and berberine for NIR808-triggered drug delivery to the brain for the treatment of AD (350). *In vivo* studies demonstrated the efficacy of these NPs in improving the memory function in AD mouse models (350). As of 1 January 2025, 182 clinical trials were identified, assessing 138 drugs for AD treatment (351). While only the APH-1105 nanoformulation, which is administered intranasally, has been studied in clinical trials. APH-1105 is a nanoformulation of a potent analogue of Bryostatin 1 (352), which is a modulator of α-secretase (353), an enzyme that cleaves the APP into a more soluble compound that is more easily cleared from the brain and does not lead to the formation of insoluble amyloid plaques (354).

5.6 NOVEL NPs FOR THE TREATMENT OF PARKINSON'S DISEASE

One of the characteristics of PD is the formation of α-synuclein (α-syn) aggregates in the brain. α-syn aggregates may lead to the formation of Lewy bodies, which are associated with the cognitive decline observed in PD (355). It was hypothesised that reducing α-syn levels in the brain could be a therapeutic approach for the treatment of PD (356). Therefore, glucose-and trehalose-functionalised carbonised polymer dots (GT-PCDs@pDNA) were developed and loaded with plasmid DNA (pDNA) to inhibit the synthesis of α-syn by silencing *SNCA* gene (355). The NPs were surface decorated with glucose to facilitate crossing the BBB via glucose transporter-1. *In vivo* studies demonstrated the accumulation of GT-PCDs@pDNA in the brain following i.v. administration. These NPs improved behavioural symptoms such as increased walking distance and speed. GT-PCDs@pDNA

NPs also reduced TNF-α levels in the brain tissues of PD mouse models, compared to untreated diseased animals. To promote the innate neuronal regenerative mechanisms in PD, a hydrogel formulation was developed, which contained polydopamine-coated polyurethane NPs (357). These NPs were employed to crosslink a glycol chitosan solution to formulate a biocompatible and biodegradable hydrogel. The hydrogel formulation significantly increased the gene expression of IL-10 in J774A.1 macrophages. This hydrogel was injected into the dopaminergic neurons in the substantia nigra of PD rat models by stereotaxic injections. The aim was to promote neuronal stem cell proliferation in the CNS. In addition to the hydrogel administration, the treatment was combined with acupuncture. This approach significantly improved behavioural measurements in PD rats, such as symmetrical forelimb usage (357). Auranofin (AUF) is an FDA-approved anti-rheumatic agent, which also showed neuroprotective effects (358). In a recent work, to enhance delivery of AUF into the CNS for the treatment of PD, AUF was loaded into hybrid chitosan-lipid NPs and was orally administered to rats prior to the administration of rotenone (a PD-inducing agent) (359). The NPs were not surface decorated with a BBB targeting ligand, and their positively charged zeta potential (42 mV) was considered sufficient to interact with the negatively charged surface of the BBB. The AUF-containing NPs protected against neurodegeneration induced by rotenone administration and increased the total distance travelled in the treatment group compared to untreated PD rat models (359).

To target neuroinflammation in PD, gold nanoclusters were decorated either with D-GSH or L-GSH (360). GSH contains three amino acids: cysteine, glutamate, and glycine. GSH has two distinct chiral enantiomeric forms: L-GSH and D-GSH due to the chiral centres of both cysteine and glutamate. D-GSH or L-GSH gold nanoclusters (2 nm size) were incubated with Zinc nitrate hexahydrate ($Zn(NO_3)_2.6H_2O$) and 2-methylimidazole to achieve an NP size of 90 nm (360). These were designated as L-Au-ZIF and D-Au-ZIF NPs. Interestingly, L-Au-ZIF NPs accumulated more in the brain than D-Au-ZIF NPs in PD rat models. However, mouse PD models receiving D-Au-ZIF NPs intravenously showed improved behavioural evaluations, such as staying longer on a rotary rod than L-Au-ZIF NPs (360).

Piperine is the major component of *Piper retrofractum* and *Piper nigrum* with potential therapeutic effects in AD (361) and PD (362). Piperine was dissolved in a solution containing Tween 80 and normal saline in the ratio of 1:25 and administered orally (362). It was hypothesised that poor aqueous solubility of piperine (0.04 mg/mL) would limit brain uptake and therapeutic effects on the CNS (363). Therefore, piperine NPs were formulated by applying flash nano-precipitation via a four-stream Multi-Inlet Vortex Mixer system. The mixing system contained four arms; each composed of a syringe pump and a syringe. Two arms contained ethanolic solutions of piperine and Eudragit L 100-55, and two arms contained aqueous solutions of Poloxamer 188. The solutions were mixed in a chamber to induce NP precipitation. Optimization of the process led to the formation of piperine NPs with an average size of 225 nm and an encapsulation efficiency of 94%. Piperine NPs achieved an aqueous solubility of 52 mg/mL, and oral administration of piperine NPs led to significantly higher blood concentrations compared to piperine suspension in rats (363).

5.7 NOVEL NPs FOR THE TREATMENT OF TBI

TBI is a significant public health issue and is a leading cause of mortality both in youth and the elderly. It has been referred to as a "silent epidemic". TBI has been considered as a non-genetic and non-ageing-related risk factor for AD. This is mainly due to the neuronal loss in TBI, which may trigger the development of AD. Therefore, it has been suggested that NPs encapsulating siRNAs for the treatment of TBI, such as targeting APP and beta-site amyloid precursor protein cleaving enzyme 1 (*BACE1*) genes (364). TBI is associated with two types of injuries: primary injury and secondary injury. The primary injury is related to the initial and direct impact of an external force. The primary injury itself can be either focal or diffuse (Figure 5.6). In the focal brain injury, a specific region of the brain is damaged (Figure 5.6A), typically where the external force impacts.

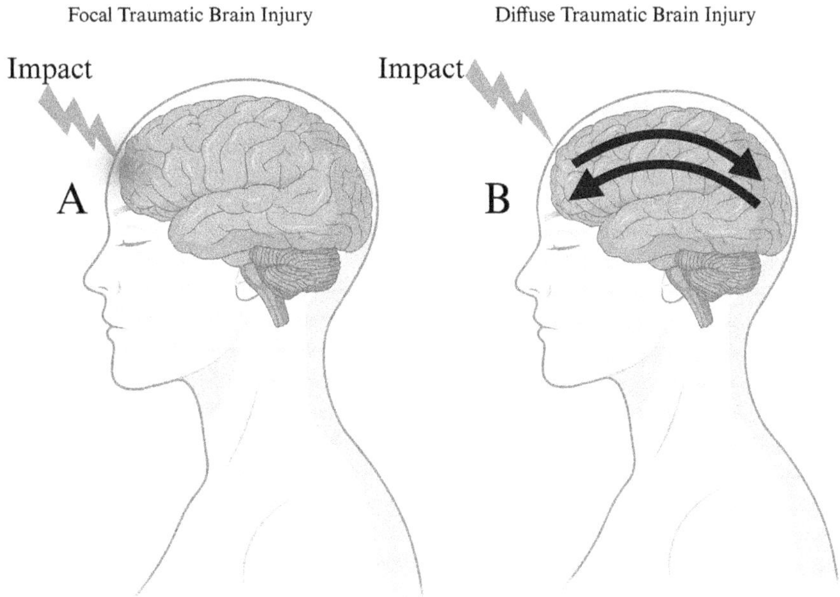

FIGURE 5.6 TBI can be categorised as focal or diffuse depending on the location and spread of the damage. A: Focal TBI causes injury in one specific area, often from direct impacts like bumps or skull fractures. B: Diffuse TBI results in widespread, non-localised brain damage, commonly seen in acceleration-deceleration injuries leading to diffuse axonal injury. Created in Biorender (Toronto, ON M5V 2J1, Canada).

In primary diffuse injury, the damage is spread in the brain tissue, for example, when the brain is shaken (Figure 5.6B). The secondary damage/secondary brain injury (SBI) is the consequence of the primary damage, which is delayed and can be the result of neuroinflammation development or hydrocephalus (cerebral oedema). The delayed development of secondary damage provides a window of opportunity to implement suitable treatments to avoid further CNS injuries following TBI. Mitochondrial dysfunction is categorised as an SBI occurring in injured neurons and microglia, which may be prevented by compounds such as cyclosporine A. However, cyclosporine exhibits poor aqueous solubility; therefore, a nanocrystal formulation was developed, and the NPs were coated with platelet and microglia membranes (CsA-NC@M-$_{PB}$). These NPs accumulated in the brains of mice with mild TBI (mTBI) following i.v. administration, which significantly reduced TNF-α levels in the injured brain and improved memory function of the mouse models (365). mTBI generally goes undetected. This is when a mild mechanical stress is transferred to the brain. However, mTBI may lead to a long-lasting neuronal functional deficit. To diagnose mTBI, $NaGdF_4$ NPs were developed, which were visualised by MRI. $NaGdF_4$ NPs were administered intravenously and allowed for precise identification of microbleeding following mTBI in a mouse model (366).

To restore mitochondrial function in TBI, oxidised carbon nanozymes (OCNs) were developed, which were derived from single-wall carbon nanotubes. When mitochondrial function is impaired, the glycolytic pathway becomes the source of energy for the cell, a process that occurs in the cytoplasm. The aim of employing OCNs was to promote oxidation of NADH to NAD+ in the cytoplasm to enhance the rate of glycolytic energy production, leading to increased lactate generation in the cell. Coating OCNs with PEG led to the formation of PEGylated hydrophilic carbon clusters (PEG-HCCs). Administration of PEG-HCCs to bEnd.3 brain endothelial cells enhanced ATP production via both mitochondrial and glycolytic pathways. To reduce the size of the haemorrhage at the injury site, OCNs were conjugated with deferoxamine, an iron chelator. Deferoxamine-conjugated oxidized carbon nanozymes (DEFPEG-cOACs) were administered intraperitoneally to TBI rat models, and the haemorrhage area was considerably smaller in the treatment group compared to the vehicle-treated TBI models (367).

To inhibit neuroinflammation as well as brain oedema, corticosteroids may be used for the TBI treatment (368). To reduce drug side effects and increase drug accessibility to the brain, an injectable nanocomposite hydrogel was developed, which was composed of a polysaccharide matrix (carboxymethyl chitosan and oxidized dextran) hydrogel with 156-nm mesoporous PDA NPs loaded with dexamethasone dispersed in the hydrogel. The NP-loaded hydrogel formulation was injected into the surgical side of the TBI mouse model. Treated TBI mouse models exhibited reduced levels of proinflammatory cytokines such as IL-1β and TNF-α in the brain as well as significantly improved learning ability compared to the untreated TBI mouse models (369).

To suppress neuroinflammation following TBI, engineered extracellular vesicles (EVs) were developed from human adipose-derived mesenchymal stem cells via transfection with lentivirus (370). The EVs contained upregulated miRNAs such as miRNA-21-5p, which regulates (down-regulation of) highly expressed inflammatory cytokines such as TNF-α and NF-κB signalling in brain microvascular endothelial cells (371). Nasal administration of the EVs (size: 118 nm) significantly reduced NF-κB protein expression in the BV2 microglia cell line, thereby reducing the levels of proinflammatory cytokines such as TNF-α and IL-1β in the brains of TBI mouse models (370). In another study, circular DNA and siRNA were complexed to form sub-100 nm NPs (RNPs). Loading RNPs with TNF-α siRNA reduced TNF-α mRNA expression in LPS-activated BV2 cells. TNF-α siRNA-loaded RNPs were administered intravenously to TBI mouse models at a dose of 4 mg/kg. The RNPs were coated with PEG, and increased BBB permeability contributed to the penetration of these NPs into the brain. TNF-α-siRNA RNP-treated TBI mouse models exhibited reduced TNF-α mRNA expression in the brain compared to the PBS-treated injured mice (372). To reduce secondary damage following TBI, which is caused by the infiltration of peripheral immune cells into the damaged regions of the brain, researchers investigated PLGA NPs. It was shown that i.v. administration of at least 1 mg of 500 nm PLGA NPs (exhibiting immunomodulatory properties) was required per TBI-mouse model to observe a statistically significant improvement in motor functions (373).

EVs derived from human induced pluripotent stem cells (hiPSC-sEVs) have shown neuroprotective effects. The EVs were loaded with a wide range of microRNAs and non-coding RNAs. hiPSC-sEVs demonstrated the ability to be internalised into damaged neuronal cells and prevent cell-death from acute damage such as TBI (374). Since EVs must be stored at −20°C to −80°C to maintain their therapeutic efficacy, these conditions limit their use in rural clinics and on the battlefield. To allow storage at ambient conditions, EVs were lyophilised with the protection of trehalose. The lyophilised EVs maintained their original size (157 nm) after two months storage at room temperature, as well as therapeutic efficacy, which was tested by a scratch assay (375).

The ROS levels increase in the brain following TBI due to disruptions to oxygen homeostasis. To scavenge the excessive ROS, curcumin-based NPs were developed. To achieve this, a hydrophobic core composed of poly (propylene sulphide) and curcumin was formed, which was surface-decorated with lecithin and DSPE-PEG2000-CAQK (376). The inclusion of CAQK was to facilitate crossing the BBB following i.v. administration. The NPs (C-PPS/C) were formed by a single nanoprecipitation process. The curcumin loading capacity was 77%. The poly (propylene sulphide) polymer contains sulphide in each monomer, which can react with oxygen in an oxidation process. The reaction converts poly (propylene sulphide) to poly (propylene sulphone), a hydrophilic compound, which triggers the release of encapsulated hydrophobic drug molecules (377). Indeed, it was shown that the release of curcumin increased from C-PPS/C NPs in the presence of H_2O_2, which was not observed from control NP formulations in which the hydrophobic core was formed from PLGA. Proinflammatory cytokine levels were significantly reduced in the brain following i.v. administration of C-PPS/C NPs at a dose of 10 mg/kg (based on curcumin content) 6 h after TBI in mouse models. In addition, the treatment significantly increased the memory function of the treated TBI mouse models compared to the untreated TBI control group (376). In another approach to target ROS generation following TBI, Tempol-loaded polyurethane nanocapsules were developed. Tempol (4-hydroxy-2,2,6,6-tetramethylpiperidin-1-oxyl) exhibits strong antioxidant properties. The NPs were not surface-decorated with BBB targeting ligands, and BBB disruption following TBI was considered sufficient for targeting the injured brain. The NPs contained 77.61 ± 3.04 μg/mg of Tempol, with a size of 141.4 ± 6.13 nm, and a zeta potential of −11.9 ± 3.54 mV (378).

Baicalein is a traditional Chinese medicine active ingredient, which has shown neuroprotective properties. However, baicalein has poor water solubility and low bioavailability, as well as rapid *in vivo* elimination. Therefore, baicalein-loaded liposomes were developed with an average size of 158 nm and a zeta potential of −28 mV. The baicalein-loaded liposomes were administered orally at a dose range of 50–200 mg/kg. The treatment at a dose of 200 mg/kg improved the BBB integrity and motor function (reducing the forelimb asymmetry score). Although baicalein-loaded liposomes did not have surface brain-targeting ligands, the liposomes used pathways such as the disrupted BBB integrity to access the injured brain (379).

Researchers have become more interested in cell membrane-coated NPs due to the presence of unique proteins, polysaccharides, and lipid layers, which could promote their therapeutic efficacy, such as avoiding elimination by the immune system (380). To prevent ROS-associated secondary injury following TBI, Mn_3O_4 NPs were surface-coated with neutrophil membranes overexpressing BBB targeting proteins such as CXCR4. Mn_3O_4 NPs were chosen as the core NPs due to their excellent ROS-scavenging ability. The cell-membrane-coated Mn_3O_4 NPs (NCM@MP) had an average particle size of 79 nm and a zeta potential of −7.54 mV. NCM@MP were administered intravenously to TBI mouse models, which relieved brain oedema and efficiently improved learning ability compared to the untreated animals (381).

To deliver therapeutic proteins into the brain for the treatment of TBI, a BBB-crossing biomimetic nanocarrier system was developed. The cores of NPs were formulated by mixing the therapeutic proteins (protamine and catalase) with hyaluronic acid. Protamine is a polycationic peptide with a Mw in the range of 4000–5000 Da (382). Hyaluronic acid is a negatively charged polysaccharide. Therefore, the idea was that these two oppositely charged ingredients would electrostatically form complexes with a wide range of proteins. This initial mixture leads to the formation of positively charged NPs that encapsulate therapeutic proteins. Then the core NPs were coated with anionic liposomes composed of dioleoylphosphatydic acid and 1,2-dimyristoyl-sn-glycero-3-phosphocholine. Subsequently, the liposomal NPs were incubated with ApoE3 to allow surface adsorption of ApoE3 on the liposomal NPs, to achieve BBB traversing NPs (CAT-HA-PRTM-rHDL NPs) and these were administered via tail i.v. injections to TBI mouse models. CAT-HA-PRTM-rHDL NPs significantly reduced $IBA1^+$ expression (a sign of activated microglia) in the cortex and hippocampus of the TBI mouse models. In addition, treatment with CAT-HA-PRTM-rHDL NPs significantly improved the memory function of the TBI mouse models compared to the untreated TBI mouse models, as evaluated by the Morris water maze test (383).

Intracranial hypertension (ICH) is a common characteristic of severe TBI (384). Hyperosmolar therapy (including mannitol or hypertonic saline) is the first line of treatment (385). There are other approaches to reducing ICH, such as CSF drainage. When these minimally invasive methods are ineffective, the surgery may be applied to reduce ICH. Surgery is also applied to remove intracerebral hematomas, as part of ICH management. To precisely identify the location of hematomas, silver-indium-selenide quantum dots (core) were doped with aluminium and zinc sulphide to form a shell, and the NPs were surface-coated with PEG (to escape rapid clearance by the reticuloendothelial system). The quantum dots (denoted as AISe@ZnS:Al QDs) had an average diameter of 3 nm and a zeta potential of −7 mV. AISe@ZnS:Al QDs were excited by 675 nm laser irradiation and had a near-infrared emission at 820 nm. AISe@ZnS:Al QDs were administered intravenously at a dose of 333 mg/kg to TBI mouse models 3 h post-injury. The mouse models had TBI induced by dropping a weight of 20 g from a 10-cm distance above the head, with each mouse being hit on the dura mater vertically, causing a brain injury. The injury created intracranial bleeding, which was detected at 5 min and reached a maximum at 30 min post-injection. The AISe@ZnS:Al QDs allowed researchers to distinguish between superficial and intracranial hemorrhage and detect the occurrence of secondary intracranial hemorrhage. This guided the surgical evacuation of the intracranial hematomas. The AISe@ZnS:Al QDs did not have a BBB-targeting ligand, and the compromised BBB following TBI was sufficient to allow penetration of the NPs into the damaged regions of the brain (386).

Following TBI, the expression of vascular cell adhesion molecule-1 (VCAM-1) increased on the BBB in TBI mouse models. This allowed a 10-fold increase in the uptake of VCAM-1 surface-decorated liposomes following i.v. administration. These observations indicate that the BBB becomes permeable following TBI, for sizes around 4 nm, but for targeting larger-sized NPs, a BBB-targeting ligand should be considered (387).

Fibronectin type III domain-containing 5 (FNDC5) is a glycosylated transmembrane protein, which is widely distributed in high-energy-demanding body compartments such as the heart, liver, brain, and skeletal muscle. FNDC5 is proteolytically cleaved to yield irisin, the fibronectin III component of FNDC5 (388). It was found that exogenous irisin markedly increased mitochondrial content and mitochondrial biogenesis in hepatic ischemia-reperfusion injury (389). Interestingly, it was observed that irisin serum levels were significantly correlated with the neurological outcomes of TBI patients, that is, the higher serum irisin levels led to good/excellent recovery with a return to baseline functional status (390). Furthermore, the overexpression of FNDC5 led to the suppression of oxidative stress (measured by dihydroethidium staining) and reduced neuronal apoptosis (determined by TUNEL staining of the cortex) in TBI mouse models. The overexpression of FNDC5 was achieved by stereotaxic injection of the AAV9 virus expressing FNDC5 into the brains of the mice, which were subjected to brain trauma 14 days after the virus injection. This observation encouraged the researchers to develop RVG29-surface-decorated 200 nm PLGA/PEG NPs loaded with irisin for the delivery of irisin to TBI mouse models overexpressing FNDC5 (390). TBI mouse models treated with irisin-loaded NPs (a single dose of 100 μg/kg given intravenously every day after TBI induction) showed significantly improved motor and memory functions compared to untreated TBI mouse models. These functions became similar to those of healthy mice for memory functions after 5 days of treatment, and after 14 days for motor functions (390).

S100B is a biomarker for assessing the severity and potential intracranial injury of mild TBI. Serum S100B levels are routinely measured to evaluate mild TBI injury levels in patients (391). The electrochemiluminescence immunoassay (ECLIA) is intended for use on Elecsys and Cobas e immunoassay analysers for S100B serum measurements, as well as ELISA kits (PS842, Beyotime). To reduce the cost and simplify the S100B serum measurements, S100B monoclonal antibody-functionalised gold NPs were developed. Resonance Rayleigh scattering was employed for the detection of S100B. The results from the NP formulations were in good agreement with the conventional ELISA kit measurements for samples obtained from TBI patients (392).

5.8 NOVEL NPs FOR THE TREATMENT OF IS

When part of a cerebral artery is blocked, that is where the stroke happens, which leads to the death of surrounding neurons within a few minutes. This region is called the ischemic core, and there is a peripheral zone around the ischemic core that is called the ischemic penumbra. This zone is also ischemic, but there are viable cells for a few hours that receive some but not sufficient oxygen and nutrients from collateral arteries. The oxidative stress increases in the penumbra zone, and levels of oxidative stress biomarkers such as F2-isoprostanes also increase in the plasma of patients with acute IS (393). Evaluating these biomarkers is important, as they are signs of salvageable brain tissue before irreversible energy failure and cell death occur in the penumbra zone. In addition, peripheral macrophages (monocyte-derived macrophages) infiltrate into the infarct zone, which contributes to behavioural recovery (394). To accelerate recovery from IS, liposome formulations were developed with membranes composed of 1-palmitoyl-2-oleyl-sn-glycero-3-phosphocholine (POPC), cholesterol, and macrophage membranes. The liposomes were loaded with ceria NPs (which have ROS scavenging properties) and annexin 1-derived peptide $Ac_{2\text{-}26}$ (Ac-AMVSEFLKQAWFIENEEQEYVQTVK (395)) to modulate microglial/macrophage polarization for reducing cerebral inflammation. The macrophage-mimetic liposomes were administered intravenously to mouse models of IS. The liposome formulation improved memory and behavioural functions, and reduced the infarct volume and ischemic penumbra TNF-α, IL-6, and IL-1β cytokine levels in the treated models compared to those treated with PBS (396).

The ROS formation has been demonstrated within the infarct region (7 mm × 5 mm) of cerebral IS in animal models (397). It was found that the occlusion of the middle cerebral artery led to a steady accumulation of ROS in the region, with the levels becoming higher than control readings after 80 min of the occlusion. However, there was a sudden burst in the ROS levels as soon as reperfusion occurred (the removal of the occlusion). The ROS levels rapidly decreased but still were much higher than the control levels at two hours post-reperfusion. Therefore, the removal of accumulated ROS has been a target for NP formulations (397). A ROS triggered, hyaluronic acid-based hydrogel was developed, which was a mixture of hyaluronic acid and platelet lysate. The latter is part of platelet-rich plasma (PRP) that contains an abundance of growth factors, including vascular endothelial growth factor (VEGF), platelet-derived growth factor, and basic fibroblast growth factor. These biological compounds contribute to angiogenesis, such as the creation of bypass blood vessels around the occlusion region (neovascularization). The hydrogel also contained triglycerol monostearate NPs loaded with docosahexaenoic acid (DHA). These NPs were enzymatically cleaved by overexpressed matrix metalloproteinases at the cerebral infarct site, leading to the release of DHA, which prevented neuroinflammation (by reducing brain TNF-α levels), reduced ROS, and promoted neurogenesis (by promoting the migration of neuronal stem cells and neuronal progenitor cells to the injured area). The hydrogel was formed using B-PEG_{500}-B (B denotes for 3-aminobenzeneboronic) as a crosslinker between hyaluronic acid molecules. The hydrogel was injected into the infarct cavity. Treated IS mouse models displayed significant improvements in motor coordination and balance at 28 days post-treatment compared to untreated mouse models (398).

Pituitary adenylate cyclase-activating polypeptide (PACAP38) is a neuroprotective peptide (a 38-residue peptide HSDGIFTDSYSRYRKQMAVKKYLAAVLGKRYKQRVKNK-NH_2 (399, 400)) that modulates microglia toward a pro-reparative phenotype (producing anti-inflammatory cytokine), but it has a short *in vivo* half-life, making systemic delivery of this peptide inefficient. Therefore, PACAP38 was loaded into PLGA NPs and dispersed in a hyaluronic acid-based hydrogel. The hydrogel was injected stereotaxically onto the surface of the brain close to the infarct region. The PACAP38-hydrogel formulation reduced the infarct volume compared to untreated and injured stroke-mouse models. In addition, the PACAP38-hydrogel formulation reduced Iba-1 expression (an indication of activated microglia) at the infarct regions. This study also found that the expression of Iba-1 was reduced at the core of the stroke, but more significantly at regions greater than 200 μm distant from the infarct core. Interestingly, the expression of S100B increased in the stroke microenvironment, and PCAP38 treatment reduced the expression of this biomarker. Stroke-injured mice showed improvements in motor functions upon treatment with the PACAP38-hydrogel formulation compared to the untreated stroke-mouse models (401).

Edaravone is a free radical scavenger that was investigated for the clinical treatment of IS (402). However, liver and kidney dysfunctions have been associated with edaravone (403). Due to the lack of a positive benefit-risk balance, Mitsubishi Tanabe Pharma GmbH withdrew the application for Marketing Authorisation of Radicava, intended for the treatment of ALS, from the European market. Edaravone is a 30 mg/100 mL solution for i.v. infusion only. However, a new formulation of edaravone has been approved for the treatment of IS which is known as edaravone dexborneol. This novel formulation is comprised of two active ingredients: edaravone and (+)-borneol, with synergistic antioxidant and anti-inflammatory effects (404). A recent Phase III clinical trial found good functional outcomes from a 14-day edaravone dexborneol infusion treatment (once every 12 h) versus edaravone alone at day 90 after receiving the treatment (404). Another clinical trial found that average functional independence measures were higher for the edaravone dexborneol-treated group compared to the placebo group, which received 100 mL of normal saline twice a day (all patients received standard medical management according to the guidelines); however, differences did not reach statistical significance (405). It should be noted that 46% to 51% of patients with acute IS and successful reperfusion by endovascular treatment achieved functional independence by day 90, but there were patients who achieved reperfusion without regaining functional independence. This represents an unmet need (405). To reduce off-target effects of edaravone, a NP platform was developed with dextran-carbonyldiimidazole-hydroxymethylphenylboronic acid pinacol ester loading edaravone (DEX-CDIPBE-ED). The choice of the phenylborate ester bond was to provide an ROS sensitivity to the NPs, allowing them to release edaravone in an elevated-oxidative stress microenvironment. These NPs had an average diameter of 70 nm and were administered intravenously to rat models of IS. DEX-CDIPBE-ED NPs at a dose of 2 mg/kg significantly improved the motor functions of IS rat models, compared to the untreated IS rat models. The DEX-CDIPBE-ED NPs were not decorated with BBB-targeting ligands, but it was assumed that diseases such as IS would increase BBB permeability (406).

In another approach, CeO_2 NPs were decorated on Zn-dipsacoside B (DB) NPs to form CZB NPs (107 nm), which formed a complex with anti-C5a aptamer on the NP surface (407). These NPs were also loaded with deferoxamine, an FDA-approved iron-chelator, with preclinical observations of reducing brain ischemic injury (408) and reducing mitochondrial free radicals in IS (409). Then the nanocomposite NPs were surfaces-covered with RVG29-attached-HL-60 cell membranes (RH@CZB-D-A nanoplatform). Dipsacoside B is an extract from the traditional Chinese medicine *Dipsacus asper Wall*, known for improving cognitive functions (410). An anti-C5a aptamer was employed to block C5a on microglia to achieve an anti-inflammatory property. Cell membranes of differentiated and stimulated HL-60 cells were employed due to overexpression of LFA-1 (CD11a/CD18), which helps targeting ICAM-1 on the BBB. To further ensure targeting of the BBB, the cell membranes were co-precipitated with RVG29-PEG-DSPE. The RH@CZB-D-A NPs were administered intravenously every day to MCAO mice with 1 h postreperfusion. The RH@CZB-D-A NPs at a high dose of 30 mg/kg (administered at 1, 25, 49 h post IS, followed by 20 mg/kg daily) significantly reduced infarct volume on day 3 of the treatment compared to IS mouse models treated with saline. Furthermore, RH@CZB-D-A NPs significantly reduced IL-1β levels in the brains of mouse models at the dose of 30 mg/kg, and significantly improved memory functions (evaluated by tests such as the Morris water maze) at a dose of 20 mg/kg compared to saline-treated mouse models (407). This is a very complex formulation and certainly demonstrates *in vivo* efficacy; however, taking these complex systems to clinical trials would be challenging as each step of the formulation should be validated and quantified for achieving consistent clinical responses.

To target ROS generated in the IS, lipid-based NPs were developed which contained DSPE-PEG2000-T7 in the formulation and were loaded with YHV984 (T7-LNP@YHV984). YHV984 was employed to inhibit Hv1 voltage-gated proton channel in microglia, which is required for full activity of NADPH oxidase (NOX) by compensating for loss of NOX–exported charge to generate ROS in IS (411). Therefore, YHV984 alleviates neuroinflammation post-IS. T7 peptide (a brain-targeting ligand)-conjugated lipid allowed convenient mixing with other lipids in the formulation such as Dlin-MC3-DMA. Interestingly, T7-LNP@YHV984 NPs further reduced ROS levels in OGD (deprived of oxygen and glucose to raise cellular ROS levels)-treated BV2 cells compared to the cells treated with YHV984 alone. The T7-LNP@YHV984 NPs were administered intravenously to tMCAO mice (transient middle cerebral artery occlusion) and significantly reduced infarct volume 24 h after induction of IS at an equivalent YHV984 dose of 1 mg/kg, as well as improved survival and memory functions compared to a control group (7 days post-IS). The IS mouse models treated with T7-LNP@YHV984 showed fewer activated microglia (evaluated by Iba-1 expression) in the damaged brain slices compared to a control group (412).

To target the thrombus (blood clot) in the brain artery, disc-shaped PLGA NPs were manufactured with surface-conjugated tissue plasminogen activator (tPA). These discoidal NPs (length 260 nm, width 100 nm) were administered intravenously to MCAO mouse models (single shot at a dose of 1 mg/kg tPA equivalent), and the NPs accumulated in the brain. The discoidal NPs increased blood flow in the cerebral artery 3 h post treatment with the NPs. That was sufficient for mouse models to gain weight close to normal mice 14 days post-treatment. In this work, an MCAO mouse model was created by the formation of mouse blood-clot *ex vivo* and then implanting the blood clot (thrombus) inside the internal carotid artery, which led to the reduction of >70% blood flow in the right cerebral artery compared to blood flow before surgery. In acute IS, at least 87% blockage of blood flow is involved (413).

To target the ischemic region and penumbra of IS, CZ-M13 NPs were developed, which were formulated by surface decoration of bacteriophage M13, decorated with $Ce_{0.9}Zr_{0.1}$-O_2 nanozyme (ultrasmall NPs mimicking superoxide dismutase enzyme). Bacteriophage M13 is a virus and predator for *E. coli*, but harmless to neuronal cells. *In vitro* studies showed M1-polarised microglia actively phagocytised CZ-M13 NPs. To avoid off-target effects, a nasal formulation of CZ-M13 NPs was developed. Therefore, the aim was to deliver CZ-M13 NPs to the ischemic regions of the IS, using M1-microglia as transporters. Indeed, CZ-M13 NPs accumulated in the core region of the stroke in MCAO mouse stroke models (created by a photochemical technique (414): irradiating the right sensorimotor cortex of mice injected intravenously with Rose Bengal at a dose of 30 mg/kg) after daily nasal administration for 7 days. It was shown that the infarct volume and IL-6 levels were significantly reduced in the infarct hemisphere of CZ-M13-treated IS mouse models compared to those models treated with saline. The treated mice also showed significantly improved motor functions (improved limb coordination) compared to the injured mice treated with saline (415). To further improve drug delivery to the brain using nasal formulations, an ultrasonic nasal drop formulation was developed to promote neuronal repair and regeneration following cerebral IS. The formulation contained liposomes loaded with Timosaponin BII, a natural anti-inflammatory drug (416). The core of the liposomes also contained barium titanate (BTO) NPs. BTO NPs had a tetragonal shape with a 50 nm diameter, which provided mechanical strength to the liposomes. These liposomes (LTO@TB) were loaded into hyaluronic acid microspheres with abundant surface negative charges (MS@LTO@TB). To enhance brain drug delivery, focused ultrasonic stimulation was applied to the nasal area. It was shown that applying ultrasound to BV2 cells treated with LPS and MS@LTO@TB significantly reduced the number of $CD86^+$ cells and increased $CD206^+$ cells compared to those cells without ultrasound treatment or LPS-activated BV2 cells, indicating strong anti-inflammatory properties of the formulation and the treatment. An aqueous formulation of MS@LTO@TB was administered intranasally to I/R (MCAO/reperfusion) mouse models. The MCAO was established by inserting a suture into the external carotid artery, and the blockage was removed after 1 h of ischemia. Applying ultrasound with MS@LTO@TB improved memory functions of ischemic/reperfusion mouse models compared to animals administered MS@LTO@TB only, although the difference did not reach statistical significance level. However, both MS@LTO@TB + ultrasound and MS@LTO@TB-only treatments significantly improved the memory functions of ischemic/reperfusion animal models compared to untreated animal models after daily administration for 14 days (417).

Following IS as well as ROS, the brain levels of iron (ferrous ions (Fe^{2+})) increase (418, 419) at the core and penumbra (420), a phenomenon known as ferroptosis, which leads to lipid peroxidation and cell membrane rupture (421). To reduce brain iron levels after IS, molecularly imprinted NPs (MINPs) were developed with high selectivity and chelation specificity towards iron. MINPs were manufactured using $FeSO_4$ molecules as a template, which were mixed with acrylamide as a core polymer, ethylene glycol dimethylacrylate as a cross-linker, and azodiisobutyronitrile as an initiator. Following the polymerization, $FeSO_4$ was embedded inside the generated NPs. Washing the NPs removed $FeSO_4$, and a mesoporous structure was developed (suggested by BET surface area analysis). MINPs demonstrated time-dependent iron (Fe^{2+}) absorption. The MINPs were administered intravenously to I/R models (after 2 hours, the occlusion was removed) at a dose of 5 mg/kg daily for 7 days. The MINPs migrated to the IS region of the brain by transport methods such as hitchhiking on circulating neutrophils. It was shown that neuronal injury (measured by NeuN immunofluorescence staining) and proinflammatory cytokine levels (IL-1β, TNF-α, and IL-6) were significantly reduced in the ischemic brain of the IS mouse models treated with MINPs compared to the untreated IS mouse models. In addition, MINP-treated I/R mice showed improved memory functions (evaluated by the Morris Water Maze test) compared to the untreated mouse models (422). Table 5.1 summarises other examples of novel approaches for the treatment of IS.

5.9 NOVEL NPs FOR THE TREATMENT OF GBM

Bevacizumab is a VEGF-neutralizing antibody which is administered systemically to prevent GBM growth by inhibiting angiogenesis, a key feature of GBM. The therapeutic efficacy of bevacizumab is severely limited due to its limited ability of bevacizumab to cross the BBB. To enhance the brain bioavailability of bevacizumab, a nasal formulation of PLGA NPs was developed. These NPs were loaded with bevacizumab and dichloroacetate (a glycolysis inhibitor (432)). The NPs (BDNP) were administered intranasally to nude mouse models, which were orthotopically inoculated either with 12O150 GBM patient-derived neurospheres (more aggressive than U-251 MG cells) or U251MG-Luc tumour cells, at a dose of BDNP corresponding to 2.5 mg/kg of bevacizumab and 7.5 mg/kg of dichloroacetate. The treatment significantly improved the median survival of the animals, more than 130 days for U251MG-Luc mouse models compared to PBS treatment (93 days) and for 12O150 GBM mouse models (53 days) compared to those treated with PBS (47 days). Interestingly, the surface of the NPs was conjugated with AP-9 peptide (YKLPGHHHHYRP) to target CD14 on the GBM tumour cells, but this approach did not increase the accumulation of the NPs in the brains of U-251MG-bearing mouse models (433).

To promote activation of the immune system against GBM (anti-GBM immunity), sonodynamic therapy was combined with brain-targeting NPs loaded with resiquimod (434). Sonodynamic therapy is emerging as a novel technique for the treatment of GBM. Sonodynamic therapy employs FUS directed into the tumour environment to induce cell apoptosis via upregulation of intracellular ROS (435). The core of the NPs was made of carboxymethyl-β-cyclodextrin (CM-β-CD) conjugated with 5,10,15,20-Tetrakis(4-aminophenyl) porphyrin (TAPP), which carried resiquimod within the CM-β-CD (called Stars NV). The aim of TAPP was to increase the efficacy of sonodynamic therapy (a sonosensitizer agent). The surface of the NPs was conjugated with T7 brain-targeting peptide (KHAIYPRH). The sonodynamic NPs were administered intravenously to intracranial GL261 GBM-bearing mice at a dose of 10 mg/kg on days 0, 3, 6, 9, and 12 (with day 6 being tumour inoculation day), with ultrasound irradiation applied 2 h post-administration at a frequency of 1.0 MHz, a power density of 1 W/cm^2, and a duration of 5 min (434). The treatment increased survival of the animal models, improved memory and motor functions of the animals, and elevated the M1/M2 tumour-associated macrophages (TAM) ratio by 31-fold (434).

In another approach to induce immunogenic cell death in relation to immunotherapy of GBM, the C-X-C motif chemokine ligand-12 (CXCL12)/C-X-C motif chemokine receptor-4 (CXCR4) axis was blocked by AMD@iNP$_{DBCO}$ (112 nm) and CPI@iNP$_{N3}$ (108 nm) micelles (436). It is believed that CXCL12/CXCR4 signalling (chemokine CXCL12 and its receptor CXCR4) plays a key role in the development of resistance to temozolomide (TMZ) therapy, and the blockade of this interaction sensitises GBM cells to TMZ (437). Furthermore, blocking CXCL12/CXCR4 signalling with FDA-approved plerixafor (a CXCR4 antagonist also known as AMD3100) decreased immunosuppression in metastatic breast cancers in mouse models (438). Similar observation of immunosuppression alleviation (reduction in M2 macrophages) was made when blocking the CXCL12/CXCR4 signalling pathway with AMD3100 in GBM treatment (439). The AMD@iNP$_{DBCO}$

TABLE 5.1
Summary of Other NP Formulation for the Treatment of Ischemic Brain.

NP Formulation (Size, Charge)	API	Target	Route of Admin.	Main Outcomes Following Treatment	Ref
Liposomes loaded with sulfonated chitosan and containing activated HL-60 neutrophil cell membrane (118 nm, −13.2 mV)	angiogenic properties of sulfonated chitosan was employed	The inflammatory microenvironment within the ischemic core	i.v.	• Liposomes targeted an inflammatory model zone (lower limb inflammation) • $\uparrow$ Memory and motor functions of MCAO/reperfusion (1 h) mouse models.	(423)
PDA-based NPs loaded with idebenone, surface coated with red blood cell membranes (to improve blood compatibility) and surface conjugated with DNase I to navigate NPs to the IS lesion (100 nm, −20 mV)	Idebenone, a potent antioxidant to protect mitochondria and restore cellular ATP generation	Ischemia-reperfusion areas rich in neutrophils	i.v.	• $\downarrow$ Infarct volume 24 post-treatment in MCAO/reperfusion (1.5 h) mouse models	(424)
EVs derived from hypoxia-conditioned human neuronal stem cells loaded with miR-125a-5p (117 nm, −28 mV)	miR-125a-5p	Brain as general	i.v. received only one injection at the time of reperfusion	• $\uparrow$ Polarization of anti-inflammatory M2 microglia in MCAO/reperfusion (2 h) rat models • $\uparrow$ Motor functional behavioural recovery on day 7 of treatment • $\downarrow$ Brain levels of IL-1β, TNF-α, IL-6 and $\uparrow$ IL-10 and IL-4 compared to untreated IS models	(425)
Triple drug nanocarriers made of ginsenoside Rb1, 3-n-butylphthalide, and probucol (115 nm, −13 mV)	Ginsenoside Rb1, 3-n-butylphthalide, and probucol	• Tween 80 as well as surface adsorption of serum apolipoproteins to bind lipoprotein receptors on the BBB were employed to cross the BBB • Target was the ischemic region of the brain	i.v.	• NPs indeed targeted the IS region of the brain in the MCAO/reperfusion (1 h) mouse model • Significantly $\downarrow$ infarct volume after 7 days of the treatment compared to mouse models treated with PBS	(426)

(Continued)

TABLE 5.1 *(Continued)*
Summary of Other NP Formulation for the Treatment of Ischemic Brain.

NP Formulation (Size, Charge)	API	Target	Route of Admin.	Main Outcomes Following Treatment	Ref
Prussian blue nanozyme-based NPs loaded with aspirin and surface modified with brain-targeting T7 peptide-conjugated to human serum albumin, A@HPB@THSA (153 nm, −11 mV)	Prussian blue nanozyme to scavenge ROS and aspirin to inhibit platelet aggregation	• To reduce neurovascular unit dysfunction in the IS region • T7 peptide was used to facilitate crossing the BBB	Single i.v. shot at the time of reperfusion at a dose of 2 mg/kg	• Significantly ↓ infarct volume and number of M1 microglia, • ↑ Number of M2 microglia, after 22.5 h of the treatment in Rat MCAO/ reperfusion (1.5 h) models compared to untreated models	(427)
Lipid NPs (170 nm, −3 mV)	circSCMH1 (promoting motor function recovery)	• To repair neurovascular units in the ischemic region • To improve sensorimotor function recovery • To comprehensively explore therapeutic efficacy of circSCMH1	Nasal administration was the main route, but i.v. was also investigated, which NPs also accumulated in the ischemic region	• A photothrombotic stroke mouse model was developed (no reperfusion) • Single nasal administration at 24 h post ischemic induction • A dose of 120 μg/kg NPs ↑motor function, memory function (28 days post IS), expression of myelin basic protein • ↓ iNOS in the ischemic region	(428)
Chitosan NPs (250 nm)	Glycine and proline (neuroprotective amino acids)	• Ischemic region	i.v. 1 h after reperfusion	• ↓ Infarct volume in rat models of MCAO/ reperfusion (1 h) at a dose of 10 mg/kg NPs	(429)
Celery seed derived LNPs (60 nm, −12 mV)	Varius lipids, apigenin and luteolin	• Neuroinflammation • Neuronal apoptosis	i.v. 2 h post reperfusion, and also on days 2, 4, and 6 at a dose range of 5-15 mg/kg	• NPs crossed the BBB *in vitro* • ↓ infarct volume in MCAO/reperfusion (1 h) mouse model at both doses of 10 and 15 mg/kg • ↓ Brain levels of TNF-α, IL-1β## and IL-6 • ↑ Memory and motor functions of IS animal models	(430)
Liposomes loaded with Deferoxamine and surface coated with isolated platelet membranes (137 nm, −20 mV)	Deferoxamine	Targeting ischemic region	i.v. 25 mg/kg equivalent to deferoxamine at the beginning of the reperfusion	• NPs targeted the ischemic region in MCAO/ reperfusion (1 h) mouse model • NPs ↓ excessive iron in stroke legions	(431)

micelles contained AMD3100, and the CPI@iNP$_{N3}$ micelles carried CPI-444 (an adenosine 2A receptor inhibitor). It should be noted that the expression of the adenosine 2A receptor on immune cells (including regulatory T cells) and activation of this receptor by adenosine (generated via excess ATP produced during immunogenic cell death) contributes to the immunosuppressive effects seen in the tumour microenvironment (440). The AMD@iNP$_{DBCO}$ micelles were made of a PEG-PLGA and iRGD-PLGA-PEG polymer mixture, with iRGD mediating targeting and crossing the BBB. These micelles also contained poly(2-azepane ethyl methacrylate)-DBCO, a pH-responsive compound, which exposed DBCO on the surface of the AMD@iNP$_{DBCO}$ micelles within the acidic environment of the tumour. This caused linking of the AMD@iNP$_{DBCO}$ micelles with the CPI@iNP$_{N3}$ micelles in the tumour environment, creating large clusters that would not be internalised by the cells. The CPI@iNP$_{N3}$ micelles also contained the brain-targeting iRGD peptide (c(CRGDKGPDC)). A combinational formulation of AMD@iNP$_{DBCO}$ and CPI@iNP$_{N3}$ micelles (identical doses of 5 mg/kg for each NP formulation) was administered intravenously to GL261-Luc cell-GBM-bearing mice in three doses on days 10, 15, and 20 post-tumour implantation. The mice were also subjected to X-ray radiotherapy on days 11, 12, 13, 14, 16, 17, 18, and 19 post-tumour implantations with a 2-Gy irradiation dose each time (a total of 16 Gy dose). The treatment significantly prevented tumour growth compared to GBM mouse models treated with PBS, leading to increased survival of the treated mice from 30 days for the PBS group to 60 days for the treatment group. Furthermore, the number of cells expressing $CD3^+CD8^+$ (an indication of immunologically active cells) significantly increased in the brain of treated animals compared to those mice treated with PBS. In addition, the plasma levels of pro-inflammatory cytokines such as TNF-α and IFN-γ were significantly increased in the treated group compared to the control group (treated with PBS) (436).

The resistance to TMZ treatment has been considered as one of the main factors for the recurrence of GBM. It was shown that the overexpression of BRD4, a key protein in DNA repair, occurred in the GBM tissue of intracranial U251 or GL261 tumour-bearing mice by TMZ treatment (441). It was suggested that overexpression of BRD4 may play a key role in the development of TMZ treatment resistance by providing several DNA repair pathways in GBM cells. To down-regulate BRD4 expression in GBM cells, a PROteolysis-Targeting Chimeras (PROTAC) platform was employed. In the PROTAC platform (442), a small bifunctional drug molecule is employed that binds the target protein at one end and binds to a native cellular control system at the other end to initiate a mechanism leading to protein degradation in cells (Figure 5.7). In simple terms, a protein of interest (POI) is tagged with multiple ubiquitin molecules (443) to make the protein nonfunctional, leading to the protein degradation by mechanisms in place within the cell. More specifically, bifunctional PROTAC molecules bind to the POI at one end while the other end binds to an E3 ligase to form a ternary complex. The recruited E3 ligase then recognises a ubiquitin-E2 enzyme complex and facilitates transfer of the ubiquitin molecule to the POI. The ternary complex dissociates, the ubiquitylated POI is removed by a proteasome, and the PROTAC is released and ready to bind another POI (considering that the interaction with the POI is non-covalent). There are several developed PROTAC molecules such as ARV-771, which mediates the degradation of bromodomain and extra-terminal (BET) proteins including BRD4 (444). In a study, ARV-771 was conjugated to a lipid molecule (1-palmitoyl-2-hydroxy-sn-glycero-3-phosphocholine) via an $^1O_2$-cleavable thioacetal ketone bond, which was cleaved within the acidic environment of GBM. The ARV-771-lipid conjugate was used in the formulation of liposomes which were loaded with TMZ. The liposome formulation also contained U251MG-R cell membranes to cross the BBB and specifically target GBM cells (441). The liposome formulation was designated as M@TP, which was administered intravenously to U251MG-R orthotopic GBM mouse models. The treated mouse models showed longer survival and inhibited tumour volume growth (441). In another study, ARV-771 was conjugated to thioacetal ketone (ARV-771-TK) and the compound was conjugated to octa-ethylene (ethylene glycol) to enhance the water solubility of ARV-771-TK (OEG_8-TK-ARV-771, OT7) (445). OT7 was then loaded into acid and light-sensitive self-assembled

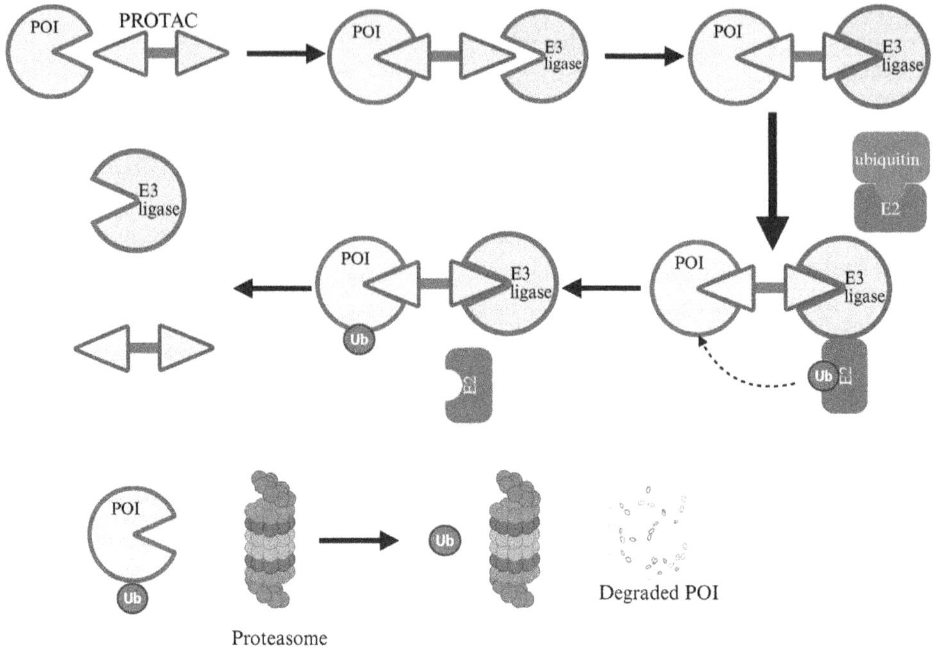

FIGURE 5.7 Schematic presentation of proteolysis-targeting chimaera (PROTAC) technology. This small molecule functions by the recruitment of an active E3 ligase to tag a protein of interest (POI) for disposal. Bifunctional PROTAC small molecules bind to the POI at one end while the other end binds to the E3 ligase to induce a complex. This activates the recruited E3 ligase, which mediates the transfer of Ubiquitin (Ub) from an E2 enzyme to the POI. The complex dissociates, and E3 ligase and PROTAC are released, leaving an ubiquitylated POI. This attracts the attention of a proteasome, which degrades the POI efficiently. The PROTAC can bind to another POI if the interaction with the POI is non-covalent and the PROTAC remains functional.

pyropheophorbide a (PPa) conjugated poly (ethylene glycol)-block-poly(diisopropanol aminoethyl methacrylate co-hydroxyl methacrylate) (PEG-b-PDPA-PPa, PDPA-PPa) NPs (PPT7 NPs), PPa was the light-sensitive component, which releases singlet oxygen ($^1O_2$) upon irradiation. To bypass the BBB and enable direct delivery of ARV-771 to GBM tumour cells, a microneedle patch was developed with needles made of two layers: an inner layer and an outer layer. The outer layer contained PPT7 (loaded with OT7), while the inner layer contained bovine serum albumin-coated manganese oxide NPs (BM NPs). Upon insertion of the microneedle into the GBM tumour during surgical implantation and removal of the backing layer (substrate layer), only the needles remained in the tumour. The outer layer of the needles dissolved in the tumour and released PPT7 NPs. It was assumed that PPT7 NPs would dissociate in the acidic environment of lysosomes following cellular uptake and release of OT7. PPT7 NPs also released OT7 and ARV-771 following irradiation with a 671-nm laser. As singlet oxygen ($^1O_2$) was essential for the release of ARV771 from OT7, BM NPs were gradually released from the inner layer of the microneedle and facilitated the formation of O_2 from accumulated H_2O_2 within the tumour, which also caused the generation of $^1O_2$ (categorised as ROS (446)), leading to further degradation of OT7 and therefore greater release of ARV771. *In vitro* studies showed complete exhaustion of the microneedle (release of all NPs) within 12 h. Therefore, this treatment was intended for the eradication of existing GBM cells. The administration of the microneedle implant improved survival in both subcutaneous and orthotopic (in the brain) GBM mouse models (445). This is an interesting approach, but the questions are how deep the NPs can migrate. Also, human tumours would be much larger than tumours induced in mice; therefore, the pins of microneedles would need to be much longer. Furthermore, when a patient undergoes surgery, the entire tumour may be removed; therefore, there may not be an opportunity to insert an implant into the tumour.

To induce immunogenic cell death, paclitaxel (PTX) NPs were developed. These were composed of PTX (90%) and indocyanine green (10%). PTX NPs were coated with dendritic cell membranes (TAA), which were conjugated with lymphocyte function-associated antigen-1 (LFA-1), very late antigen-4 (VLA-4), αCTLA-4, and αPD1. VLA-4 and LFA-1 contributed to crossing the BBB and targeting tumour cells, while αCTLA-4 and αPD1 helped promote T cell-mediated immune responses. PTX-NPs were administered intravenously at a dose of 5 mg/kg equivalent to PTX, to GL261-Luc GBM mice every two days. The treated mice showed improved survival and an increased population of $CD8^+$ presenting T cells compared to mice treated with PBS. PTX NPs also showed an increased survival rate compared to mice treated with TMZ (447). In another approach to stimulate the immune system for the suppression of GBM cells, virus-like particles (VLPs) were developed. The core of VLPs contained a modified version of an immunogenic protein from hepatitis B virus. Part of this protein contained a peptide (FSHHNIIRL), which is known as $EphA2_{671-679}$ and is expressed on the surface of tumour cells (448). The proteins self-assembled to form NPs with negative surface charges, which allowed for coating them with Ca^{2+}. The resulting NPs were further coated with phosphate (PO_4^{3-}) to achieve mineralised (CaP) NPs (HBc VLPs) with enhanced mucosal adhesion properties. A nasal formulation of HBc VLPs was administered to orthotopic GL261 tumour-bearing C57BL/6 mice on three occasions (Day 7, 12, and 17 post-brain tumour cell inoculation). The aim of the study was to promote an immune response via HBc VLPs in the brain against GBM cells. The survival of the animal models increased in the treatment group compared to PBS-treated animals. Furthermore, the population of M1 macrophages significantly increased in the brains of the treated animals compared to control animals treated with PBS (449).

To address multi-barriers in the treatment of GBM, Trojanbots were developed. The core of the NPs was made of gelatine with surface-adsorbed doxorubicin (GeDOX NPs). These NPs were surface-coated with *E. coli* membrane by extruding the bacteria through a 200 nm polycarbonate membrane first to harvest the cell membrane and then mixing it with GeDOX NPs. This step was followed by covalently conjugating catalase enzyme to the bacterial membrane on the surface of the GeDOX NPs (designated as CatNbots). The presence of the bacterial membrane on the surface of the NPs promoted uptake of the CatNbots by neutrophils and crossing of an *in vitro* model of the BBB. Furthermore, CatNbots were able to move in the media (propulsion) in the presence of H_2O_2, which helped with the penetration of CatNbots into the tumour, where there are higher levels of H_2O_2. These NPs were administered intravenously to GL261 glioma-bearing mice, and the NPs accumulated in the tumour, most likely via neutrophils that phagocytosed the NPs while they were in the blood. The idea was that while the neutrophils were carrying CatNbots (Trojanbots), they would arrive at the tumour environment, release CatNbots, and these would relocate deep into the tumour because of the H_2O_2 gradient and deliver DOX to most of the tumour cells. Indeed, the administration of CatNbots increased the survival of the animal models compared to those treated with PBS (450).

The recurrence of GBM has been observed in patients with time periods as short as six months (451). Radiotherapy resistance (radioresistance) may contribute to GBM recurrence (452). It was found that the brain microenvironment selectively enhances the radioresistance of GBM cells (453). Within the tumour, the hypoxic regions may host radioresistant GBM cells (452) as well as other factors such as glioma stem cells (454). Furthermore, the tumour hypoxic conditions stimulate the release of GSH, which inactivates ROS generated during radiotherapy. The detection of hypoxic regions of GBM by minimally invasive methods may allow for targeting of radiotherapy to these specific regions of the tumour. To achieve this target, 2-nitroimidazole was employed as a hypoxia-targeting ligand, which was conjugated to the surface of PEG-coated superparamagnetic iron oxide NPs (SPION-PG-NI). These NPs exhibited MRI contrast properties. SPION-PG-NI were administered intravenously to subcutaneous glioblastoma-bearing mice and successfully accumulated in the hypoxic regions of the tumour, which were visualised by MRI and confirmed by immunohistological staining using pimonidazole. The treatment prevented the increase in the tumour volume in the mice compared to the saline group. Unexpectedly, SPION-PG-NI further reduced ROS levels upon irradiation (X-ray), which caused cytotoxicity towards tumour cells. Therefore, the radiotherapy exerted antitumour effects by increasing ROS levels in normoxic regions and significantly reducing ROS levels below healthy levels in the hypoxic regions (potentially radioresistant areas) (455).

To facilitate successful lysosomal escape for RNA therapy, bioengineered protein nanocarriers were developed. RNA therapy (e.g., siRNA and miRNA) presents a promising strategy for the treatment of gene-associated diseases such as GBM. Small interfering RNA is negatively charged, double-stranded RNA with 21–23 base pairs, a length of 7–8 nm, and a molecular weight of 13.3 kDa. siRNA exerts its therapeutic effects by binding to a specific mRNA in the cytoplasm and suppressing the expression of a target protein. Due to the charge of siRNA, this type of compound cannot cross the cell membrane; therefore, an NP platform may be employed to neutralise/mask the negative charges, thus enabling cellular uptake. The cellular uptake typically occurs via endocytosis and the formation of intracellular vesicles (456). There are five different pathways of cellular uptake: clathrin-dependent, caveolin-dependent, clathrin- and the caveolin- independent, phagocytosis, and micropinocytosis. When the cellular uptake occurs via the clathrin-dependent pathway, the intracellular vesicles (100-500 nm) are transported within the cell with the help of intracellular actin filaments and join with early endosomes with a typical pH of 6.5 (457). Depending on the NP, they can escape from the early endosomes via typical mechanisms such as membrane fusion, membrane destabilization, particle swelling, and osmotic rupture, causing the release of the NPs into the cytoplasm (Figure 5.8). If this fails, the early endosomes are converted into late endosomes (pH 5.5), which fuse with lysosomes to form endolysosomal vesicles,

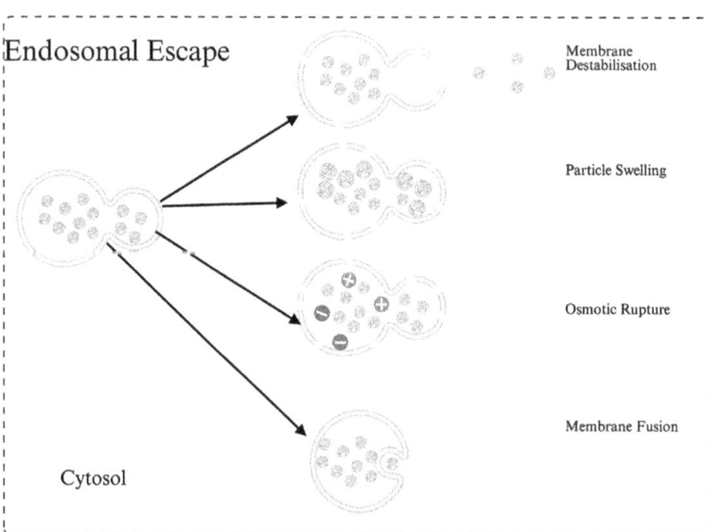

FIGURE 5.8 Schematic presentation of endosomal escape of NPs and release into the cytoplasm by: Membrane fusion, Membrane destabilization, particle swelling, and osmotic rupture.

where the pH drops to 4.5, and there are plenty of proteases, lipases, phosphatases, and nucleases. These enzymes work to digest the contents of the endolysosomal vesicles, such as ferritin, to extract useful ingredients such as iron (458) for cellular use, including in the mitochondria. The remaining materials are then expelled into the extracellular space. Therefore, NPs (carrying siRNA) should escape from the early endosomes to prevent the degradation of siRNA later in the endosomal stage. Ferritin is an interesting protein as it is hollow with the capacity of storing up to 4,500 iron atoms (458). Ferritin binds to transferrin receptor 1 (459), which is also expressed on the BBB (460), promoting uptake of ferritin into the brain (461). Ferritin may be considered an excellent drug delivery system to the brain due to its uniform size and biocompatibility. However, ferritin molecules require modification to enhance their nucleic acid capacity, as the cavity of the ferritin molecules carries negative charges. Therefore, mutants of the ferritin protein molecule were produced (462). The ferritin mutant was disassembled into 24 subunits, each positively charged by reducing the pH to pH 1-2. This was followed by electrostatic complex formation with negatively charged nucleic acids, and then increasing the pH to pH 7–8 re-assembled the ferritin monomers, forming the final nanocarriers (462). Early studies suggested that the loading capacity of mutant ferritin nanocarriers was six to eight nucleic acids per carrier. Therefore, when mutant ferritin was internalised inside the lysosomes, the acidic environment would disassemble the ferritin protein and release the monomers and eventually nucleic acids such as siRNA. To ensure full disassembly of mutant ferritin proteins at weak acids (similar to early endosome pH) and enlargement of the protein internal cavity, the ferritin proteins were further modified and truncated (tHFn+) (463). The idea of endosomal escape was based on the disassembling of ferritin-siRNA carriers at early endosomal stages and the exposure of positive charges to rupture endosomes and release the subunits. Confocal laser scanning microscopy demonstrated that Cy5.5-labelled tHFn+NPs accumulated inside endosomes after 30 min of treatment of U87MG cells with the NPs. Therefore, Cy5.5 labelled tHFn+ NPs were never enriched inside lysosomes. siRNA loaded tHFn+ NPs (siTERT@tHFN+ NPs) reduced the expression of telomerase reverse transcriptase (*TERT*) by more than 50% in U87MG cells at a 200 nM dose. Interestingly, lipid NPs loaded with the same siRNA achieved similar silencing efficiency in the cells, while these could not cross an *in vitro* model of the BBB or accumulate in the brains of Balb/c-nude mice bearing orthotopic glioma when the mice received the NPs via i.v. injections. Furthermore, mice bearing orthotopic U87MG-lu-mCherry glioma cells received siTERT@tHFN+ NPs at a dose of 1 mg/kg of the siRNA every 2 days via i.v. injections. The growth of the tumour and the expression of TERT protein in the brain were significantly reduced after 18 days of tumour induction. This study is explained here in more detail to highlight the steps that perhaps should be taken when GBMs are targeted by NPs carrying nucleic acids (463). Table 5.2 summaries other examples of novel approaches for the treatment of GBM.

5.10 CONCLUSION

Despite progress, challenges persist: CPPs often lack tissue specificity, exhibit suboptimal pharmacokinetics, and face hurdles in clinical translation. Strategies to address these include combining CPPs with NPs or designing stimuli-responsive systems (e.g., pH- or enzyme-triggered release) to boost targeting and bioavailability. While over 25 CPP-based therapies are in clinical trials, none have yet received FDA approval, largely due to insufficient *in vivo* data on stability, immunogenicity, and long-term safety. In summary, novel NPs and CPPs are advancing CNS drug delivery by offering a versatile, low-toxicity platform, with ongoing research focused on refining specificity and overcoming biological barriers to unlock their full therapeutic potential.

TABLE 5.2
Summary of Other NP Formulation for the Treatment of GBM.

Formulation	API	Target Molecular Mechanism	Route of Admin.	Main Outcomes	Ref
Antibody conjugated Niosomes	Bevacizumab or its Fab fragment conjugated to the surface of niosomes	Targeting VEGF	NA (*in vitro*)	• NP size in the range of 100–200 nm • Neutral surface charge • The NPs demonstrated cytotoxicity against GBM U87 cells but not to HUVEC cells (a model of healthy cells)	(464)
Thioether-bridged fluorescence carbon dots with surface-conjugated dopamine quinone ligand (CDs@S-DAQ)	Imaging technique for sensitive and specific recognition of GBM	Sensitive sensing of glutathione as its levels increase to 10 mM in tumour cells	NA (*in vitro*)	• NP size less than 10 nm • Fluorescence intensity increased by increasing GSH concentration with 550 nm UV-vis excitation wavelength and emission at 625 nm • GBM (U87) cells internalised CDs@S-DAQ and demonstrated red fluorescence after 3 h of incubation upon irradiation with a 550 nm excitation wavelength	(465)
Graphitic carbon nitride, ($g\text{-}C_3N_4$)	NP themselves	Generating ROS to depolarise mitochondrial cell membrane integrity	NA (*in vitro*)	• NP size was 300 nm • IC_{50} of 4.7 μg/mL for NPs, and 9.3 μg/mL for TMZ in LN229 cell line • Minimal toxicity against non-cancerous cells (MEF)	(466)
Albumin NPs	Palbociclib	Inhibition of CDK4 and CDK6	i.v.	• NP size was 221 nm and 10.6 mV zeta potential • The surface of NPs was conjugated with polysorbate 80 to cross the BBB. • The IC_{50} of the NPs was 0.6 μM against SH-SY5Y. • The NPs showed higher AUC in different regions of the brain compared to the API alone	(467)

Pegylated chitosan NPs with surface-conjugated lactoferrin to target GBM cells	Thymoquinone	To induce apoptosis in GBM cells	IN	• NP size was 130 nm and 21 mV zeta potential • The IC_{50} of the NPs was 31 μg/mL against U87 MG cells • NPs delivered more amounts of thymoquinone to the brain compared to the API suspension	(468)
Porphyrin-based metallic with a core formed with Yttrium and Terbium, and the core was surface coated with sodium Gd NPs	NP themselves	To generate ROS upon photoactivation (X-ray) and also trigger immune cell death	i.v.	• The surface of NPs was coated with M1-polarised microglia • NP size was 100 nm and −5 mV zeta potential • NPs accumulated in the brain 24 h post i.v. injection to orthotopic GL261-luc^+ GBM bearing mice • The survival of the GBM mice treated with NPs increased compared to the mice treated with PBS • Injection of IL-12 with NPs increased infiltration of $CD3^+$ and $CD8^+$ T cells into the tumour	(469)
Self-assembled luteolin NPs within a PVA hydrogel	Luteolin	To inhibit epithelial mesenchymal transition pathway and prevent GMB recurrence via the β-catenin/Cyclin D1 pathway	The hydrogel containing NPs was deposited in the location of the brain where GL261-luc cells implanted	• NP size was 187 nm and −15 mV zeta potential • The NPs showed IC^{50} around 16 μg/mL against GL261 cells • The growth of tumour was monitored using IVIS • The treatment inhibited the growth of the tumour 25 days after tumour inoculation	(470)

(Continued)

TABLE 5.2 ***(Continued)***
Summary of Other NP Formulation for the Treatment of GBM.

Formulation	API	Target Molecular Mechanism	Route of Admin.	Main Outcomes	Ref
Liposomes with surface conjugated ginsenoside F2	FTY20 (a structurally modified sphingosine, and a natural product extracted from *Cordyceps sinensis*)	• To induce ferroptosis in GBM cells via the AMPK-mTOR-GPX4 Pathway • Ginsenoside F2 was employed to target glucose transporters on the BBB and GBM cells	i.v. to orthotopic mouse models of GBM inoculated with LN229-Luc cells allowing to monito the tumour growth using an IVIS	• NP size was 123-nm and -15 mV zeta potential • DiR-loaded NPs accumulated in the tumour • The treatment inhibited the growth of the tumour 22 days after tumour inoculation • ↑ Survival of treated mouse models compared to the mice that received PBS, which received the NPs every 4 days from the 10th day after implementation of LN299-Luc cells	(471)
Self-assembled NPs of PiDPA copolymer (poly ethylene glycol methyl ether-2-(diisopropylamino) ethyl methacrylate	NPs loaded with oxaliplatin as API, Gd as MRI contrast agent and Cy5 was conjugated to PiDPA as optical imaging agent	To induce cell death by ribosome synthesis stress	Via CED to orthotopic GL261 tumour-bearing mice on the 5th day of tumour implantation and received treatment every 5 days after that	• NP size was 113 nm and −26 mV zeta potential • NPs disassembled at pH 6 and released Gd and oxaliplatin • Both MRI and near-infrared fluorescence imaging demonstrated distribution of the formulation in the tumour • ↑ Survival of treated mouse models compared to those mice received normal saline	(472)
SPIONs with transferrin surface conjugated	Carmustine as API and transferrin was used to target GBM cells	• API to induce cytotoxicity by DNA crosslinking in GBM cells • Iron-oxide NPs to generate ROS and destroy GBM cells as well as triggering M1 polarization in the tumour associated macrophages	IN to GL261-Luc orthotopic glioma-bearing mouse model	• NP size was 42 nm and -35 mV zeta potential • The NPs showed IC_{50} of 2.56 μg/mL GL261 cells • The tumour became invisible in the treated mice 28 days after tumour cell implantation • ↑ Survival of the NP treated mice compared to saline-treated mice • ↑ population of CD86 expressing cells in the brain (excluding the tumour) of the NP treated mice compared to the saline treated animals	(473)

REFERENCES

1. Collaborators GBDNSD. Global, regional, and national burden of disorders affecting the nervous system, 1990-2021: A systematic analysis for the global burden of disease study 2021. Lancet Neurol. 2024;23(4):344–81.
2. Castillo-Gonzalez J, Gonzalez-Rey E. Beyond wrecking a wall: Revisiting the concept of blood-brain barrier breakdown in ischemic stroke. Neural Regen Res. 2025;20(7):1944–56.
3. Huang N, Huang Y, Deng Z, Qi S, Zhang W, Liu Y et al. Blood-brain barrier dysfunction in epilepsy: Mechanisms, therapeutic strategies and future orientation (review). Int J Mol Med. 2025;56(3).
4. Kim S, Jung UJ, Kim SR. The crucial role of the blood-brain barrier in neurodegenerative diseases: Mechanisms of disruption and therapeutic implications. J Clin Med. 2025;14(2).
5. Profaci CP, Munji RN, Pulido RS, Daneman R. The blood-brain barrier in health and disease: Important unanswered questions. J Exp Med. 2020;217(4).
6. Sweeney MD, Sagare AP, Zlokovic BV. Blood-brain barrier breakdown in Alzheimer disease and other neurodegenerative disorders. Nat Rev Neurol. 2018;14(3):133–50.
7. Zhang F, Zhang J, Wang X, Han M, Fei Y, Wang J. Blood-brain barrier disruption in schizophrenia: Insights, mechanisms, and future directions. Int J Mol Sci. 2025;26(3):873.
8. Zhao Z, Nelson AR, Betsholtz C, Zlokovic BV. Establishment and dysfunction of the blood-brain barrier. Cell. 2015;163(5):1064–78.
9. Alahmari A. Blood-brain barrier overview: Structural and functional correlation. Neural Plast. 2021;2021:6564585.
10. Lacoste B, Prat A, Freitas-Andrade M, Gu C. The blood-brain barrier: Composition, properties, and roles in brain health. Cold Spring Harb Perspect Biol. 2025;17(5).
11. Rust R, Yin H, Achon Buil B, Sagare AP, Kisler K. The blood-brain barrier: A help and a hindrance. Brain. 2025;148(7):2262–82.
12. Arora S, Bajaj T, Kumar J, Goyal M, Singh A, Singh C. Recent advances in delivery of peptide and protein therapeutics to the brain. J Pharmacol Exp Ther. 2024;388(1):54–66.
13. Ghorai SM, Deep A, Magoo D, Gupta C, Gupta N. Cell-penetrating and targeted peptides delivery systems as potential pharmaceutical carriers for enhanced delivery across the blood-brain barrier (BBB). Pharmaceutics. 2023;15(7).
14. Lin J, Yu Z, Gao X. Advanced noninvasive strategies for the brain delivery of therapeutic proteins and peptides. ACS Nano. 2024;18(34):22752–79.
15. Porro G, Basile M, Xie Z, Tuveri GM, Battaglia G, Lopes CDF. A new era in brain drug delivery: Integrating multivalency and computational optimisation for blood-brain barrier permeation. Adv Drug Deliv Rev. 2025;224:115637.
16. Pedder JH, Sonabend AM, Cearns MD, Michael BD, Zakaria R, Heimberger AB et al. Crossing the blood-brain barrier: Emerging therapeutic strategies for neurological disease. Lancet Neurol. 2025;24(3):246–60.
17. Stockwell J, Abdi N, Lu X, Maheshwari O, Taghibiglou C. Novel central nervous system drug delivery systems. Chem Biol Drug Des. 2014;83(5):507–20.
18. Frankel AD, Pabo CO. Cellular uptake of the tat protein from human immunodeficiency virus. Cell. 1988;55(6):1189–93.
19. Green M, Loewenstein PM. Autonomous functional domains of chemically synthesized human immunodeficiency virus tat trans-activator protein. Cell. 1988;55(6):1179–88.
20. Park J, Ryu J, Kim KA, Lee HJ, Bahn JH, Han K et al. Mutational analysis of a human immunodeficiency virus type 1 Tat protein transduction domain which is required for delivery of an exogenous protein into mammalian cells. J Gen Virol. 2002;83(Pt 5):1173–81.
21. Vives E, Brodin P, Lebleu B. A truncated HIV-1 Tat protein basic domain rapidly translocates through the plasma membrane and accumulates in the cell nucleus. J Biol Chem. 1997;272(25):16010–7.
22. Joliot A, Pernelle C, Deagostini-Bazin H, Prochiantz A. Antennapedia homeobox peptide regulates neural morphogenesis. Proc Natl Acad Sci U S A. 1991;88(5):1864–8.
23. Derossi D, Joliot AH, Chassaing G, Prochiantz A. The third helix of the Antennapedia homeodomain translocates through biological membranes. J Biol Chem. 1994;269(14):10444–50.
24. Bohmova E, Machova D, Pechar M, Pola R, Venclikova K, Janouskova O et al. Cell-penetrating peptides: A useful tool for the delivery of various cargoes into cells. Physiol Res. 2018;67(Suppl 2):S267–S79.
25. Shoari A, Tooyserkani R, Tahmasebi M, Lowik D. Delivery of various cargos into cancer cells and tissues via cell-penetrating peptides: A review of the last decade. Pharmaceutics. 2021;13(9):1391.

26. Arif M, Kabir M, Ahmed S, Khan A, Ge F, Khelifi A et al. DeepCPPred: A deep learning framework for the discrimination of cell-penetrating peptides and their uptake efficiencies. IEEE/ACM Trans Comput Biol Bioinform. 2022;19(5):2749–59.
27. de Oliveira ECL, da Costa KS, Taube PS, Lima AH, Junior CSS. Biological membrane-penetrating peptides: Computational prediction and applications. Front Cell Infect Microbiol. 2022;12:838259.
28. de Oliveira ECL, Santana K, Josino L, Lima ELAH, de Souza de Sales Junior C. Predicting cell-penetrating peptides using machine learning algorithms and navigating in their chemical space. Sci Rep. 2021;11(1):7628.
29. Gu ZF, Hao YD, Wang TY, Cai PL, Zhang Y, Deng KJ et al. Prediction of blood-brain barrier penetrating peptides based on data augmentation with Augur. BMC Biol. 2024;22(1):86.
30. Hallbrink M, Karelson M. Prediction of cell-penetrating peptides. Methods Mol Biol. 2015;1324:39–58.
31. Imre A, Balogh B, Mandity I. GraphCPP: The new state-of-the-art method for cell-penetrating peptide prediction via graph neural networks. Br J Pharmacol. 2025;182(3):495–509.
32. Kumar N, Du Z, Li Y. pLM4CPPs: Protein language model-based predictor for cell penetrating peptides. J Chem Inf Model. 2025;65(3):1128–39.
33. Manavalan B, Subramaniyam S, Shin TH, Kim MO, Lee G. Machine-learning-based prediction of cell-penetrating peptides and their uptake efficiency with improved accuracy. J Proteome Res. 2018;17(8):2715–26.
34. Park H, Park JH, Kim MS, Cho K, Shin JM. In silico screening and optimization of cell-penetrating peptides using deep learning methods. Biomolecules. 2023;13(3).
35. Ramasundaram M, Sohn H, Madhavan T. A bird's-eye view of the biological mechanism and machine learning prediction approaches for cell-penetrating peptides. Front Artif Intell. 2024;7:1497307.
36. Serebrennikova M, Grafskaia E, Maltsev D, Ivanova K, Bashkirov P, Kornilov F et al. TriplEP-CPP: Algorithm for predicting the properties of peptide sequences. Int J Mol Sci. 2024;25(13).
37. Shi K, Xiong Y, Wang Y, Deng Y, Wang W, Jing B et al. PractiCPP: a deep learning approach tailored for extremely imbalanced datasets in cell-penetrating peptide prediction. Bioinformatics. 2024;40(2):Btae058.
38. Su R, Hu J, Zou Q, Manavalan B, Wei L. Empirical comparison and analysis of web-based cell-penetrating peptide prediction tools. Brief Bioinform. 2020;21(2):408–20.
39. Sutcliffe R, Doherty CPA, Morgan HP, Dunne NJ, McCarthy HO. Strategies for the design of biomimetic cell-penetrating peptides using AI-driven in silico tools for drug delivery. Biomater Adv. 2025;169:214153.
40. Wei HH, Yang W, Tang H, Lin H. The development of machine learning methods in cell-penetrating peptides identification: A brief review. Curr Drug Metab. 2019;20(3):217–23.
41. Wolfe JM, Fadzen CM, Choo ZN, Holden RL, Yao M, Hanson GJ et al. Machine learning to predict cell-penetrating peptides for antisense delivery. ACS Cent Sci. 2018;4(4):512–20.
42. Falanga A, Bellavita R, Braccia S, Galdiero S. Hydrophobicity: The door to drug delivery. J Pept Sci. 2024;30(5):e3558.
43. Derakhshankhah H, Jafari S. Cell penetrating peptides: A concise review with emphasis on biomedical applications. Biomed Pharmacother. 2018;108:1090–6.
44. Behzadipour Y, Hemmati S. Covalent conjugation and non-covalent complexation strategies for intracellular delivery of proteins using cell-penetrating peptides. Biomed Pharmacother. 2024;176:116910.
45. Gori A, Lodigiani G, Colombarolli SG, Bergamaschi G, Vitali A. Cell penetrating peptides: Classification, mechanisms, methods of study, and applications. ChemMedChem. 2023;18(17):e202300236.
46. Sun Z, Huang J, Fishelson Z, Wang C, Zhang S. Cell-penetrating peptide-based delivery of macromolecular drugs: Development, strategies, and progress. Biomedicines. 2023;11(7).
47. Moreno-Vargas LM, Prada-Gracia D. Exploring the chemical features and biomedical relevance of cell-penetrating peptides. Int J Mol Sci. 2024;26(1).
48. Zakany F, Mandity IM, Varga Z, Panyi G, Nagy P, Kovacs T. Effect of the lipid landscape on the efficacy of cell-penetrating peptides. Cells. 2023;12(13):1700.
49. Berillo D, Yeskendir A, Zharkinbekov Z, Raziyeva K, Saparov A. Peptide-based drug delivery systems. Medicina (Kaunas). 2021;57(11).
50. Culkins C, Adomanis R, Phan N, Robinson B, Slaton E, Lothrop E et al. Unlocking the gates: Therapeutic agents for noninvasive drug delivery across the blood-brain barrier. Mol Pharm. 2024;21(11):5430–54.
51. Guidotti G, Brambilla L, Rossi D. Cell-penetrating peptides: from basic research to clinics. Trends Pharmacol Sci. 2017;38(4):406–24.
52. Prades R, Teixido M, Oller-Salvia B. New trends in brain shuttle peptides. Mol Pharm. 2025;22(3):1100–9.

53. Hardingham GE. Coupling of the NMDA receptor to neuroprotective and neurodestructive events. Biochem Soc Trans. 2009;37(Pt 6):1147–60.
54. Hardingham GE, Bading H. Synaptic versus extrasynaptic NMDA receptor signalling: implications for neurodegenerative disorders. Nat Rev Neurosci. 2010;11(10):682–96.
55. Lai TW, Shyu WC, Wang YT. Stroke intervention pathways: NMDA receptors and beyond. Trends Mol Med. 2011;17(5):266–75.
56. Chytla A, Gajdzik-Nowak W, Olszewska P, Biernatowska A, Sikorski AF, Czogalla A. Not just another scaffolding protein family: The multifaceted MPPs. Molecules. 2020;25(21).
57. Levy AM, Gomez-Puertas P, Tumer Z. Neurodevelopmental disorders associated with PSD-95 and its interaction partners. Int J Mol Sci. 2022;23(8).
58. Won S, Levy JM, Nicoll RA, Roche KW. MAGUKs: multifaceted synaptic organizers. Curr Opin Neurobiol. 2017;43:94–101.
59. Lee HJ, Zheng JJ. PDZ domains and their binding partners: structure, specificity, and modification. Cell Commun Signal. 2010;8(8).
60. Cui H, Hayashi A, Sun HS, Belmares MP, Cobey C, Phan T et al. PDZ protein interactions underlying NMDA receptor-mediated excitotoxicity and neuroprotection by PSD-95 inhibitors. J Neurosci. 2007;27(37):9901–15.
61. Ballarin B, Tymianski M. Discovery and development of NA-1 for the treatment of acute ischemic stroke. Acta Pharmacol Sin. 2018;39(5):661–8.
62. Aarts M, Liu Y, Liu L, Besshoh S, Arundine M, Gurd JW et al. Treatment of ischemic brain damage by perturbing NMDA receptor- PSD-95 protein interactions. Science. 2002;298(5594):846–50.
63. Li Y, Zhang J. Animal models of stroke. Animal Model Exp Med. 2021;4(3):204–19.
64. Liu F, McCullough LD. Middle cerebral artery occlusion model in rodents: methods and potential pitfalls. J Biomed Biotechnol. 2011;2011:464701.
65. Srejic LR, Hutchison WD, Aarts MM. Uncoupling PSD-95 interactions leads to rapid recovery of cortical function after focal stroke. J Cereb Blood Flow Metab. 2013;33(12):1937–43.
66. Sun HS, Doucette TA, Liu Y, Fang Y, Teves L, Aarts M et al. Effectiveness of PSD95 inhibitors in permanent and transient focal ischemia in the rat. Stroke. 2008;39(9):2544–53.
67. Li R, Huang Y, Chen L, Zhou H, Zhang M, Chang L et al. Targeted delivery of intranasally administered nanoparticles-mediated neuroprotective peptide NR2B9c to brain and neuron for treatment of ischemic stroke. Nanomedicine. 2019;18:380–90.
68. Cook DJ, Teves L, Tymianski M. A translational paradigm for the preclinical evaluation of the stroke neuroprotectant Tat-NR2B9c in gyrencephalic nonhuman primates. Sci Transl Med. 2012;4(154):154ra33.
69. Cook DJ, Teves L, Tymianski M. Treatment of stroke with a PSD-95 inhibitor in the gyrencephalic primate brain. Nature. 2012;483(7388):213–7.
70. Fisher M, Feuerstein G, Howells DW, Hurn PD, Kent TA, Savitz SI et al. Update of the stroke therapy academic industry roundtable preclinical recommendations. Stroke. 2009;40(6):2244–50.
71. Fisher M, Savitz SI. Pharmacological brain cytoprotection in acute ischaemic stroke - renewed hope in the reperfusion era. Nat Rev Neurol. 2022;18(4):193–202.
72. Hill MD, Martin RH, Mikulis D, Wong JH, Silver FL, Terbrugge KG et al. Safety and efficacy of NA-1 in patients with iatrogenic stroke after endovascular aneurysm repair (ENACT): A phase 2, randomised, double-blind, placebo-controlled trial. Lancet Neurol. 2012;11(11):942–50.
73. Hill MD, Goyal M, Menon BK, Nogueira RG, McTaggart RA, Demchuk AM et al. Efficacy and safety of nerinetide for the treatment of acute ischaemic stroke (ESCAPE-NA1): A multicentre, double-blind, randomised controlled trial. Lancet. 2020;395(10227):878–87.
74. Mayor-Nunez D, Ji Z, Sun X, Teves L, Garman JD, Tymianski M. Plasmin-resistant PSD-95 inhibitors resolve effect-modifying drug-drug interactions between alteplase and nerinetide in acute stroke. Sci Transl Med. 2021;13(588).
75. Tymianski M, Hill MD, Goyal M, Christenson J, Menon BK, Swartz RH et al. Safety and efficacy of nerinetide in patients with acute ischaemic stroke enrolled in the early window: A post-hoc meta-analysis of individual patient data from three randomised trials. Lancet Neurol. 2025;24(3):208–17.
76. Ospel JM, Adams C, Tymianski M, Goyal M, Hill M. Health economic impact of nerinetide in addition to mechanical thrombectomy without concurrent alteplase. Interv Neuroradiol. 2023: 15910199231193455.
77. Fladt J, Guo J, Specht JL, Wang M, Chan LL, McTaggart R et al. Infarct evolution on MR-DWI after thrombectomy in acute stroke patients randomized to nerinetide or placebo: The REPERFUSE-NA1 study. Neurology. 2024;102(2):e207976.

78. Ospel JM, Goyal M, Menon BK, Almekhlafi MA, Zerna C, Nogueira RG et al. Factors influencing nerinetide effect on infarct volume in patients without alteplase in the randomized ESCAPE-NA1 trial. Stroke. 2025;56(1):14–21.
79. Goyal M, Menon BK, Ospel J, Almekhlafi M, Zerna C, Nogueira R et al. Factors influencing nerinetide effect on clinical outcome in patients without alteplase treatment in the ESCAPE-NA1 trial. J Stroke. 2025;27(1):95–101.
80. Joundi RA, Smith EE, Ganesh A, Nogueira RG, McTaggart RA, Demchuk AM et al. Time from hospital arrival until endovascular thrombectomy and patient-reported outcomes in acute ischemic stroke. JAMA Neurol. 2024;81(7):752–61.
81. Pensato U, Bosshart SL, Stebner A, Dowlatshahi D, Bang OY, Sahlas DJ et al. Comparison of noncontrast computed tomography, multiphase computed tomography angiography, and computed tomography perfusion to assess infarct growth rate in acute stroke. Stroke. 2025;56(3):657–66.
82. Christenson J, Hill MD, Swartz RH, Adams C, Benavente O, Casaubon LK et al. Efficacy and safety of intravenous nerinetide initiated by paramedics in the field for acute cerebral ischaemia within 3 h of symptom onset (FRONTIER): A phase 2, multicentre, randomised, double-blind, placebo-controlled study. Lancet. 2025;405(10478):571–82.
83. Hill MD, Goyal M, Demchuk AM, Menon BK, Field TS, Guest WC et al. Efficacy and safety of nerinetide in acute ischaemic stroke in patients undergoing endovascular thrombectomy without previous thrombolysis (ESCAPE-NEXT): A multicentre, double-blind, randomised controlled trial. Lancet. 2025;405(10478):560–70.
84. Benali F, Fladt J, Jaroenngarmsamer T, Bala F, Singh N, Ospel JM et al. Association of brain atrophy with functional outcome and recovery trajectories after thrombectomy: Post hoc analysis of the ESCAPE-NA1 Trial. Neurology. 2023;101(15):e1521–e30.
85. Ospel JM, Rinkel L, Ganesh A, Demchuk A, Joshi M, Poppe A et al. Influence of infarct morphology and patterns on cognitive outcomes after endovascular thrombectomy. Stroke. 2024;55(5):1349–58.
86. Xu B, Xiao AJ, Chen W, Turlova E, Liu R, Barszczyk A et al. Neuroprotective effects of a PSD-95 inhibitor in neonatal hypoxic-ischemic brain injury. Mol Neurobiol. 2016;53(9):5962–70.
87. Zhu YB, Jia GL, Wang JW, Ye XY, Lu JH, Chen JL et al. Activation of CaMKII and GluR1 by the PSD-95-GluN2B coupling-dependent phosphorylation of GluN2B in the spinal cord in a rat model of type-2 diabetic neuropathic pain. J Neuropathol Exp Neurol. 2020;79(7):800–8.
88. D'Mello R, Marchand F, Pezet S, McMahon SB, Dickenson AH. Perturbing PSD-95 interactions with NR2B-subtype receptors attenuates spinal nociceptive plasticity and neuropathic pain. Mol Ther. 2011;19(10):1780–92.
89. Zhou HH, Zhang L, Zhang HX, Xu BR, Zhang JP, Zhou YJ et al. Tat-HA-NR2B9c attenuate oxaliplatin-induced neuropathic pain. Exp Neurol. 2019;311:80–7.
90. Wang Z, Chen Z, Yang J, Yang Z, Yin J, Duan X et al. Treatment of secondary brain injury by perturbing postsynaptic density protein-95-NMDA receptor interaction after intracerebral hemorrhage in rats. J Cereb Blood Flow Metab. 2019;39(8):1588–601.
91. Dykstra CM, Ratnam M, Gurd JW. Neuroprotection after status epilepticus by targeting protein interactions with postsynaptic density protein 95. J Neuropathol Exp Neurol. 2009;68(7):823–31.
92. Tse K, Hammond D, Simpson D, Beynon RJ, Beamer E, Tymianski M et al. The impact of postsynaptic density 95 blocking peptide (Tat-NR2B9c) and an iNOS inhibitor (1400W) on proteomic profile of the hippocampus in C57BL/6J mouse model of kainate-induced epileptogenesis. J Neurosci Res. 2019;97(11):1378–92.
93. Wu P, Zhao H, Gou X, Wu X, Zhang S, Deng G et al. Targeted delivery of polypeptide nanoparticle for treatment of traumatic brain injury. Int J Nanomedicine. 2019;14:4059–69.
94. Zhang X, Zhang R, Wu J. Inhibition of the NR2B-PSD95 interaction exerts neuroprotective effects on retinal ischemia-reperfusion injury. Neuroscience. 2022;490:89–99.
95. Zhang H, Xu L, He Y, Zhang Z, Zhang J, Yu Q et al. Tat-NR2B9c attenuates oxidative stress via inhibition of PSD95-NR2B-nNOS complex after subarachnoid hemorrhage in rats. Neuropharmacology. 2024;251:109905.
96. Smaga I, Wydra K, Witek K, Surowka P, Suder A, Pieniazek R et al. Intravenous administration of Tat-NR2B9c peptide, a PSD95 inhibitor, attenuates reinstatement of cocaine-seeking behavior in rats. Behav Brain Res. 2022;416:113537.
97. Lee YR, Chen M, Pandolfi PP. The functions and regulation of the PTEN tumour suppressor: New modes and prospects. Nat Rev Mol Cell Biol. 2018;19(9):547–62.
98. Ning K, Pei L, Liao M, Liu B, Zhang Y, Jiang W et al. Dual neuroprotective signaling mediated by downregulating two distinct phosphatase activities of PTEN. J Neurosci. 2004;24(16):4052–60.

99. Chen Z, Trotman LC, Shaffer D, Lin HK, Dotan ZA, Niki M et al. Crucial role of p53-dependent cellular senescence in suppression of Pten-deficient tumorigenesis. Nature. 2005;436(7051):725–30.
100. Trotman LC, Alimonti A, Scaglioni PP, Koutcher JA, Cordon-Cardo C, Pandolfi PP. Identification of a tumour suppressor network opposing nuclear Akt function. Nature. 2006;441(7092):523–7.
101. Trotman LC, Wang X, Alimonti A, Chen Z, Teruya-Feldstein J, Yang H et al. Ubiquitination regulates PTEN nuclear import and tumor suppression. Cell. 2007;128(1):141–56.
102. Wang X, Trotman LC, Koppie T, Alimonti A, Chen Z, Gao Z et al. NEDD4-1 is a proto-oncogenic ubiquitin ligase for PTEN. Cell. 2007;128(1):129–39.
103. Zhang S, Taghibiglou C, Girling K, Dong Z, Lin SZ, Lee W et al. Critical role of increased PTEN nuclear translocation in excitotoxic and ischemic neuronal injuries. J Neurosci. 2013;33(18):7997–8008.
104. Zhao J, Yin L, Jiang L, Hou L, He L, Zhang C. PTEN nuclear translocation enhances neuronal injury after hypoxia-ischemia via modulation of the nuclear factor-kappaB signaling pathway. Aging (Albany NY). 2021;13(12):16165–77.
105. Zheng Y, Gu H, Kong Y. Targeting PTEN in ischemic stroke: From molecular mechanisms to therapeutic potentials. Exp Neurol. 2025;383:115023.
106. Wang Z, Jia S, Zhang Q, Wang Y, Huang B, Zheng L. LC-MS/MS assay for the determination of tat-K13, a novel interfering peptide for the treatment of ischemic stroke, in human plasma and its application to a pharmacokinetics study. Biomed Chromatogr. 2021;35(7):e5095.
107. Wang Z, Jin Y, Shen Q, Zhuo L, Liu R, Di X et al. First-in-human safety, tolerability, and pharmacokinetics of SY-007, a prolonged action neuroprotective drug for ischemic stroke, in healthy Chinese subjects. Eur J Pharm Sci. 2022;170:106104.
108. Sekar S, Marks WN, Gopalakrishnan V, Greba Q, Snutch TP, Howland JG et al. Evidence for altered insulin signalling in the brains of genetic absence epilepsy rats from Strasbourg. Clin Exp Pharmacol Physiol. 2020;47(9):1530–6.
109. Nisha TD, Gohil B, Modak AS, Singh NT, Mukherjee C et al. Downregulation of Pten improves Huntington's disease phenotype by reducing Htt aggregates and Cell death. Mol Neurobiol. 2025;62(7):8752–67.
110. Sekar S, Taghibiglou C. Elevated nuclear phosphatase and tensin homolog (PTEN) and altered insulin signaling in substantia nigral region of patients with Parkinson's disease. Neurosci Lett. 2018;666:139–43.
111. Shimano H, Sato R. SREBP-regulated lipid metabolism: convergent physiology - divergent pathophysiology. Nat Rev Endocrinol. 2017;13(12):710–30.
112. Taghibiglou C, Martin HG, Rose JK, Ivanova N, Lin CH, Lau HL et al. Essential role of SBP-1 activation in oxygen deprivation induced lipid accumulation and increase in body width/length ratio in Caenorhabditis elegans. FEBS Lett. 2009;583(4):831–4.
113. Liu S, Li X, Fan P, Gu Y, Yang A, Wang W et al. The potential role of transcription factor sterol regulatory element binding proteins (SREBPs) in Alzheimer's disease. Biomed Pharmacother. 2024;180:117575.
114. Xu X, Jin W, Chang R, Ding X. Research progress of SREBP and its role in the pathogenesis of autoimmune rheumatic diseases. Front Immunol. 2024;15:1398921.
115. Taghibiglou C, Martin HG, Lai TW, Cho T, Prasad S, Kojic L et al. Role of NMDA receptor-dependent activation of SREBP1 in excitotoxic and ischemic neuronal injuries. Nat Med. 2009;15(12):1399–406.
116. Taghibiglou C, Lu J, Mackenzie IR, Wang YT, Cashman NR. Sterol regulatory element binding protein-1 (SREBP1) activation in motor neurons in excitotoxicity and amyotrophic lateral sclerosis (ALS): Indip, a potential therapeutic peptide. Biochem Biophys Res Commun. 2011;413(2):159–63.
117. Adejare A. Drug discovery approaches for the treatment of neurodegenerative disorders: Alzheimer's disease. Amsterdam; Boston: Academic Press is an imprint of Elsevier; 2017. xvii, 290 pages p.
118. Sekar S, Omran E, Gopalakrishnan V, Howland JG, Snutch TP, Taghibiglou C. Elevated sterol regulatory elementary binding protein 1 and GluA2 levels in the hippocampal nuclear fraction of Genetic Absence Epilepsy Rats from Strasbourg. Epilepsy Res. 2017;136:1–4.
119. Makgoo L, Mosebi S, Mbita Z. The role of death-associated protein kinase-1 in cell homeostasis-related processes. Genes (Basel). 2023;14(6):1274.
120. Xu LZ, Li BQ, Li FY, Li Y, Qin W, Zhao Y et al. NMDA receptor GluN2B subunit is involved in excitotoxicity mediated by death-associated protein kinase 1 in Alzheimer's Disease. J Alzheimers Dis. 2023;91(2):877–93.
121. Gan CL, Zou Y, Xia Y, Zhang T, Chen D, Lan G et al. Inhibition of death-associated protein kinase 1 protects against epileptic seizures in mice. Int J Biol Sci. 2021;17(9):2356–66.
122. Li SX, Han Y, Xu LZ, Yuan K, Zhang RX, Sun CY et al. Uncoupling DAPK1 from NMDA receptor GluN2B subunit exerts rapid antidepressant-like effects. Mol Psychiatry. 2018;23(3):597–608.

123. Zhang T, Kim BM, Lee TH. Death-associated protein kinase 1 as a therapeutic target for Alzheimer's disease. Transl Neurodegener. 2024;13(1):4.
124. Tu W, Xu X, Peng L, Zhong X, Zhang W, Soundarapandian MM et al. DAPK1 interaction with NMDA receptor NR2B subunits mediates brain damage in stroke. Cell. 2010;140(2):222–34.
125. Tu G, Fu T, Yang F, Yao L, Xue W, Zhu F. Prediction of GluN2B-CT(1290-1310)/DAPK1 interaction by protein(-)peptide docking and molecular dynamics simulation. Molecules. 2018;23(11):3018.
126. Schmidt ME, Caron NS, Aly AE, Lemarie FL, Dal Cengio L, Ko Y et al. DAPK1 Promotes extrasynaptic GluN2B phosphorylation and striatal spine instability in the YAC128 mouse model of Huntington disease. Front Cell Neurosci. 2020;14:590569.
127. Liu Y, Stockwell BR, Jiang X, Gu W. p53-regulated non-apoptotic cell death pathways and their relevance in cancer and other diseases. Nat Rev Mol Cell Biol. 2025;26(8):600–14.
128. Ho CJ, Lin RW, Zhu WH, Wen TK, Hu CJ, Lee YL et al. Transcription-independent and -dependent p53-mediated apoptosis in response to genotoxic and non-genotoxic stress. Cell Death Discov. 2019;5:131.
129. Pei L, Shang Y, Jin H, Wang S, Wei N, Yan H et al. DAPK1-p53 interaction converges necrotic and apoptotic pathways of ischemic neuronal death. J Neurosci. 2014;34(19):6546–56.
130. Zhao J, Liu X, Blayney A, Zhang Y, Gandy L, Mirsky PO et al. Intrinsically disordered N-terminal domain (NTD) of p53 interacts with mitochondrial PTP regulator cyclophilin D. J Mol Biol. 2022;434(9):167552.
131. Wang X, Pei L, Yan H, Wang Z, Wei N, Wang S et al. Intervention of death-associated protein kinase 1-p53 interaction exerts the therapeutic effects against stroke. Stroke. 2014;45(10):3089–91.
132. Yang H, Yang H, Lu Y, Ma S, Liu Y, Jia G et al. Targeted delivery of extracellular matrix protected against neurologic defects after focal ischemia reperfusion in rats. J Stroke Cerebrovasc Dis. 2015;24(1):154–62.
133. Golubovskaya VM, Finch R, Zheng M, Kurenova EV, Cance WG. The 7-amino-acid site in the proline-rich region of the N-terminal domain of p53 is involved in the interaction with FAK and is critical for p53 functioning. Biochem J. 2008;411(1):151–60.
134. Jiang L, Ma Y, Wang J, Tao X, Wei D. The transduction of His-TAT-p53 fusion protein into the human osteogenic sarcoma cell line (Saos-2) and its influence on cell cycle arrest and apoptosis. Mol Biol Rep. 2008;35(1):1–8.
135. Ryu J, Lee HJ, Kim KA, Lee JY, Lee KS, Park J et al. Intracellular delivery of p53 fused to the basic domain of HIV-1 Tat. Mol Cells. 2004;17(2):353–9.
136. Snyder EL, Meade BR, Saenz CC, Dowdy SF. Treatment of terminal peritoneal carcinomatosis by a transducible p53-activating peptide. PLoS Biol. 2004;2(2):E36.
137. Vitali R, Cesi V, Tanno B, Ferrari-Amorotti G, Dominici C, Calabretta B et al. Activation of p53-dependent responses in tumor cells treated with a PARC-interacting peptide. Biochem Biophys Res Commun. 2008;368(2):350–6.
138. Yan H, Liu N, Zhao Z, Zhang X, Xu H, Shao B et al. Expression and purification of human TAT-p53 fusion protein in Pichia pastoris and its influence on HepG2 cell apoptosis. Biotechnol Lett. 2012;34(7):1217–23.
139. Illanes-Gonzalez J, Flores-Munoz C, Vitureira N, Ardiles AO. Pannexin 1 channels: A bridge between synaptic plasticity and learning and memory processes. Neurosci Biobehav Rev. 2025;174:106173.
140. Koval M, Schug WJ, Isakson BE. Pharmacology of pannexin channels. Curr Opin Pharmacol. 2023;69:102359.
141. Saez JC, Ocaranza FJ, Prieto-Villalobos J, Orellana JA. Connexin and pannexin hemichannels: Broad-spectrum players in neuroinflammatory signaling. J Neurochem. 2025;169(9):e70237.
142. Yeung AK, Patil CS, Jackson MF. Pannexin-1 in the CNS: Emerging concepts in health and disease. J Neurochem. 2020;154(5):468–85.
143. Ruan Z, Orozco IJ, Du J, Lu W. Structures of human pannexin 1 reveal ion pathways and mechanism of gating. Nature. 2020;584(7822):646–51.
144. Weilinger NL, Tang PL, Thompson RJ. Anoxia-induced NMDA receptor activation opens pannexin channels via Src family kinases. J Neurosci. 2012;32(36):12579–88.
145. Weilinger NL, Lohman AW, Rakai BD, Ma EM, Bialecki J, Maslieieva V et al. Metabotropic NMDA receptor signaling couples Src family kinases to pannexin-1 during excitotoxicity. Nat Neurosci. 2016;19(3):432–42.
146. Man HY, Lin JW, Ju WH, Ahmadian G, Liu L, Becker LE et al. Regulation of AMPA receptor-mediated synaptic transmission by clathrin-dependent receptor internalization. Neuron. 2000;25(3):649–62.

147. Ahmadian G, Ju W, Liu L, Wyszynski M, Lee SH, Dunah AW et al. Tyrosine phosphorylation of GluR2 is required for insulin-stimulated AMPA receptor endocytosis and LTD. EMBO J. 2004;23(5):1040–50.
148. Fox CJ, Russell K, Titterness AK, Wang YT, Christie BR. Tyrosine phosphorylation of the GluR2 subunit is required for long-term depression of synaptic efficacy in young animals in vivo. Hippocampus. 2007;17(8):600–5.
149. Lin XJ, Zhang JJ, Yu LC. GluR2-3Y Inhibits the acquisition and reinstatement of morphine-induced conditioned place preference in rats. Neurosci Bull. 2016;32(2):177–82.
150. Zhang J, Liu Z, Liu X, Wang X, Yu L. Intravenous injection of GluR2-3Y inhibits repeated morphine-primed reinstatement of drug seeking in rats. Brain Sci. 2023;13(4):590.
151. Yu YJ, Chang CH, Gean PW. AMPA receptor endocytosis in the amygdala is involved in the disrupted reconsolidation of Methamphetamine-associated contextual memory. Neurobiol Learn Mem. 2013;103:72–81.
152. Zhang W, Zhang X, Lei M, Zhang D, Qin G, Zhou J et al. Dopamine D2 receptor activation blocks GluA2/ROS positive feedback loop to alienate chronic-migraine-associated pain sensitization. Antioxidants (Basel). 2024;13(6):725.
153. Ashourpour F, Jafari A, Babaei P. Chronic administration of Tat-GluR23Y ameliorates cognitive dysfunction targeting CREB signaling in rats with amyloid beta neurotoxicity. Metab Brain Dis. 2021;36(4):701–9.
154. Ashourpour F, Jafari A, Babaei P. Co-treatment of AMPA endocytosis inhibitor and GluN2B antagonist facilitate consolidation and retrieval of memory impaired by beta amyloid peptide. Int J Neurosci. 2022;132(7):714–23.
155. Eyvani K, Letafatkar N, Babaei P. AMPA receptors endocytosis inhibition attenuates cognition deficit via c-Fos/BDNF signaling in amyloid beta neurotoxicity. Exp Aging Res. 2025;51(3):303–15.
156. Lopez J, Gamache K, Schneider R, Nader K. Memory retrieval requires ongoing protein synthesis and NMDA receptor activity-mediated AMPA receptor trafficking. J Neurosci. 2015;35(6):2465–75.
157. Pereyra M, de Landeta AB, Dalto JF, Katche C, Medina JH. AMPA receptor expression requirement during long-term memory retrieval and its association with mTORC1 signaling. Mol Neurobiol. 2021;58(4):1711–22.
158. Cazakoff BN, Howland JG. AMPA receptor endocytosis in rat perirhinal cortex underlies retrieval of object memory. Learn Mem. 2011;18(11):688–92.
159. Ferrara NC, Jarome TJ, Cullen PK, Orsi SA, Kwapis JL, Trask S et al. GluR2 endocytosis-dependent protein degradation in the amygdala mediates memory updating. Sci Rep. 2019;9(1):5180.
160. Hardt O, Nader K, Wang YT. GluA2-dependent AMPA receptor endocytosis and the decay of early and late long-term potentiation: Possible mechanisms for forgetting of short- and long-term memories. Philos Trans R Soc Lond B Biol Sci. 2014;369(1633):20130141.
161. Babaei P, Javer S, Abedinzade M. Therapeutic effects of combined and chronic treatment of Tat-GluA23y and D-Serine on cognitive dysfunction in postmenopausal rats. Exp Aging Res. 2024;50(5):633–51.
162. Migues PV, Liu L, Archbold GE, Einarsson EO, Wong J, Bonasia K et al. Blocking synaptic removal of GluA2-containing AMPA receptors prevents the natural forgetting of long-term memories. J Neurosci. 2016;36(12):3481–94.
163. Ashby DM, Dias C, Aleksandrova LR, Lapish CC, Wang YT, Phillips AG. Disruption of long-term depression potentiates latent inhibition: Key role for central nucleus of the amygdala. Int J Neuropsychopharmacol. 2021;24(7):580–91.
164. Liu TY, Cheng Y, Qin XY, Yu LC. Pharmacologically inhibiting GluR2 internalization alleviates neuropathic pain. Neurosci Bull. 2015;31(5):611–6.
165. Bai Y, Zhou L, Wu X, Dong Z. D-serine enhances fear extinction by increasing GluA2-containing AMPA receptor endocytosis. Behav Brain Res. 2014;270:223–7.
166. Marcondes LA, de CMJ, Nachtigall EG, Narvaes RF, Izquierdo I, Furini CRG. PKMzeta maintains remote contextual fear memory by inhibiting GluA2-dependent AMPA receptor endocytosis in the prelimbic cortex. Neuroscience. 2022;497:97–106.
167. Migues PV, Hardt O, Finnie P, Wang YW, Nader K. The maintenance of long-term memory in the hippocampus depends on the interaction between N-ethylmaleimide-sensitive factor and GluA2. Hippocampus. 2014;24(9):1112–9.
168. Outram AR, Brown MW, Warburton EC, Barker GRI. A critical role for long-term potentiation mechanisms in the maintenance of object recognition memory in perirhinal cortex revealed by the infusion of zeta inhibitory pseudosubstrate. Front Behav Neurosci. 2022;16:970291.

169. Wang W, Tan T, Yu Y, Huang Z, Du Y, Han H et al. Inhibition of AMPAR endocytosis alleviates pentobarbital-induced spatial memory deficits and synaptic depression. Behav Brain Res. 2018;339:66–72.
170. Yi L, Tian Q, Xu B, Pang Y, Dong Z. Targeting the PS1-ELK1 protein-protein interaction with a peptide-based inhibitor reduces Abeta production and alleviates memory decline in Alzheimer's disease. Int J Biol Macromol. 2025;334(Pt 2):149074.
171. Catterall WA. Voltage-gated sodium channels at 60: structure, function and pathophysiology. J Physiol. 2012;590(11):2577–89.
172. Wang J, Ou SW, Wang YJ. Distribution and function of voltage-gated sodium channels in the nervous system. Channels (Austin). 2017;11(6):534–54.
173. Wisedchaisri G, Gamal El-Din TM. Druggability of voltage-gated sodium channels-exploring old and new drug receptor sites. Front Pharmacol. 2022;13:858348.
174. Mahling R, Hegyi B, Cullen ER, Cho TM, Rodriques AR, Fossier L et al. De novo design of a peptide modulator to reverse sodium channel dysfunction linked to cardiac arrhythmias and epilepsy. Cell. 2025;188(22):6170–85e19.
175. Karpova A, Hiesinger PR, Kuijpers M, Albrecht A, Kirstein J, Andres-Alonso M et al. Neuronal autophagy in the control of synapse function. Neuron. 2025;113(7):974–90.
176. Palabiyik AA. The role of Bcl-2 in controlling the transition between autophagy and apoptosis (review). Mol Med Rep. 2025;32(1).
177. Palmer JE, Wilson N, Son SM, Obrocki P, Wrobel L, Rob M et al. Autophagy, aging, and age-related neurodegeneration. Neuron. 2025;113(1):29–48.
178. Shoji-Kawata S, Sumpter R, Leveno M, Campbell GR, Zou Z, Kinch L et al. Identification of a candidate therapeutic autophagy-inducing peptide. Nature. 2013;494(7436):201–6.
179. Mishra E, Thakur MK. Tat-Beclin-1 ameliorates memory by improving neuronal cytoarchitecture and mitigating mitochondrial dysfunction in scopolamine-induced amnesic male mice. ACS Pharmacol Transl Sci. 2024;7(11):3462–75.
180. De Risi M, Torromino G, Tufano M, Moriceau S, Pignataro A, Rivagorda M et al. Mechanisms by which autophagy regulates memory capacity in ageing. Aging Cell. 2020;19(9):e13189.
181. Hylin MJ, Zhao J, Tangavelou K, Rozas NS, Hood KN, MacGowan JS et al. A role for autophagy in long-term spatial memory formation in male rodents. J Neurosci Res. 2018;96(3):416–26.
182. Tam TH, Zhang W, Tu Y, Hicks JL, Farcas S, Kim D et al. Pain hypersensitivity is dependent on autophagy protein Beclin 1 in males but not females. Cell Rep. 2024;43(6):114293.
183. Mathew B, Chennakesavalu M, Sharma M, Torres LA, Stelman CR, Tran S et al. Autophagy and post-ischemic conditioning in retinal ischemia. Autophagy. 2021;17(6):1479–99.
184. Forte M, Bianchi F, Cotugno M, Marchitti S, De Falco E, Raffa S et al. Pharmacological restoration of autophagy reduces hypertension-related stroke occurrence. Autophagy. 2020;16(8):1468–81.
185. Alder J, Fujioka W, Giarratana A, Wissocki J, Thakkar K, Vuong P et al. Genetic and pharmacological intervention of the p75NTR pathway alters morphological and behavioural recovery following traumatic brain injury in mice. Brain Inj. 2016;30(1):48–65.
186. Head BP, Patel HH, Niesman IR, Drummond JC, Roth DM, Patel PM. Inhibition of p75 neurotrophin receptor attenuates isoflurane-mediated neuronal apoptosis in the neonatal central nervous system. Anesthesiology. 2009;110(4):813–25.
187. Lemkuil BP, Head BP, Pearn ML, Patel HH, Drummond JC, Patel PM. Isoflurane neurotoxicity is mediated by p75NTR-RhoA activation and actin depolymerization. Anesthesiology. 2011;114(1):49–57.
188. Pearn ML, Hu Y, Niesman IR, Patel HH, Drummond JC, Roth DM et al. Propofol neurotoxicity is mediated by p75 neurotrophin receptor activation. Anesthesiology. 2012;116(2):352–61.
189. Scheller A, Oehlke J, Wiesner B, Dathe M, Krause E, Beyermann M et al. Structural requirements for cellular uptake of alpha-helical amphipathic peptides. J Pept Sci. 1999;5(4):185–94.
190. Sebastiani A, Golz C, Werner C, Schafer MK, Engelhard K, Thal SC. Proneurotrophin binding to P75 neurotrophin receptor (P75ntr) is essential for brain lesion formation and functional impairment after experimental traumatic brain injury. J Neurotrauma. 2015;32(20):1599–607.
191. Buhusi M, Etheredge C, Granholm AC, Buhusi CV. Increased hippocampal ProBDNF contributes to memory impairments in aged mice. Front Aging Neurosci. 2017;9:284.
192. Sun W, Mei Y, Li X, Yang Y, An L. Maternal immune activation-induced proBDNF-mediated neural information processing dysfunction at hippocampal CA3-CA1 synapses associated with memory deficits in offspring. Front Cell Dev Biol. 2022;10:1018586.
193. Sheng R, Zhang TT, Felice VD, Qin T, Qin ZH, Smith CD et al. Preconditioning stimuli induce autophagy via sphingosine kinase 2 in mouse cortical neurons. J Biol Chem. 2014;289(30):20845–57.

194. Wacker BK, Freie AB, Perfater JL, Gidday JM. Junctional protein regulation by sphingosine kinase 2 contributes to blood-brain barrier protection in hypoxic preconditioning-induced cerebral ischemic tolerance. J Cereb Blood Flow Metab. 2012;32(6):1014–23.
195. Wacker BK, Park TS, Gidday JM. Hypoxic preconditioning-induced cerebral ischemic tolerance: role of microvascular sphingosine kinase 2. Stroke. 2009;40(10):3342–8.
196. Yung LM, Wei Y, Qin T, Wang Y, Smith CD, Waeber C. Sphingosine kinase 2 mediates cerebral preconditioning and protects the mouse brain against ischemic injury. Stroke. 2012;43(1):199–204.
197. Liu H, Toman RE, Goparaju SK, Maceyka M, Nava VE, Sankala H et al. Sphingosine kinase type 2 is a putative BH3-only protein that induces apoptosis. J Biol Chem. 2003;278(41):40330–6.
198. Song DD, Zhang TT, Chen JL, Xia YF, Qin ZH, Waeber C et al. Sphingosine kinase 2 activates autophagy and protects neurons against ischemic injury through interaction with Bcl-2 via its putative BH3 domain. Cell Death Dis. 2017;8(7):e2912.
199. Chen JL, Wang XX, Chen L, Tang J, Xia YF, Qian K et al. A sphingosine kinase 2-mimicking TAT-peptide protects neurons against ischemia-reperfusion injury by activating BNIP3-mediated mitophagy. Neuropharmacology. 2020;181:108326.
200. Wang XX, Li M, Xu XW, Zhao WB, Jin YM, Li LL et al. BNIP3-mediated mitophagy attenuates hypoxic-ischemic brain damage in neonatal rats by inhibiting ferroptosis through P62-KEAP1-NRF2 pathway activation to maintain iron and redox homeostasis. Acta Pharmacol Sin. 2025;46(1):33–51.
201. Bolhassani A, Jafarzade BS, Mardani G. In vitro and in vivo delivery of therapeutic proteins using cell penetrating peptides. Peptides. 2017;87:50–63.
202. Cerrato CP, Langel U. An update on cell-penetrating peptides with intracellular organelle targeting. Expert Opin Drug Deliv. 2022;19(2):133–46.
203. Greene LA, Zhou Q, Siegelin MD, Angelastro JM. Targeting transcription factors ATF5, CEBPB and CEBPD with cell-penetrating peptides to treat brain and other cancers. Cells. 2023;12(4).
204. Krautwald S, Dewitz C, Fandrich F, Kunzendorf U. Inhibition of regulated cell death by cell-penetrating peptides. Cell Mol Life Sci. 2016;73(11–12):2269–84.
205. Lichtenstein M, Zabit S, Hauser N, Farouz S, Melloul O, Hirbawi J et al. TAT for enzyme/protein delivery to restore or destroy cell activity in human diseases. Life (Basel). 2021;11(9):924.
206. Margus H, Padari K, Pooga M. Insights into cell entry and intracellular trafficking of peptide and protein drugs provided by electron microscopy. Adv Drug Deliv Rev. 2013;65(8):1031–8.
207. Parrasia S, Szabo I, Zoratti M, Biasutto L. Peptides as pharmacological carriers to the brain: Promises, shortcomings and challenges. Mol Pharm. 2022;19(11):3700–29.
208. Pirhaghi M, Mamashli F, Moosavi-Movahedi F, Arghavani P, Amiri A, Davaeil B et al. Cell-penetrating peptides: Promising therapeutics and drug-delivery systems for neurodegenerative diseases. Mol Pharm. 2024;21(5):2097–117.
209. Ramsey JD, Flynn NH. Cell-penetrating peptides transport therapeutics into cells. Pharmacol Ther. 2015;154:78–86.
210. Reissmann S, Filatova MP. New generation of cell-penetrating peptides: Functionality and potential clinical application. J Pept Sci. 2021;27(5):e3300.
211. Xie J, Bi Y, Zhang H, Dong S, Teng L, Lee RJ et al. Cell-penetrating peptides in diagnosis and treatment of human diseases: From preclinical research to clinical application. Front Pharmacol. 2020;11:697.
212. Bolton SJ, Jones DN, Darker JG, Eggleston DS, Hunter AJ, Walsh FS. Cellular uptake and spread of the cell-permeable peptide penetratin in adult rat brain. Eur J Neurosci. 2000;12(8):2847–55.
213. Spiegel S, Joly S, Meli I, Chan A, Pernet V. Modulation of antibody transport in the brain and spinal cord through the intranasal pathway. Neurotherapeutics. 2025;22:e00606.
214. Gothwal A, Muolokwu CE, Chaulagain B, Mahanta AK, Singh J. Intranasal delivery of pApoE2 via penetratin-mannose multi-functionalized chitosan polymeric micelles to the brain: Reduced total tau and phosphorylated tau burden in transgenic Alzheimer's mouse model. Int J Biol Macromol. 2025;310(Pt 4):143542.
215. Khafagy ES, Kamei N, Fujiwara Y, Okumura H, Yuasa T, Kato M et al. Systemic and brain delivery of leptin via intranasal coadministration with cell-penetrating peptides and its therapeutic potential for obesity. J Control Release. 2020;319:397–406.
216. Kamei N, Hashimoto A, Tanaka E, Murata K, Yamaguchi M, Yokoyama N et al. Therapeutic effects of anti-amyloid beta antibody after intravenous injection and efficient nose-to-brain delivery in Alzheimer's disease mouse model. Drug Deliv Transl Res. 2022;12(11):2667–77.
217. Kamei N, Shingaki T, Kanayama Y, Tanaka M, Zochi R, Hasegawa K et al. Visualization and quantitative assessment of the brain distribution of insulin through nose-to-brain delivery based on the cell-penetrating peptide noncovalent strategy. Mol Pharm. 2016;13(3):1004–11.

218. Kamei N, Suwabe S, Arime K, Bando H, Murata K, Yamaguchi M et al. Investigation of the transport pathways associated with enhanced brain delivery of peptide drugs by intranasal coadministration with penetratin. Pharmaceutics. 2021;13(11):1745.
219. Kamei N, Takeda-Morishita M. Brain delivery of insulin boosted by intranasal coadministration with cell-penetrating peptides. J Control Release. 2015;197:105–10.
220. Kamei N, Tanaka M, Choi H, Okada N, Ikeda T, Itokazu R et al. Effect of an enhanced nose-to-brain delivery of insulin on mild and progressive memory loss in the senescence-accelerated mouse. Mol Pharm. 2017;14(3):916–27.
221. Wang Y, Liu W, Sun Y, Dong X. Transthyretin-penetratin: A potent fusion protein inhibitor against Alzheimer's amyloid-beta fibrillogenesis with high blood brain barrier crossing capability. Bioconjug Chem. 2024;35(3):419–31.
222. Ritchie DL, Smith C. Pathological spectrum of sporadic Creutzfeldt-Jakob disease. Pathology. 2025;57(2):196–206.
223. Udayakumar S, Girigoswami A, Girigoswami K. A review on current theories and potential therapies for prion diseases. Mol Biol Rep. 2025;52(1):674.
224. Skrlj N, Drevensek G, Hudoklin S, Romih R, Curin Serbec V, Dolinar M. Recombinant single-chain antibody with the Trojan peptide penetratin positioned in the linker region enables cargo transfer across the blood-brain barrier. Appl Biochem Biotechnol. 2013;169(1):159–69.
225. Spencer B, Williams S, Rockenstein E, Valera E, Xin W, Mante M et al. alpha-synuclein conformational antibodies fused to penetratin are effective in models of Lewy body disease. Ann Clin Transl Neurol. 2016;3(8):588–606.
226. Mihardja SS, Gao D, Sievers RE, Fang Q, Feng J, Wang J et al. Targeted in vivo extracellular matrix formation promotes neovascularization in a rodent model of myocardial infarction. PLoS One. 2010;5(4):e10384.
227. Patel S, Tsang J, Harbers GM, Healy KE, Li S. Regulation of endothelial cell function by GRGDSP peptide grafted on interpenetrating polymers. J Biomed Mater Res A. 2007;83(2):423–33.
228. Harreither E, Rydberg HA, Amand HL, Jadhav V, Fliedl L, Benda C et al. Characterization of a novel cell penetrating peptide derived from human Oct4. Cell Regen. 2014;3(1):2.
229. Chen W, Wang YJ. Multifaceted roles of OCT4 in tumor microenvironment: biology and therapeutic implications. Oncogene. 2025;44(18):1213–29.
230. Lofgren K, Wahlstrom A, Lundberg P, Langel U, Graslund A, Bedecs K. Antiprion properties of prion protein-derived cell-penetrating peptides. FASEB J. 2008;22(7):2177–84.
231. Westergren I, Johansson BB. Altering the blood-brain barrier in the rat by intracarotid infusion of polycations: A comparison between protamine, poly-L-lysine and poly-L-arginine. Acta Physiol Scand. 1993;149(1):99–104.
232. Matsui H, Tomizawa K, Lu YF, Matsushita M. Protein Therapy: in vivo protein transduction by polyarginine (11R) PTD and subcellular targeting delivery. Curr Protein Pept Sci. 2003;4(2):151–7.
233. Wong S, Brown AD, Abrahams AB, Nurzak AN, Eltaher HM, Sykes DA et al. A modified cell-penetrating peptide enhances insulin and oxytocin delivery across an RPMI 2650 nasal epithelial cell barrier in vitro. Pharmaceutics. 2024;16(10):1267.
234. Cheng YS, Chen ZT, Liao TY, Lin C, Shen HC, Wang YH et al. An intranasally delivered peptide drug ameliorates cognitive decline in Alzheimer transgenic mice. EMBO Mol Med. 2017;9(5):703–15.
235. Marshall J, Wong KY, Rupasinghe CN, Tiwari R, Zhao X, Berberoglu ED et al. Inhibition of N-methyl-D-aspartate-induced retinal neuronal death by polyarginine peptides is linked to the attenuation of stress-induced hyperpolarization of the inner mitochondrial membrane potential. J Biol Chem. 2015;290(36):22030–48.
236. Komin A, Bogorad MI, Lin R, Cui H, Searson PC, Hristova K. A peptide for transcellular cargo delivery: Structure-function relationship and mechanism of action. J Control Release. 2020;324:633–43.
237. Gotanda Y, Wei FY, Harada H, Ohta K, Nakamura KI, Tomizawa K et al. Efficient transduction of 11 poly-arginine peptide in an ischemic lesion of mouse brain. J Stroke Cerebrovasc Dis. 2014;23(8):2023–30.
238. Pham W, Zhao BQ, Lo EH, Medarova Z, Rosen B, Moore A. Crossing the blood-brain barrier: a potential application of myristoylated polyarginine for in vivo neuroimaging. Neuroimage. 2005;28(1):287–92.
239. Kumar M, Medarova Z, Pantazopoulos P, Dai G, Moore A. Novel membrane-permeable contrast agent for brain tumor detection by MRI. Magn Reson Med. 2010;63(3):617–24.
240. Szelenyi ER, Navarrete JS, Murry AD, Zhang Y, Girven KS, Kuo L et al. An arginine-rich nuclear localization signal (ArgiNLS) strategy for streamlined image segmentation of single cells. Proc Natl Acad Sci U S A. 2024;121(32):e2320250121.

241. Liu Y, Ran R, Chen J, Kuang Q, Tang J, Mei L et al. Paclitaxel loaded liposomes decorated with a multifunctional tandem peptide for glioma targeting. Biomaterials. 2014;35(17):4835–47.
242. Liu Y, Mei L, Xu C, Yu Q, Shi K, Zhang L et al. Dual receptor recognizing cell penetrating peptide for selective targeting, efficient intratumoral diffusion and synthesized anti-glioma therapy. Theranostics. 2016;6(2):177–91.
243. Matsushita M, Tomizawa K, Moriwaki A, Li ST, Terada H, Matsui H. A high-efficiency protein transduction system demonstrating the role of PKA in long-lasting long-term potentiation. J Neurosci. 2001;21(16):6000–7.
244. Mozafari M, Md Hashim SN, Ahmad Amin Noordin KB, Zainal SA, Azlina A. Nuclear factor of activated T cells (NFAT) proteins as targeted molecules in diseases: A narrative review. Cureus. 2024;16(12):e75844.
245. Kamei N, Yamaoka A, Fukuyama Y, Itokazu R, Takeda-Morishita M. Noncovalent Strategy with cell-penetrating peptides to facilitate the brain delivery of insulin through the blood-brain barrier. Biol Pharm Bull. 2018;41(4):546–54.
246. Akita T, Oda Y, Kimura R, Nagai M, Tezuka A, Shimamura M et al. Involvement of trigeminal axons in nose-to-brain delivery of glucagon-like peptide-2 derivative. J Control Release. 2022;351:573–80.
247. Sasaki-Hamada S, Funane T, Nakao Y, Sasaki R, Nagai M, Ueta Y et al. Intranasal administration of neuromedin U derivatives containing cell-penetrating peptides and a penetration-accelerating sequence induced memory improvements in mice. Peptides. 2018;99:241–6.
248. Ellert-Miklaszewska A, Szymczyk A, Poleszak K, Kaminska B. Delivery of the VIVIT peptide to human glioma cells to interfere with calcineurin-NFAT signaling. Molecules. 2021;26(16).
249. Batulu H, Du GJ, Li DZ, Sailike D, Fan YH, Geng D. Effect of poly-arginine R18 on neurocyte cell growth via autophagy in traumatic brain injury. Exp Ther Med. 2019;17(5):4109–15.
250. Chiu LS, Anderton RS, Clark VW, Cross JL, Knuckey NW, Meloni BP. Effect of polyarginine peptide R18D following a traumatic brain injury in Sprague-Dawley Rats. Curr Ther Res Clin Exp. 2020;92:100584.
251. Chiu LS, Anderton RS, Cross JL, Clark VW, Edwards AB, Knuckey NW et al. Assessment of R18, COG1410, and APP96-110 in excitotoxicity and traumatic brain injury. Transl Neurosci. 2017;8:147–57.
252. Edwards AB, Anderton RS, Knuckey NW, Meloni BP. Assessment of therapeutic window for poly-arginine-18D (R18D) in a P7 rat model of perinatal hypoxic-ischaemic encephalopathy. J Neurosci Res. 2018;96(11):1816–26.
253. Edwards AB, Cross JL, Anderton RS, Knuckey NW, Meloni BP. Poly-arginine R18 and R18D (D-enantiomer) peptides reduce infarct volume and improves behavioural outcomes following perinatal hypoxic-ischaemic encephalopathy in the P7 rat. Mol Brain. 2018;11(1):8.
254. MacDougall G, Anderton RS, Edwards AB, Knuckey NW, Meloni BP. The neuroprotective peptide poly-arginine-12 (R12) reduces cell surface levels of NMDA NR2B receptor subunit in cortical neurons; investigation into the involvement of endocytic mechanisms. J Mol Neurosci. 2017;61(2):235–46.
255. MacDougall G, Anderton RS, Mastaglia FL, Knuckey NW, Meloni BP. Proteomic analysis of cortical neuronal cultures treated with poly-arginine peptide-18 (R18) and exposed to glutamic acid excitotoxicity. Mol Brain. 2019;12(1):66.
256. MacDougall G, Anderton RS, Trimble A, Mastaglia FL, Knuckey NW, Meloni BP. Poly-arginine-18 (R18) confers neuroprotection through glutamate receptor modulation, intracellular calcium reduction, and preservation of mitochondrial function. Molecules. 2020;25(13):2977.
257. Meloni BP, Chen Y, Harrison KA, Nashed JY, Blacker DJ, South SM et al. Poly-arginine peptide-18 (R18) reduces brain injury and improves functional outcomes in a nonhuman primate stroke model. Neurotherapeutics. 2020;17(2):627–34.
258. Milani D, Bakeberg MC, Cross JL, Clark VW, Anderton RS, Blacker DJ et al. Comparison of neuroprotective efficacy of poly-arginine R18 and R18D (D-enantiomer) peptides following permanent middle cerebral artery occlusion in the Wistar rat and in vitro toxicity studies. PLoS One. 2018;13(3):e0193884.
259. Milani D, Clark VW, Cross JL, Anderton RS, Knuckey NW, Meloni BP. Poly-arginine peptides reduce infarct volume in a permanent middle cerebral artery rat stroke model. BMC Neurosci. 2016;17(1):19.
260. Milani D, Clark VW, Feindel KW, Blacker DJ, Bynevelt M, Edwards AB et al. Comparative assessment of the proteolytic stability and impact of poly-arginine peptides R18 and R18D on infarct growth and penumbral tissue preservation following middle cerebral artery occlusion in the Sprague Dawley Rat. Neurochem Res. 2021;46(5):1166–76.
261. Milani D, Cross JL, Anderton RS, Blacker DJ, Knuckey NW, Meloni BP. Delayed 2-h post-stroke administration of R18 and NA-1 (TAT-NR2B9c) peptides after permanent and/or transient middle cerebral artery occlusion in the rat. Brain Res Bull. 2017;135:62–8.

262. Tahmasbi F, Madani Neishaboori A, Mardani M, Toloui A, Komlakh K, Azizi Y et al. Efficacy of polyarginine peptides in the treatment of stroke: A systematic review and meta-analysis. Brain Behav. 2023;13(1):e2858.
263. Meloni BP, Craig AJ, Milech N, Hopkins RM, Watt PM, Knuckey NW. The neuroprotective efficacy of cell-penetrating peptides TAT, penetratin, Arg-9, and Pep-1 in glutamic acid, kainic acid, and in vitro ischemia injury models using primary cortical neuronal cultures. Cell Mol Neurobiol. 2014;34(2):173–81.
264. Morris MC, Depollier J, Mery J, Heitz F, Divita G. A peptide carrier for the delivery of biologically active proteins into mammalian cells. Nat Biotechnol. 2001;19(12):1173–6.
265. Bajinka O, Aikins AR, Jallow L. Peptidyl-prolyl isomerase PIN1 as a pivotal regulator of cancer progression and therapy. Pathol Res Pract. 2025;275:156221.
266. Park JH, Shin MJ, Youn GS, Yeo HJ, Yeo EJ, Kwon HJ et al. PEP-1-PIN1 promotes hippocampal neuronal cell survival by inhibiting cellular ROS and MAPK phosphorylation. Biomedicines. 2024;12(10):2532.
267. Choi YJ, Kim DW, Shin MJ, Yeo HJ, Yeo EJ, Lee LR et al. PEP-1-GLRX1 reduces dopaminergic neuronal cell loss by modulating MAPK and apoptosis signaling in Parkinson's disease. Molecules. 2021;26(11):3329.
268. Ryu EJ, Kim DW, Shin MJ, Jo HS, Park JH, Cho SB et al. PEP-1-glutaredoxin 1 protects against hippocampal neuronal cell damage from oxidative stress via regulation of MAPK and apoptotic signaling pathways. Mol Med Rep. 2018;18(2):2216–28.
269. Sohn EJ, Kim DW, Kim MJ, Jeong HJ, Shin MJ, Ahn EH et al. PEP-1-metallothionein-III protein ameliorates the oxidative stress-induced neuronal cell death and brain ischemic insults. Biochim Biophys Acta. 2012;1820(10):1647–55.
270. Zhang YE, Fu SZ, Li XQ, Chen P, Wang JL, Che J et al. PEP-1-SOD1 protects brain from ischemic insult following asphyxial cardiac arrest in rats. Resuscitation. 2011;82(8):1081–6.
271. Kim MJ, Park M, Kim DW, Shin MJ, Son O, Jo HS et al. Transduced PEP-1-PON1 proteins regulate microglial activation and dopaminergic neuronal death in a Parkinson's disease model. Biomaterials. 2015;64:45–56.
272. Eom SA, Kim DW, Shin MJ, Ahn EH, Chung SY, Sohn EJ et al. Protective effects of PEP-1-Catalase on stress-induced cellular toxicity and MPTP-induced Parkinson's disease. BMB Rep. 2015;48(7):395–400.
273. Kim DW, Jeong HJ, Kang HW, Shin MJ, Sohn EJ, Kim MJ et al. Transduced human PEP-1-catalase fusion protein attenuates ischemic neuronal damage. Free Radic Biol Med. 2009;47(7):941–52.
274. Choi YJ, Shin MJ, Youn GS, Park JH, Yeo HJ, Yeo EJ et al. Protective effects of PEP-1-GSTA2 protein in hippocampal neuronal cell damage induced by oxidative stress. Int J Mol Sci. 2023;24(3):2767.
275. Liu Y, Wang S, Luo S, Li Z, Liang F, Zhu Y et al. Intravenous PEP-1-GDNF is protective after focal cerebral ischemia in rats. Neurosci Lett. 2016;617:150–5.
276. An JJ, Lee YP, Kim SY, Lee SH, Lee MJ, Jeong MS et al. Transduced human PEP-1-heat shock protein 27 efficiently protects against brain ischemic insult. FEBS J. 2008;275(6):1296–308.
277. Lee YP, Kim DW, Kang HW, Hwang JH, Jeong HJ, Sohn EJ et al. PEP-1-heat shock protein 27 protects from neuronal damage in cells and in a Parkinson's disease mouse model. FEBS J. 2012;279(11):1929–42.
278. Jung HY, Kim DW, Yim HS, Yoo DY, Kim JW, Won MH et al. Heme oxygenase-1 protects neurons from ischemic damage by upregulating expression of Cu,Zn-superoxide dismutase, catalase, and brain-derived neurotrophic factor in the rabbit spinal cord. Neurochem Res. 2016;41(4):869–79.
279. Youn JK, Kim DW, Kim ST, Park SY, Yeo EJ, Choi YJ et al. PEP-1-HO-1 prevents MPTP-induced degeneration of dopaminergic neurons in a Parkinson's disease mouse model. BMB Rep. 2014;47(10):569–74.
280. Hahn KR, Kwon HJ, Moon SM, Kim W, Hwang IK, Kim DW et al. Therapeutic potential of cell-permeable PEP-1-Srxn1 in mitigating oxidative and ischemic damage in the hippocampus. Neurochem Int. 2025;187:105988.
281. Cho JH, Hwang IK, Yoo KY, Kim SY, Kim DW, Kwon YG et al. Effective delivery of Pep-1-cargo protein into ischemic neurons and long-term neuroprotection of Pep-1-SOD1 against ischemic injury in the gerbil hippocampus. Neurochem Int. 2008;52(4–5):659–68.
282. Choi JH, Kim DW, Yoo DY, Jeong HJ, Kim W, Jung HY et al. Repeated administration of PEP-1-Cu,Zn-superoxide dismutase and PEP-1-peroxiredoxin-2 to senescent mice induced by D-galactose improves the hippocampal functions. Neurochem Res. 2013;38(10):2046–55.
283. Eum WS, Kim DW, Hwang IK, Yoo KY, Kang TC, Jang SH et al. In vivo protein transduction: Biologically active intact pep-1-superoxide dismutase fusion protein efficiently protects against ischemic insult. Free Radic Biol Med. 2004;37(10):1656–69.
284. Jia J, Chen F, Wu Y. Recombinant PEP-1-SOD1 improves functional recovery after neural stem cell transplantation in rats with traumatic brain injury. Exp Ther Med. 2018;15(3):2929–35.

285. Kim W, Kim DW, Yoo DY, Chung JY, Hwang IK, Won MH et al. Neuroprotective effects of PEP-1-Cu,Zn-SOD against ischemic neuronal damage in the rabbit spinal cord. Neurochem Res. 2012;37(2):307–13.
286. Yoo DY, Shin BN, Kim IH, Kim W, Kim DW, Yoo KY et al. Effects of Cu,Zn-superoxide dismutase on cell proliferation and neuroblast differentiation in the mouse dentate gyrus. Neurochem Res. 2012;37(2):261–7.
287. Yoon DK, Hwang IK, Yoo KY, Lee YB, Lee JJ, Kim JH et al. Comparison of alpha-synuclein immunoreactivity and protein levels in ischemic hippocampal CA1 region between adult and aged gerbils and correlation with Cu,Zn-superoxide dismutase. Neurosci Res. 2006;55(4):434–41.
288. Yune TY, Lee JY, Jiang MH, Kim DW, Choi SY, Oh TH. Systemic administration of PEP-1-SOD1 fusion protein improves functional recovery by inhibition of neuronal cell death after spinal cord injury. Free Radic Biol Med. 2008;45(8):1190–200.
289. Kim YN, Jung HY, Eum WS, Kim DW, Shin MJ, Ahn EH et al. Neuroprotective effects of PEP-1-carbonyl reductase 1 against oxidative-stress-induced ischemic neuronal cell damage. Free Radic Biol Med. 2014;69:181–96.
290. Sohn EJ, Shin MJ, Kim DW, Son O, Jo HS, Cho SB et al. PEP-1-GSTpi protein enhanced hippocampal neuronal cell survival after oxidative damage. BMB Rep. 2016;49(7):382–7.
291. Kim W, Kim DW, Shin BN, Yoo DY, Nam SM, Kim MJ et al. PEP-1-frataxin significantly increases cell proliferation and neuroblast differentiation by reducing lipid peroxidation in the mouse dentate gyrus. Neurochem Res. 2011;36(12):2452–8.
292. Ahn EH, Kim DW, Shin MJ, Kim YN, Kim HR, Woo SJ et al. PEP-1-ribosomal protein S3 protects dopaminergic neurons in an MPTP-induced Parkinson's disease mouse model. Free Radic Biol Med. 2013;55:36–45.
293. Ahn EH, Kim DW, Shin MJ, Kwon SW, Kim YN, Kim DS et al. Chlorogenic acid improves neuroprotective effect of PEP-1-ribosomal protein S3 against ischemic insult. Exp Neurobiol. 2011;20(4):169–75.
294. Hwang IK, Yoo KY, Kim DW, Kim SY, Park JH, Ryoo ZY et al. Ischemia-induced ribosomal protein S3 expressional changes and the neuroprotective effect against experimental cerebral ischemic damage. J Neurosci Res. 2008;86(8):1823–35.
295. Lee YP, Kim DW, Lee MJ, Jeong MS, Kim SY, Lee SH et al. Human brain pyridoxal-5'-phosphate phosphatase (PLPP):protein transduction of PEP-1-PLPP into PC12 cells. BMB Rep. 2008;41(5):408–13.
296. Jeong HJ, Yoo DY, Kim DW, Yeo HJ, Cho SB, Hyeon J et al. Neuroprotective effect of PEP-1-peroxiredoxin2 on CA1 regions in the hippocampus against ischemic insult. Biochim Biophys Acta. 2014;1840(7):2321–30.
297. Ahn EH, Kim DW, Shin MJ, Kim HR, Kim SM, Woo SJ et al. PEP-1-PEA-15 protects against toxin-induced neuronal damage in a mouse model of Parkinson's disease. Biochim Biophys Acta. 2014;1840(6):1686–700.
298. Jung HY, Cho SB, Kim W, Yoo DY, Won MH, Choi GM et al. Phosphatidylethanolamine-binding protein 1 protects CA1 neurons against ischemic damage via ERK-CREB signaling in Mongolian. gerbil. 2018;8(11):1370.
299. Kim W, Cho SB, Jung HY, Yoo DY, Oh JK, Choi GM et al. Phosphatidylethanolamine-binding protein 1 ameliorates ischemia-induced inflammation and neuronal damage in the rabbit spinal cord. Cells. 2019;8(11).
300. Zhang DD, Jin C, Zhang YT, Gan XD, Zou MJ, Wang YY et al. A novel IL-1RA-PEP fusion protein alleviates blood-brain barrier disruption after ischemia-reperfusion in male rats. J Neuroinflammation. 2018;15(1):16.
301. Hahn KR, Kwon HJ, Yoon YS, Kim DW, Hwang IK. Phosphoglycerate kinase 1 protects against ischemic damage in the gerbil hippocampus. Aging (Albany NY). 2022;14(22):8886–99.
302. Wang X, Zhang Q, Lv L, Fu J, Jiang Y, Xin H et al. Glioma and microenvironment dual targeted nanocarrier for improved antiglioblastoma efficacy. Drug Deliv. 2017;24(1):1401–9.
303. Kim MJ, Kim DW, Yoo KY, Sohn EJ, Jeong HJ, Kang HW et al. Protective effects of transduced PEP-1-Frataxin protein on oxidative stress-induced neuronal cell death. J Neurol Sci. 2010;298(1–2):64–9.
304. Jung HY, Kwon HJ, Kim W, Hahn KR, Moon SM, Yoon YS et al. The neuroprotective effects of phosphoglycerate mutase 5 are mediated by decreasing oxidative stress in HT22 hippocampal cells and gerbil hippocampus. Neurochem Int. 2022;157:105346.
305. Choi SH, Kim DW, Kim SY, An JJ, Lee SH, Choi HS et al. Transduced human copper chaperone for Cu,Zn-SOD (PEP-1-CCS) protects against neuronal cell death. Mol Cells. 2005;20(3):401–8.

306. Yoo DY, Kim DW, Kim MJ, Choi JH, Jung HY, Nam SM et al. Sodium butyrate, a histone deacetylase inhibitor, ameliorates SIRT2-induced memory impairment, reduction of cell proliferation, and neuroblast differentiation in the dentate gyrus. Neurol Res. 2015;37(1):69–76.
307. Chang JC, Chao YC, Chang HS, Wu YL, Chang HJ, Lin YS et al. Intranasal delivery of mitochondria for treatment of Parkinson's disease model rats lesioned with 6-hydroxydopamine. Sci Rep. 2021;11(1):10597.
308. Kim MH, Maeng HJ, Yu KH, Lee KR, Tsuruo T, Kim DD et al. Evidence of carrier-mediated transport in the penetration of donepezil into the rat brain. J Pharm Sci. 2010;99(3):1548–66.
309. Baehr C, Reichel V, Fricker G. Choroid plexus epithelial monolayers – a cell culture model from porcine brain. Cerebrospinal Fluid Res. 2006;3(1):13.
310. Ghersi-Egea J-F, Strazielle N, Catala M, Silva-Vargas V, Doetsch F, Engelhardt B. Molecular anatomy and functions of the choroidal blood-cerebrospinal fluid barrier in health and disease. Acta Neuropathol. 2018;135(3):337–61.
311. Choi JD, Moon Y, Kim HJ, Yim Y, Lee S, Moon WJ. Choroid plexus volume and permeability at brain MRI within the Alzheimer disease clinical spectrum. Radiology. 2022;304(3):635–45.
312. Althubaity N, Schubert J, Martins D, Yousaf T, Nettis MA, Mondelli V et al. Choroid plexus enlargement is associated with neuroinflammation and reduction of blood brain barrier permeability in depression. NeuroImage: Clinical. 2022;33:102926.
313. Shrestha B, Paul D, Pachter JS. Alterations in tight junction protein and IgG permeability accompany leukocyte extravasation across the choroid plexus during neuroinflammation. J Neuropathol Exp Neurol. 2014;73(11):1047–61.
314. Sevigny J, Chiao P, Bussiere T, Weinreb PH, Williams L, Maier M et al. The antibody aducanumab reduces Abeta plaques in Alzheimer's disease. Nature. 2016;537(7618):50–6.
315. Blennow K, Fredman P, Wallin A, Gottfries CG, Karlsson I, Långström G et al. Protein analysis in cerebrospinal fluid. II. Reference values derived from healthy individuals 18-88 years of age. Eur Neurol. 1993;33(2):129–33.
316. Mueller S, Kline C, Stoller S, Lundy S, Christopher L, Reddy AT et al. PNOC015: Repeated convection-enhanced delivery of MTX110 (aqueous panobinostat) in children with newly diagnosed diffuse intrinsic pontine glioma. Neuro Oncol. 2023;25(11):2074–86.
317. Ruan Z, Zeng L, Zhang J, Qin H, Huang Z, Yan H et al. Intrathecal nivolumab and IL-2 for treatment of leptomeningeal metastases in EGFR-mutated lung adenocarcinoma. Lung Cancer. 2025;204:108586.
318. Xu Q, Huang K, Tong L, Yan D, Zhao Y, Yang H et al. Intrathecal pemetrexed efficacy and cerebrospinal fluid tumor marker response in refractory leptomeningeal metastasis of non-small-cell lung cancer: A single-arm phase II trial. BMC Med. 2025;23(1):301.
319. Muir KW, Bulters D, Willmot M, Sprigg N, Dixit A, Ward N et al. Intracerebral implantation of human neural stem cells and motor recovery after stroke: Multicentre prospective single-arm study (PISCES-2). J Neurol, Neurosurg Psychiatry. 2020;91(4):396–401.
320. Rezai AR, D'Haese P-F, Finomore V, Carpenter J, Ranjan M, Wilhelmsen K et al. Ultrasound blood–brain barrier opening and aducanumab in Alzheimer's disease. New England Journal of Medicine. 2024;390(1):55–62.
321. Evans RW. Complications of lumbar puncture. Neurol Clin. 1998;16(1):83–105.
322. Hollingworth M, Zacharoulis S. Infusion-related side-effects during convection enhanced delivery for brainstem-diffuse midline glioma/diffuse intrinsic pontine glioma. J Neurooncol. 2022;159(2):417–24.
323. Vogel M, Köck P, Strasser J, Kalbermatten C, Binder H, Dürsteler KM et al. Nasal opioid agonist treatment in patients with severe opioid dependence: a case series. Eur Addict Res. 2022;28(1):80–6.
324. Vogel M, Meyer M, Westenberg JN, Kormann A, Simon O, Salim Hassan Fadlelseed R et al. Safety and feasibility of intranasal heroin-assisted treatment: 4-week preliminary findings from a Swiss multicentre observational study. Harm Reduct J. 2023;20(1):2.
325. Sun Z-F, Gao X, Pinto JM, He Y, Yang Q, Tian J et al. Morphological evaluation using MRI of the olfactory filaments (fila) in a post-traumatic olfactory rat model. World J Otorhinolaryngol Head Neck Surg. 2018;4(1):50–6.
326. Soriano RM, Solares CA, Pradilla G, DelGaudio JM. Endoscopic study of the distribution of olfactory filaments: A cadaveric study. Am J Rhinol Allergy. 2021;35(2):226–33.
327. Kurihara S, Tei M, Hata J, Mori E, Fujioka M, Matsuwaki Y et al. MRI tractography reveals the human olfactory nerve map connecting the olfactory epithelium and olfactory bulb. Commun Biol. 2022;5(1):843.
328. Marín-Padilla M. The human brain intracerebral microvascular system: development and structure. Front Neuroanat. 2012;6:38.

329. Al-Zahaby SA, Farag MR, Alagawany M, Taha HSA, Varoni MV, Crescenzo G et al. Zinc oxide nanoparticles (ZnO-NPs) induce cytotoxicity in the zebrafish olfactory organs via activating oxidative stress and apoptosis at the ultrastructure and genetic levels. Animals (Basel). 2023;13(18).
330. Clementino AR, Pellegrini G, Banella S, Colombo G, Cantù L, Sonvico F et al. Structure and fate of nanoparticles designed for the nasal delivery of poorly soluble drugs. Mol Pharm. 2021;18(8):3132–46.
331. Kiemle Trindade IE, Oliveira Camargo Gomes A, Martins Sampaio-Teixeira AC, Kiemle Trindade SH. Adult nasal volumes assessed by acoustic rhinometry. Braz J Otorhinolaryngol. 2007;73(1):32–9.
332. Formica ML, Real DA, Picchio ML, Catlin E, Donnelly RF, Paredes AJ. On a highway to the brain: A review on nose-to-brain drug delivery using nanoparticles. Applied Materials Today. 2022;29:101631.
333. Zee-Cheng RK, Cheng CC. Delivery of anticancer drugs. Methods Find Exp Clin Pharmacol. 1989;11(7–8):439–529.
334. Schröder U, Sabel BA. Nanoparticles, a drug carrier system to pass the blood-brain barrier, permit central analgesic effects of i.v. dalargin injections. Brain Res. 1996;710(1–2):121–4.
335. You H, Zhang S, Zhang Y, Chen Q, Wu Y, Zhou Z et al. Engineered bacterial outer membrane vesicles-based doxorubicin and CD47-siRNA co-delivery nanoplatform overcomes immune resistance to potentiate the immunotherapy of glioblastoma. Adv Materials (Deerfield Beach, Fla). 2025;37(15):e2418053.
336. Krienke C, Kolb L, Diken E, Streuber M, Kirchhoff S, Bukur T et al. A noninflammatory mRNA vaccine for treatment of experimental autoimmune encephalomyelitis. Science. 2021;371(6525):145–53.
337. Phan NM, Nguyen TL, Min DK, Kim J. Mesoporous polydopamine nanoparticle-based tolerogenic vaccine induces antigen-specific immune tolerance to prevent and treat autoimmune multiple sclerosis. Biomaterials. 2025;316:122997.
338. Zhang R, Yao X, Li Q, Li X, Ma Q, Huang W et al. Self-assembled nanoparticles of rapamycin prodrugs for the treatment of multiple sclerosis. J Colloid Interface Sci. 2025;683:448–59.
339. Wang H, Shang J, He Z, Zheng M, Jia H, Zhang Y et al. Dual peptide nanoparticle platform for enhanced antigen-specific immune tolerance for the treatment of experimental autoimmune encephalomyelitis. Biomater Sci. 2022;10(14):3878–91.
340. Tuohy VK, Lu Z, Sobel RA, Laursen RA, Lees MB. Identification of an encephalitogenic determinant of myelin proteolipid protein for SJL mice. J Immunol. (Baltimore, Md: 1950). 1989;142(5):1523–7.
341. Miller SD, Karpus WJ. Experimental autoimmune encephalomyelitis in the mouse. Curr Protoc Immunol. 2007;77: 15.1.1–15.1.18.
342. Morris AH, Hughes KR, Oakes RS, Cai MM, Miller SD, Irani DN et al. Engineered immunological niches to monitor disease activity and treatment efficacy in relapsing multiple sclerosis. Nat Commun. 2020;11(1):3871.
343. Shi C, Zhang J, Wang H, Chen C, Han M, Gao L et al. Trojan horse nanocapsule enabled in situ modulation of the phenotypic conversion of Th17 cells to treg cells for the treatment of multiple sclerosis in mice. Adv Materials (Deerfield Beach, Fla). 2023;35(11):e2210262.
344. Montes M, Zhang X, Berthelot L, Laplaud D-A, Brouard S, Jin J et al. Oligoclonal myelin-reactive T-cell infiltrates derived from multiple sclerosis lesions are enriched in Th17 cells. Clin Immunol. 2009;130(2):133–44.
345. Moser T, Akgün K, Proschmann U, Sellner J, Ziemssen T. The role of TH17 cells in multiple sclerosis: Therapeutic implications. Autoimmun Rev. 2020;19(10):102647.
346. Lerma-Martin C, Badia-i-Mompel P, Ramirez Flores RO, Sekol P, Schäfer PSL, Riedl CJ et al. Cell type mapping reveals tissue niches and interactions in subcortical multiple sclerosis lesions. Nature Neuroscience. 2024;27(12):2354–65.
347. Rezaeimanesh N, Rafiee P, Saeedi R, Khosravian P, Sahraian MA, Eskandarieh S et al. The effect of crocin-selenium nanoparticles on the cognition and oxidative stress markers of multiple sclerosis patients: A randomized triple-blinded placebo-controlled clinical trial. Biometals. 2024;37(2):305–19.
348. Wang X, Chen S, Xia X, Du Y, Wei Y, Yang W et al. Lysosome-targeting protein degradation through endocytosis pathway triggered by polyvalent nano-chimera for AD therapy. Adv Mater. 2025;37(5):2411061.
349. Du B, Zou Q, Wang X, Wang H, Yang X, Wang Q et al. Multi-targeted engineered hybrid exosomes as Aβ nanoscavengers and inflammatory modulators for multi-pathway intervention in Alzheimer's disease. Biomaterials. 2025;322:123403.
350. Lv Y, Asghar S, Ye P, Dong H, Hu R, Yuan C et al. Near-infrared light-triggered methylene blue/berberine co-loaded polydopamine nanoparticles modified with sialic acid for Alzheimer's disease therapy. J Colloid Interface Sci. 2025;698:138083.

351. Cummings JL, Zhou Y, Lee G, Zhong K, Fonseca J, Leisgang-Osse AM et al. Alzheimer's disease drug development pipeline: 2025. Alzheimers Dement Transl Res Clin Interv. 2025;11(2):e70098.
352. Panwar A, Khan MI, Kumar R, Kumar R, Rai SK, Kumar A. Emerging Novel therapeutic approaches for the treatment of Alzheimer's disease. Adv Alzheimers Dis. 2024;13(3):65–94.
353. Cummings J, Zhou Y, Lee G, Zhong K, Fonseca J, Cheng F. Alzheimer's disease drug development pipeline: 2024. Alzheimers Dement Transl Res Clin Interv. 2024;10(2):e12465.
354. Taléns-Visconti R, de Julián-Ortiz JV, Vila-Busó O, Diez-Sales O, Nácher A. Intranasal drug administration in Alzheimer-type dementia: Towards clinical applications. Pharmaceutics. 2023;15(5):1399.
355. Fan TS, Liu SC, Wu RM. Alpha-synuclein and cognitive decline in Parkinson disease. Life (Basel). 2021;11(11).
356. Zhai L, Gao Y, Yang H, Wang H, Liao B, Cheng Y et al. A ROS-Responsive nanoparticle for nuclear gene delivery and autophagy restoration in Parkinson's disease therapy. Biomaterials. 2025;321:123345.
357. Chen T-Y, Xu J, Tai C-H, Wen T-K, Hsu S-h. Biodegradable, electroconductive self-healing hydrogel based on polydopamine-coated polyurethane nano-crosslinker for Parkinson's disease therapy. Biomaterials. 2025;320:123268.
358. Kushawaha SK, Ashawat MS, Baldi A. Auranofin-loaded PLGA nanoparticles alleviate cognitive deficit induced by streptozotocin in rats model: modulation of oxidative stress, neuroinflammatory markers, and neurotransmitters. Naunyn-Schmiedeberg's Arch Pharmacol. 2024;397(12):10031–47.
359. Soni D, Garg Y, Upadhayay S, Bhatia A, Basir B, Singh SK et al. Auranofin-loaded chitosan-lipid hybrid nanoparticle protects against rotenone model of Parkinson's disease via modulation of GSK-3β/Nrf2/HO-1 signaling. Eur J Pharmacol. 2025;998:177523.
360. Chen J, Xu G, Shen R, Xu J, Lu C, Li X et al. Communications among neurocytes in Parkinson's disease regulated by differential metabolism and blood-brain barrier traversing of chiral gold cluster-MOF integrated nanoparticles. Advanced Science. 2025;12(23):2500026.
361. Nazifi M, Oryan S, Esfahani DE, Ashrafpoor M. The functional effects of piperine and piperine plus donepezil on hippocampal synaptic plasticity impairment in rat model of Alzheimer's disease. Life Sci. 2021;265:118802.
362. Shrivastava P, Vaibhav K, Tabassum R, Khan A, Ishrat T, Khan MM et al. Anti-apoptotic and anti-inflammatory effect of piperine on 6-OHDA induced Parkinson's rat model. J Nutr Biochem. 2013;24(4):680–7.
363. Li J, Leung SSY, Chan EHY, Jiang C, Ho ETY, Zuo Z. Significantly increased aqueous solubility of piperine via nanoparticle formulation serves as the most critical factor for its brain uptake enhancement. Int J Nanomed. 2025;20:3945–59.
364. Jin J, Zhang H, Lu Q, Tian L, Yao S, Lai F et al. Nanocarrier-mediated siRNA delivery: A new approach for the treatment of traumatic brain injury-related Alzheimer's disease. Neural Regen Res. 2025;20(9):2538–55.
365. Zhou Y, Qiao S, Zhang L, Liu M, Ji Q, Zhang B et al. Hybrid membrane-coated Cyclosporine A nanocrystals preventing secondary brain injury via alleviating neuroinflammatory and oxidative stress. J Control Release. 2025;383:113795.
366. Li W, Jiao M, Li Y, Wu S, Zhang P, Luo X et al. Hypersensitive multiparametric MRA for mild traumatic brain injury: simultaneous imaging of hemorrhagic foci, vascular integrity, and collaterals. Analytical chemistry. 2025;97(25):13021–30.
367. Mouli K, Liopo AV, McHugh EA, Underwood E, Zhao J, Dash PK et al. Oxidized carbon nanoparticles enhance cellular energetics with application to injured brain. Adv Healthc Mater. 2025;14(8):e2401629.
368. Wang Y, Huang L, Li J, Duan J, Pan X, Menon BK et al. Efficacy and safety of corticosteroids for stroke and traumatic brain injury: A systematic review and meta-analysis. Syst Rev. 2025;14(1):54.
369. Zhang B, Bai M, Yang M, Wang Y, Chen X, Liu B et al. Injectable nanocomposite hydrogel for localized precision delivery of dexamethasone after traumatic brain injury: Dual modulation of neuroinflammation and blood-brain barrier restoration. J Transl Med. 2025;23(1):579.
370. Li P, Sun S, Zhu X, Liu X, Yin R, Chen Y et al. Intranasal delivery of engineered extracellular vesicles promotes neurofunctional recovery in traumatic brain injury. J Nanobiotechnol. 2025;23(1):229.
371. Ge X, Huang S, Gao H, Han Z, Chen F, Zhang S et al. miR-21-5p alleviates leakage of injured brain microvascular endothelial barrier in vitro through suppressing inflammation and apoptosis. Brain Res. 2016;1650:31–40.
372. Han S, Yoo W, Carton O, Joo J, Kwon EJ. PEGylated multimeric RNA nanoparticles for siRNA delivery in traumatic brain injury. Small (Weinheim an der Bergstrasse, Germany). 2025;21(10):e2405806.

373. Bertossi R, Kurz JE, McGuire T, Peng CY, Kessler JA. Intravenous immunomodulatory nanoparticles prevent secondary damage after traumatic brain injury. J Neurotrauma. 2025;42(1–2):94–106.
374. Barilani M, Peli V, Manzini P, Pistoni C, Rusconi F, Pinatel EM et al. Extracellular vesicles from human induced pluripotent stem cells exhibit a unique MicroRNA and CircRNA signature. Int J Biol Sci. 2024;20(15):6255–78.
375. Jones B, Patel R, Wang B, Evans-Nguyen T, Patel NA. Lyophilized small extracellular vesicles (sEVs) derived from human adipose stem cells maintain efficacy to promote healing in neuronal injuries. Biomedicines. 2025;13(2).
376. Fu X, Zhang Y, Chen G, Mao G, Tang J, Xu J et al. Responsive nanoparticles synergize with Curcumin to break the "reactive oxygen Species-Neuroinflammation" vicious cycle, enhancing traumatic brain injury outcomes. J Nanobiotechnol. 2025;23(1):172.
377. Gupta MK, Meyer TA, Nelson CE, Duvall CL. Poly(PS-b-DMA) micelles for reactive oxygen species triggered drug release. J Control Release. 2012;162(3):591–8.
378. Ale T, Ale T, Baker KJ, Zuniga KM, Hutcheson J, Lavik E. Delivery of tempol from polyurethane nanocapsules to address oxidative stress post-injury. Bioconjugate chemistry. 2025;36(2):146–51.
379. Yuan D, Xu W, Raza F, Zafar H, Guan S, Wang Z et al. Baicalein long-circulating liposomes improve histological and functional outcomes after traumatic brain injury. Adv Pharmacol Pharm Sci. 2024;2024:2869332.
380. Zhang M, Xiang C, Niu R, He X, Luo W, Liu W et al. Liposomes as versatile agents for the management of traumatic and nontraumatic central nervous system disorders: drug stability, targeting efficiency, and safety. Neural Regen Res. 2025;20(7):1883–99.
381. Li H, Sun D, Zhao Z, Fang J, Li M, Lv C et al. Neutrophil membrane-derived nanoparticles protect traumatic brain injury via inhibiting calcium overload and scavenging ROS. J Nanobiotechnol. 2024;22(1):477.
382. Ruseska I, Fresacher K, Petschacher C, Zimmer A. Use of protamine in nanopharmaceuticals—A review. Nanomaterials. 2021;11(6):1508.
383. Huang J, Fu Y, Wang A, Shi K, Peng Y, Yi Y et al. Brain delivery of protein therapeutics by cell matrix-inspired biomimetic nanocarrier. Adv Materials (Deerfield Beach, Fla). 2024;36(31):e2405323.
384. Vrettou CS, Mentzelopoulos SD. Second- and third-tier therapies for severe traumatic brain injury. J Clin Med. 2022;11(16):4790.
385. Patel S, Maria-Rios J, Parikh A, Okorie ON. Diagnosis and management of elevated intracranial pressure in the emergency department. Int J Emerg Med. 2023;16(1):72.
386. Jiao M, Li X, Liu H, Cai P, Yang X, McHugh KJ et al. Aqueous grown quantum dots with robust near-infrared fluorescence for integrated traumatic brain injury diagnosis and surgical monitoring. ACS Nano. 2024;18(29):19038–53.
387. Omo-Lamai S, Nong J, Savalia K, Kelley BJ, Wu J, Esteves-Reyes S et al. Targeting of nanoparticles to the cerebral vasculature after traumatic brain injury. PLoS One. 2024;19(6):e0297451.
388. Young MF, Valaris S, Wrann CD. A role for FNDC5/Irisin in the beneficial effects of exercise on the brain and in neurodegenerative diseases. Prog Cardiovasc Dis. 2019;62(2):172–8.
389. Bi J, Zhang J, Ren Y, Du Z, Li Q, Wang Y et al. Irisin alleviates liver ischemia-reperfusion injury by inhibiting excessive mitochondrial fission, promoting mitochondrial biogenesis and decreasing oxidative stress. Redox Biol. 2019;20:296–306.
390. Ge Y, Wu X, Cai Y, Hu Q, Wang J, Zhang S et al. FNDC5 prevents oxidative stress and neuronal apoptosis after traumatic brain injury through SIRT3-dependent regulation of mitochondrial quality control. Cell Death Dis. 2024;15(5):364.
391. Oris C, Kahouadji S, Durif J, Bouvier D, Sapin V. S100B, actor and biomarker of mild traumatic brain injury. Int J Mol Sci. 2023;24(7).
392. Tiantian W, Yonghui W, Junbo L. Antibody-labeled gold nanoparticle based resonance Rayleigh scattering detection of S100B. Anal Methods. 2024;16(19):3074–80.
393. Lorenzano S, Rost NS, Khan M, Li H, Batista LM, Chutinet A et al. Early molecular oxidative stress biomarkers of ischemic penumbra in acute stroke. Neurology. 2019;93(13):e1288–e98.
394. Wattananit S, Tornero D, Graubardt N, Memanishvili T, Monni E, Tatarishvili J et al. Monocyte-derived macrophages contribute to spontaneous long-term functional recovery after stroke in mice. J Neurosci. 2016;36(15):4182–95.
395. Maderna P, Yona S, Perretti M, Godson C. Modulation of phagocytosis of apoptotic neutrophils by supernatant from dexamethasone-treated macrophages and annexin-derived peptide Ac2–261. J Immunol. 2005;174(6):3727–33.
396. Bai L, Li N, Li H, Dai Y, Bai Y, Ma F et al. Macrophage-mimetic liposomes co-delivering ceria and Ac2-26 peptide for penumbra protection in ischemic stroke. J Nanobiotechnol. 2025;23(1):488.

397. Peters O, Back T, Lindauer U, Busch C, Megow D, Dreier J et al. Increased formation of reactive oxygen species after permanent and reversible middle cerebral artery occlusion in the rat. Journal of Cerebral Blood Flow & Metabolism. 1998;18(2):196–205.
398. Zhang W, Liu Y, Wu Y, Wang Z, Liu X, Hu Q et al. ROS-triggered biomimetic hydrogel soft scaffold for ischemic stroke repair. Biomaterials. 2025;319:123217.
399. Vaudry D, Falluel-Morel A, Bourgault S, Basille M, Burel D, Wurtz O et al. Pituitary adenylate cyclase-activating polypeptide and its receptors: 20 years after the discovery. Pharmacol Rev. 2009;61(3):283–357.
400. Denes V, Geck P, Mester A, Gabriel R. Pituitary adenylate cyclase-activating polypeptide: 30 years in research spotlight and 600 million years in service. Journal of Clinical Medicine. 2019;8(9):1488.
401. Ho E, Li DX, Cui H, Akbar D, Siu R, Morshead CM, et al. Immunomodulation with local, sustained delivery of pituitary adenylate cyclase activating polypeptide results in improved functional recovery in stroke-injured mice. Adv Healthc Mater. 2025;14::e2500765.
402. Chen C, Li M, Lin L, Chen S, Chen Y, Hong L. Clinical effects and safety of edaravone in treatment of acute ischaemic stroke: A meta-analysis of randomized controlled trials. J Clin Pharm Ther. 2021;46(4):907–17.
403. Abe M, Kaizu K, Matsumoto K. A case report of acute renal failure and fulminant hepatitis associated with edaravone administration in a cerebral infarction patient. Ther Apher Dial. 2007;11(3):235–40.
404. Xu J, Wang A, Meng X, Yalkun G, Xu A, Gao Z et al. Edaravone dexborneol versus edaravone alone for the treatment of acute ischemic stroke. Stroke. 2021;52(3):772–80.
405. Chen H-S, Zhao Z-A, Shen X-Y, Qiu S-Q, Cui Y, Qiu J et al. Edaravone dexborneol for ischemic stroke with sufficient recanalization after thrombectomy: a randomized phase II trial. Nat Commun. 2025;16(1):2393.
406. Zhou Y, Chen F, Lin Y, Wang Q, Hong Y, Zhao S et al. ROS-response nanoparticles for edaravone delivery: Pathological-activated neuroprotection against ischemic-reperfusion injury in ischemic stroke. Adv Healthc Mater. 2025;14(25):e2501962.
407. Li XS, Li HY, Peng D, Liu HR, Wu L, Wang KJ, et al. A biomimetic multifunctional nanoplatform with precise targeting-penetration and efficient synergistic therapy for ischemic stroke. ACS Appl Mater Interfaces. 2025;17:39925–54.
408. Palmer C, Roberts RL, Bero C. Deferoxamine posttreatment reduces ischemic brain injury in neonatal rats. Stroke. 1994;25(5):1039–45.
409. Im DS, Jeon JW, Lee JS, Won SJ, Cho SI, Lee YB et al. Role of the NMDA receptor and iron on free radical production and brain damage following transient middle cerebral artery occlusion. Brain Res. 2012;1455:114–23.
410. Peng S, Sun T, Yang D, Zhao H, Lin L, Xia B et al. Dipsacoside B ameliorates cognitive impairment in sepsis-associated encephalopathy by reducing Th17 cell infiltration and neuroinflammation. Biochem Pharmacol. 2024;227:116428.
411. Wu LJ, Wu G, Akhavan Sharif MR, Baker A, Jia Y, Fahey FH et al. The voltage-gated proton channel Hv1 enhances brain damage from ischemic stroke. Nat Neurosci. 2012;15(4):565–73.
412. Yang Z, Jin L, Li L, Wu Y, Liu W, Feng X et al. Brain targeted lipid nanoparticles with Hv1 inhibitors alleviate neuroinflammation post-ischemic stroke. J Nanobiotechnol. 2025;23(1):464.
413. Liu C, Wang J, Wang Z, Fan X, Pan J, Guo X et al. Discoidal nanoparticles exploit thrombus-induced shear gradients to enhance site-specific thrombolysis in stroke. Acta Biomater. 2025;201:471–84.
414. Talley Watts L, Zheng W, Garling RJ, Frohlich VC, Lechleiter JD. Rose bengal photothrombosis by confocal optical imaging in vivo: A model of single vessel stroke. J Vis Exp. 2015;100:e52794.
415. Zhu Z, Jin L, Wang Q, Shi H, Cheng K, Mao Z. Inhalable Ce nanozyme-backpacked phage aims at ischemic cerebral injury by M1-microglia hitchhiking. Adv Materials (Deerfield Beach, Fla). 2025;37(27):e2419903.
416. Lu WQ, Qiu Y, Li TJ, Tao X, Sun LN, Chen WS. Timosaponin B-II inhibits pro-inflammatory cytokine induction by lipopolysaccharide in BV2 cells. Arch Pharm Research. 2009;32(9):1301–8.
417. Zhu N, Wang F, Han Z, Ling S, Wong DW, Ye S et al. Noninvasive method for achieving the regeneration of damaged nerves via ultrasonic nasal drops. Bioact Mater. 2025;49:342–61.
418. Tuo QZ, Lei P, Jackman KA, Li XL, Xiong H, Li XL et al. Tau-mediated iron export prevents ferroptotic damage after ischemic stroke. Mol Psychiatry. 2017;22(11):1520–30.
419. Kondo Y, Ogawa N, Asanuma M, Ota Z, Mori A. Regional differences in late-onset iron deposition, ferritin, transferrin, astrocyte proliferation, and microglial activation after transient forebrain ischemia in rat brain. J Cerebral Blood Flow Metab. 1995;15(2):216–26.

420. Millerot-Serrurot E, Bertrand N, Mossiat C, Faure P, Prigent-Tessier A, Garnier P et al. Temporal changes in free iron levels after brain ischemia relevance to the timing of iron chelation therapy in stroke. Neurochem Intl. 2008;52(8):1442–8.
421. Hu X, Bao Y, Li M, Zhang W, Chen C. The role of ferroptosis and its mechanism in ischemic stroke. Exp Neurol. 2024;372:114630.
422. Lv R, Li F, Liu Y, Song M, Yuan J, Zhang G et al. Molecularly imprinted nanoparticles hitchhiking on neutrophils for precise treatment of ischemic stroke. J Colloid Interface Sci. 2025;689:137246.
423. Gao Z, Wang X, Zhang W, Wang J, Liu C. Inflammatory oriented nanospheres-reconstructed extracellular matrix in ischemic stroke. ACS Nano. 2025;19(24):22065–78.
424. Zhang T, Liu P, Shen W, Li C, Zhao Z, Wu Y et al. DNase I-mediated chemotactic nanoparticles for NETs targeting and microenvironment remodeling treatment of acute ischemic stroke. Advanced Science. 2025;12(34):e03689.
425. Liu Q, Ma H, Liao J, Zhu Z, Chen H, Sun D et al. MiR-125a-5p in extracellular vesicles of neural stem cells acts as a crosstalk signal modulating neuroinflammatory microenvironment to alleviate cerebral ischemia-reperfusion injury. Theranostics. 2025;15(14):7064–89.
426. Guo Y, Jia M, He Q, Zeng J, Zhang Y, Gao Y et al. A carrier-free triple-drug co-assembled nanoformulation for synergistic treatment of cerebral ischemia-reperfusion injury. ACS Appl Mater Interfaces. 2025;17(23):33618–32.
427. Song Y, Wang L, Peng T, Shan L, Wan B, Tang M et al. Brain-targeting biomimetic nanozyme enhances neuroprotection in ischemic stroke by remodeling the neurovascular unit. Journal of controlled release: official journal of the Controlled Release Society. 2025;382:113750.
428. Jia Y, Xu L, Leng S, Sun Y, Huang X, Wang Y et al. Nose-to-brain delivery of circular RNA SCMH1-loaded lipid nanoparticles for ischemic stroke therapy. Adv Materials (Deerfield Beach, Fla). 2025;37(19):e2500598.
429. Gao J, Wang X, Wan Q. One nanoparticle delivers two different neuroprotective amino acids into ischemic brain and protects against neuronal death in rat cerebral ischemia injury. Mol Cell Neurosci. 2025;133:104009.
430. Han D, Zhang J, Li D, Wang C. Celery seed derived reconstituted lipid nanoparticles as an innate neuron-targeted neuroprotective nanomedicine for ischemic stroke treatment. J Nanobiotechnol. 2025;23(1):298.
431. Wang P, Lv X, Tian S, Yang W, Feng M, Chang S et al. Platelet membrane-based nanoparticles for targeted delivery of deferoxamine to alleviate brain injury induced by ischemic stroke. Int J Nanomed. 2025;20:7533–48.
432. Tataranni T, Piccoli C. Dichloroacetate (DCA) and cancer: An overview towards clinical applications. Oxid Med Cell Longevity. 2019;2019:8201079.
433. Pacheco C, de Dios O, Ramiréz-González MA, Martins C, Fialho SL, Baltazar F et al. Multimodal nanoparticles co-delivering bevacizumab and dichloroacetate for dual targeting of neoangiogenesis and hyperglycolysis in glioblastoma treatment. Journal of controlled release: official journal of the Controlled Release Society. 2025;384:113931.
434. Cheng W, Yang J, Pan Y, Qu H, Duan Z, Wu J et al. Noninvasive activation of local and systemic immunity with a sequential-targeting sonodynamic nanovaccine to treat glioblastoma. ACS Nano. 2025;19(30):27804–24.
435. Keenlyside A, Marples T, Gao Z, Hu H, Nicely LG, Nogales J et al. Development and optimisation of in vitro sonodynamic therapy for glioblastoma. Sci Rep. 2023;13(1):20215.
436. Wei R, Xie K, Li T, Lin W, Zhao Y, Li J et al. Immunity/metabolism dual-regulation via an acidity-triggered bioorthogonal assembly nanoplatform enhances glioblastoma immunotherapy by targeting CXCL12/CXCR4 and adenosine-A2AR pathways. Biomaterials. 2025;319:123216.
437. Chiang IT, Liu YC, Liu HS, Ali AAA, Chou SY, Hsu TI et al. Regorafenib reverses temozolomide-induced CXCL12/CXCR4 signaling and triggers apoptosis mechanism in glioblastoma. Neurotherapeutics. 2022;19(2):616–34.
438. Chen IX, Chauhan VP, Posada J, Ng MR, Wu MW, Adstamongkonkul P et al. Blocking CXCR4 alleviates desmoplasia, increases T-lymphocyte infiltration, and improves immunotherapy in metastatic breast cancer. Proceedings of the National Academy of Sciences. 2019;116(10):4558–66.
439. Wei R, Li J, Lin W, Pang X, Yang H, Lai S et al. Nanoparticle-mediated blockade of CXCL12/CXCR4 signaling enhances glioblastoma immunotherapy: Monitoring early responses with MRI radiomics. Acta Biomater. 2024;177:414–30.
440. Allard B, Beavis PA, Darcy PK, Stagg J. Immunosuppressive activities of adenosine in cancer. Curr Opin Pharmacol. 2016;29:7–16.

441. Xu Q, Hu X, Ullah I, Lin T, Liu Y, Dong H et al. Biomimetic hybrid PROTAC nanovesicles block multiple DNA repair pathways to overcome temozolomide resistance against orthotopic glioblastoma. Adv Materials (Deerfield Beach, Fla). 2025;37(29):e2504253.
442. Lai AC, Crews CM. Induced protein degradation: an emerging drug discovery paradigm. Nat Rev Drug Discov. 2017;16(2):101–14.
443. Yang Q, Zhao J, Chen D, Wang Y. E3 ubiquitin ligases: styles, structures and functions. Mol Biomed. 2021;2(1):23.
444. Raina K, Lu J, Qian Y, Altieri M, Gordon D, Rossi AMK et al. PROTAC-induced BET protein degradation as a therapy for castration-resistant prostate cancer. Proc Natl Acad Sci U S A. 2016; 113(26):7124–9.
445. Jiang X, Lai Y, Xia W, Yang W, Wang J, Pan J et al. Self-oxygenating PROTAC microneedle for spatiotemporally-confined protein degradation and enhanced glioblastoma therapy. Adv Materials (Deerfield Beach, Fla). 2025;37(13):e2411869.
446. Maurya PK. Chapter 10 - animal biotechnology as a tool to understand and fight aging. In: Verma AS, Singh A, editors. Animal Biotechnology. San Diego: Academic Press; 2014. p. 177–91.
447. Kuang L, Han M, Wu X, Deng Z, Liu T, Yin Y et al. Starting the engine and releasing the brakes of T-cell responses: A biomimetic dendritic cell nanoplatform for improved glioblastoma immunotherapy. ACS Nano. 2025;19(23):21365–84.
448. Hatano M, Kuwashima N, Tatsumi T, Dusak JE, Nishimura F, Reilly KM et al. Vaccination with EphA2-derived T cell-epitopes promotes immunity against both EphA2-expressing and EphA2-negative tumors. J Transl Med. 2004;2(1):40.
449. Wang C, Feng W, Li J, Wang J, Liu L, Ye SH et al. Enhanced nano-vaccine utilizing biomineralized virus-like particles for efficient glioblastoma immunotherapy via the nose-to-brain delivery pathway. ACS Nano. 2025;19(22):21154–68.
450. Gao Y, Mao M, Li Y, Xuan M, Wu Y, He Q. A self-directed Trojanbot-enzymatic nanobot in neutrobot for active target therapy of glioblastoma. Nature communications. 2025;16(1):5263.
451. Ho W-M, Chen C-Y, Chiang T-W, Chuang T-J. A longer time to relapse is associated with a larger increase in differences between paired primary and recurrent IDH wild-type glioblastomas at both the transcriptomic and genomic levels. Acta neuropathologica communications. 2024;12(1):77.
452. Chédeville AL, Madureira PA. The role of hypoxia in glioblastoma radiotherapy resistance. Cancers. 2021;13(3):542.
453. Jamal M, Rath BH, Tsang PS, Camphausen K, Tofilon PJ. The brain microenvironment preferentially enhances the radioresistance of CD133(+) glioblastoma stem-like cells. Neoplasia. 2012;14(2):150–8.
454. Ali MY, Oliva CR, Noman ASM, Allen BG, Goswami PC, Zakharia Y et al. Radioresistance in glioblastoma and the development of radiosensitizers. Cancers. 2020;12(9).
455. Yoshino Y, Yoshino F, Aoki I, Mori Y, Suzuki G, Tsuji S et al. 2-Nitroimidazole-functionalized superparamagnetic iron oxide nanoparticles detect hypoxic regions of glioblastomas on MRI and improve radiotherapy efficacy. ACS Nano. 2025;19(13):12762–76.
456. Donahue ND, Acar H, Wilhelm S. Concepts of nanoparticle cellular uptake, intracellular trafficking, and kinetics in nanomedicine. Adv Drug Del Rev. 2019;143:68–96.
457. Toscano F, Torres-Arias M. Nanoparticles cellular uptake, trafficking, activation, toxicity and in vitro evaluation. Curr Res Immunol. 2023;4:100073.
458. Antileo E, Garri C, Tapia V, Muñoz JP, Chiong M, Nualart F et al. Endocytic pathway of exogenous iron-loaded ferritin in intestinal epithelial (Caco-2) cells. Am J Physiol Gastrointest Liver Physiol. 2013;304(7):G655–61.
459. Li L, Fang CJ, Ryan JC, Niemi EC, Lebrón JA, Björkman PJ et al. Binding and uptake of H-ferritin are mediated by human transferrin receptor-1. Proc Natl Acad Sci U S A. 2010;107(8):3505–10.
460. Johnsen KB, Burkhart A, Thomsen LB, Andresen TL, Moos T. Targeting the transferrin receptor for brain drug delivery. Prog Neurobiol. 2019;181:101665.
461. Fan K, Jia X, Zhou M, Wang K, Conde J, He J et al. Ferritin nanocarrier traverses the blood brain barrier and kills glioma. ACS Nano. 2018;12(5):4105–15.
462. Zhang B, Chen X, Tang G, Zhang R, Li J, Sun G et al. Constructing a nanocage-based universal carrier for delivering TLR-activating nucleic acids to enhance antitumor immunotherapy. Nano Today. 2022;46:101564.
463. Jin Y, Zhang B, Li J, Guo Z, Zhang C, Chen X et al. Bioengineered protein nanocarrier facilitating siRNA escape from lysosomes for targeted RNAi therapy in glioblastoma. Sci Adv. 2025; 11(8):eadr9266.

464. Cakir N, Oncel H, Ozkan A, Bicak D, Akgun Bas S, Mustafaoglu N. Developing Immunoniosomes (INs): Antibody and Fab conjugations of niosomal nanoparticles via UV-NBS and EDC/NHS chemistry for treating glioblastoma cells. Int J Pharm X. 2025;10:100367.
465. Wei N, Xiao Y, Jin Y, Zhang L, Chen P, Du S. Thioether-bridged carbon dots-dopamine quinone conjugate probe for glutathione sensing and discrimination between glioblastoma and normal cells. Talanta. 2025;294:128249.
466. Alonso AV, Murugesan A, Gogoi R, Chandrabose S, Abass KS, Sharma V et al. Graphitic carbon nitride nanoparticle: g-C(3)N(4) synthesis, characterization, and its biological activity against glioblastoma. Eur J Pharmacol. 2025;1003:177999.
467. Kurawattimath V, Wilson B, Geetha KM, Divekar K. Surface functionalized albumin nanoparticles of palbociclib with amplified brain delivery for treating brain glioblastoma. Sci Rep. 2025;15(1):28433.
468. Trivedi S, Agade R, Belgamwar V. Engineering of dual-functionalized intranasal nanovesicles embedded with thymoquinone for targeted modulation of the PI3K/AKT pathway in glioblastoma therapy. Mol Pharm. 2025;22(8):5075–93.
469. Wang L, Xu D, Hu X, Quan R, Lu D, Li Z et al. Cascade reaction-driven biomimetic scintillant/metal-organic frameworks for X-ray triggered combinational therapy against glioma. Mater Today Bio. 2025;33:102069.
470. Zhou L, Zhao Q, Gu L, Tian R, Li Y, Xiong X. Application of PVA hydrogel loaded with luteolin nanoparticles in anti EMT treatment after GBM. Mater Today Bio. 2025;33:101956.
471. Xu H, Wu Z, Tang J, Gan Y, Li J, Yu Y et al. Ginsenoside F2-modified liposomes delivering FTY720 enhance glioblastoma targeting and antitumor activity via ferroptosis. Phytomedicine. 2025;144:156917.
472. He Q, Zhang S, Wang Y, Gao Y, Han H. An optical/magnetic imaging nanoplatform-mediated convection-enhanced delivery of oxaliplatin for glioma treatment. Nanoscale. 2025;17(30):17636–46.
473. Wang H, Ren Q, Wu G, Kong J, Jin X, Huang P et al. Nose-to-brain delivery of transferrin-modified carmustine-loaded iron nanoparticles for enhanced glioblastoma treatment. J Drug Target. 2025:1880–1891.

6 Exosomes for Drug Delivery to the Brain

Mosavar Farahani

6.1 INTRODUCTION

Exosomes are nanosized, naturally occurring extracellular vesicles (EVs) secreted by cells, serving as vital mediators of intercellular communication. These vesicles are particularly valuable in therapeutic applications because of their distinctive capacity to transport and deliver a wide range of biomolecules.

One of the current strategies for drug delivery to the brain involves designing vehicles that can cross the blood–brain barrier (BBB), using information found on the surface of EVs. EVs, particularly small EVs (sEVs) play a crucial role in modulating signal transmission throughout the brain in physiological conditions and are released by various types of neural cells into biofluids, which helps to deliver to therapeutic targets in neurological disorders. Under pathological conditions, these EVs can contribute to the dissemination of misfolded or pathogenic proteins across distinct brain regions.

This chapter explores the therapeutic potential of EVs in the context of neurological disorders, focusing on three major neurodegenerative diseases, Alzheimer's disease (AD), Parkinson's disease (PD), and multiple sclerosis (MS), as well as the brain cancer glioblastoma (GBM). We begin by reviewing the fundamental characteristics of EVs, including their heterogeneity, biogenesis, and roles in both physiological and pathological conditions. The discussion then proceeds to their emerging applications in neurological diseases, emphasising their use as drug delivery systems. This includes the stages of EV-mediated drug delivery, their capacity to overcome the BBB, current progress in preclinical and clinical studies, and prospects for their successful therapeutic implementation.

6.2 EXTRACELLULAR VESICLES AND THEIR BIOGENESIS

EVs are nano-to-micro-sized lipid-bilayer membrane-enclosed vesicles, that are released by all cell types. Cells communicate with each other by secreting signalling molecules, such as proteins, lipids, and nucleic acids. Cells can package these signalling molecules into EVs to avoid rapid degradation and escape immune surveillance among other purposes. This allows for local and long-distance intercellular communication. EVs are released under both normal and pathological conditions in all tissues and bodily fluids [1,2]. They are crucial in maintaining physiological equilibrium and homeostasis and they influence pathological processes by participating in intercellular signalling [3]. Although cellular communication typically involves secreted factors and direct cell contact, the secretion and uptake of EVs have been described as an additional layer of intercellular communication [4]. EVs are involved in intercellular communication. Once they have been released from parental cells, they carry their biological cargoes into bodily fluids. Their subsequent recognition and uptake result in alterations to recipient cell behaviour [5].

6.2.1 Extracellular Vesicles Biogenesis

Advances in isolation and characterisation techniques have enabled the identification of various distinct EV subtypes. EV types are named based on their method of origin, functional characteristics, or physical appearance. The International Society for Extracellular Vesicles (ISEV) has played a crucial

DOI: 10.1201/9781003482406-6

TABLE 6.1
Classification of Extracellular Vesicles (EVs). Based on their Size and Biogenesis [1,2]

	Extracellular Vesicles Subtypes Characteristics		
	Exosomes	**Microvesicles**	**Apoptotic Bodies**
Size	30–150 nm	150–1000 nm	50–5,000 nm
Biogenesis/Release	Fusion of **multivesicular bodies** with the plasma membrane, via ESCRT- dependent or ESCRT-independent mechanisms of cargo sorting, released upon fusion of MVEs with the plasma membrane	Outward blebbing of the **plasma membrane**	Released from apoptotic cells upon activation of apoptosis-related transduction pathways
Commonly used markers	ESCRT complex proteins, CD9, CD63, CD81, Alix, Flotillin-1, ESCRT-3, TSG101	Annexin A1, ARF6, integrins, selectins, tissue factor	Phosphatidylserine, Annexin V
Pathways	ESCRT-dependent Tetraspanin-dependent Stimuli-dependent	Ca^{2+}- dependent Stimuli- and cell dependent	Apoptosis related

role in advancing the field of EV research, by providing guidelines and recommendations for EV classification [6]. Their guidelines categorise EVs into three main subtypes based on their biogenesis pathways (Table 6.1) [4]. Exosomes, which are generated by inward budding of endosomal compartments (multivesicular bodies––MVBs) that later fuse with the plasma membrane (Figure 6.1).

1. Microvesicles (also known as microparticles), which arise from the outward budding of the plasma membrane
2. Apoptotic bodies are generated by fragmentation of cells undergoing apoptosis

Establishing consistent terminology remains a major challenge in the field, as highlighted by the ISEV MISEV2023 guidelines, which emphasize the use of 'extracellular vesicle' as a generic term and recommend precise definitions based on measurable characteristics such as origin, size, and molecular markers. Some of the protein cargo of microvesicles overlaps with those of exosomes, which makes definitive identification difficult. Furthermore, isolating highly purified preparations of different types of EVs is challenging, and there is heterogeneity even within each class. The size of microvesicles is more heterogeneous with diameters ranging from 100 nm to 1μm [6]

While EVs are the preferred umbrella term in the current nomenclature, the term "exosomes" remains prevalent in the literature, particularly in studies investigating their role in drug delivery. In this chapter, "exosomes" and "EVs" may be used interchangeably, with emphasis placed on their applications in therapeutic delivery systems. As most of the referenced articles focus on exosomes, especially in the context of overcoming the BBB, the term "exosomes" will be used more frequently throughout.

Exosome biogenesis begins within endosomes, where inward budding of the limiting membrane of late endosomes leads to the formation of intraluminal vesicles (ILVs) within multivesicular endosomes (MVEs). This budding event results in the incorporation of endocytosed transmembrane proteins into ILVs, maintaining their signalling orientation- exposing the ectodomain outward and enclosing the cytoplasmic tail and associated cargo internally. Upon fusion of MVEs with the plasma membrane, ILVs are released into the extracellular space as exosomes, typically ranging from 30 to 150 nm in diameter [1]. The formation and cargo sorting of ILVs are largely governed by the endosomal sorting complex required for transport (ESCRT). Additionally, tetraspanin proteins such as CD63, CD81, and CD9 are highly enriched in exosomes and are widely utilised as molecular

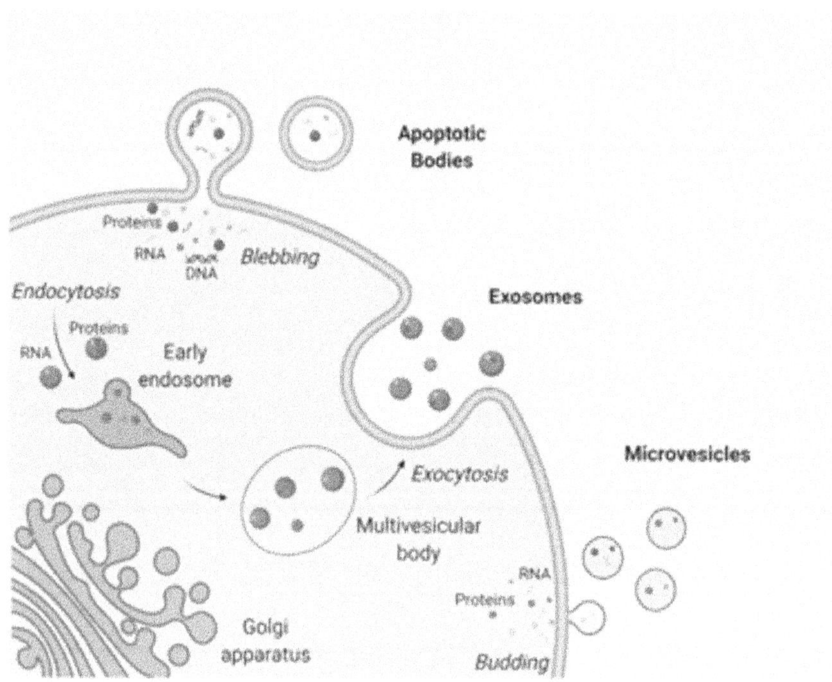

FIGURE 6.1 Biogenesis of extracellular vesicle (EV). Exosomes originate as intraluminal vesicles within multivesicular bodies (MVBs) and are released into the extracellular space when these MVBs fuse with the plasma membrane via exocytosis. In contrast, microvesicles are formed by the direct outward budding and shedding of the plasma membrane. Apoptotic bodies are generated during apoptosis as the dying cell fragments into membrane-bound vesicles. Figure is taken from Dang et al. [7] with permission.

markers [8]. Members of the RAB family of small GTPases are key regulators of vesicular trafficking, including exosome secretion. Disruption of RAB function has been shown to significantly impair exosome release [9].

6.2.2 Composition of Exosomes

Exosomes, a subset of SEVs, are enriched with a diverse array of proteins, lipids and nucleic acids that reflect their complex biological functions. They play critical roles in both maintaining physiological balance and driving disease processes. Their unique cargo, ability to cross barriers, and cell-specific signatures position them as powerful tools for diagnostics, prognostics, and targeted therapies––particularly in the context of neurological diseases [5]. Enclosed by a lipid bilayer, exosomes carry membrane trafficking proteins (e.g. GTPases, Annexins, Flotillin, Rabs, Dynamin, and Syntaxin), cytoskeletal elements (e.g. Actin, Tubulin, and Cofilin), and ESCRT-associated proteins (e.g. Alix and TSG-101), which are critical for vesicle formation and structural integrity. Surface markers include tetraspanins (CD9, CD63, CD81, etc.), adhesion molecules (e.g. Integrins), MHC molecules (class I and II) and signalling receptors (Figure 6.2). Exosomes also contain heat shock proteins, cytokines, growth factors and nucleic acids (mRNA, miRNA, non-coding RNA, DNA), which are vital for their roles in intercellular communication, immune modulation, and disease progression. Lipid components such as cholesterol, ceramides, phospholipids, and glycoproteins contribute to membrane stability and functional specificity [8, 10]. Exosomes carry bioactive cargo that is selectively packaged through regulated sorting mechanisms. While they reflect the molecular profile of their cells of origin, exosomes are not merely passive snapshots. Instead, they are enriched with

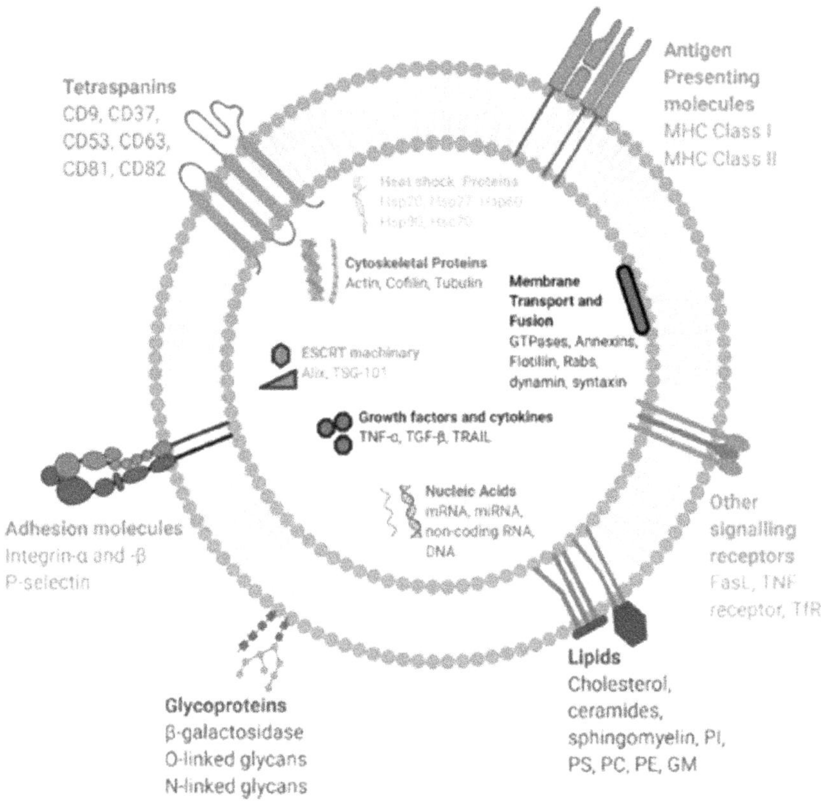

FIGURE 6.2 Schematic representation of an exosome and its biological composition. Exosomes are composed of a phospholipid bilayer membrane enclosing intraluminal fluid with cytoplasmic origin. They contain biomolecules from their cell of origin, which include, other than lipids, several types of proteins (e.g. involved in cell adhesion, as well as other transmembrane and intraluminal proteins with various functions) and nucleic acids. Figure taken from Gurung et al. [10] with permission.

specific biomolecules that confer distinct functional roles in both physiological and pathological contexts [11].

6.2.3 Physiological and Pathological Roles of Exosomes

Exosomes circulate in a variety of biological fluids including plasma, cerebrospinal fluid (CSF), urine, saliva, breast milk, and cell culture media. Their molecular composition is reflective of the cell of origin, making them promising biomarkers for diagnostic and prognostic applications [12]. In physiological contexts, exosomes maintain homeostasis by supporting tissue repair, immune balance, and neural communication. For example, in the central nervous system (CNS), exosomes facilitate neurogenesis, synaptic plasticity, and myelin maintenance. They are secreted by neurons, oligodendrocytes, astrocytes, and microglia, each contributing distinct molecular cargos (such as myelin proteins, cytokines, or receptor subunits) that support normal brain function [11, 13].

However, under pathological conditions, this communication network can be hijacked. Dysregulated exosomal signalling has been implicated in cancer progression, cardiovascular dysfunction, and immune-mediated diseases. In the nervous system, exosomes may contribute to neurodegenerative and neuroinflammatory conditions by disseminating misfolded proteins (e.g. tau and α-synuclein) or inflammatory mediators, thereby exacerbating disease states [5, 13].

A key feature of exosomes is their ability to cross biological barriers, notably the BBB. This property underlies both their diagnostic potential and their promise as drug delivery vehicles. Studies have demonstrated that exosomes, can traverse the BBB under both physiological and pathological conditions. For instance, tumour-derived exosomes have been shown to disrupt BBB integrity via miRNA-mediated actin cytoskeleton remodelling [14]. Additionally, the BBB becomes increasingly permeable in many neurological diseases, such as Alzheimer's or MS, which further facilitates EV transfer between the CNS and the periphery [15]. Notably, engineered exosomes have also been shown to deliver therapeutic cargoes across the BBB, supporting their utility in CNS drug delivery. This has been successfully demonstrated using modified exosomes to deliver siRNA to neurons *in vivo*, highlighting their translational potential [16].

6.2.4 Isolation of Exosomes

There are various methods by which EVs can be isolated, depending on the source material (e.g. choice of biofluid) and the downstream assays to be performed (Table 6.2). Differential ultracentrifugation is typically used when studying samples, such as conditioned cell-culture media, usually due to the larger volumes of material available for processing. Isolating neural-specific EVs is especially challenging, especially when using them for biomarker discovery, since how many EVs cross the BBB remains unknown, and access to the primary tissue is not amenable for routine testing. Analysis of EVs isolated from CSF is possible and has been used in biomarker discovery studies for a few neurological diseases. EVs can be isolated from blood (serum and plasma), CSF, urine, and other biofluids. An early example is the identification of mRNA and miRNA signatures from GBM-derived extracellular vesicles, which were detectable in peripheral circulation [17]. Since then, several published studies have reported EV, associated protein, and genetic biomarkers for several neurological diseases, which will be discussed below [5]. Differential ultracentrifugation (also known as density gradient centrifugation) has been used for many years, however it is labour-intensive and can result in considerable sEV loss [6]. Ultrafiltration is a fast, sterile, and scalable approach with a high RNA yield- however, it can result in low purity, potential sEV loss due to membrane attachment and shear stress that may degrade vesicles. sEV precipitation is a straightforward and scalable method that preserves bioactivity, though it requires long processing

TABLE 6.2
Classification of Common Methods of EVs Isolation. EVs can Be Isolated from Various Biofluid, Cells Cultured *In Vitro* and the Collection of Conditioned Medium to Enrich for EVs [2, 18]

Technique	Principle of the Method	Comments
Ultracentrifugation	Size-based techniques	Classic, time-consuming
Co-precipitation	Changing the solubility and dispersibility of particles by using hydrophilic polymers	High yield, simple operation, suitable for large samples—but low purity
Size exclusion chromatography	Separation based on the negative Zeta potential of the EVs surface	Integrity preserved
Immunoaffinity enrichment	Specific binding of capture molecules to the EV surface markers	Highly specific
Microfluidics-based technologies	Mainly chip technology designed according to EVs physical properties	Biocompatibility requires low sample volumes and high purity
Ultrafiltration	Using filtration membranes, the separation is based on particle size	Sterile, fast, no special equipment required, highest exosomal RNA yield

times and has a risk of co-precipitating contaminants. Immunological separation is highly specific, fast, and requires a smaller sample volume, but its high reagent costs and low yield present challenges. Lastly, microfluidics-based technologies offer high purity, precision, and cost efficiency, yet they are limited by low sample capacity and an inability to distinguish vesicles with similar surface markers (Table 6.2).

6.2.5 Characterisation of Exosomes

Isolated EVs can be characterised and quantified using different methods, including flow cytometry, western blot, and electron microscopy, or atomic force microscopy for characterisation nanoparticle tracking for EV quantification (Figure 6.3). One way to characterise exosomes is to check their sizes, shape, and surface morphology.

6.2.6 Advantages and Challenges of Using Exosomes for Drug Delivery

Exosomes are lipid bilayer-enclosed vesicles that protect therapeutic cargo, such as proteins and nucleic acids, while facilitating efficient cellular uptake. Their endogenous origin contributes to excellent biocompatibility, stability, low immunogenicity, and minimal toxicity, making them highly promising candidates for drug delivery, particularly in targeting the brain [20].

One of the major advantages of exosomes is their ability to be engineered through surface modifications. Surface engineering can enhance targeting specificity, reduce off-target effects, and improve therapeutic efficacy. Exosomes also demonstrate organ-specific accumulation, extended plasma half-lives, and the ability to evade immune clearance, partly due to surface markers like CD47. These features allow them to bypass the reticuloendothelial system (RES) and prolong circulation time. Additionally, their membranes can be functionalised with native, exogenous, or engineered ligands to improve uptake by target cells and enhance delivery efficiency. Their inherent ability to cross the BBB further highlights their potential in treating CNS disorders [21]. Despite these advantages, several challenges limit the clinical translation of exosome-based therapies. The efficiency of cargo transfer, especially for mRNA, miRNA, and proteins, remains suboptimal, affecting therapeutic consistency. Exosome heterogeneity, along with variability in isolation, purification, and preparation methods, poses significant barriers to standardisation and reproducibility. Although surface features like CD47 help avoid immune detection, exosomes are still subject to rapid clearance, often requiring additional modifications to extend their circulation. Long-term storage is another challenge, as maintaining exosome integrity and bioactivity requires specialised conditions [22].

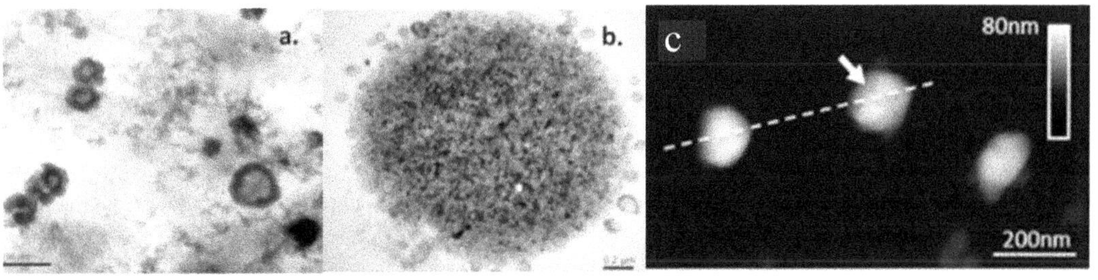

FIGURE 6.3 TEM and AFM characterisation of exosomes. (a) Transmission electron microscopy (TEM) image of exosomes enriched from lymphocyte cell culture. Exosomes are visible as small 50–100 nm vesicles with bi-layered membranes (scale bar: 100 nm). (b) Exosomes immuno-isolated using magnetic beads coated with anti-HLA antibody. The TEM image shows a bead coated with exosomes (scale bar: 0.2 μm). Figure taken from Farahani et al. [19]. (c) Atomic force microscopy (AFM) topographic image of exosomes immobilised on a mica surface using PeakForce tapping mode (Multimode 8, Bruker).

While exosomes exhibit intrinsic stability in circulation due to their negative surface charge and CD47 expression, clearance kinetics can vary widely. Some studies report rapid elimination following intravenous administration, which may be influenced by the cell source, isolation techniques, or specific surface protein and lipid profiles. These mechanisms are not yet fully understood. Moreover, exosomes may possess innate targeting capabilities based on their lipid and protein composition. For example, progenitor cell-derived exosomes have been shown to deliver bioactive cargo that promotes angiogenesis, tissue repair, and immune modulation, highlighting their dual role as drug carriers and potential agents for cellular therapy [21].

Successful clinical translation of exosome-based therapies requires overcoming several challenges. These include scaling up production and isolation processes, establishing standardised protocols, and deepening our understanding of exosome biogenesis, cargo sorting, subpopulations, and internalisation pathways. Such insights are essential for refining strategies that enhance the therapeutic utility of exosomes.

The following section will examine specific examples of exosome applications in the treatment of brain diseases.

6.2.6.1 Exosomes in Neurological Diseases

Exosomes are increasingly recognised as key players in both the pathology and potential treatment of major brain disorders, including AD, PD, MS, and GBM. Within the CNS, exosomes contribute to essential functions such as synaptic plasticity, axonal transport, neuroprotection, and the clearance of cellular debris. They facilitate communication between neurons and glial cells and influence immune responses, showing both pro-inflammatory and anti-inflammatory effects depending on their source and molecular cargo [20].

In physiological conditions, exosomes help maintain brain homeostasis. However, in neurological diseases, they can act as carriers of pathological signals. For example, stressed astrocytes may release exosomes loaded with neurotoxic amyloid-β or pro-inflammatory molecules that accelerate neurodegeneration [23]. In AD and PD, exosomes have been implicated in the spread of misfolded proteins such as amyloid-β, tau, and α-synuclein, hallmarks of these diseases [24].

Beyond their pathological roles, exosomes are being actively investigated for their diagnostic and therapeutic potential. Because their molecular cargo reflects the state of their parent cells, exosomes offer a promising avenue for biomarker discovery in neurodegenerative conditions. Their ability to cross the BBB makes them especially valuable for CNS drug delivery. Engineered exosomes have demonstrated success in transporting therapeutic molecules across the BBB, reducing inflammation, and slowing disease progression in preclinical models [5].

Antibodies, enabling selective interaction with receptors expressed on brain endothelial cells or neurons. Common strategies include conjugation of ligands like rabies virus glycoprotein (RVG) peptide or transferrin to facilitate receptor-mediated uptake [11]. In addition to surface functionalisation, exosomes can be loaded with diverse therapeutic cargos, such as small molecules, proteins, or nucleic acids, including microRNAs (miRNAs), using techniques like electroporation, sonication, or genetic engineering of donor cells. miRNAs delivered via exosomes have shown significant promise in modulating gene expression and neuroinflammatory pathways, offering neuroprotective and regenerative effects in conditions such as AD, stroke, and neuropathic pain. These modifications not only improve targeting specificity but also increase the efficiency of drug delivery, providing a platform for treating neurological disorders and brain tumours while minimising off-target effects [25].

Clinical trials are now exploring how EVs can be harnessed for both diagnosis and treatment, with several studies focused on brain-related conditions such as AD, PD, stroke, and GBM (clinicaltrials.gov). Importantly, exosomes play dual roles in neurological diseases. On one hand, they may contribute to disease progression by spreading toxic proteins and amplifying inflammation. On the other, they hold promise as tools for diagnosis and therapy. Unlocking their full potential will require deeper insights into their biogenesis, cargo selection, and targeting mechanisms. These discoveries will be critical for advancing exosome-based strategies in the treatment of neurodegenerative disorders.

6.2.6.2 Exosomes in Alzheimer's Disease (AD)

AD is the most prevalent neurodegenerative disorder, characterised by progressive cognitive decline driven by the accumulation of amyloid-β (Aβ) peptides, hyperphosphorylated tau proteins, and widespread neurodegeneration. Extracellular Aβ plaques and intracellular neurofibrillary tangles composed of tau protein spread throughout the brain, exacerbating neuronal damage and dysfunction [26,27]. Despite decades of research, a definitive cure remains elusive, emphasising the urgent need for innovative therapeutic strategies.

Exosomes have emerged as promising candidates in AD therapy due to their low immunogenicity, ability to cross the BBB, and capacity to carry therapeutic biomolecules. In rodent models, mesenchymal stem cell-derived extracellular vesicles (MSC-EVs) have been shown to reduce neuroinflammation and oxidative stress markers, including IL-1β, TNF-α, Iba1, and HO-1, while increasing neurogenesis [28–30]. Exosomes derived from CNS-resident cells also demonstrate neuroprotective effects. For instance, repeated intravenous administration of endothelial cell-derived EVs has reduced Aβ deposition and improved cognitive performance in AD mouse models [31]. Similarly, intracerebroventricular injections of neural stem cell-derived EVs (NSC-EVs) have enhanced memory and protected synaptic integrity [32].

Additionally, exosomes function not only as passive carriers of therapeutic agents, but they also play an active role in the clearance of amyloid-β peptides. In vitro studies have revealed that MSC-derived exosomes contain neprilysin (NEP), a membrane-bound metallopeptidase essential for degrading Aβ peptides. NEP levels decline with age and in AD, potentially contributing to Aβ accumulation. *In vivo*, exosomes may act as scavengers by binding soluble Aβ and facilitating its delivery to microglia for degradation [33]. Beyond therapy, exosomes are being explored as non-invasive diagnostic biomarkers. Blood-derived exosomes enriched for CNS origin carry proteins such as Aβ42, total tau (t-tau), and phosphorylated tau (p-T181-tau and p-S396-tau), with profiles that mirror CSF findings. This suggests their utility in early diagnosis and monitoring of AD progression [34]. Several proof-of-concept studies have demonstrated the potential of engineered exosomes for targeted drug delivery in AD. For example, Alvarez-Erviti et al. [16] used RVG-targeted exosomes to deliver siRNA across the BBB, successfully silencing the BACE1 gene, a key enzyme in Aβ production [35]. The researchers showed that human MSC-derived exosomes contribute to Aβ degradation in vitro, while neuroblastoma-derived exosomes have enhanced Aβ clearance in AD mouse models [36]. Collectively, these findings highlight the dual role of exosomes in AD, as therapeutic agents capable of modulating disease pathology and as diagnostic tools offering insights into disease progression. Table 6.3 summarises key data from the reviewed studies.

6.2.6.3 Exosomes in Parkinson's Disease (PD)

PD is the second most prevalent neurodegenerative disorder worldwide, characterised by the progressive degeneration of dopaminergic neurons in the substantia nigra and the pathological accumulation of misfolded α-synuclein aggregates, known as Lewy bodies [37,38]. These hallmark features disrupt cellular homeostasis, leading to dysregulated proteostasis, mitochondrial dysfunction, oxidative stress, and chronic neuroinflammation. Emerging evidence highlights the critical role of exosomes, in the intercellular transmission of α-synuclein. These vesicles facilitate the spread of both monomeric and oligomeric α-synuclein species, thereby amplifying protein aggregation and contributing to disease progression. Moreover, exosomes secreted by activated microglia and astrocytes exacerbate neuroinflammatory responses, further accelerating neuronal damage. Lysosomal dysfunction has also been implicated in PD pathogenesis. Disruption of lysosomal degradation pathways impairs intracellular protein clearance and enhances the release of α-synuclein-laden exosomes, reinforcing the cycle of neurodegeneration [39,40].

TABLE 6.3
Exosomes as a Treatment in Models of Neurological Disorders. Summary of EV-based Therapeutic Strategies in Preclinical Models of Neurological Diseases

Disease	Material	Via	Model	Main Findings	Reference
Alzheimer's Disease (AD)	RVG targeted exosomes	Injection	Mice	exosome-mediated siRNA delivery was demonstrated by mRNA and protein knockdown of BACE1	[16]
	Human Mesenchymal stem cells EVs	NA	*In vitro*	Lysis of beta amyloid peptide	[35]
	Neuroblastoma derived exosomes	Injection	Mice	Enhanced beta amyloid clearance	[36]
	Mesenchymal-stem-cells derived EVs with tyrosine phosphatase-2	Intravenous Injection	Mice	Relieve oxidative stress, decreased Aβ deposition, and improved cognitive function	[46]
	Mannose-coated EVs	Injection	Mice	Microglial targeting enhanced Aβ uptake and lysosomal clearance, improving cognition	[47]
Parkinson's Disease (PD)	RVG-exosomes loaded with α-Syn siRNA	Intravenous Injection	Mice	Downregulation of alpha synuclein	[41]
	Bone marrow-derived macrophages	Intranasal injection	Mice	Significant neuroprotective effects of exosome-based catalase formulations	[42]
	Dopamine-loaded exosomes	Intravenous	Mice	Improve distribution of dopamine & disease phenotype without toxicity	[43]
	Implanting engineered HEK-293T exosomes	Intranasal injection	Mice	Delivering catalase mRNA via designer exosomes with reduced neurotoxicity and neuroinflammation	[44]
	EVs derived stem cells	Intranasal injection	Rat	Enhanced rat gait parameters and increased tyrosine hydroxylase expression in the substantia nigra	[45]
	EVs derived human teeth stem cells	Intranasal	Rat	Reversed memory impairments, with effects lasting up to 6 days post-treatment	[48]
	miR-124-3p-loaded EVs	Intracerebroventricularly administered	Mice	Protected dopaminergic neurons and reversing motor deficits	[49]

Multiple sclerosis (MS)	IFNγ-stimulated dendritic cell exosomes	Intranasal injection	Rat	Increased CNS myelination *in vivo*	[50]
	EVs derived from human adipose tissue mesenchymal stem cells	Intravenous	Theiler's murine encephalomyelitis virus (TMEV)-induced demyelinating disease (TMEV-IDD)	Decreasing neuroinflammation. Reduction in brain atrophy. Improvement in motor deficits	[51]
	Human MSCs EVs	Intravenous	EAE mouse	IFNγ stimulation alleviated EAE severity, reducing clinical scores, demyelination, and neuroinflammation while increasing spinal cord Tregs.	[52]
	BM mesenchymal stem cells-derived exosomes	Intravenous	EAE rat	Attenuation of inflammation and demyelination of the CNS in rats by regulating the polarisation of microglia	[53]
	Placental Mesenchymal Stem Cell-Derived EVs	Intravenous	EAE murine	Enhancement of Myelin Regeneration, reduced DNA damage in oligodendroglia populations, and increased myelination within the spinal cord of treated mice	[54]
	Engineered exosomes carrying montelukast and PDGFA	Intranasal administration	Cuprizone mouse model	Engineered exosomes mitigate demyelination, promote differentiation of oligodendrocyte progenitor cells, enhance remyelination, and reduce axonal injury	[55]
	Engineered exosomes loaded with brain-derived neurotrophic factor	Intranasal administration	Mice	Improvement in demyelination	[56]
	Resveratrol-loaded exosomes	Intranasal administration	EAE mice	reduced CNS and peripheral inflammation	[57]
	Transferrin-loaded EVs isolated from mouse plasma	Intranasal administration	Mice	Promoting oligodendrocyte progenitor cells differentiation, and accelerating remyelination	[58]
Glioblastoma	T7- peptide decorated exosomes	Intravenous	Rat	T7-exosomes with Antisense miRNA oligonucleotides against miR-21 reduced the miR-21 level in the glioblastoma and induced the expression of PDCD4 and PTEN in tumours, resulting in reduction of tumour sizes.	[59]
	Oligopeptide modified exosome loaded with doxorubicin (Pep2-Exos-DOX)	Intravenous	Rat	Pep2-Exos-DOX possessed significant anti-cancer activity against glioblastoma	[60]
	Yeast cytosine deaminase: uracilphosphoribosyl transferase (yCD::UPRT-MSC)-exosomes	Intraperitoneally, Subcutaneously, Intranasally	Rat	yCD::UPRT-MSC exosomes, in the presence of the prodrug 5-fluorocytosine, inhibited the growth of rat C6 glioblastoma cells	[61]

Given their dual role as mediators of pathology and potential therapeutic carriers, exosomes have got significant interest in PD research. Innovative studies have demonstrated that engineered exosomes can deliver therapeutic cargoes with high specificity and minimal immunogenicity. For instance, RVG-tagged exosomes loaded with α-synuclein-targeting siRNA effectively downregulate its expression [41]. Similarly, bone marrow-derived macrophage exosomes delivering catalase confer neuroprotection [42], while dopamine-loaded exosomes enhance dopamine distribution and improve motor deficits without inducing toxicity [43]. Additionally, exosomes derived from HEK-293T cells carrying catalase mRNA and stem cell-derived EVs have shown promise in reducing neuroinflammation and improving motor function in preclinical models [44,45] (see Table 6.3).

6.2.6.4 Exosomes as Therapeutic Tools in Multiple Sclerosis

MS is a chronic, immune-mediated demyelinating disorder of the CNS, primarily affecting young adults, with a higher prevalence among females. The disease is characterised by multifocal demyelinating lesions, chronic neuroinflammation, astrogliosis, and progressive axonal degeneration. Although the precise aetiology remains elusive, the pathogenesis is driven by the activation of autoreactive T and B lymphocytes that target myelin antigens. This immune response leads to sustained inflammation, disruption of the BBB, and subsequent neurodegeneration [62–65].

While current immunomodulatory therapies aim to suppress inflammation and reduce relapse rates, they offer limited efficacy in promoting remyelination or neuronal repair and are often associated with adverse effects. Consequently, there is growing interest in novel, cell-free therapeutic strategies that can modulate immune responses and support CNS regeneration. Recent studies have highlighted the dual role of EVs in the pathophysiology of MS. CNS-resident cells such as astrocytes, neurons, oligodendrocytes, microglia, and mast cells secrete EVs that can either exacerbate or mitigate disease processes. Under pathological conditions, exosomes can carry pro-inflammatory mediators, including microRNAs like miR-34a, which contribute to neuroinflammation and neuronal injury [66].

Astrocyte-derived EVs have been shown to propagate neurotoxicity by transferring pathogenic molecules to neurons, while endothelial-derived EVs facilitate BBB disruption, enhancing immune cell infiltration into the CNS. Additionally, B-cell-derived extracellular EVs have been shown to exert cytotoxic effects on oligodendrocytes, potentially contributing to demyelination in MS. *In vitro* studies demonstrated that EV-enriched fractions from B cells of MS patients induce oligodendrocyte death, whereas those from healthy controls do not, suggesting a pathogenic role for altered B cell secretome in MS [67]. On the contrary, EVs derived from mesenchymal stem cells have been reported to enhance oligodendrocyte precursor differentiation and promote remyelination, highlighting the multifaceted roles of EVs as both drivers of pathology and mediators of repair in MS [52].

Preclinical models have demonstrated the therapeutic promise of mesenchymal stem cell-derived EVs (MSC-EVs) in modulating immune responses and promoting remyelination. In experimental autoimmune encephalomyelitis (EAE) models, intravenous administration of MSC-EVs has been shown to downregulate pro-inflammatory cytokines (e.g. IL-1β, TNF-α), upregulate anti-inflammatory cytokines such as IL-10, reduce demyelination, and enhance oligodendrocyte maturation [66]. Intranasal delivery of EVs has also been associated with increased regulatory T cell populations, which are essential for maintaining immune tolerance [68]. Further supporting this therapeutic potential, Pusic et al. [72] demonstrated that exosomes derived from IFNγ-stimulated dendritic cells can promote CNS myelination *in vivo*. Similarly, Riazifar et al. [52] reported that IFNγ stimulation of EVs reduces EAE severity and neuroinflammation. EVs from other sources, such as periodontal ligament stem cells and oligodendrocytes, have also shown efficacy in restoring immune tolerance and reducing disease severity in EAE models [69]. Collectively, these findings suggest that EV-based therapies represent a promising, cell-free approach for MS treatment, capable of both reducing neuroinflammation and promoting CNS repair [58]. A summary of the therapeutic potential of various EV sources and delivery strategies in MS models is provided in Table 6.3.

6.2.6.5 Exosomes in Glioblastomas

Gliomas are a heterogeneous group of primary brain tumours composed of malignant glial cells alongside non-neoplastic elements such as neurons, astrocytes, microglia, macrophages, and vascular cells. This cellular diversity contributes to a complex tumour microenvironment, where dynamic interactions between tumour and stromal cells drive disease progression and therapeutic resistance. In gliomas, exosomal cargo and function are often reprogrammed to support tumour growth, invasion, and immune evasion [70].

These vesicles facilitate the transfer of oncogenic signals, contributing to the malignancy's adaptability and aggressiveness. GBM, the most aggressive glioma subtype, is particularly challenging due to its high heterogeneity, resistance to standard therapies, and frequent recurrence. These factors highlight the need for novel therapeutic strategies. Exosomes have emerged as promising tools in this context, both as diagnostic biomarkers via liquid biopsy and as targeted delivery vehicles for therapeutic agents.

Pioneering work by Alvarez-Erviti et al. [16] demonstrated that exosomes engineered with RVG peptides could deliver siRNA to brain cells following systemic administration, establishing a foundation for brain-targeted exosome therapies. Building on this, recent preclinical studies have explored diverse engineering strategies for GBM treatment [71,72]. As outlined in Table 6.3, several innovative strategies have demonstrated the therapeutic potential of engineered exosomes in glioblastoma models. For instance, T7 peptide-decorated exosomes delivering antisense oligonucleotides targeting miR-21 have been shown to significantly reduce tumour size [59]. Similarly, exosomes modified with oligopeptides and loaded with doxorubicin have exhibited potent anti-tumour effects [60, 71]. Another promising approach involves the use of mesenchymal stem cell (MSC)–derived exosomes obtained from cells genetically engineered to express the yeast *cytosine deaminase–uracil phosphoribosyl transferase* (yCD::UPRT) fusion gene. These exosomes can transfer the mRNA of the suicide gene to glioblastoma cells, enabling the intracellular conversion of the non-toxic prodrug 5-fluorocytosine (5-FC) into the active chemotherapeutic agent 5-fluorouracil (5-FU). This strategy allows for the localised activation of chemotherapy within the tumour microenvironment, thereby enhancing antitumour efficacy while potentially minimising systemic toxicity, as demonstrated in preclinical models [61]. These findings underscore the therapeutic versatility of engineered exosomes, offering innovative solutions to overcome key challenges in GBM treatment, including BBB penetration, tumour specificity, and drug resistance. Further details are provided in Table 6.3.

Table 6.3 presents various studies demonstrating the application of EVs, including exosomes, for the treatment of AD, PD, MS, and Glioblastoma. Details include the type of EVs or exosome formulation, delivery method, experimental model (*in vitro* or *in vivo*), and main therapeutic findings. Notable outcomes include targeted gene silencing, enhanced clearance of pathological proteins, neuroprotection, reduced neuroinflammation, improved myelination, and tumour suppression (References correspond to the original studies detailing each therapeutic approach).

6.3 CONCLUSION AND FUTURE PROSPECT

Exosomes represent a promising frontier in drug delivery for neurological diseases, offering key advantages such as low immunogenicity, efficient cellular uptake, and an inherent ability to cross the BBB. Their natural composition and biological functions enable targeted delivery of therapeutic cargo, positioning them ahead of many synthetic nanocarriers. However, significant translational hurdles remain, including scalable manufacturing, standardised isolation and characterisation protocols, and limited brain accumulation following systemic administration. To address these challenges, emerging engineering strategies, such as surface modification and genetic tailoring, are being explored to enhance targeting specificity, prolong circulation, and improve therapeutic efficacy.

In conclusion, while obstacles persist, rapid advances in exosome research and bioengineering continue to unlock new possibilities. Future success will depend on robust manufacturing pipelines, harmonised characterisation standards, and comprehensive safety assessments to support clinical translation. With sustained innovation and rigorous validation, exosomes hold immense potential as a transformative platform for precision medicine, enabling targeted and personalised therapies for diverse neurological disorders.

REFERENCES

1. D.K. Jeppesen, Q. Zhang, J.L. Franklin, R.J. Coffey, Extracellular vesicles and nanoparticles: Emerging complexities, Trends Cell Biol 33 (2023) 667–681. https://doi.org/10.1016/j.tcb.2023.01.002.
2. M. Colombo, G. Raposo, C. Théry, Biogenesis, secretion, and intercellular interactions of exosomes and other extracellular vesicles, Annu Rev Cell Dev Biol 30 (2014) 255–289. https://doi.org/10.1146/annurev-cellbio-101512-122326.
3. G. Raposo, P.D. Stahl, Extracellular vesicles – on the cusp of a new language in the biological sciences, Extracell Vesicles Circ Nucl Acids 4 (2023) 240–254. https://doi.org/10.20517/evcna.2023.18.
4. J. Rädler, D. Gupta, A. Zickler, S.E. Andaloussi, Exploiting the biogenesis of extracellular vesicles for bioengineering and therapeutic cargo loading, Mol Ther 31 (2023) 1231–1250. https://doi.org/10.1016/j.ymthe.2023.02.013.
5. A.F. Hill, Extracellular vesicles and neurodegenerative diseases, J Neurosci 39 (2019) 9269–9273. https://doi.org/10.1523/jneurosci.0147-18.2019.
6. J.A. Welsh, D.C.I. Goberdhan, L. O'Driscoll, E.I. Buzas, C. Blenkiron, B. Bussolati, H. Cai, D. Di Vizio, T.A.P. Driedonks, U. Erdbrügger, J.M. Falcon-Perez, Q.L. Fu, A.F. Hill, M. Lenassi, S.K. Lim, M.G. Mahoney, S. Mohanty, A. Möller, R. Nieuwland, T. Ochiya, S. Sahoo, A.C. Torrecilhas, L. Zheng, A. Zijlstra, S. Abuelreich, R. Bagabas, P. Bergese, E.M. Bridges, M. Brucale, D. Burger, R.P. Carney, E. Cocucci, R. Crescitelli, E. Hanser, A.L. Harris, N.J. Haughey, A. Hendrix, A.R. Ivanov, T. Jovanovic-Talisman, N.A. Kruh-Garcia, V. Ku'ulei-Lyn Faustino, D. Kyburz, C. Lässer, K.M. Lennon, J. Lötvall, A.L. Maddox, E.S. Martens-Uzunova, R.R. Mizenko, L.A. Newman, A. Ridolfi, E. Rohde, T. Rojalin, A. Rowland, A. Saftics, U.S. Sandau, J.A. Saugstad, F. Shekari, S. Swift, D. Ter-Ovanesyan, J.P. Tosar, Z. Useckaite, F. Valle, Z. Varga, E. van der Pol, M.J.C. van Herwijnen, M.H.M. Wauben, A.M. Wehman, S. Williams, A. Zendrini, A.J. Zimmerman, C. Théry, K.W. Witwer, Minimal information for studies of extracellular vesicles (MISEV2023): From basic to advanced approaches, J Extracell Vesicles 13 (2024) e12404. https://doi.org/10.1002/jev2.12404.
7. X.T.T. Dang, J.M. Kavishka, D.X. Zhang, M. Pirisinu, M.T.N. Le, Extracellular vesicles as an efficient and versatile system for drug delivery, Cells 9 (2020) 2191. https://doi.org/10.3390/cells9102191.
8. G. van Niel, D.R.F. Carter, A. Clayton, D.W. Lambert, G. Raposo, P. Vader, Challenges and directions in studying cell-cell communication by extracellular vesicles, Nat Rev Mol Cell Biol 23 (2022) 369–382. https://doi.org/10.1038/s41580-022-00460-3.
9. L. Blanc, M. Vidal, New insights into the function of rab GTPases in the context of exosomal secretion, Small GTPases 9 (2018) 95–106. https://doi.org/10.1080/21541248.2016.1264352.
10. S. Gurung, D. Perocheau, L. Touramanidou, J. Baruteau, The exosome journey: From biogenesis to uptake and intracellular signalling, Cell Commun Signal 19 (2021) 47. https://doi.org/10.1186/s12964-021-00730-1.
11. H. Zeng, S. Guo, X. Ren, Z. Wu, S. Liu, X. Yao, Current strategies for exosome cargo loading and targeting delivery, Cells 12 (2023) 1416. https://doi.org/10.3390/cells12101416.
12. D. Yu, Y. Li, M. Wang, J. Gu, W. Xu, H. Cai, X. Fang, X. Zhang, Exosomes as a new frontier of cancer liquid biopsy, Mol Cancer 21 (2022) 56. https://doi.org/10.1186/s12943-022-01509-9.
13. C. Caruso Bavisotto, F. Scalia, A. Marino Gammazza, D. Carlisi, F. Bucchieri, E. Conway de Macario, A.J.L. Macario, F. Cappello, C. Campanella, Extracellular vesicle-mediated Cell–Cell communication in the nervous system: Focus on neurological diseases, Int J Mol Sci 20 (2019) 434. https://doi.org/10.3390/ijms20020434.
14. N. Tominaga, N. Kosaka, M. Ono, T. Katsuda, Y. Yoshioka, K. Tamura, J. Lötvall, H. Nakagama, T. Ochiya, Brain metastatic cancer cells release microRNA-181c-containing extracellular vesicles capable of destructing blood–brain barrier, Nat Commun 6 (2015) 6716. https://doi.org/10.1038/ncomms7716.
15. K. Ridder, S. Keller, M. Dams, A.-K. Rupp, J. Schlaudraff, D. Del Turco, J. Starmann, J. Macas, D. Karpova, K. Devraj, C. Depboylu, B. Landfried, B. Arnold, K.H. Plate, G. Höglinger, H. Sültmann, P. Altevogt, S. Momma, Extracellular vesicle-mediated transfer of genetic information between the hematopoietic system and the brain in response to inflammation, PLoS Biol 12 (2014) e1001874. https://doi.org/10.1371/journal.pbio.1001874.

16. L. Alvarez-Erviti, Y. Seow, H. Yin, C. Betts, S. Lakhal, M.J.A. Wood, Delivery of siRNA to the mouse brain by systemic injection of targeted exosomes, Nat Biotechnol 29 (2011) 341–345. https://doi.org/10.1038/nbt.1807.
17. J. Skog, T. Würdinger, S. van Rijn, D.H. Meijer, L. Gainche, W.T. Curry, B.S. Carter, A.M. Krichevsky, X.O. Breakefield, Glioblastoma microvesicles transport RNA and proteins that promote tumour growth and provide diagnostic biomarkers, Nat Cell Biol 10 (2008) 1470–1476. https://doi.org/10.1038/ncb1800.
18. E. Willms, C. Cabañas, I. Mäger, M.J.A. Wood, P. Vader, Extracellular vesicle heterogeneity: Subpopulations, isolation techniques, and diverse functions in cancer progression, Front Immunol 9 (2018) 738. https://doi.org/10.3389/fimmu.2018.00738.
19. M. Farahani, C. Rubbi, L. Liu, J.R. Slupsky, N. Kalakonda, CLL exosomes modulate the transcriptome and behaviour of recipient stromal cells and are selectively enriched in miR-202-3p, PLoS One 10 (2015) e0141429. https://doi.org/10.1371/journal.pone.0141429.
20. S.L. Reed, A. Escayg, Extracellular vesicles in the treatment of neurological disorders, Neurobiol Dis 157 (2021) 105445. https://doi.org/10.1016/j.nbd.2021.105445.
21. O.M. Elsharkasy, J.Z. Nordin, D.W. Hagey, O.G. de Jong, R.M. Schiffelers, S.E. Andaloussi, P. Vader, Extracellular vesicles as drug delivery systems: Why and how? Adv Drug Deliv Rev 159 (2020) 332–343. https://doi.org/10.1016/j.addr.2020.04.004.
22. P. Shirmast, M.A. Shahri, A. Brent, A. Idris, N.A.J. McMillan, Delivering therapeutic RNA into the brain using extracellular vesicles, Mol Ther Nucleic Acids 35 (2024) 102373. https://doi.org/10.1016/j.omtn.2024.102373.
23. S. Busatto, G. Morad, P. Guo, M.A. Moses, The role of extracellular vesicles in the physiological and pathological regulation of the blood-brain barrier, FASEB Bioadv 3 (2021) 665–675. https://doi.org/10.1096/fba.2021-00045.
24. J. Chen, C. Tian, X. Xiong, Y. Yang, J. Zhang, Extracellular vesicles: new horizons in neurodegeneration, EBioMedicine 113 (2025) 105605. https://doi.org/10.1016/j.ebiom.2025.105605.
25. T. Premchandani, A. Tatode, J. Taksande, M. Umekar, M. Qutub, U.M. Hussain, P. Singanwad, Engineered exosomes as smart drug carriers: Overcoming biological barriers in CNS and cancer therapy, Drugs and Drug Candidates 4 (2025) 19. https://doi.org/10.3390/ddc4020019.
26. C. Reitz, C. Brayne, R. Mayeux, Epidemiology of Alzheimer disease, Nat Rev Neurol 7 (2011) 137–152. https://doi.org/10.1038/nrneurol.2011.2.
27. C.R. Jack, H.J. Wiste, T.M. Therneau, S.D. Weigand, D.S. Knopman, M.M. Mielke, V.J. Lowe, P. Vemuri, M.M. Machulda, C.G. Schwarz, J.L. Gunter, M.L. Senjem, J. Graff-Radford, D.T. Jones, R.O. Roberts, W.A. Rocca, R.C. Petersen, Associations of Amyloid, Tau, and neurodegeneration biomarker profiles with rates of memory decline among individuals without dementia, JAMA 321 (2019) 2316. https://doi.org/10.1001/jama.2019.7437.
28. E. Reza-Zaldivar, M. Hernández-Sapiéns, Y. Gutiérrez-Mercado, S. Sandoval-Ávila, U. Gomez-Pinedo, A. Márquez-Aguirre, E. Vázquez-Méndez, E. Padilla-Camberos, A. Canales-Aguirre, Mesenchymal stem cell-derived exosomes promote neurogenesis and cognitive function recovery in a mouse model of Alzheimer's disease, Neural Regen Res 14 (2019) 1626. https://doi.org/10.4103/1673-5374.255978.
29. M. Guo, Z. Yin, F. Chen, P. Lei, Mesenchymal stem cell-derived exosome: A promising alternative in the therapy of Alzheimer's disease, Alzheimers Res Ther 12 (2020) 109. https://doi.org/10.1186/s13195-020-00670-x
30. Y. Ge, J. Wu, L. Zhang, N. Huang, Y. Luo, A new strategy for the regulation of neuroinflammation: Exosomes derived from mesenchymal stem cells, Cell Mol Neurobiol 44 (2024) 24. https://doi.org/10.1007/s10571-024-01460-x.
31. J. Pan, R. He, Q. Huo, Y. Shi, L. Zhao, Brain microvascular endothelial cell derived exosomes potently ameliorate cognitive dysfunction by enhancing the clearance of Aβ through up-regulation of P-gp in mouse model of AD, Neurochem Res 45 (2020) 2161–2172. https://doi.org/10.1007/s11064-020-03076-1.
32. B. Li, J. Liu, G. Gu, X. Han, Q. Zhang, W. Zhang, Impact of neural stem cell-derived extracellular vesicles on mitochondrial dysfunction, sirtuin 1 level, and synaptic deficits in Alzheimer's disease, J Neurochem 154 (2020) 502–518. https://doi.org/10.1111/jnc.15001.
33. K. Yuyama, Y. Igarashi, Exosomes as carriers of Alzheimer's Amyloid-ß, Front Neurosci 11 (2017). https://doi.org/10.3389/fnins.2017.00229.
34. S. Hornung, S. Dutta, G. Bitan, CNS-derived blood exosomes as a promising source of biomarkers: Opportunities and challenges, Front Mol Neurosci 13 (2020). https://doi.org/10.3389/fnmol.2020.00038.
35. T. Katsuda, R. Tsuchiya, N. Kosaka, Y. Yoshioka, K. Takagaki, K. Oki, F. Takeshita, Y. Sakai, M. Kuroda, T. Ochiya, Human adipose tissue-derived mesenchymal stem cells secrete functional neprilysin-bound exosomes, Sci Rep 3 (2013) 1197. https://doi.org/10.1038/srep01197.

36. K. Yuyama, H. Sun, S. Sakai, S. Mitsutake, M. Okada, H. Tahara, J. Furukawa, N. Fujitani, Y. Shinohara, Y. Igarashi, Decreased Amyloid-β pathologies by intracerebral loading of glycosphingolipid-enriched exosomes in Alzheimer model mice, J Biol Chem 289 (2014) 24488–24498. https://doi.org/10.1074/jbc.M114.577213.
37. J. Jankovic, Parkinson's disease: Clinical features and diagnosis, J Neurol Neurosurg Psychiatry 79 (2008) 368–376. https://doi.org/10.1136/jnnp.2007.131045.
38. L.V. Kalia, A.E. Lang, Parkinson's disease, The Lancet 386 (2015) 896–912. https://doi.org/10.1016/S0140-6736(14)61393-3.
39. E.M. Rocha, B. De Miranda, L.H. Sanders, Alpha-synuclein: Pathology, mitochondrial dysfunction and neuroinflammation in Parkinson's disease, Neurobiol Dis 109 (2018) 249–257. https://doi.org/10.1016/j.nbd.2017.04.004.
40. S. Sarkar, S. Patranabis, Emerging role of extracellular vesicles in intercellular communication in the brain: Implications for neurodegenerative diseases and therapeutics, Cell Biochem Biophys (2024). https://doi.org/10.1007/s12013-024-01221-z.
41. J.M. Cooper, P.B.O. Wiklander, J.Z. Nordin, R. Al-Shawi, M.J. Wood, M. Vithlani, A.H.V. Schapira, J.P. Simons, S. El-Andaloussi, L. Alvarez-Erviti, Systemic exosomal siRNA delivery reduced alpha-synuclein aggregates in brains of transgenic mice, Movement Disorders 29 (2014) 1476–1485. https://doi.org/10.1002/mds.25978.
42. M.J. Haney, N.L. Klyachko, Y. Zhao, R. Gupta, E.G. Plotnikova, Z. He, T. Patel, A. Piroyan, M. Sokolsky, A.V. Kabanov, E.V. Batrakova, Exosomes as drug delivery vehicles for Parkinson's disease therapy, J Controlled Release 207 (2015) 18–30. https://doi.org/10.1016/j.jconrel.2015.03.033.
43. M. Qu, Q. Lin, L. Huang, Y. Fu, L. Wang, S. He, Y. Fu, S. Yang, Z. Zhang, L. Zhang, X. Sun, Dopamine-loaded blood exosomes targeted to brain for better treatment of Parkinson's disease, J Controlled Release 287 (2018) 156–166. https://doi.org/10.1016/j.jconrel.2018.08.035.
44. R. Kojima, D. Bojar, G. Rizzi, G.C. Hamri, M.D. El-Baba, P. Saxena, S. Ausländer, K.R. Tan, M. Fussenegger, Designer exosomes produced by implanted cells intracerebrally deliver therapeutic cargo for Parkinson's disease treatment, Nat Commun 9 (2018) 1305. https://doi.org/10.1038/s41467-018-03733-8.
45. K. Narbute, V. Piļipenko, J. Pupure, Z. Dzirkale, U. Jonavičė, V. Tunaitis, K. Kriaučiūnaitė, A. Jarmalavičiūtė, B. Jansone, V. Kluša, A. Pivoriūnas, Intranasal administration of extracellular vesicles derived from human teeth stem cells improves motor symptoms and normalizes tyrosine hydroxylase expression in the substantia Nigra and Striatum of the 6-hydroxydopamine-treated rats, Stem Cells Transl Med 8 (2019) 490–499. https://doi.org/10.1002/sctm.18-0162.
46. F. Xu, Y. Wu, Q. Yang, Y. Cheng, J. Xu, Y. Zhang, H. Dai, B. Wang, Q. Ma, Y. Chen, F. Lin, C. Wang, Engineered extracellular vesicles with SHP2 high expression promote mitophagy for Alzheimer's disease treatment, Advanced Materials 34 (2022). https://doi.org/10.1002/adma.202207107.
47. Y. Hao, C. Su, X. Liu, H. Sui, Y. Shi, L. Zhao, Bioengineered microglia-targeted exosomes facilitate Aβ clearance via enhancing activity of microglial lysosome for promoting cognitive recovery in Alzheimer's disease, Biomaterials Advances 136 (2022) 212770. https://doi.org/10.1016/j.bioadv.2022.212770.
48. K. Narbute, V. Pilipenko, J. Pupure, T. Klinovičs, J. Auders, U. Jonavičė, K. Kriaučiūnaitė, A. Pivoriūnas, V. Kluša, Time-dependent memory and gait improvement by intranasally-administered extracellular vesicles in Parkinson's disease model rats, Cell Mol Neurobiol 41 (2021) 605–613. https://doi.org/10.1007/s10571-020-00865-8.
49. M. Esteves, R. Abreu, H. Fernandes, C. Serra-Almeida, A.T. Martins, M. Barão, A.C. Cristóvão, C. Saraiva, R. Ferreira, L. Ferreira, L. Bernardino, MicroRNA-124-3p-enriched small extracellular vesicles as a therapeutic approach for Parkinson's disease, Molecular Therapy 30 (2022) 3176–3192. https://doi.org/10.1016/j.ymthe.2022.06.003.
50. A.D. Pusic, K.M. Pusic, B.L.L. Clayton, R.P. Kraig, IFNγ-stimulated dendritic cell exosomes as a potential therapeutic for remyelination, J Neuroimmunol 266 (2014) 12–23. https://doi.org/10.1016/j.jneuroim.2013.10.014.
51. F. Laso-García, J. Ramos-Cejudo, F.J. Carrillo-Salinas, L. Otero-Ortega, A. Feliú, M. Gómez-de Frutos, M. Mecha, E. Díez-Tejedor, C. Guaza, M. Gutiérrez-Fernández, Therapeutic potential of extracellular vesicles derived from human mesenchymal stem cells in a model of progressive multiple sclerosis, PLoS One 13 (2018) e0202590. https://doi.org/10.1371/journal.pone.0202590.
52. M. Riazifar, M.R. Mohammadi, E.J. Pone, A. Yeri, C. Lässer, A.I. Segaliny, L.L. McIntyre, G.V. Shelke, E. Hutchins, A. Hamamoto, E.N. Calle, R. Crescitelli, W. Liao, V. Pham, Y. Yin, J. Jayaraman, J.R.T.

Lakey, C.M. Walsh, K. Van Keuren-Jensen, J. Lotvall, W. Zhao, Stem Cell-Derived Exosomes as Nanotherapeutics for Autoimmune and Neurodegenerative Disorders, ACS Nano 13 (2019) 6670–6688. https://doi.org/10.1021/acsnano.9b01004.
53. Z. Li, F. Liu, X. He, X. Yang, F. Shan, J. Feng, Exosomes derived from mesenchymal stem cells attenuate inflammation and demyelination of the central nervous system in EAE rats by regulating the polarization of microglia, Int Immunopharmacol 67 (2019) 268–280. https://doi.org/10.1016/j.intimp.2018.12.001.
54. K. Clark, S. Zhang, S. Barthe, P. Kumar, C. Pivetti, N. Kreutzberg, C. Reed, Y. Wang, Z. Paxton, D. Farmer, F. Guo, A. Wang, Placental mesenchymal stem cell-derived extracellular vesicles promote myelin regeneration in an animal model of multiple sclerosis, Cells 8 (2019) 1497. https://doi.org/10.3390/cells8121497.
55. Y. Xiao, Y. Zhang, Y.-H. Gao, Z.-H. Zhao, J. He, R. Gao, Y.-X. Guo, L.-B. Wang, X. Li, A targeted extracellular vesicles loaded with montelukast in the treatment of demyelinating diseases, Biochem Biophys Res Commun 594 (2022) 31–37. https://doi.org/10.1016/j.bbrc.2022.01.051.
56. Y. Zhai, Q. Wang, Z. Zhu, Y. Hao, F. Han, J. Hong, W. Zheng, S. Ma, L. Yang, G. Cheng, High-efficiency brain-targeted intranasal delivery of BDNF mediated by engineered exosomes to promote remyelination, Biomater Sci 10 (2022) 5707–5718. https://doi.org/10.1039/D2BM00518B.
57. X. Zheng, K. Sun, Y. Liu, X. Yin, H. Zhu, F. Yu, W. Zhao, Resveratrol-loaded macrophage exosomes alleviate multiple sclerosis through targeting microglia, J Controlled Release 353 (2023) 675–684. https://doi.org/10.1016/j.jconrel.2022.12.026.
58. V. Mattera, F. Occhiuzzi, J. Correale, J.M. Pasquini, Remyelinating effect driven by transferrin-loaded extracellular vesicles, Glia 72 (2024) 338–361. https://doi.org/10.1002/glia.24478.
59. G. Kim, M. Kim, Y. Lee, J.W. Byun, D.W. Hwang, M. Lee, Systemic delivery of microRNA-21 antisense oligonucleotides to the brain using T7-peptide decorated exosomes, J Controlled Release 317 (2020) 273–281. https://doi.org/10.1016/j.jconrel.2019.11.009.
60. Y. Wang, Y. Huo, C. Zhao, H. Liu, Y. Shao, C. Zhu, L. An, X. Chen, Z. Chen, Engineered exosomes with enhanced stability and delivery efficiency for glioblastoma therapy, J Controlled Release 368 (2024) 170–183. https://doi.org/10.1016/j.jconrel.2024.02.015.
61. M. Tibensky, J. Jakubechova, U. Altanerova, A. Pastorakova, B. Rychly, L. Baciak, B. Mravec, C. Altaner, Gene-directed enzyme/prodrug therapy of rat brain tumor mediated by human mesenchymal stem cell suicide gene extracellular vesicles in vitro and in vivo, Cancers (Basel) 14 (2022) 735. https://doi.org/10.3390/cancers14030735.
62. J. Correale, M.I. Gaitán, M.C. Ysrraelit, M.P. Fiol, Progressive multiple sclerosis: from pathogenic mechanisms to treatment, Brain (2016) aww258. https://doi.org/10.1093/brain/aww258.
63. R. Martin, M. Sospedra, M. Rosito, B. Engelhardt, Current multiple sclerosis treatments have improved our understanding of MS autoimmune pathogenesis, Eur J Immunol 46 (2016) 2078–2090. https://doi.org/10.1002/eji.201646485.
64. A.L. Greenfield, S.L. Hauser, B-cell therapy for multiple sclerosis: Entering an era, Ann Neurol 83 (2018) 13–26. https://doi.org/10.1002/ana.25119.
65. N. Tang, Exosomes in multiple sclerosis and Alzheimer's disease – Adversary and ally, Biomed J 47 (2024) 100665. https://doi.org/10.1016/j.bj.2023.100665.
66. S. Li, Z. Lei, T. Sun, The role of microRNAs in neurodegenerative diseases: a review, Cell Biol Toxicol 39 (2023) 53–83. https://doi.org/10.1007/s10565-022-09761-x.
67. J.A. Benjamins, L. Nedelkoska, H. Touil, P.M. Stemmer, N.J. Carruthers, B.P. Jena, A.R. Naik, A. Bar-Or, R.P. Lisak, Exosome-enriched fractions from MS B cells induce oligodendrocyte death, Neurol Neuroimmunol Neuroinflamm 6 (2019). https://doi.org/10.1212/NXI.0000000000000550.
68. A. Fathollahi, S.M. Hashemi, M. Haji Molla Hoseini, S. Tavakoli, E. Farahani, F. Yeganeh, Intranasal administration of small extracellular vesicles derived from mesenchymal stem cells ameliorated the experimental autoimmune encephalomyelitis, Int Immunopharmacol 90 (2021) 107207. https://doi.org/10.1016/j.intimp.2020.107207.
69. G. Casella, J. Rasouli, A. Boehm, W. Zhang, D. Xiao, L.L.W. Ishikawa, R. Thome, X. Li, D. Hwang, P. Porazzi, S. Molugu, H.-Y. Tang, G.-X. Zhang, B. Ciric, A. Rostami, Oligodendrocyte-derived extracellular vesicles as antigen-specific therapy for autoimmune neuroinflammation in mice, Sci Transl Med 12 (2020). https://doi.org/10.1126/scitranslmed.aba0599.
70. A. Balakrishnan, S. Roy, T. Fleming, H.S. Leong, C. Schuurmans, The emerging role of extracellular vesicles in the glioma microenvironment: Biogenesis and clinical relevance, Cancers (Basel) 12 (2020). https://doi.org/10.3390/cancers12071964.

71. Y. Jiao, L. Yang, R. Wang, G. Song, J. Fu, J. Wang, N. Gao, H. Wang, Drug delivery across the blood–brain barrier: A new strategy for the treatment of neurological diseases, Pharmaceutics 16 (2024) 1611. https://doi.org/10.3390/pharmaceutics16121611.
72. S. Singh, D. Dey, D. Barik, I. Mohapatra, S. Kim, M. Sharma, S. Prasad, P. Wang, A. Singh, G. Singh, Glioblastoma at the crossroads: Current understanding and future therapeutic horizons, Signal Transduct Target Ther 10 (2025) 213. https://doi.org/10.1038/s41392-025-02299-4.

7 Targeting the Brain with Monoclonal Antibodies

Shreyansh Chauhan, Jai Naik, and Mayur Patel

7.1 INTRODUCTION

Monoclonal antibodies (mAbs) have emerged as promising therapeutic agents due to their exceptionally high affinity, specificity, and ability to target pathological molecules with minimal off-target effects. These biologics have shown substantial therapeutic efficacy across a range of diseases, such as neurodegenerative conditions, cancers, and autoimmune disorders. Despite their broad application in treating different pathological conditions, their utility in treating central nervous system (CNS) disorders remains significantly restricted. The restrictive nature of the blood-brain barrier (BBB) prevents maximum macromolecules, such as mAbs, from efficiently entering the brain parenchyma. This biological barrier closely regulates CNS homeostasis, allowing only select molecules to cross while excluding therapeutic agents, thereby limiting their efficacy at the target site. Even when mAbs are administered directly into the cerebrospinal fluid (CSF), their rapid clearance through CSF outflow pathways considerably decreases their half-life within the CNS, limiting their therapeutic exposure (1,2). Despite these barriers, the therapeutic potential of mAbs in CNS diseases is evident from recent Food and Drug Administration (FDA)-approved treatments. For instance, Aduhelm® (aducanumab) and Leqembi® (lecanemab) have demonstrated efficacy in lowering amyloid-beta plaques in Alzheimer's disease (AD), although achieving effective brain concentrations required high systemic doses, which have been associated with adverse effects, including cerebral amyloid angiopathy-related edema and BBB disruption (2–4). In neuro-oncology, the application of mAbs for primary and metastatic brain tumors has been explored due to the pressing need for effective therapies for CNS malignancies. The unique immunosuppressive microenvironment of CNS tumors, coupled with the physiological and physical challenges posed by the BBB, complicates therapeutic interventions. Nevertheless, advancements in antibody engineering, such as the development of fully human and humanized antibodies, have provided effective tools for against these malignancies, paving the way for improved outcomes in this field (5–8). The financial strain of mAb therapies emphasizes the importance of developing effective delivery strategies. For instance, the high yearly expense of therapies like Aduhelm® (aducanumab), about $56,000 per patient, is a direct consequence of the high systemic doses required to overcome inadequate BBB penetration (9–12). Enhancing the delivery of mAbs has the potential to increase therapeutic efficacy, while improving patient accessibility and reducing treatment costs. To overcome these challenges, various innovative strategies have been developed to improve mAb delivery across the BBB and their bioavailability in the CNS. These strategies include receptor-mediated transcytosis (RMT), which ultilize endogenous transport systems to ferry mAbs across the BBB, focused ultrasound (FUS), which temporarily disrupts the BBB utilizing microbubbles (MBs) and nanotechnology-based systems, like nanoparticle conjugates, which increase brain targeting and retention. Moreover, intranasal delivery provides a non-invasive alternative to bypass the BBB, with advances in modulating CSF-to-brain transport pathways being explored to optimize therapeutic exposure (2,9,11).

Targeting the brain with mAbs shows a unique set of opportunities and challenges, reflecting the difficulty of delivering therapeutics across the highly selective BBB. Exploring advanced innovative strategies to improve the CNS bioavailability of mAbs emphasizes the transformative potential of these therapies in overcoming long-standing barriers in brain-targeted drug delivery.

DOI: 10.1201/9781003482406-7

These advancements open new opportunities for resolving critical unmet needs in the treatment of CNS malignancies, neurodegenerative diseases, and other neurological disorders.

7.2 THE BBB

The BBB functions as a highly selective and dynamic interface between the systemic circulation and the brain parenchyma, tightly regulating the translocation of ions, cellular elements, and molecules. It plays a vital role in providing a defense mechanism against inflammation, pathogenic invasion, toxic substances, injury, and various neuropathological conditions (9,13,14). Anatomically, the BBB is composed of specialized brain microvascular endothelial cells (BECs) that form the lining of cerebral microvasculature. These BECs, along with pericytes, astrocytic end-feet processes, and the basement membrane (acellular component), collectively constitute the neurovascular unit. This intricate system assures the selective permeability of the BBB. The BECs are distinct from peripheral endothelial cells due to their tight junctions formed by proteins such as occludins, junction adhesion molecules, and claudins. These tight junctions produce a polarizing abluminal and luminal compartment, a robust paracellular barrier, and limit transcellular transport through reduced pinocytosis and the activity of efflux transporters. Consequently, the BBB closely regulates the movement of ions and nutrients into the brain while preventing the entry of potentially harmful substances (2,9,13,15,16). The adjacent non-endothelial cellular components of the neurovascular unit, including astrocytes and pericytes, play vital roles (17–20). Pericytes are embedded within the vascular basement membrane and extend elongated processes along the abluminal surface of the endothelium. They contribute to immune cell infiltration, angiogenesis, and cerebral blood flow regulation (18). Astrocytes, with their polarized end-feet processes that surround the microvasculature, express channels such as aquaporin-4 (AQP4), regulating the diffusion of solutes and water between the blood and brain (17). The basement membrane (acellular component) produced by astrocytes, pericytes, and BECs ensures structural support and contributes to BBB integrity. It comprises extracellular matrix proteins, such as heparan sulfate proteoglycans, nidogens, collagen IV isoforms, and laminins (2,6,9,13–21).

The BBB is essential to preserving the composition of CSF and brain interstitial fluid, which ensures neurons' integrative functions. Accurate synaptic transmission, temporal and spatial summation, and general neural function rely on this homeostasis. The BBB supplies vital nutrients while preventing the entry of neurotoxins, microorganisms, leukocytes, red blood cells, plasma components, and other foreign substances. Any breach in the BBB can cause the entry of neurotoxins, adversely affecting neuronal and synaptic functions, potentially leading to neuronal damage and death. The BBB also stabilizes ionic compositions in the brain, essential for synaptic and axonal signaling, preventing fluctuations after physiological activities like eating or exercise. It separates centrally and peripherally acting neurotransmitters and facilitates the clearance of larger molecules and metabolites, preserving the brain's microenvironment. While the BBB is critical for brain protection, its stringent permeability presents a significant obstacle to therapeutic and diagnostic interventions (2,9,13–16). Approximately 98% of small-molecule drugs and most macromolecular therapeutics cannot penetrate the CNS due to the BBB's restrictive nature (2,22,23). The unique properties of BECs absence of fenestrations, low transcytosis rates, abundance of mitochondria, and expression of efflux transporters, further limit drug delivery. Tight junctions, adherens junctions, and desmosomes between BECs severely impede paracellular transport, making innovative strategies essential to bypass the BBB for effective CNS therapies (2,6,9,13–16). Figure 7.1 shows a schematic representation of the BBB, highlighting its structure and the key cellular components involved.

7.2.1 THE CHALLENGE OF CROSSING THE BBB

The BBB is a highly selective semipermeable structure that poses significant challenges to the delivery of mAbs to the brain. These challenges arise primarily due to the large size of mAbs and the

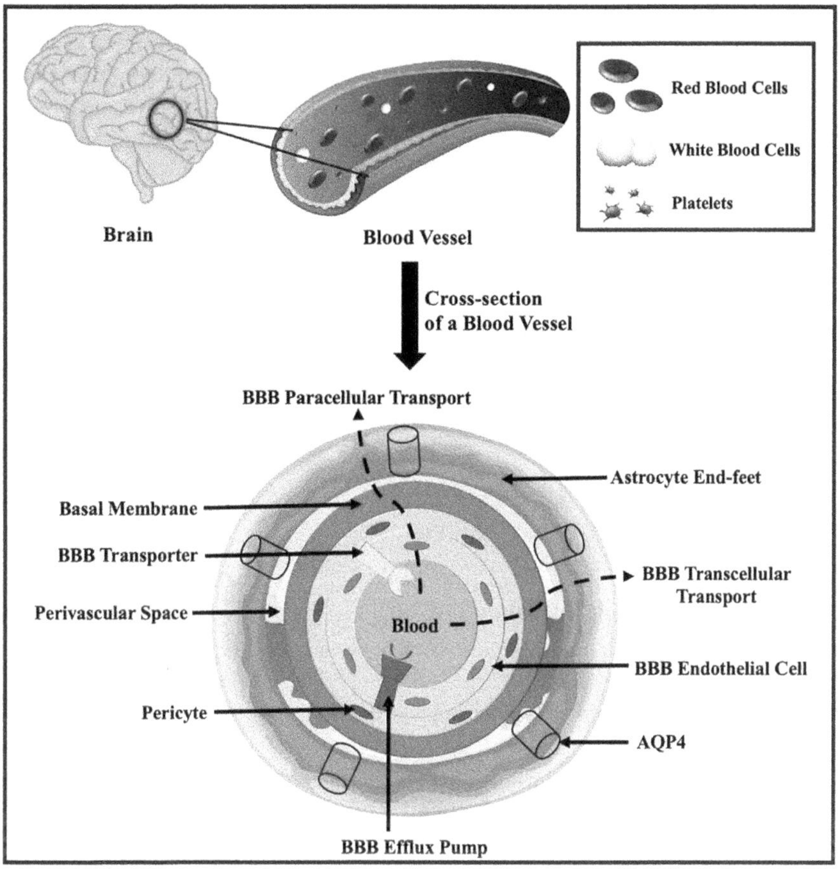

FIGURE 7.1 The BBB structure and key cellular components

restrictive nature of the BBB, which prevents most substances, including therapeutic agents, from freely entering the brain. Studies using radiolabeled IgG antibodies have shown that less than 0.05% of the administered dose reaches the brain parenchyma within 2 hours of intravenous or intraperitoneal administration (9). This limited permeability is attributed to the tight junctions between endothelial cells, low pinocytic activity, and the presence of efflux transporters that actively remove substances from the brain. Although small amounts of IgG can cross the BBB, the mechanisms governing this process remain elusive. It has been proposed that the neonatal Fc receptor (FcRn), expressed on the surface of BECs and the choroid plexus, may facilitate the transcytosis of IgG across the BBB (9,24,25). However, studies using *in vitro* models derived from human induced pluripotent stem cells suggest that IgG transport may occur independently of FcRn through nonspecific fluid-phase endocytosis, such as micropinocytosis (9,26,27). Once IgG traversed the BBB, it was quickly removed from the brain *via* reverse transcytosis, which was mediated by FcRn that lowered the residence time of therapeutic antibodies within the brain and further made it more difficult to use them for treating CNS disorders. (9,28–30).

In neuro-oncologic conditions like gliomas, the integrity of the BBB is interrupted, allowing macromolecules, such as mAbs, to penetrate tumor tissues. The infiltrative components of gliomas remain protected by intact BBB regions, posing a persistent challenge for antibody-based therapies (8,31–33). Further, the degree of BBB disruption differs with disease progression, and the time period of antibody administration can substantially affect the therapeutic outcomes. These factors emphasize the difficulty of delivering mAbs efficiently to CNS tumors. To overcome these challenges, researchers have investigated different strategies to improve the delivery of antibodies across the BBB. Recently, they have engineered bispecific antibodies that target receptors associated with BBB transport, like the transferrin receptor (TfR), which has shown potential by exploiting TfR-mediated endocytosis for increased brain penetration. Other strategies, such as FUS, utilize temporarily disrupting the BBB, conjugating antibodies in nanoparticles designed to traverse the BBB, and modifying the Fc domain of antibodies to prevent interactions with FcRn (8).

Preclinical and clinical studies provide evidence that mAbs have the potential to exert therapeutic effects on the CNS. For example, in AD, peripherally administered antibodies against amyloid-beta have localized to plaques and facilitated their clearance, though the exact mechanism remains disputed. Similarly, paraneoplastic antibodies are produced peripherally in some neurological conditions, which can cross the BBB and bind specifically to the brain parenchyma, demonstrating certain natural permeability for specific antibodies. In the case of gliomas, radiolabeled antibody-drug conjugates have shown selective tumor localization, further demonstrating that mAbs can accumulate in intracranial sites under certain conditions (8,34,35).

7.2.2 Blood-CSF Fluid Barrier and the Challenge Associated with Crossing the BCSFB

The choroid plexus forms a critical component of the Blood-CSF Fluid Barrier (BCSFB), creating an additional interface within the CNS to regulate molecular exchange. Structurally, the CP comprises a richly vascularized stroma containing porous capillaries and specialized epithelial cells. Strategically located within each of the four cerebral ventricles, the CP is primarily responsible for CSF production. Through active transport mechanisms, the CP facilitates the transfer of nutrients, metabolic by-products, and ions between the bloodstream and the CSF, thereby preserving the chemical homeostasis essential for CNS function. Following its secretion by CP epithelial cells, CSF flows from the lateral ventricles into the third ventricle *via* the interventricular foramina and subsequently passes through the cerebral aqueduct to reach the fourth ventricle. From there, CSF circulates within the subarachnoid space surrounding the brain and spinal cord, as well as through the central canal of the spinal cord. The subarachnoid space is demarcated by epithelial-like arachnoid barrier cells, which separate the CSF from the fenestrated vasculature of the dura mater. Consequently, arachnoid barrier cells also contribute to the structural and functional integrity of the BCSFB. Further, analogous to the BECs of the BBB, the CP epithelial cells and arachnoid barrier cells are interconnected by tight and adherens junctions, which limit paracellular diffusion into the CSF. This selective restriction allows for the regulation of solute transport via specific cellular transporters, maintaining a controlled biochemical environment across the barrier. Moreover, BCSFB cells exhibit a broad repertoire of transport proteins, which are often asymmetrically localized between the apical and basolateral membranes to fine-tune chemical homeostasis. In addition, similar to the BECs of the BBB, CP epithelial cells are enriched in mitochondria to support the high energy demands associated with active transepithelial transport processes (2,36).

To assess drug exposure in the brain, CSF concentrations are frequently used as a surrogate marker for widespread brain penetration. However, this approach can lead to an overestimation of mAb concentrations within the brain parenchyma. Multiple studies have demonstrated that molecules administered directly into the CSF undergo rapid clearance and show limited diffusion into the brain tissue (1,2,37). Consequently, while CSF antibody concentrations may reflect transport across the BCSFB, they do not necessarily provide an accurate estimation of mAb deposition within the brain or predict therapeutic efficacy. Supporting evidence for mAb transport across the BCSFB includes observations of its comparatively greater permeability relative to the BBB. Although the BCSFB and BBB possess distinct transporter expression profiles, influencing their selective permeability, the BCSFB has been shown to be inherently more permeable than the BBB (2,38). This enhanced permeability is characterized by increased leakage of plasma proteins across the BCSFB and a lesser electrical resistance of the epithelial barrier cells (2,38,39)

7.3 THE MABS

The mAbs are large, heterodimeric glycoproteins with a molecular weight of approximately 150 kDa. They are composed of two identical heavy chains (50 kDa each) and two identical light chains (25 kDa each), linked together by interchain disulfide bonds to form a characteristic Y-shaped structure. Each heavy chain contains three constant domains (C_H1, C_H2, and C_H3) and one variable domain (VH), while each light chain consists of one constant domain (CL) and one variable domain (VL). The paratope (antigen-binding site) was formed through the intricate intertwining of the variable domains (VL and VH). Within these domains, three complementarity-determining regions (CDRs) play a vital role in defining the antibodies' binding affinity and specificity for its target epitope. Based on the sequences of their heavy chains, antibodies are categorized into five main classes: IgA, IgD, IgE, IgG, and IgM. Among these, IgG antibodies are the most widely utilized in therapeutic applications due to their high stability, relatively straightforward production, and extended circulating half-life. Further, the IgG is categorized into four subtypes: IgG1, IgG2, IgG3, and IgG4. The crystallizable fragment region of IgG mediates effector functions by interacting with components of the immune system, including Fc gamma receptors and C1q on immune cells. These interactions enable processes such as antibody-dependent cellular cytotoxicity (ADCC) and complement-dependent cytotoxicity (CDC). Furthermore, IgG subtypes, such as IgG3 and IgG1, are highly effective activators of the complement pathway. Where IgG1, in particular, shows higher efficacy in promoting ADCC, making it the ideal subtype for cancer therapeutics where cell lysis is required. In contrast, IgG4 and IgG2 demonstrated decreased effector functions, which can be further lowered through engineering, making them preferred for conditions where CDC or ADCC are undesirable (40–50).

The structural diversity of antibodies, specifically in their variable regions, allows them to bind to a variety of antigens with minimal off-target effects and high specificity. This specificity is controlled by the unique sequence of CDRs, which are tailored for binding distinct epitopes. Moreover, the Fc region of mAbs governs their interactions, distribution, and elimination with immune effector cells, influencing their therapeutic and pharmacokinetic efficacy. These attributes, coupled with the ability to develop their structure for reduced immunogenicity or enhanced performance, established mAbs as a vital component of modern therapeutics (41,42).

7.3.1 Categories of mAbs

7.3.1.1 Fully Human mAbs

The development of fully human mAbs has been accomplished through advanced techniques, including transgenic mouse models and phage display. These techniques involve isolating single cells derived from immunized or naïve human donors or immunoglobulin genes from peripheral blood lymphocytes. These genes are utilized to construct combinatorial libraries, which are then expressed in bacterial systems to identify and isolate colonies producing functional antibodies. Fully human mAbs, devoid of murine components, demonstrate minimal immunogenicity and improved pharmacokinetic profiles, including slower plasma clearance. Notable examples include erenumab, used for migraine prophylaxis, and ofatumumab, approved for multiple sclerosis (MS). These fully human antibodies represent a significant advancement in mAb development, offering superior therapeutic potential with reduced adverse immunogenic responses (11,42).

7.3.1.2 Humanized Antibodies

Advancements in molecular biology have facilitated the engineering and development of humanized mAbs, which are approximately 90% human-derived in composition, with only about 10% murine protein. These antibodies exhibit markedly reduced immunogenicity compared to their chimeric counterparts. Humanized antibodies are engineered by grafting CDRs critical for antigen binding from murine antibodies into human immunoglobulin frameworks. This process eliminates non-binding murine sequences and replaces them with human counterparts. Additionally, selective amino acid substitutions in the constant regions, derived from the parental murine antibody, can further optimize functionality. Humanized mAbs preserve the antigen-binding affinity and specificity of their original murine predecessors while minimizing immunogenicity and acquiring desired biological properties. The majority of therapeutic mAbs currently used or under development for neurological disorders are humanized (11,51,52).

7.3.1.3 Chimeric Antibodies

The clinical limitations of murine antibodies prompted the development of chimeric mAbs, which incorporate both human and murine components. By replacing the Fc region of the murine antibody with the human constant region, chimeric antibodies were created to reduce immunogenicity while retaining antigen-binding specificity. These antibodies contain approximately 34% murine protein, localized to the variable regions, resulting in a significantly lower incidence of human anti-mouse antibodies (HAMAs) responses compared to murine mAbs. Additionally, chimeric mAbs improved pharmacokinetics, enhanced cellular cytotoxicity, exhibit a broader range of antigen specificities, and have a longer half-life. Infliximab and rituximab are examples of chimeric mAbs, which are currently approved for use in neurological conditions (11,42).

7.3.1.4 Murine Antibodies

These antibodies are derived solely from mouse proteins and represent the original class of mAbs developed. Due to their murine origin, these antibodies are recognized as allogeneic proteins by the human immune system, often triggering the formation of HAMAs within 2–3 weeks of administration. HAMAs frequently neutralize murine antibodies, leading to their rapid inactivation or altered pharmacokinetics, which accelerates plasma clearance. These immunogenic limitations have rendered murine mAbs unsuitable for long-term clinical use, and none are currently utilized in neurology (11,53,54). Figure 7.2 schematically illustrates the distribution of human and murine-derived components in fully human, humanized, chimeric, and murine mAbs.

7.3.2 Mechanisms Facilitating Antibody Transport Across the BBB

Over the past few decades, several transport mechanisms have been identified as key pathways enabling macromolecule passage across the BBB. Under normal physiological conditions, only about 0.1% of circulating antibodies can penetrate the brain, highlighting the formidable challenge posed by the BBB. To address this, brain-targeting mAbs have emerged as a transformative class of therapeutics designed to overcome the physiological barrier posed by the BBB and to engage disease-specific targets within the CNS. These mAbs exert their therapeutic efficacy primarily by leveraging specialized transport pathways, including carrier-mediated transport (CMT), adsorptive-mediated endocytosis (AMT), and receptor-mediated transcytosis (RMT). Each of these mechanisms offers unique advantages in enhancing the delivery and targeting efficiency of therapeutic antibodies, paving the way for improved treatment outcomes in several neurological disorders (11,55).

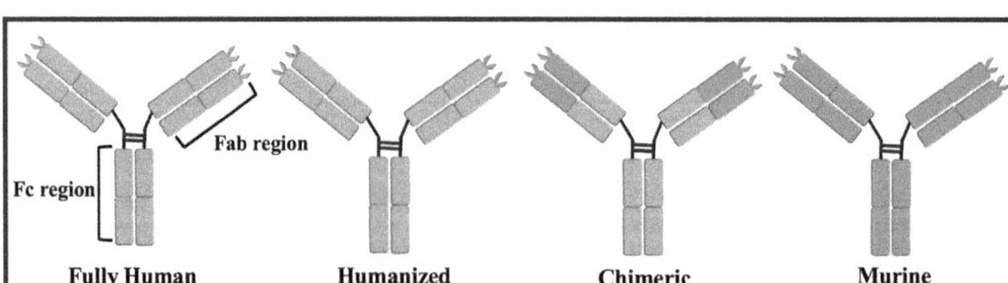

FIGURE 7.2 Types of monoclonal antibodies

7.3.2.1 Carrier-Mediated Transport

The CMT mechanism facilitates the rapid transfer of small molecules such as hormones, vitamins, amino acids, and glucose across the BBB. This transport is saturable due to finite carrier availability and plays a crucial role in maintaining CNS homeostasis by enabling bidirectional exchange. Notable carriers include CLUT1, which transports LAT1, mannose, and glucose, responsible for the transport of amino acids (neutral). Although conjugating therapeutic molecules to their analogues or endogenous ligands offers a theoretical strategy for brain delivery *via* CMT. However, this approach is difficult to transport large molecules such as antibodies due to the carrier's preference for high stereoselectivity and small substrates (55–58).

7.3.2.2 Adsorptive-Mediated Endocytosis

The AMT mechanism depends on electrostatic interactions among circulating cationic molecules and the negatively charged luminal surface of BBB endothelial cells, leading to internalization of the positively charged molecules. To enhance brain uptake, proteins like IgG and albumin have been chemically modified by introducing cationic groups onto their surfaces. Studies have shown that cationization, achieved by covalently attaching primary amine groups to antibodies, significantly promotes their uptake into the brain *via* AMT. Although AMT possesses a high transport capacity, it exhibits poor specificity and low affinity, as cationized molecules can also interact with negatively charged membranes of peripheral tissues, resulting in non-proportional brain targeting. Therefore, the non-selective nature of AMT must be carefully considered when designing brain-targeted therapeutic antibodies (55–59).

7.3.2.3 Receptor-Mediated Transcytosis

RMT has emerged as one of the most promising strategies for antibody delivery to the brain. This pathway involves three major receptor classes: lipid transporters (e.g., low-density lipoprotein receptor-related protein 1, LRP1), insulin transporters (e.g., insulin receptor (IR)), and iron transporters (e.g., TfR). In RMT, the ligand binds to its receptor, the ligand–receptor complex undergoes endocytosis, and the cargo is subsequently exocytosed across the abluminal membrane into the brain parenchyma. However, high-affinity antibodies targeting RMT receptors may be redirected to lysosomal degradation pathways following internalization. To overcome this limitation, optimizing the ligand–receptor binding affinity has proven to be a successful strategy to enhance transcytosis efficiency (55–57,59).

7.3.3 Marketed mAbs for Brain-Targeted Therapies

Several mAbs have received regulatory approval for the treatment of brain-related diseases, including MS, AD, and brain tumors such as neuroblastoma and glioblastoma. These therapeutic antibodies, encompassing fully human, humanized, chimeric, and murine types, have received approval from major regulatory agencies such as the EMA, FDA, and others for clinical use across various indications (11).

Notably, not all mAbs targeting neurological diseases are required to cross the BBB to exert their therapeutic effects. For instance, Tysabri (natalizumab), approved for MS, exerts its peripheral action by inhibiting the infiltration of activated immune cells into the CNS through the blockade of immune cell adhesion to the endothelial cells of the BBB (60). Conversely, certain mAbs directly targeting pathological entities within the brain must cross the BBB to achieve therapeutic efficacy. Moreover, prominent examples include lecanemab and aducanumab, both directed against amyloid-beta plaques, which have been approved for AD treatment and are presumed to require BBB traversal to mediate amyloid clearance (61). Several mAbs currently used in neurology were originally developed for rheumatological disorders, such as tocilizumab or hematological malignancies, such as rituximab, ofatumumab, and alemtuzumab. Others, however, have been specifically engineered for neurological applications, such as ocrelizumab for MS and various mAbs approved for migraine prophylaxis (11). Table 7.1 represents the currently marketed and under-development mAbs for CNS disorders.

TABLE 7.1
Currently Marketed and Under-Development Mabs for CNS Disorders.

Disease	Name	Type	Target	Action	Route	Approval Status/ Phase	References
AD	LY3372993 (Remternetug)	IgG1 (Human)	Amyloid-β (Aβ)	Binds amyloid plaques; under investigation	Intravenous (IV)	Phase 3	(2,59,62,63)
	Aducanumab (Aduhelm)	IgG1 (Human)	Aβ	Removes amyloid plaques	IV	2021	(2,59,62,63)
	ABBV-8E12	Human mAb	Tau protein	Binds tau protein	IV	Phase 2	(2,59,62,63)
	AL002	Human mAb	TREM-2 receptor	Activates microglial response	IV	Phase 1	(2,59,62,63)
	BIB092	Human mAb	Tau protein	Binds tau protein	IV	Phase 2	(2,59,62,63)
	AL003	Human mAb	SIGLEC-3	Modulates immune response	IV	Phase 1	(2,59,62,63)
	Zagotenemab	Human mAb	Tau protein	Binds aggregated tau protein	IV	Phase 2	(2,59,62,63)
	JNJ-63733657	Human mAb	Tau protein	Binds tau protein	IV	Phase 1	(2,59,62,63)
	Lecanemab (Leqembi)	IgG1 (Humanized)	Aβ	Binds soluble and insoluble Aβ	IV	2023	(2,59,62,63)
	Crenezumab	IgG4 (Humanized)	Aβ	Binds oligomeric forms of Aβ	IV	Phase 3	(2,59,62,63)
	Semorinemab	IgG4 (Humanized)	Tau protein	Binds N-terminal tau protein	IV	Phase 2	(2,59,62,63)
	Solanezumab	IgG1 (Humanized)	Monomers	Binds soluble monomeric Aβ	IV	Phase 3	(2,59,62,63)
	Gantenerumab	Human IgG1	Aβ	Binds aggregated Aβ	IV	Phase 3	(2,59,62,63)
	E2814	IgG1 (Humanized)	Tau protein	Binds tau protein to prevent aggregation	IV	Phase 2/3	(2,59,62,63)
Frontotemporal Dementia	AL001 (Latozinemab)	IgG1 (Human)	Sortilin	Modulates progranulin levels	IV	Phase 3	(2,59,62,63)
Glioblastoma	Biotin-coupled BC-4 + Avidin + [^{90}Y]-Biotin	Human mAb	Tenascin	Pre-targeted radioimmunotherapy	IV	Phase 1/2	(2,59,62,63)
	^{125}I-mAb 425	Human mAb	Epidermal growth factor receptor (EGFR)	Radiolabeled antibody targeting EGFR	IV	Phase 2	(2,59,62,63)
	^{131}I-BC-2 mAb	Human mAb	Tenascin	Radiolabeled antibody targeting tenascin	IV	Phase 2	(2,59,62,63)
	^{131}I-chTNT-1/B MAb	Human mAb	DNA-histone H1 complex	Radiolabeled antibody targeting necrotic tumor cells	IV	Phase 1/2	(2,59,62,63)
	At-labeled 81C6 mAb	Human mAb	Tenascin	Radiolabeled antibody targeting tenascin	IV	Phase 1/2	(2,59,62,63)
	[^{188}Re]-labeled Nimotuzumab	Humanized mAb	EGFR	Radiolabeled antibody targeting EGFR	IV	Phase 1	(2,59,62,63)
	Depatuxizumab mafodotin	IgG1 ADC	EGFR	Antibody-drug conjugate targeting EGFR	IV	Phase 2b/3	(2,59,62,63)

Neuroblastoma	^{131}I-omburtamab	Murine mAb	BT-H3	Radiolabeled antibody targeting B7-H3	IV	Phase 2/3	(2,59,62,63)
	Dinutuximab (Unituxin)	IgG1 (Chimeric)	GD2	Targets GD2 on neuroblastoma cells	IV	2015	(2,59,62,63)
MS	Divozilimab (Ivlizi)	IgG1 (Humanized)	CD20	Depletes B cells	IV	2023	(2,59,62,63)
	Ublituximab (Briumvi)	IgG1 (Humanized)	CD20	Depletes B cells	IV	2022	(2,59,62,63)
	Daclizumab (Zinbryta)	IgG1 (Humanized)	CD25	Blocks high-affinity IL-2 receptor	Subcutaneous (SC)	2016	(2,59,62,63)
	Ocrelizumab (Ocrevu)	IgG1 (Humanized)	CD20	Depletes B cells	IV	2017	(2,59,62,63)
	Natalizumab (Tysabri)	IgG4 (Humanized)	α4β1 integrin (CD49d)	Inhibits lymphocyte entry into brain parenchyma	IV	2014	(2,59,62,63)
	Alemtuzumab (Lemtrada)	IgG1 (Humanized)	CD52	Depletes CD52+ T and B cells	IV	2014	(2,59,62,63)
Neurological Diseases (General)	Vyepti (Eptinezumab)	IgG1 (Humanized)	Calcitonin gene-related peptide (CGRP) ligand	High selectivity bind CGRP isoforms b and a	IV	2020	(11,64)
	Actemra (Tocilizumab)	IgG1 (Humanized)	Interleukin-6 receptor (IL-6R)	IL-6R signaling blockade	IV	2020	(11,65–67)
	Enspryng (Satralizumab)	IgG2 (Humanized)	IL-6R	IL-6R signaling blockade	SC	2020	(11,68,69)
	Kesimpta (Ofatumumab)	IgG1 (Fully human)	CD20	Reduces B cells	SC	2020	(11,70,71)
	Uplizna (Inebilizumab)	IgG1 (Humanized)	CD19	Reduces plasmablasts/plasma cells and B cells	IV	2020	(11,72,73)
	Soliris (Eculizumab)	IgG2/4 (Humanized)	C5 complement protein	Inhibits terminal C5 complement pathway	IV	2019	(11,74–78)
	Emgality (Galcanezumab)	IgG4 (Humanized)	CGRP ligand	Binds CGRP and prevents biological activity	SC	2018	(11,79–84)
	Ajovy (Fremanezumab)	IgG2 (Humanized)	CGRP ligand	Specifically binds CGRP isoforms a and b	SC	2018	(11,85,86)
	Aimovig (Erenumab)	IgG2 (Fully human)	CGRP receptor	Reversibly and competitively binds CGRP receptor	SC	2018	(11,87–91)
	Avastin (Bevacizumab)	IgG1 (Humanized)	Vascular endothelial growth factor (VEGF)	Inhibits angiogenesis	IV	2009	(11,92–95)

7.4 CURRENT APPROACH FOR DELIVERING MAB UPTAKE TO THE BRAIN

Despite significant advances in the delivery of mAbs to the brain, the clinical development of brain-targeting mAbs continues to face substantial challenges. For example, several candidate mAbs engineered to promote axonal remyelination in MS patients, including anti-Nogo-A, sHIgM22, anti-LINGO-1 (opicinumab), and VX15/2503, have failed to demonstrate efficacy in clinical trials (96–99). Similarly, the phase II clinical trial of the anti-Tau mAb (8E12) in AD patients was discontinued due to a lack of therapeutic benefit (100). In many instances, limited brain delivery resulting from the restrictive nature of the BBB has been identified as a critical factor contributing to these clinical failures (2). To address the challenges posed by the BBB and enhance mAb delivery to the brain, researchers have explored several innovative strategies. These include bispecific monoclonal antibodies (bsAbs), FUS, conjugating antibodies with nanoparticles, and other delivery approaches (Figure 7.3 and Table 7.2) (2,9,11). In addition to improving transport across the BBB, novel approaches are being explored to enhance mAb retention within the brain, enabling gradual accumulation and sustained therapeutic concentrations over time. Although no mAb delivery method has yet received FDA approval specifically for brain-targeting therapies, several emerging strategies have demonstrated remarkable potential. Continued innovation and refinement of these

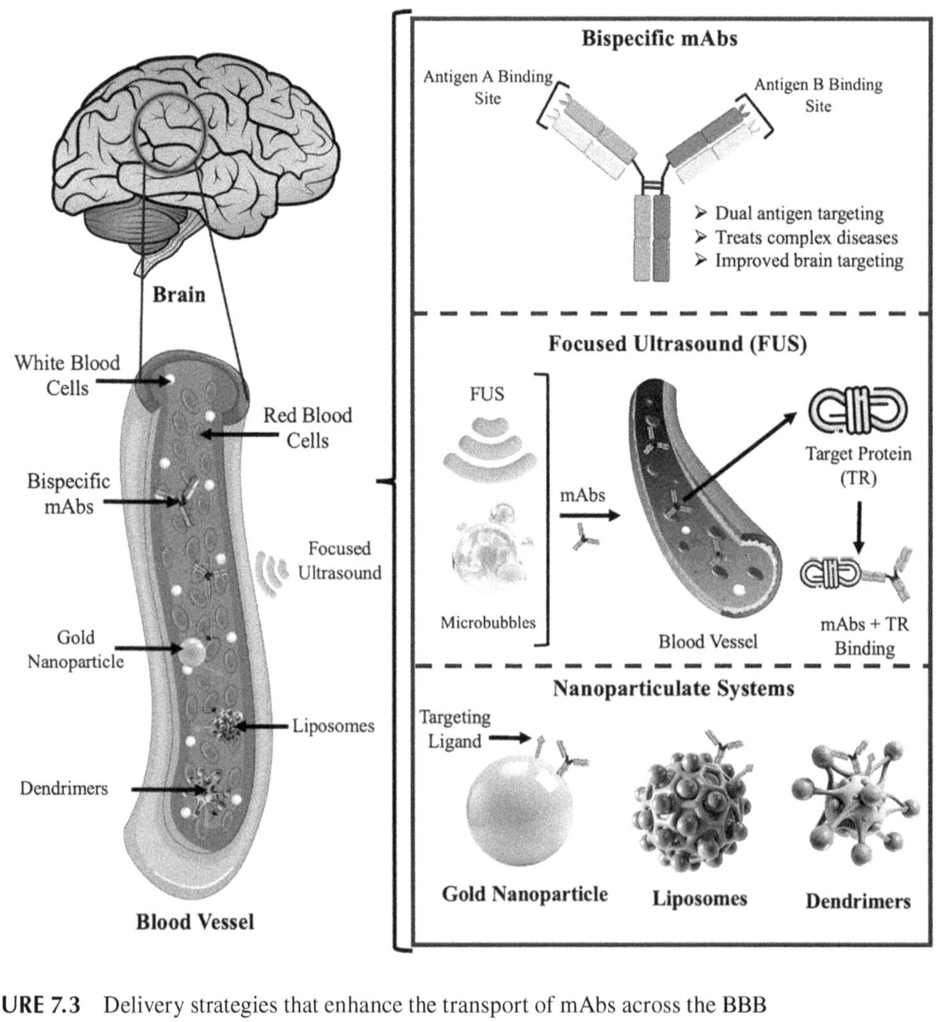

FIGURE 7.3 Delivery strategies that enhance the transport of mAbs across the BBB

TABLE 7.2
Delivery Strategies that Enhance the Transport of Therapeutic Antibodies Across the BBB.

Delivery Strategy	Antibody	Target	Disease	Mode of Transport	References
bsAbs	Anti-β-site amyloid precursor protein-cleaving enzyme 1 (Anti-BACE1) mAb	BACE1	AD	RMT (TfR)	(101)
	Anti-Aβ antibody, 3D6 scFv, IgG	Aβ	AD	RMT (TfR)	(102)
	Anti-Aβ mAb, 13C3	Protofibrillar Aβ	AD	RMT (TfR)	(103)
	Anti-Aβ mAb	Aβ peptide	AD	RMT (TfR)	(104–107)
FUS	Herceptin®/ Trastuzumab	Human epidermal growth factor receptor (HER2)	Breast cancer	Not determined (ND)	(108)
	Anti-α-synuclein mAb	α-synuclein	Parkinson's disease	ND	(109)
	Anti-tau mAb (RNF5)	Tau (all isoforms)	Fronto-temporal lobar degeneration	ND	(110)
	Anti-tau antibody, IgG, Fab, RN2N scFv	Tau (isoforms with two N-terminal domains)	Fronto-temporal lobar degeneration	ND	(111,112)
	Anti-Aβ mAb, A07/2a	pGlu3 Aβ	AD	ND	(113)
	Anti-Aβ polyclonal (Rabbit sera)	Aβ	AD	ND	(114)
	Aduhelm®/ Aducanumab	Oligomeric Aβ	AD	ND	(115)
	Avastin®/ Bevacizumab	VEGF	Malignant glioma	ND	(116)
	Anti-Aβ mAb	Aβ	AD	ND	(117,118)
Gold Nanoparticle	Anti-BACE1 mAb	BACE1	AD	RMT (TfR)	(119)
Liposome	Trastuzumab	HER2 receptor	Brain malignancies (*in vitro* BBB model)	RMT (TfR)	(120)
	Cetuximab	EGFR	Brain malignancies (*in vitro* BBB model)	RMT (TfR)	(120)
	Anti-CD133 mAb	Glioblastoma stem cells	Glioblastoma	RMT (LRP-1)	(121)
Dendrimer	Anti-NKG2a mAb	NK cell receptor	Glioblastoma	RMT (caveolae-mediated endocytosis)	(122)

delivery platforms hold significant promise for advancing the treatment of a wide range of neurological disorders (2).

7.4.1 bsAbs for Brain Delivery

The bsAbs are engineered to simultaneously recognize and bind two distinct epitopes, endowing them with multifunctional capabilities beyond the scope of conventional mAbs. This dual specificity enables spatially and temporally restricted effects, which cannot be replicated by administering a combination of two monospecific antibodies. The versatility of bsAbs lies in their ability to mediate diverse biological functions, such as delivering ligands to particular cell populations, promoting receptor internalization, redirecting effector immune cells to defined target cells, or concurrently blocking multiple signaling pathways. Importantly, they can also facilitate translocation across physiological barriers, such as the BBB, a critical challenge in CNS therapeutics. The development of bsAbs specifically targeting TfR and IR aims to enhance BBB penetration and therapeutic efficacy. One of the earliest examples was a tetravalent antibody, where a bivalent anti-human insulin receptor (HIR) antibody was amalgamated with two single-chain variable fragments (scFv) targeting anti-Aβ. However, recent studies showed that high-affinity binding to TfR or IR could lead to excessive accumulation and degradation within BECs, limiting transport into brain tissue (2,9,11). To address this, Yu et al. demonstrated that decreasing the affinity of antibodies for TfR significantly improved brain uptake. Their engineered monovalent bispecific antibody targeting β-secretase (BACE-1) and TfR enhanced BBB transport compared to a monospecific anti-BACE-1 mAb, with monovalent binding minimizing lysosomal degradation by altering trafficking away from lysosomes (123). Though no bsAbs are currently approved for neurological disorders, preclinical models demonstrate their potential. For instance, a bispecific construct combining an LDLR-binding domain and an epitope that enhances alpha-secretase activity showed neuroprotective activity in a mouse model of AD (124). Similarly, dual targeting of angiopoietin-2 (Ang-2) and translocator protein, or Ang-2 and VEGF, has yielded survival benefits in preclinical glioblastoma models (125). These successes underscore the potential of bsAbs to revolutionize treatment approaches across disease domains, including neurodegenerative disorders and brain tumors.

7.4.1.1 TfR-Targeting Bispecific Antibodies

TfR mediates the transcytosis of iron-bound transferrin into the brain and most studied RMT receptor for brain delivery (126). Advancements in antibody engineering have facilitated the development of numerous strategies employing anti-TfR antibodies to enable the transport of antibodies across the BBB. Various TfR-targeting tools are utilized for brain targeting, including antibody clones such as scFvs, sFabs, 8D3, OX26, variable new antigen receptors, and peptides (GYR) (127). The efficiency of transcytosis appears to be significantly influenced by the interaction between the TfR and bsAbs. Modulating the antibody's affinity for TfR at the acidic lysosomal pH (~5.5) can reduce lysosomal degradation, thereby facilitating enhanced transcytosis and improving the subsequent delivery of the antibody into the brain parenchyma. (9). Further, the use of the 8D3 antibody (anti-TfR), which exhibits bivalent binding to TfR, has been shown to result in sub-optimal transcytosis. Consequently, the development of bi-specific antibodies with TfR binding (monovalent) has significantly enhanced transcytosis efficiency. For example, attachment of scFv 8D3 (anti-TfR) to the C-terminal light chain of mAb158 (anti-Aβ) or the use of anti-TfR Fab fusion to the Fc domain of anti-Aβ mAbs led to improved monovalent binding and transcytosis without peripheral Fc-FcγR receptor engagement (104–106). Furthermore, Roche initiated a clinical trial, Phase 1b/2a (NCT04639050), utilizing this monovalent binding technology to deliver anti-Aβ mAb (gantenerumab) in Alzheimer's patients (9). Structure-based mutations introduced into the CDRs of the anti-TfR variable domains to reduce TfR affinity have also been shown to enhance brain exposure (103). Additionally, reducing the size of the bispecific antibody, as shown by faster brain parenchymal delivery of 56 kDa scFv fusions compared to full-size IgG fusions (210 kDa), provided additional benefits (102). Instead of employing an anti-TfR antibody, Denali Therapeutics Inc. developed a small segment of the human Fc domain to interact with the TfR, termed transport vehicles (TVs). This was accomplished by identifying a 9-amino acid sequence within the IgG1 (human) CH3 domain that could be altered without disrupting the native function of the Fc domain. Randomization of amino acid sequence generated a library of variants, and subsequent screening enabled the selection of constant region clones capable of binding to the extracellular domain of TfR (human). These TVs can be further engineered by fusing them to two Fab arms, either with identical epitopes (to mimic a conventional IgG antibody) or with distinct epitopes (to mimic bsAbs), thus forming antibody TVs (101). Moreover, a clinical trial (NCT03568175), evaluating the efficacy and safety of JR-141, was studies have demonstrated effective brain delivery and improvements in neurocognitive function, along with mild to moderate drug-related adverse events. Additionally, a clinical trial with Phase I/II (NCT04251026) is ongoing to assess enzyme replacement therapy in the treatment of Hunter syndrome by delivering DNL310, a fusion protein comprising the enzyme TVs and iduronate-2-sulfatase (127).

7.4.1.2 IR-Targeting Bispecific Antibodies

The IR is another promising target for RMT-mediated brain delivery. They are expressed on the luminal surface of BECs, mediate the transport of insulin from the bloodstream into the brain. Similar to the TfR, though less prominently, the IR has been exploited to enable the delivery of biotherapeutic agents into the brain parenchyma (128,129). Currently, ArmaGen Technologies is developing a humanized anti-insulin receptor monoclonal antibody (HIRMAb) and its fusion proteins for the treatment of lysosomal storage disorders (128). A clinical trial with Phase I/II is investigating valanafusp alpha, a fusion of HIRMAb with a lysosomal enzyme, which demonstrated successful delivery to both the CNS and peripheral tissues. This dual targeting is attributed to the antibody's affinity for both the TR and mannose-6-phosphate receptor. The clinical data also indicated cognitive stabilization in pediatric patients with severe mucopolysaccharidosis (type I) (clinical trials NCT03071341 and NCT03053089). Moreover, dose-dependent hypersensitivity reactions were observed in some non-human primates during preclinical studies involving antibodies targeting the IR. To address this issue, a clinical trial with Phase II (NCT03053089) was conducted where laronidase was fused to an insulin receptor-binding IgG and administered *via* IV infusion for the treatment of Hurler syndrome (mucopolysaccharidosis type I). The resulting fusion protein, termed AGT-181, was engineered to bind to IR on BECs, thereby promoting transcytosis across the BBB and enabling CNS) delivery of laronidase. Overall, the therapy was found to be safe and well tolerated. While some participants developed hypoglycemia, these drawbacks were successfully controlled with the co-administration of dextrose (9,127).

7.4.2 FUS

The FUS is an emerging non-invasive technique that transiently increases BBB permeability, facilitating enhanced transcellular and paracellular transport of therapeutic agents into the brain parenchyma. This method requires the IV administration of gas-filled lipid vesicles, known as MBs, before the application of FUS. Upon entering circulation and encountering the ultrasound field, MBs undergo oscillation, generating mechanical forces on the vascular endothelium. This mechanical activity transiently disrupts tight junctions, thereby stimulating vesicle-mediated transcytosis and enhancing paracellular transport, ultimately facilitating the extravasation of nanoparticles, antibodies, and other therapeutic agents into the brain (9,130–132). Subsequent studies revealed that the combination of FUS with IV administration of MBs can reduce the acoustic parameter required to achieve BBB opening without causing the risk of tissue damage. Several studies have explored the extent of BBB opening induced by FUS combined with MBs using fluorescently labeled dextran molecules. A concurrent study discovered that a 0.57-MPa acoustic pressure permitted the transport of 70 kDa and 3 kDa dextrans across the BBB, whereas a larger dextrans size (2000 kDa) required a higher pressure (0.84 MPa). However, pressures at or above 0.84 MPa increased the risk of inertial cavitation and subsequent microhemorrhage formation. Furthermore, studies with liposomes (55–200 nm) demonstrated that an increased MBs dose (from 0.1 to 0.5 μL/g) substantially improved the delivery of larger nanoparticles, emphasizing that both acoustic parameters and MBs dosing critically influence the extent of BBB opening (2,133,134). Moreover, preclinical studies of FUS-mediated BBB opening have successfully enhanced the delivery of therapeutic mAbs, like trastuzumab, bevacizumab, and interleukin-12, and proteins for the treatment of malignant gliomas and metastatic brain tumors. Bevacizumab (immunotherapeutic), a mAb approved by the FDA for several cancer types, such as glioblastoma, showed significantly improved brain penetration when administered with FUS. In mice, FUS increased brain levels of bevacizumab by 5.7 to 56.7-fold compared to controls without FUS, leading to suppressed tumor growth and a 135% increase in median survival time, compared to a 48% increase with bevacizumab treatment alone (9,108,116,135). Currently, A clinical trial (NCT04446416) is underway to assess the efficacy and safety of FUS-assisted bevacizumab delivery in patients with recurrent glioblastoma (9). FUS has also been utilized to facilitate mAb delivery in neurodegenerative disease models, including AD and parkinson's disease. In parkinson's disease (transgenic mouse model), FUS-enhanced delivery of anti-α-synuclein antibodies reduced pathological aggregation more effectively than antibody administration alone (109). Alzheimer's pathology, characterized by extracellular Aβ plaques and intracellular tau neurofibrillary tangles, has been a primary target. Raymond et al. first demonstrated a 2.7 ± 1.2-fold higher anti-Aβ antibody uptake into brain parenchyma following FUS (114). Subsequently, Jordão et al. showed that repeated FUS applications significantly reduced plaque load after delivering an anti-Aβ antibody. Similar reductions were observed in APP/PS1dE9 mice and rabbit models. FUS-mediated delivery of an anti-pyroglutamate-3 Aβ antibody led to a 5.5-fold brain accumulation compared to the antibody alone, with corresponding improvements in spatial learning and memory (113).

7.4.2.1 Scanning Ultrasound

Conversely to FUS, which uses a single sonication site, SUS applies multiple sonication sites across the entire brain. This method is designed to improve the concentration and distribution of therapeutic molecules throughout the brain. When combined with MBs (termed SUS+MB), this approach substantially increases the brain concentrations of diverse antibody formats, including Fab fragments (~50 kDa), full-length IgGs (~155 kDa), and scFv (~27 kDa). In tau transgenic pR5 mice, SUS+MB enhanced the brain concentration of these antibodies by 30-, 19-, and 20-fold, respectively. Weekly SUS+MB treatments combined with tau-specific scFv administration not only elevated antibody delivery by ~11-fold but also promoted neuronal uptake, facilitating tau clearance and improving behavioral outcomes. More recently, the delivery of an anti-Aβ antibody analogous to Aduhelm® (aducanumab) using SUS+MB resulted in improved plaque clearance and better cognitive performance compared to either treatment alone (9,110–112,115,136,137).

7.4.3 Nanoparticulate Systems for Brain Delivery

Various nanoparticulate systems have recently gained attention as a promising strategy for enabling drug delivery across the BBB into the CNS. Nanoscale carrier systems can be engineered to conjugate a wide range of molecules, including proteins, peptides, mAbs, nucleic acids, and small-molecule drugs. These particles can be finely tuned in terms of size, shape, surface charge, hydrophobicity, and chemical composition to achieve controlled drug release, protection from biological degradation, and site-specific targeting. Among the different nanoparticles studied, metallic nanoparticles (MNPs), liposomes, and dendrimers have shown particular promise for crossing the BBB, with the capability to either load or conjugate therapeutic or targeting antibodies. Notably, many current studies focus on enhancing BBB penetration and target specificity through the conjugation of antibodies or nanobodies to nanoparticles (9,138–141).

7.4.3.1 Metallic Nanoparticles

MNPs represent a distinct class of nanocarriers characterized by their small size, high density, and rigid, spherical structure. Unlike other nanoparticle systems, such as liposomes or polymeric nanoparticles, which typically encapsulate therapeutic agents within a core, MNO are composed of solid metal, including silver, iron, and gold, or metallic allotropes of non-metals, and primarily rely on surface conjugation for drug delivery. Therapeutic molecules, including small drugs, peptides, or targeting ligands, are commonly attached to the external surface of MNPs to achieve site-specific delivery. Despite their solid structure, MNPs have demonstrated the ability to traverse the BBB via passive transmembrane diffusion or RMT, depending on their surface modifications and size. These unique properties make metallic nanoparticles an attractive platform for neurological applications, particularly for imaging, targeted drug delivery, and theranostic strategies in CNS disorders. Gold nanoparticles (AuNPs) have gained considerable attention as brain-targeted nanocarriers due to their tunable surface properties, biocompatibility, and efficient transport across the BBB (2,9,138,139,142). In a recent study, AuNPs, coupled with monovalent antibodies targeting both the BACE1 and TfR, demonstrated successful translocation across *in vitro* and *in vivo* BBB models *via* TfR-mediated transcytosis. Notably, the use of monovalent antibodies proved superior to bivalent antibody, particularly at low degrees of surface functionalization (<10 antibodies per AuNP), resulting in a 3-fold higher brain parenchymal uptake. This suggests that minimizing receptor crosslinking may enhance BBB penetration and intracellular trafficking. In addition to facilitating BBB transport, BACE1, a critical enzyme involved in amyloid-beta production was identified as a promising therapeutic target for AD (9,119). Although the therapeutic potential of these AuNP formulations was not fully explored in the study, the dual-targeting approach presents a compelling strategy that warrants further investigation in future therapeutic development.

7.4.3.2 Liposomes

Liposomes are self-assembled, spherical vesicular nanostructures composed of one or more phospholipid bilayers surrounding an aqueous core. Their unique architecture allows for the simultaneous encapsulation of both hydrophilic and hydrophobic therapeutic agents. Due to the structural similarity of their lipid bilayer to natural biological membranes, liposomes exhibit excellent membrane permeation capabilities, along with favorable biocompatibility and biodegradability. These properties have facilitated their successful clinical application in the treatment of various diseases (9,138,140). However, despite these advantages, the presence of a lipid bilayer is insufficient to enable efficient crossing of the BBB, a major hurdle in CNS drug delivery. To overcome this limitation, RMT has emerged as a widely studied strategy to facilitate BBB penetration, with the TfR being a primary target (120,121,143). Preclinical studies have demonstrated that liposomes co-loaded with simvastatin (cholesterol-lowering agent) and gefitinib (anti-cancer drug), and functionalized with T12 (a peptide targeting TfR) along with a nanobody selectively targeting the tumor-associated macrophage PD-L1, successfully traversed the BBB, engaged the TfR, and selectively targeted tumour-associated macrophages in a mouse model of intracranial non-small cell lung cancer (143). Similarly, H-ferritin-based liposomes, leveraging the intrinsic selectivity of H-ferritin for TfR, have been utilized to deliver therapeutic mAbs, including cetuximab and trastuzumab, across an *in vitro* human BBB model (hCMEC/D3), while upholding their inherent biofunctional characteristics (120). Although TfR is a well-characterized target for BBB crossing, alternative receptors such as the low-density lipoprotein receptor-related protein 1 (LRP1) offer even more favorable expression profiles, particularly with high expression on BECs (144). For instance, liposomes conjugated with angiopep-2 (a ligand for LRP1) and anti-CD133 mAbs for glioblastoma stem cell targeting demonstrated enhanced BBB permeability, improved site-specific delivery, prolonged survival, and negligible systemic toxicity in murine models (121). Furthermore, immunoliposomes coated with a TfR-specific antibody (OX26) and encapsulating oxaliplatin significantly increased brain oxaliplatin concentrations, achieving a threefold enhancement without requiring the complete transport of the intact liposome across the BBB, highlighting the nuanced mechanisms by which modified liposomal systems can optimize brain drug delivery (127).

7.4.3.3 Dendrimers

Dendrimers are a distinctive class of polymeric nanoparticles characterized by their radially symmetrical, highly branched, tree-like architecture emanating from a central core. Due to their compact and organized structure, dendrimers are typically much smaller than liposomes, often comparable in size to small AuNPs (10 nm). Unlike liposomes, dendrimers do not encapsulate therapeutic agents in a central aqueous core. Instead, drugs and targeting moieties are typically conjugated to the multiple reactive surface groups on their branched periphery (145). Various strategies have been explored to facilitate dendrimer transport across the BBB. These include olfactory nerve-mediated delivery (146), direct carotid artery injection (147), and less invasive approaches such as CMT and endocytosis-based uptake (148). Among the different dendrimer types, polyamidoamine (PAMAM) dendrimers have shown the most promise in CNS applications (147,148). Recent studies have reported successful delivery of PAMAM dendrimers conjugated with siLSINCT5, a non-coding RNA that suppresses tumor growth and glioma metastasis. These dendrimers were further functionalized with the tLyp-1 cell-penetrating peptide to enhance cellular uptake and with anti-NKG2A mAbs to stimulate anti-tumor immune responses by activating T and NK cells. Following IV administration in glioma-bearing mice, these multifunctional dendrimers demonstrated efficient BBB crossing *via* caveolae-mediated endocytosis, selective tumor targeting, reduced glioma growth and migration, and prolonged survival (122).

7.4.4 Alternative Strategies for Brain Delivery

7.4.4.1 BBB-Modulating Peptides

BBB-modulating peptides (BBBMPs) are a promising strategy to enhance paracellular permeability of the BBB involving the use of small cadherin-derived peptides. These peptides transiently disrupt protein–protein interactions between endothelial cells, effectively loosening tight junctions and enhancing the porosity of the BBB. The *in vitro* studies have shown that BBBM peptides inhibit calcium-dependent cell reaggregation (149), reduce transepithelial electrical resistance (150,151), and promote paracellular transport of markers such as ^{14}C-mannitol across tight junction-forming cell monolayers (150). *In vivo*, BBBM peptides have successfully enhanced brain delivery of a wide range of molecules in rodent models, including larger macromolecules, such as IgG mAb (150 kDa), albumin (67 kDa), IRDye800CW PEG (25 kDa), and brain-Derived Neurotrophic Factor (BDNF, 13 kDa), and small molecules (e.g., medium-sized peptides, gadopentetic acid, mannitol, anticancer agents). These enhancements have been quantitatively confirmed using imaging modalities like radioactive tracing, mass spectrometry, Near-Infrared Fluorescence (NIRF), and Magnetic Resonance Imaging (MRI) (2). Notably, co-administration of IgG mAbs with BBBM peptides in mice led to a 2-4 fold increase in brain antibody accumulation compared to mAb delivery alone. The enhancement was peptide-dependent, with cyclic peptides demonstrating superior efficacy over linear analogs. Furthermore, repeated administration of BBBM peptides following a single mAb injection resulted in higher brain deposition than a single co-dose, suggesting a cumulative permeation effect (152,153). While research findings indicate superior therapeutic outcomes using BBBM-enhanced delivery of BDNF in mouse models of AD and MS, similar evaluations for antibody therapies remain unexplored. Importantly, no significant safety concerns have been reported to date. Repeated peptide dosing in mice did not cause motor abnormalities, weight loss, astrogliosis, or neuroinflammation (154,155). Moreover, unlike the osmotic BBBD method, BBBM peptides did not affect cerebral blood flow

(156). Although BBBM-mediated delivery allows albumin penetration, which is known to be potentially neurotoxic, minimal toxicity has been observed. This could be attributed to the transient and size-restricted nature of the BBB opening. For instance, permeability was observed for larger molecules (Galbumin MW ~65 kDa) was about 40 min and for small molecules (Gd-DTPA MW ~500 Da), it was between 1 to 4 h. Further, BBBMs also exhibit molecular size selectivity, permitting transport of IgG mAbs (150 kDa) but not fibronectin (220 kDa), thereby potentially minimizing the entry of unwanted blood proteins into the CNS. Despite these advantages, further long-term safety studies, including dose escalation experiments, are necessary to fully assess potential adverse effects (2).

7.4.4.2 Enhancing Antibody Retention in the Brain

The mAbs demonstrated to undergo quick efflux following their delivery into the CNS, thereby limiting their therapeutic window and efficacy. To address this, strategies aimed at enhancing mAb retention within the brain are a promising solution to improve brain exposure. Although this area remains underexplored, one pivotal study has shown that binding of mAbs to neural targets within the CNS extracellular matrix can substantially increase their retention and cumulative concentration in the brain. Further, the study describes the brain exposure of an anti-myelin oligodendrocyte glycoprotein (anti-MOG) mAb, which particularly targets a brain-specific antigen, which was evaluated with anti-TfR mAb (a molecule known to traverse the BBB *via* non-targeting control mAb and RMT). While the anti-TfR mAb shows an increase in initial brain accumulation within the first 4 days, its levels decline quickly because of the systemic clearance, as TfR is abundantly expressed throughout peripheral tissues. In contrast, the anti-MOG mAb demonstrated prolonged CNS retention and improved brain concentrations over 10 days. This extended exposure was linked to its target engagement with the MOG protein, allowing the antibody to evade quick efflux and slowly accumulate in the brain despite low initial penetration. These results show the potential of target-specific retention as a powerful strategy to improve therapeutic efficacy and brain exposure of mAbs in the treatment of neurological disorders (1,2,157).

7.4.4.3 Osmotic BBB Opening

The osmotic BBB opening (OBBBO) shows (1970s) the transient and reversible disruption of the BBB in rabbits through intracarotid infusion of hyperosmolar electrolyte solutions (2). This strategy, also known as BBB disruption, works by inducing an osmotic withdrawal of water from BBB endothelial cells. This process results in cell separation and shrinkage of tight junctions, thereby improving paracellular permeability. Notably, this increased permeability is size-dependent, with smaller molecules traversing the barrier more readily than larger molecules. Despite its non-specific nature, BBB disruption has shown the capacity to significantly enhance the transport of high-molecular-weight proteins to the brain, including Fab fragments, antibodies such as IgM, IgG, and albumin. In primate models immunized against measles, BBB disruption facilitated a notable increase in endogenous IgG levels in the CNS, highlighting its potential to boost the efficacy of immunotherapies targeting brain infections. Additionally, the delivery of exogenous mAbs such as IgG was also enhanced following IV administration in rats subjected to BBB disruption. However, the clinical translation of BBB disruption is limited by safety concerns. The procedure results in a prolonged opening of the BBB, which can permit the infiltration of circulating neurotoxic plasma proteins like fibrinogen and albumin to enter brain tissue and potentially cause irreversible damage. For instance, uptake of these proteins has been associated with nerve damage and neuroinflammation. In rat models, BBB disruption has been shown to activate microglial cells, initiate a sterile inflammatory response, and alter cerebral blood flow. Moreover, clinical studies employing BBB disruption for oncologic drug delivery have reported a 13% incidence of seizures, further underscoring the need for caution and refinement in this method (2,158–161).

7.4.4.4 Intranasal Brain Delivery of Peptides and Proteins

The intranasal brain delivery has emerged as a promising method for delivering proteins and peptides directly to the brain. This approach avoids enzymatic degradation in the systemic circulation, which enhances the bioavailability of fragile peptide drugs compared to IV and IP administrations. This route has been explored for delivering molecules such as oxytocin and insulin, some of which have advanced to clinical trials for the treatment of neurodegenerative diseases, including AD and autism spectrum disorder. For successful brain delivery, molecules administered nasally must traverse the nasal epithelium and the cribriform plate, accessing the brain via pathways such as the olfactory bulb and the trigeminal nerve. To facilitate this process, various intranasal delivery enhancers have been investigated, including nanoparticle-based carriers, receptor-mediated transport mechanisms, cell-penetrating peptides, and tight junction disruptors (e.g., carnitines, ultrasound). While some of these enhancement strategies have progressed to clinical evaluation, many remain in the preclinical stage. Nevertheless, continued optimization of these delivery platforms may open new avenues for effective treatment of CNS disorders. Overall, these strategies highlight the ongoing research to overcome the challenges of brain targeting, focusing on improving BBB penetration, targeting efficiency, and reducing potential toxicity. However, most of these methods are still in the experimental or early clinical stages, and further studies are essential for their successful translation into clinical therapies for CNS diseases (2,162–164).

7.5 CONCLUSION

The development of effective delivery strategies for mAbs and other biotherapeutics across the BBB is pivotal to advancing the treatment of neurological disorders. Enhancing BBB penetration not only has the potential to improve therapeutic efficacy but also to reduce manufacturing costs, treatment frequency, and the overall healthcare burden. Among the mechanisms investigated, RMT has emerged as a particularly promising approach. The bsAbs targeting receptors such as TfR and IR have demonstrated the capacity to facilitate the transport of therapeutic antibodies across the BBB. Several RMT-based platforms have entered clinical trials for indications such as AD, MS, and glioblastoma. While TfR-targeted approaches dominate current investigations, alternative RMT receptors with greater specificity and expression in BECs warrant further exploration. Such diversification could help minimize peripheral exposure and improve brain-targeting precision. Notably, most studies to date have utilised IgG antibodies, which face inherent challenges in crossing the BBB due to their size. Consequently, smaller antibody formats such as antibody fragments scFvs, nanobodies, and engineered bsAbs are gaining interest for their potential to enhance BBB penetration and therapeutic performance. Understanding the mechanisms of mAb uptake, distribution, and clearance within the brain parenchyma is equally critical to optimizing drug exposure and pharmacodynamic outcomes (2,9,11,127).

Furthermore, combinatorial strategies, including the use of FUS to transiently disrupt the BBB and nanoparticle-based delivery systems by conjugating antibodies, offer synergistic advantages by enhancing delivery efficiency while maintaining safety. Despite encouraging preclinical and early clinical data, the translation of BBB-penetrating mAb therapies into approved treatments remains limited. Barriers, including safety concerns, inconsistent CNS delivery, and scalability challenges, have hindered widespread clinical success. Nonetheless, pharmaceutical companies and academic institutions are investing profoundly in dedicated brain-delivery platforms (2,9,11,40,48). Initiatives such as the Innovative Medicines Initiative's IM2PACT project exemplify collaborative efforts aimed at accelerating progress in this field. The mAbs are poised to become a cornerstone in the future landscape of CNS therapeutics (127). Continued innovation in antibody engineering, receptor targeting, and delivery technologies supported by robust translational research and well-designed clinical trials will be essential in transforming conceptual breakthroughs into clinical realities. As research advances, the next generation of brain-targeted mAb therapies holds strong potential to revolutionize the treatment of neurological and neurodegenerative diseases through safe, effective, and targeted delivery across the BBB.

REFERENCES

1. Noguchi Y, Kato M, Ozeki K, Ishigai M. Pharmacokinetics of an intracerebroventricularly administered antibody in rats. MAbs. 2017;9(7):1210–5.
2. Schwinghamer K, Siahaan TJ. Enhancing antibody exposure in the central nervous system: Mechanisms of uptake, clearance, and strategies for improved brain delivery. J Nanotheranostics. 2023;4(4):463–79.
3. Cummings J. Anti-amyloid monoclonal antibodies are transformative treatments that redefine Alzheimer's disease therapeutics. Drugs. 2023 May 15;83(7):569–76.
4. Gandy S, Ehrlich ME. Moving the needle on Alzheimer's disease with an anti-oligomer antibody. New Engl J Med. 2023;388(1):80–1.
5. Smith K, Garman L, Wrammert J, Zheng NY, Capra JD, Ahmed R et al. Rapid generation of fully human monoclonal antibodies specific to a vaccinating antigen. Nat Protoc. 2009 Mar 26;4(3):372–84.
6. Wu D, Chen Q, Chen X, Han F, Chen Z, Wang Y. The blood–brain barrier: Structure, regulation and drug delivery. Signal Transduct Target Ther. 2023 May 25;8(1):217.
7. Himes BT, Geiger PA, Ayasoufi K, Bhargav AG, Brown DA, Parney IF. Immunosuppression in glioblastoma: Current understanding and therapeutic implications. Front Oncol. 2021 Oct 28;11.
8. Ramapriyan R, Sun J, Curry A, Richardson LG, Ramesh T, Gaffey MA et al. The role of antibody-based therapies in neuro-oncology. Antibodies. 2023 Nov 13;12(4):74.
9. Bajracharya R, Caruso AC, Vella LJ, Nisbet RM. Current and emerging strategies for enhancing antibody delivery to the brain. Pharmaceutics. 2021 Nov 26;13(12):2014.
10. Lu RM, Hwang YC, Liu IJ, Lee CC, Tsai HZ, Li HJ et al. Development of therapeutic antibodies for the treatment of diseases. J Biomed Sci. 2020;27(1):1.
11. Gklinos P, Papadopoulou M, Stanulovic V, Mitsikostas DD, Papadopoulos D. Monoclonal antibodies as neurological therapeutics. Pharmaceuticals. 2021 Jan 26;14(2):92.
12. Liu KY, Howard R. Can we learn lessons from the FDA's approval of aducanumab? Nat Rev Neurol. 2021 Nov 17;17(11):715–22.
13. Kadry H, Noorani B, Cucullo L. A blood–brain barrier overview on structure, function, impairment, and biomarkers of integrity. Fluids Barriers CNS. 2020 Dec 18;17(1):69.
14. Pandit R, Chen L, Götz J. The blood-brain barrier: Physiology and strategies for drug delivery. Adv Drug Deliv Rev. 2020;165–166:1–14.
15. Sharif Y, Jumah F, Coplan L, Krosser A, Sharif K, Tubbs RS. Blood brain barrier: A review of its anatomy and physiology in health and disease. Clin Anat. 2018 Sep 18;31(6):812–23.
16. Barnabas W. Drug targeting strategies into the brain for treating neurological diseases. J Neurosci Methods. 2019;311:133–46.
17. Abbott NJ, Rönnbäck L, Hansson E. Astrocyte–endothelial interactions at the blood–brain barrier. Nat Rev Neurosci. 2006;7(1):41–53.
18. Armulik A, Genové G, Betsholtz C. Pericytes: Developmental, physiological, and pathological perspectives, problems, and promises. Dev Cell. 2011;21(2):193–215.
19. Schaeffer S, Iadecola C. Revisiting the neurovascular unit. Nat Neurosci. 2021;24(9):1198–209.
20. Iadecola C. The neurovascular unit coming of age: A journey through neurovascular coupling in health and disease. Neuron. 2017;96(1):17–42.
21. Xu L, Nirwane A, Yao Y. Basement membrane and blood–brain barrier. Stroke Vasc Neurol. 2019;4(2):78–82.
22. Pardridge WM. A historical review of brain drug delivery. Pharmaceutics. 2022 Jun 16;14(6):1283.
23. Pardridge WM. Blood–brain barrier delivery. Drug Discov Today. 2007;12(1–2):54–61.
24. Barar J, Rafi MA, Pourseif MM, Omidi Y. Blood-brain barrier transport machineries and targeted therapy of brain diseases. BioImpacts. 2016;6(4):225–48.
25. Schlachetzki F, Zhu C, Pardridge WM. Expression of the neonatal Fc receptor (FcRn) at the blood–brain barrier. J Neurochem. 2002 Apr 25;81(1):203–6.
26. Mantle JL, Lee KH. Immunoglobulin G transport increases in an in vitro blood–brain barrier model with amyloid-β and with neuroinflammatory cytokines. Biotechnol Bioeng. 2019;116(7):1752–61.

27. Ruano-Salguero JS, Lee KH. Antibody transcytosis across brain endothelial-like cells occurs nonspecifically and independent of FcRn. Sci Rep. 2020 Feb 28;10(1):3685.
28. Cooper PR, Ciambrone GJ, Kliwinski CM, Maze E, Johnson L, Li Q et al. Efflux of monoclonal antibodies from rat brain by neonatal fc receptor, FcRn. Brain Res. 2013;1534:13–21.
29. Deane R, Sagare A, Hamm K, Parisi M, LaRue B, Guo H et al. IgG-assisted age-dependent clearance of Alzheimer's amyloid β peptide by the blood–brain barrier neonatal fc receptor. J Neurosci. 2005 Dec 14;25(50):11495–503.
30. Zhang Y, Pardridge WM. Mediated efflux of IgG molecules from brain to blood across the blood–brain barrier. J Neuroimmunol. 2001;114(1–2):168–72.
31. Wolburg H, Noell S, Fallier-Becker P, Mack AF, Wolburg-Buchholz K. The disturbed blood–brain barrier in human glioblastoma. Mol Aspects Med. 2012;33(5–6):579–89.
32. Solar P, Hendrych M, Barak M, Valekova H, Hermanova M, Jancalek R. Blood-brain barrier alterations and edema formation in different brain mass lesions. Front Cell Neurosci. 2022 Jul 15;16.
33. Sarkaria JN, Hu LS, Parney IF, Pafundi DH, Brinkmann DH, Laack NN et al. Is the blood–brain barrier really disrupted in all glioblastomas? A critical assessment of existing clinical data. Neuro Oncol. 2018 Jan 22;20(2):184–91.
34. Dalmau J, Rosenfeld MR. Paraneoplastic syndromes of the CNS. Lancet Neurol. 2008;7(4):327–40.
35. Bard F, Cannon C, Barbour R, Burke RL, Games D, Grajeda H et al. Peripherally administered antibodies against amyloid β-peptide enter the central nervous system and reduce pathology in a mouse model of Alzheimer disease. Nat Med. 2000;6(8):916–9.
36. Spector R, Keep RF, Robert Snodgrass S, Smith QR, Johanson CE. A balanced view of choroid plexus structure and function: Focus on adult humans. Exp Neurol. 2015;267:78–86.
37. Rubenstein JL, Combs D, Rosenberg J, Levy A, McDermott M, Damon L et al. Rituximab therapy for CNS lymphomas: Targeting the leptomeningeal compartment. Blood. 2003 Jan 15;101(2):466–8.
38. Pardridge WM. CSF, blood-brain barrier, and brain drug delivery. Expert Opin Drug Deliv. 2016;13(7):963–75.
39. Reiber H. Proteins in cerebrospinal fluid and blood: Barriers, CSF flow rate and source-related dynamics. Restor Neurol Neurosci. 2003;21(3–4):79–96.
40. Tabrizi M, Bornstein GG, Suria H. Biodistribution mechanisms of therapeutic monoclonal antibodies in health and disease. AAPS J. 2010 Mar 19;12(1):33–43.
41. Castelli MS, McGonigle P, Hornby PJ. The pharmacology and therapeutic applications of monoclonal antibodies. Pharmacol Res Perspect. 2019 Dec 20;7(6).
42. Ryman JT, Meibohm B. Pharmacokinetics of monoclonal antibodies. CPT Pharmacometrics Syst Pharmacol. 2017;6(9):576–88.
43. Desjarlais JR, Lazar GA, Zhukovsky EA, Chu SY. Optimizing engagement of the immune system by anti-tumor antibodies: An engineer's perspective. Drug Discov Today. 2007;12(21–22):898–910.
44. Nimmerjahn F, Ravetch JV. Fcγ receptors: Old friends and new family members. Immunity. 2006;24(1):19–28.
45. Ravetch JV, Bolland S. IgG fc receptors. Annu Rev Immunol. 2001;19(1):275–90.
46. Breedveld F. Therapeutic monoclonal antibodies. The Lancet. 2000;355(9205):735–40.
47. Geng X, Kong X, Hu H, Chen J, Yang F, Liang H et al. Research and development of therapeutic mAbs: An analysis based on pipeline projects. Hum Vaccin Immunother. 2015;11(12):2769–76.
48. Weiner LM, Surana R, Wang S. Monoclonal antibodies: versatile platforms for cancer immunotherapy. Nat Rev Immunol. 2010;10(5):317–27.
49. Kinder M, Greenplate AR, Strohl WR, Jordan RE, Brezski RJ. An fc engineering approach that modulates antibody-dependent cytokine release without altering cell-killing functions. MAbs. 2015;7(3):494–504.
50. An Z, Forrest G, Moore R, Cukan M, Haytko P, Huang L et al. IgG2m4, an engineered antibody isotype with reduced Fc function. MAbs. 2009 Nov 27;1(6):572–9.
51. Carter PJ. Potent antibody therapeutics by design. Nat Rev Immunol. 2006;6(5):343–57.
52. Steinitz M. Three decades of human monoclonal antibodies: Past, present and future developments. Hum Antibodies. 2009 Apr 29;18(1–2):1–10.
53. Chirmule N, Jawa V, Meibohm B. Immunogenicity to therapeutic proteins: Impact on PK/PD and efficacy. AAPS J. 2012 Mar 10;14(2):296–302.
54. van den Bemt BJ, den Broeder AA, Wolbink G, Hekster YA, van Riel PL, Benraad B et al. Anti-infliximab antibodies are already detectable in most patients with rheumatoid arthritis halfway through an infusioncycle: An open-label pharmacokinetic cohort study. BMC Musculoskelet Disord. 2011 Dec 13;12(1):12.

55. Kouhi A, Pachipulusu V, Kapenstein T, Hu P, Epstein AL, Khawli LA. Brain disposition of antibody-based therapeutics: Dogma, approaches and perspectives. Int J Mol Sci. 2021 Jun 16;22(12):6442.
56. Grabrucker AM, Ruozi B, Belletti D, Pederzoli F, Forni F, Vandelli MA et al. Nanoparticle transport across the blood brain barrier. Tissue Barriers. 2016;4(1):e1153568.
57. Yu YJ, Watts RJ. Developing therapeutic antibodies for neurodegenerative disease. Neurotherapeutics. 2013;10(3):459–72.
58. Herda LM, Polo E, Kelly PM, Rocks L, Hudecz D, Dawson KA. Designing the future of nanomedicine: Current barriers to targeted brain therapeutics. Eur J Nanomed. 2014;6(3).
59. Cavaco M, Gaspar D, ARB Castanho M, Neves V. Antibodies for the treatment of brain metastases, a dream or a reality? Pharmaceutics. 2020 Jan 13;12(1):62.
60. Khoy K, Mariotte D, Defer G, Petit G, Toutirais O, Le Mauff B. Natalizumab in multiple sclerosis treatment: From biological effects to immune monitoring. Front Immunol. 2020;11:549842.
61. Chowdhury S. Monoclonal antibody treatments for Alzheimer's disease: Aducanumab and lecanemab. Discoveries. 2023 Sep 25;11(3):e173.
62. Antibodies in Late-Stage Clinical Studies [Internet]. [cited 2025 Jun 9]. 2024 Dec. Available from: https://www.antibodysociety.org/antibodies-in-late-stage-clinical-studies/
63. Antibody Therapeutics Approved or in Regulatory Review in the EU or US [Internet]. [cited 2025 Jun 9]. 2024. Available from: https://www.antibodysociety.org/resources/approved-antibodies/
64. Ashina M, Saper J, Cady R, Schaeffler BA, Biondi DM, Hirman J et al. Eptinezumab in episodic migraine: A randomized, double-blind, placebo-controlled study (PROMISE-1). Cephalalgia. 2020 Mar 19;40(3):241–54.
65. Kieseier BC, Stüve O, Dehmel T, Goebels N, Leussink VI, Mausberg AK et al. Disease amelioration with tocilizumab in a treatment-resistant patient with neuromyelitis optica. JAMA Neurol. 2013;70(3):390.
66. Araki M, Aranami T, Matsuoka T, Nakamura M, Miyake S, Yamamura T. Clinical improvement in a patient with neuromyelitis optica following therapy with the anti-IL-6 receptor monoclonal antibody tocilizumab. Mod Rheumatol. 2013 Jul 11;23(4):827–31.
67. Kaneko A. Tocilizumab in rheumatoid arthritis: Efficacy, safety and its place in therapy. Ther Adv Chronic Dis. 2013;4(1):15–21.
68. Traboulsee A, Greenberg BM, Bennett JL, Szczechowski L, Fox E, Shkrobot S et al. Safety and efficacy of satralizumab monotherapy in neuromyelitis optica spectrum disorder: A randomised, double-blind, multicentre, placebo-controlled phase 3 trial. Lancet Neurol. 2020;19(5):402–12.
69. Yamamura T, Kleiter I, Fujihara K, Palace J, Greenberg B, Zakrzewska-Pniewska B et al. Trial of satralizumab in neuromyelitis optica spectrum disorder. New Engl J Med. 2019 Nov 28;381(22):2114–24.
70. Karlin L, Coiffier B. Ofatumumab in the treatment of non-Hodgkin's lymphomas. Expert Opin Biol Ther. 2015;15(7):1085–91.
71. Hauser SL, Bar-Or A, Cohen JA, Comi G, Correale J, Coyle PK et al. Ofatumumab versus teriflunomide in multiple sclerosis. New Engl J Med. 2020;383(6):546–57.
72. Frampton JE. Inebilizumab: First approval. Drugs. 2020 Aug 29;80(12):1259–64.
73. Agius MA, Klodowska-Duda G, Maciejowski M, Potemkowski A, Li J, Patra K et al. Safety and tolerability of inebilizumab (MEDI-551), an anti-CD19 monoclonal antibody, in patients with relapsing forms of multiple sclerosis: Results from a phase 1 randomised, placebo-controlled, escalating intravenous and subcutaneous dose study. Mult Sclerosis J. 2019 Feb 16;25(2):235–45.
74. Nishimura J, ichi, Yamamoto M, Hayashi S, Ohyashiki K, Ando K, Brodsky AL et al. Genetic variants in C5 and poor response to eculizumab. New Engl J Med. 2014 Feb 13;370(7):632–9.
75. Howard JF, Utsugisawa K, Benatar M, Murai H, Barohn RJ, Illa I et al. Safety and efficacy of eculizumab in anti-acetylcholine receptor antibody-positive refractory generalised myasthenia gravis (REGAIN): A phase 3, randomised, double-blind, placebo-controlled, multicentre study. Lancet Neurol. 2017;16(12):976–86.
76. Akaishi T, Nakashima I. Efficiency of antibody therapy in demyelinating diseases. Int Immunol. 2017;29(7):327–35.
77. Xue T, Yang Y, Lu Q, Gao B, Chen Z, Wang Z. Efficacy and safety of monoclonal antibody therapy in neuromyelitis optica spectrum disorders: Evidence from randomized controlled trials. Mult Scler Relat Disord. 2020;43:102166.
78. Pittock SJ, Berthele A, Fujihara K, Kim HJ, Levy M, Palace J et al. Eculizumab in aquaporin-4–Positive neuromyelitis optica spectrum disorder. New Engl J Med. 2019 Aug 15;381(7):614–25.
79. Martinez JM, Hindiyeh N, Anglin G, Kalidas K, Hodsdon ME, Kielbasa W et al. Assessment of immunogenicity from galcanezumab phase 3 trials in patients with episodic or chronic migraine. Cephalalgia. 2020 Aug 27;40(9):978–89.

80. Camporeale A, Kudrow D, Sides R, Wang S, Van Dycke A, Selzler KJ et al. A phase 3, long-term, open-label safety study of galcanezumab in patients with migraine. BMC Neurol. 2018;18(1):188.
81. Goadsby PJ, Dodick DW, Leone M, Bardos JN, Oakes TM, Millen BA et al. Trial of galcanezumab in prevention of episodic cluster headache. New Engl J Med. 2019 Jul 11;381(2):132–41.
82. Gklinos P, Mitsikostas DD. Galcanezumab in migraine prevention: A systematic review and meta-analysis of randomized controlled trials. Ther Adv Neurol Disord. 2020 Jan 28;13.
83. Skljarevski V, Matharu M, Millen BA, Ossipov MH, Kim BK, Yang JY. Efficacy and safety of galcanezumab for the prevention of episodic migraine: Results of the EVOLVE-2 phase 3 randomized controlled clinical trial. Cephalalgia. 2018 Jul 31;38(8):1442–54.
84. Stauffer VL, Dodick DW, Zhang Q, Carter JN, Ailani J, Conley RR. Evaluation of galcanezumab for the prevention of episodic migraine. JAMA Neurol. 2018;75(9):1080.
85. Dodick DW, Silberstein SD, Bigal ME, Yeung PP, Goadsby PJ, Blankenbiller T et al. Effect of fremanezumab compared with placebo for prevention of episodic migraine. JAMA. 2018 May 15;319(19):1999.
86. Goadsby PJ, Silberstein SD, Yeung PP, Cohen JM, Ning X, Yang R et al. Long-term safety, tolerability, and efficacy of fremanezumab in migraine. Neurology. 2020;95(18).
87. Goadsby PJ, Reuter U, Hallström Y, Broessner G, Bonner JH, Zhang F et al. A controlled trial of erenumab for episodic migraine. New Engl J Med. 2017 Nov 30;377(22):2123–32.
88. Ashina M, Goadsby PJ, Reuter U, Silberstein S, Dodick DW, Xue F et al. Long-term efficacy and safety of erenumab in migraine prevention: Results from a 5-year, open-label treatment phase of a randomized clinical trial. Eur J Neurol. 2021 May 20;28(5):1716–25.
89. Dodick DW, Ashina M, Brandes JL, Kudrow D, Lanteri-Minet M, Osipova V et al. ARISE: A phase 3 randomized trial of erenumab for episodic migraine. Cephalalgia. 2018 May 22;38(6):1026–37.
90. Tepper S, Ashina M, Reuter U, Brandes JL, Doležil D, Silberstein S et al. Safety and efficacy of erenumab for preventive treatment of chronic migraine: A randomised, double-blind, placebo-controlled phase 2 trial. Lancet Neurol. 2017;16(6):425–34.
91. Ashina M, Dodick D, Goadsby PJ, Reuter U, Silberstein S, Zhang F et al. Erenumab (AMG 334) in episodic migraine. Neurology. 2017 Sep 19;89(12):1237–43.
92. Affronti M, Lou, Jackman JG, McSherry F, Herndon JE, Massey EC, Lipp E et al. Phase II study to evaluate the efficacy and safety of rilotumumab and bevacizumab in subjects with recurrent malignant glioma. Oncologist. 2018;23(8):889–e98.
93. Norden AD, Young GS, Setayesh K, Muzikansky A, Klufas R, Ross GL et al. Bevacizumab for recurrent malignant gliomas. Neurology. 2008;70(10):779–87.
94. Chamberlain MC, Johnston S. Bevacizumab for recurrent alkylator-refractory anaplastic oligodendroglioma. Cancer. 2009 Apr 15;115(8):1734–43.
95. Wick W, Gorlia T, Bendszus M, Taphoorn M, Sahm F, Harting I et al. Lomustine and bevacizumab in progressive glioblastoma. New Engl J Med. 2017 Nov 16;377(20):1954–63.
96. Fisher TL, Reilly CA, Winter LA, Pandina T, Jonason A, Scrivens M et al. Generation and preclinical characterization of an antibody specific for SEMA4D. MAbs. 2016;8(1):150–62.
97. Ciric B, Howe CL, Soldan MP, Warrington AE, Bieber AJ, Van Keulen V et al. Human monoclonal IgM antibody promotes CNS myelin repair independent of Fc function. Brain Pathology. 2003;13(4):608–16.
98. Ruggieri S, Tortorella C, Gasperini C. Anti lingo 1 (opicinumab) a new monoclonal antibody tested in relapsing remitting multiple sclerosis. Expert Rev Neurother. 2017;17(11):1081–9.
99. Ineichen BV, Plattner PS, Good N, Martin R, Linnebank M, Schwab ME. Nogo-a antibodies for progressive multiple sclerosis. CNS Drugs. 2017 Mar 19;31(3):187–98.
100. Florian H, Arnold SE, Bateman R, Braunstein JB, Budur K, Kerwin DR, et al. P1-050: Abbv-8e12, a humanized anti-tau monoclonal antibody, for treating early Alzheimer's disease: Updated design and baseline characteristics of a phase 2 study. *Alzheimers Dementia*. 2019;15(7S_Part_4):251–252.
101. Kariolis MS, Wells RC, Getz JA, Kwan W, Mahon CS, Tong R, et al. Brain delivery of therapeutic proteins using an Fc fragment blood-brain barrier transport vehicle in mice and monkeys. Sci Transl Med. 2020 May 27;12(545):1–13.
102. Faresjö R, Bonvicini G, Fang XT, Aguilar X, Sehlin D, Syvänen S. Brain pharmacokinetics of two BBB penetrating bispecific antibodies of different size. Fluids Barriers CNS. 2021;18(1):26.
103. Do TM, Capdevila C, Pradier L, Blanchard V, Lopez-Grancha M, Schussler N et al. Tetravalent bispecific tandem antibodies improve brain exposure and efficacy in an amyloid transgenic mouse model. Mol Ther Methods Clin Dev. 2020;19:58–77.

104. Hultqvist G, Syvänen S, Fang XT, Lannfelt L, Sehlin D. Bivalent brain shuttle increases antibody uptake by monovalent binding to the transferrin receptor. Theranostics. 2017;7(2):308–18.
105. Niewoehner J, Bohrmann B, Collin L, Urich E, Sade H, Maier P et al. Increased brain penetration and potency of a therapeutic antibody using a monovalent molecular shuttle. Neuron. 2014;81(1):49–60.
106. Weber F, Bohrmann B, Niewoehner J, Fischer JAA, Rueger P, Tiefenthaler G et al. Brain shuttle antibody for Alzheimer's disease with attenuated peripheral effector function due to an inverted binding mode. Cell Rep. 2018;22(1):149–62.
107. Sehlin D, Stocki P, Gustavsson T, Hultqvist G, Walsh FS, Rutkowski JL et al. Brain delivery of biologics using a cross-species reactive transferrin receptor 1 VNAR shuttle. FASEB J. 2020 Oct 11;34(10):13272–83.
108. Park EJ, Zhang YZ, Vykhodtseva N, McDannold N. Ultrasound-mediated blood-brain/blood-tumor barrier disruption improves outcomes with trastuzumab in a breast cancer brain metastasis model. J Control Release. 2012;163(3):277–84.
109. Zhang H, Sierra C, Kwon N, Karakatsani M, Burgess M, Jackson-Lewis VR, et al. Focused-Ultrasound Mediated Anti-Alpha-Synuclein Antibody Delivery for the Treatment of Parkinson's Disease. In Proceedings of the 2018 IEEE International Ultrasonics Symposium (IUS). 2018 Oct;1–4.
110. Bajracharya R, Cruz E, Götz J, Nisbet RM. Ultrasound-mediated delivery of novel tau-specific monoclonal antibody enhances brain uptake but not therapeutic efficacy. J Control Release. 2022;349:634–48.
111. Janowicz PW, Leinenga G, Götz J, Nisbet RM. Ultrasound-mediated blood-brain barrier opening enhances delivery of therapeutically relevant formats of a tau-specific antibody. Sci Rep. 2019 Jun 25;9(1):9255.
112. Nisbet RM, Van der Jeugd A, Leinenga G, Evans HT, Janowicz PW, Götz J. Combined effects of scanning ultrasound and a tau-specific single chain antibody in a tau transgenic mouse model. Brain. 2017;140(5):1220–30.
113. Sun T, Shi Q, Zhang Y, Power C, Hoesch C, Antonelli S et al. Focused ultrasound with anti-pGlu3 aβ enhances efficacy in Alzheimer's disease-like mice via recruitment of peripheral immune cells. J Control Release. 2021;336:443–56.
114. Raymond SB, Treat LH, Dewey JD, McDannold NJ, Hynynen K, Bacskai BJ. Ultrasound enhanced delivery of molecular imaging and therapeutic agents in Alzheimer's disease mouse models. PLoS One. 2008 May 14;3(5):e2175.
115. Leinenga G, Koh WK, Götz J. A comparative study of the effects of aducanumab and scanning ultrasound on amyloid plaques and behavior in the APP23 mouse model of Alzheimer disease. Alzheimers Res Ther. 2021;13(1):76.
116. Liu HL, Hsu PH, Lin CY, Huang CW, Chai WY, Chu PC et al. Focused ultrasound enhances Central nervous system delivery of bevacizumab for malignant glioma treatment. Radiology. 2016;281(1):99–108.
117. Jordão JF, Ayala-Grosso CA, Markham K, Huang Y, Chopra R, McLaurin J et al. Antibodies targeted to the brain with image-guided focused ultrasound reduces amyloid-β plaque load in the TgCRND8 mouse model of Alzheimer's disease. PLoS One. 2010 May 11;5(5):e10549.
118. Alecou T, Giannakou M, Damianou C. Amyloid β plaque reduction with antibodies crossing the blood-brain barrier, which was opened in 3 sessions of focused ultrasound in a rabbit model. J Ultrasound Med. 2017 Nov 24;36(11):2257–70.
119. Johnsen KB, Bak M, Kempen PJ, Melander F, Burkhart A, Thomsen MS et al. Antibody affinity and valency impact brain uptake of transferrin receptor-targeted gold nanoparticles. Theranostics. 2018;8(12):3416–36.
120. Rizzuto MA, Dal Magro R, Barbieri L, Pandolfi L, Sguazzini-Viscontini A, Truffi M et al. H-ferritin nanoparticle-mediated delivery of antibodies across a BBB *in vitro* model for treatment of brain malignancies. Biomater Sci. 2021;9(6):2032–42.
121. Kim JS, Shin DH, Kim JS. Dual-targeting immunoliposomes using angiopep-2 and CD133 antibody for glioblastoma stem cells. Journal of Controlled Release. 2018;269:245–57.
122. Jin Z, Piao L, Sun G, Lv C, Jing Y, Jin R. Dual functional nanoparticles efficiently across the blood–brain barrier to combat glioblastoma *via* simultaneously inhibit the PI3K pathway and NKG2A axis. J Drug Target. 2021 Mar 16;29(3):323–35.
123. Yu YJ, Zhang Y, Kenrick M, Hoyte K, Luk W, Lu Y et al. Boosting brain uptake of a therapeutic antibody by reducing its affinity for a transcytosis target. Sci Transl Med. 2011 May 25;3(84).
124. He P, Xin W, Schulz P, Sierks MR. Bispecific antibody fragment targeting APP and inducing α-site cleavage restores neuronal health in an Alzheimer's mouse model. Mol Neurobiol. 2019 Nov 30;56(11):7420–32.

125. Li J, Zhang Z, Lv L, Qiao H, Chen X, Zou C. A bispecific antibody (ScBsAbAgn-2/TSPO) target for ang-2 and TSPO resulted in therapeutic effects against glioblastomas. Biochem Biophys Res Commun. 2016;472(2):384–91.
126. Leitner DF, Connor JR. Functional roles of transferrin in the brain. Biochim Biophys Acta. 2012;1820(3):393–402.
127. Terstappen GC, Meyer AH, Bell RD, Zhang W. Strategies for delivering therapeutics across the blood–brain barrier. Nat Rev Drug Discov. 2021;20(5):362–83.
128. Boado RJ, Pardridge WM. Brain and organ uptake in the rhesus monkey in vivo of recombinant iduronidase compared to an insulin receptor antibody–iduronidase fusion protein. Mol Pharm. 2017;14(4):1271–7.
129. Giugliani R, Giugliani L, de Oliveira Poswar F, Donis KC, Corte AD, Schmidt M et al. Neurocognitive And somatic stabilization in pediatric patients with severe mucopolysaccharidosis type i after 52 weeks of intravenous brain-penetrating insulin receptor Aatibody-iduronidase fusion protein (valanafusp alpha): An open label phase 1-2 trial. Orphanet J Rare Dis. 2018;13(1):110.
130. Chen KT, Wei KC, Liu HL. Theranostic strategy of focused ultrasound induced blood-brain barrier opening for CNS disease treatment. Front Pharmacol. 2019;10.
131. Sheikov N, McDannold N, Vykhodtseva N, Jolesz F, Hynynen K. Cellular mechanisms of the blood-brain barrier opening induced by ultrasound in presence of microbubbles. Ultrasound Med Biol. 2004;30(7):979–89.
132. Sheikov N, McDannold N, Sharma S, Hynynen K. Effect of focused ultrasound applied with an ultrasound contrast agent on the tight junctional integrity of the brain microvascular endothelium. Ultrasound Med Biol. 2008;34(7):1093–104.
133. Chen H, Konofagou EE. The size of Blood–Brain barrier opening induced by focused ultrasound is dictated by the acoustic pressure. J Cereb Blood Flow Meta. 2014 Jul 30;34(7):1197–204.
134. Choi JJ, Wang S, Tung YS, Morrison B, Konofagou EE. Molecules of various pharmacologically-relevant sizes can cross the ultrasound-induced blood-brain barrier opening in vivo. Ultrasound Med Biol. 2010;36(1):58–67.
135. Chen PY, Hsieh HY, Huang CY, Lin CY, Wei KC, Liu HL. Focused ultrasound-induced blood–brain barrier opening to enhance interleukin-12 delivery for brain tumor immunotherapy: A preclinical feasibility study. J Transl Med. 2015 Dec 17;13(1):93.
136. Leinenga G, Götz J. Scanning ultrasound removes amyloid-β and restores memory in an Alzheimer's disease mouse model. Sci Transl Med. 2015 Mar 11;7(278).
137. Leinenga G, Bodea LG, Koh WK, Nisbet RM, Götz J Delivery of antibodies into the brain using focused scanning ultrasound. Journal of Visualized Experiments. 2020 Jul 18;161:e61372.
138. Yang Z, Liu ZW, Allaker RP, Reip P, Oxford J, Ahmad Z et al. A review of nanoparticle functionality and toxicity on the central nervous system. J R Soc Interface. 2010;7(suppl_4).
139. Fratoddi I, Venditti I, Cametti C, Russo MV. Gold nanoparticles and gold nanoparticle-conjugates for delivery of therapeutic molecules. Progress and challenges. J Mater Chem B. 2014;2(27):4204–20.
140. Bulbake U, Doppalapudi S, Kommineni N, Khan W. Liposomal formulations in clinical use: An updated review. Pharmaceutics. 2017 Mar 27;9(2):12.
141. Wolfram J, Zhu M, Yang Y, Shen J, Gentile E, Paolino D et al. Safety of nanoparticles in medicine. Curr Drug Targets. 2015 Nov 19;16(14):1671–81.
142. Wang L, Tang S, Yu Y, Lv Y, Wang A, Yan X et al. Intranasal delivery of temozolomide-conjugated gold nanoparticles functionalized with anti-EphA3 for glioblastoma targeting. Mol Pharm. 2021;18(3): 915–27.
143. Yin W, Zhao Y, Kang X, Zhao P, Fu X, Mo X et al. BBB-penetrating codelivery liposomes treat brain metastasis of non-small cell lung cancer with $EGFR^{T790M}$ mutation. Theranostics. 2020;10(14):6122–35.
144. Zhang W, Liu QY, Haqqani AS, Leclerc S, Liu Z, Fauteux F et al. Differential expression of receptors mediating receptor-mediated transcytosis (RMT) in brain microvessels, brain parenchyma and peripheral tissues of the mouse and the human. Fluids Barriers CNS. 2020 Dec 22;17(1):47.
145. Kumar M, Sharma P, Maheshwari R, Tekade M, Shrivastava SK, Tekade RK. Beyond the blood–brain barrier. In: Nanotechnology-Based Targeted Drug Delivery Systems for Brain Tumors. Elsevier; 2018. p. 397–437.
146. Kim ID, Shin JH, Kim SW, Choi S, Ahn J, Han PL et al. Intranasal delivery of HMGB1 siRNA confers target Gene knockdown and robust neuroprotection in the postischemic brain. Mol Ther. 2012;20(4):829–39.

147. Srinageshwar B, Peruzzaro S, Andrews M, Johnson K, Hietpas A, Clark B et al. PAMAM dendrimers cross the Blood–Brain barrier when administered through the carotid artery in C57BL/6J mice. Int J Mol Sci. 2017 Mar 14;18(3):628.
148. Sharma A, Porterfield JE, Smith E, Sharma R, Kannan S, Kannan RM. Effect of mannose targeting of hydroxyl PAMAM dendrimers on cellular and organ biodistribution in a neonatal brain injury model. Journal of Controlled Release. 2018;283:175–89.
149. Lutz KL, Siahaan TJ. Modulation of the cellular junction protein e-cadherin in bovine brain microvessel endothelial cells by cadherin peptides. Drug Deliv. 1997 Jan 27;4(3):187–93.
150. Sinaga E, Jois SDS, Avery M, Makagiansar IT, Tambunan USF, Audus KL et al. Increasing paracellular porosity by e-cadherin peptides: Discovery of bulge and groove regions in the EC1-domain of e-cadherin. Pharm Res. 2002;19(8):1170–9.
151. Makagiansar IT, Avery M, Hu Y, Audus KL, Siahaan TJ. Improving the selectivity of HAV-peptides in modulating e-cadherin-e-cadherin interactions in the intercellular junction of MDCK cell monolayers. Pharm Res. 2001;18(4):446–53.
152. Ulapane KR, Kopec BM, Siahaan TJ. *In vivo* brain delivery and brain deposition of proteins with various sizes. Mol Pharm. 2019;16(12):4878–89.
153. Ulapane KR, Kopec BM, Siahaan TJ. Improving in vivo brain delivery of monoclonal antibody using novel cyclic peptides. Pharmaceutics. 2019 Oct 31;11(11):568.
154. Kopec B, Kiptoo P, Zhao L, Rosa-Molinar E, Siahaan T. Noninvasive Brain Delivery and Efficacy of BDNF to Stimulate Neuroregeneration and Suppression of Disease Relapse in EAE Mice. Mol Pharm. 2019 Dec 31;17(2):553–561.
155. Kopec B, Zhao L, Rosa-Molinar E, Siahaan T. Non-invasive brain delivery and efficacy of BDNF in APP/PS1 transgenic mice. Med Res Arch. 2020;8(2).
156. Sajesh BV, On NH, Omar R, Alrushaid S, Kopec BM, Wang WG et al. Validation of cadherin HAV6 peptide in the transient modulation of the blood-brain barrier for the treatment of brain tumors. Pharmaceutics. 2019 Sep 17;11(9):481.
157. Nakano R, Takagi-Maeda S, Ito Y, Kishimoto S, Osato T, Noguchi K et al. A new technology for increasing therapeutic protein levels in the brain over extended periods. PLoS One. 2019 Apr 12;14(4):e0214404.
158. Burks SR, Kersch CN, Witko JA, Pagel MA, Sundby M, Muldoon LL et al. Blood–brain barrier opening by intracarotid artery hyperosmolar mannitol induces sterile inflammatory and innate immune responses. Proc Natl Acad Sci. 2021;118(18).
159. Petersen MA, Ryu JK, Chang KJ, Etxeberria A, Bardehle S, Mendiola AS et al. Fibrinogen activates BMP signaling in oligodendrocyte progenitor cells and inhibits remyelination after vascular damage. Neuron. 2017;96(5):1003–1012.e7.
160. Muldoon LL, Neuwelt EA. BR96–DOX immunoconjugate targeting of chemotherapy in brain tumor models. J Neurooncol. 2003;65(1):49–62.
161. Remsen LG, Trail PA, Hellström I, Hellström KE, Neuwelt EA. Enhanced delivery improves the efficacy of a tumor-specific doxorubicin immunoconjugate in a human brain tumor xenograft model. Neurosurgery. 2000;46(3):704–9.
162. Tanaka A, Furubayashi T, Arai M, Inoue D, Kimura S, Kiriyama A et al. Delivery of oxytocin to the brain for the treatment of autism spectrum disorder by nasal application. Mol Pharm. 2018;15(3):1105–11.
163. Samaridou E, Alonso MJ. Nose-to-brain peptide delivery – The potential of nanotechnology. Bioorg Med Chem. 2018;26(10):2888–905.
164. Meredith ME, Salameh TS, Banks WA. Intranasal delivery of proteins and peptides in the treatment of neurodegenerative diseases. AAPS J. 2015 Jul 24;17(4):780–7.

8 *In Vitro* Evaluations of Drug Delivery to the Brain

Touraj Ehtezazi

8.1 INTRODUCTION

The blood–brain barrier (BBB) is a physiological system that protects the brain from toxic compounds circulating in the bloodstream. The BBB is composed of endothelial cells, which typically encircle to form the lumen of the blood vessel. When the ends of the parts reach each other they almost overlap and the gap is filled with certain proteins such as occludin, claudins, and cadherins (reviewed in (1, 2)). The abluminal side of brain endothelial cells is covered by a basal membrane, with pericyte cells wrapped around the endothelial cells (3) and astrocyte end-foot projections in contact. There are several mechanisms that drug molecules or nanoparticles (NPs) may cross the BBB. Small molecules typically cross the BBB via diffusion through the tight junctions, while receptor-mediated BBB transversing has been employed for antibodies (4) and NPs (5). During the development of brain-targeting drug molecules and NPs, it is essential to obtain confirmation of crossing the BBB, before *in vivo* investigations. This is to minimise the number of animals that are used in pre-clinical studies. Several BBB *in vitro* models have been developed (reviewed in (2)). Here the simplest method will be discussed that is widely used, and also we observed it to be sufficient to predict BBB targeting *in vivo*.

BBB models offer numerous advantages for pharmaceutical research and basic science. They provide a relatively simple, cost-effective platform for high-throughput screening of drug candidates. Both passive diffusion and active transport mechanisms have been evaluated by examining the directional movement of compounds. The system also allows investigation of BBB dysfunction in disease states by optimised differentiation of human induced pluripotent stem cells (6) and deletion of genes such as *FOXF2*, by exposing cells to inflammatory mediators such as TNF-α and IL-1β (7), or disease-relevant conditions such as cerebral amyloid angiopathy (8).

8.2 TRANSWELL *IN VITRO* BBB MODEL

Transwell systems consist of cell culture inserts with microporous membranes that fit into standard well plates, creating two distinct compartments: an apical chamber representing the blood side, and a basolateral chamber representing the brain side (Figure 8.1).

This configuration allows investigators to culture brain endothelial cells on the membrane, recreating the physical barrier between circulation and neural tissue (9). The porous membrane typically has a pore size ranging from 0.4 to 3.0 μm, permitting molecular exchange while supporting cell growth. The most common approach involves culturing primary or immortalised brain microvascular endothelial cells such as human hCMEC/D3 (9) or mouse brain endothelial cells b.End3 (10) on the apical surface of the transwell membrane. Special media are used to culture the brain endothelial cells, for example, EndoGRO media (SCME004) supplemented with hydrocortisone, ascorbic acid, heparin sulfate, rhEGF, L-glutamine, EndoGro-LS, and 5% foetal bovine serum is used for human brain endothelial cells. A tissue culture insert is placed in each well of a 24-well plate containing 1.6 mL of pre-warmed culture media (37°C; the basolateral compartment). After 24 h of incubation, transepithelial/endothelial electrical resistance (TEER) is measured, with values above 150–200 $\Omega \cdot cm^2$ generally indicating a functional barrier. These cells form tight junctions, characteristic of the BBB, creating a selective barrier that restricts paracellular transport.

DOI: 10.1201/9781003482406-8

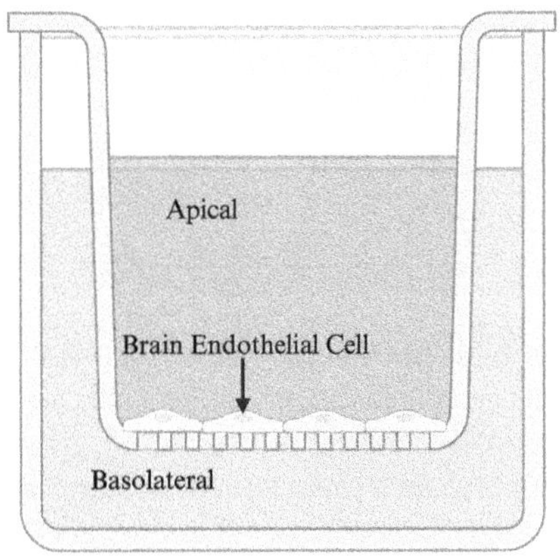

FIGURE 8.1 Schematic representation of a transwell *in vitro* BBB model, demonstrating a two-compartmental model, with an apical side representing the luminal side of the BBB, and a basolateral compartment representing the abluminal side of the BBB. The brain endothelial cells are cultured on the porous membrane of the insert.

The functionality of transwell BBB models is typically assessed through several key measurements. TEER provides real-time monitoring of tight junction formation and barrier integrity. Permeability studies involve adding test compounds to the apical chamber and measuring their appearance in the basolateral compartment over time, allowing calculation of apparent permeability coefficients. These studies help predict whether drugs can cross the BBB and reach therapeutic targets in the brain.

Recent advances have sought to address these limitations through dynamic culture systems incorporating flow and more sophisticated multi-cell type arrangements. Nevertheless, standard transwell BBB models remain valuable tools that bridge the gap between simple cell cultures and animal studies, contributing significantly to our understanding of brain barrier function and facilitating the development of neurotherapeutic agents.

8.2.1 Permeability Assessment

The permeability of the endothelial cells should be assessed for paracellular probes such as sucrose (10), as well as transcellular probes such as propranolol (10). In addition, the impermeability of the endothelial cell barrier should be evaluated using large molecular weight compounds such as FITC dextran with Mw = 70 kDa (9). To demonstrate the functionality of the BBB model, the BBB monolayer is exposed to compounds such as histamine, which would increase the permeability of the monolayer to large and charged molecules such as siRNA (9). To demonstrate that the increased BBB monolayer permeability was due to the interaction of histamine with histamine receptors, cimetidine (an H_2-receptor antagonist) may be added with histamine, when the BBB monolayer permeability should decrease to siRNA compared to the addition of histamine only (9).

8.2.2 Calculating Permeability Coefficient

The permeability coefficient (P_{app}) of an *in vitro* BBB model is calculated by plotting basolateral measurements (such as fluorescence or radioactivity) versus time. P_{app} is calculated from Equation 8.1 (11):

$$P_{app} = \frac{\Delta Q_{well}}{\Delta T \times A \times C_0} \tag{8.1}$$

where C_0 is the donor concentration in mass/cm^3, ΔT indicates time (in seconds), A denotes the surface area of the insert (barrier) in square centimetres, and ΔQ_{well} is the mass (amount) of drug transported into the basolateral compartment.

P_{app} may also be calculated from Equation 8.2:

$$P_{app} = \frac{\Delta F_{well}}{\Delta T \times A \times \Delta F_{Insert}} \times V \quad (8.2)$$

Where ΔF_{insert} is the fluorescence value measured in the insert (donor compartment/apical part) at the same time point minus the medium autofluorescence value, ΔT indicates time (in seconds), A denotes the surface area of the insert (barrier) in square centimetres, and ΔF_{well} is the fluorescence value measured in the well (receiver compartment/basolateral part) at a certain time point minus the cell culture medium autofluorescence value, and V denotes the basolateral volume in cm^3. P_{app} is expressed as cm/s (12).

8.3 LIMITATIONS OF TRANSWELL MODELS

Despite their utility, transwell models have limitations. They lack the complex three-dimensional architecture and hemodynamic forces present *in vivo*. The absence of blood flow means shear stress effects on endothelial cell function cannot be studied without additional modifications. Additionally, immortalised cell lines may not fully recapitulate all properties of primary brain endothelial cells, though they offer better reproducibility. The physiological relevance of these models may be enhanced by incorporating additional cell types such as astrocytes (13) or pericytes (14) cultured on the basolateral side of the inserts, with neurons (13) cultured in the basolateral compartment (Figure 8.2). The presence of pericytes would provide an extra level

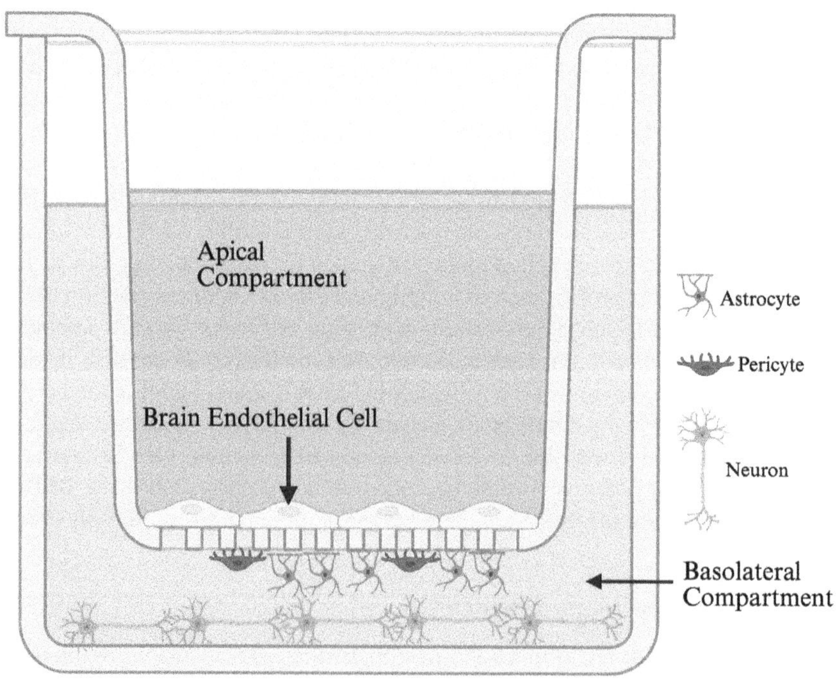

FIGURE 8.2 An enhanced transwell *in vitro* BBB model comprised of brain endothelial cells, astrocytes, pericytes, and neurons. The brain endothelial cells are cultured on the apical side of the transwell insert, while astrocytes and pericytes are cultured on the porous membrane of the transwell insert but on the basolateral (abluminal) side. Neurons are cultured on the surface of the basolateral compartment.

of communication between endothelial cells and astrocytes and potentially a prominent immune function (13). It was found that a four-cell BBB model (brain endothelial cells, astrocytes, pericytes, and neurons) exhibited consistent increase in the BBB permeability with the addition of dexamethasone to the luminal side of the BBB model (13). The expression of tight junction proteins (claudin -5 and ZO-1) increased with co-culture of brain endothelial cells with astrocytes in the presence of hydrocortisone or cAMP (15). However, in high-throughput screening, these extra cell cultures may be avoided (16).

8.4 CONCLUSIONS

The development of an *in vitro* model is essential in the process of drug development for targeting the brain. There are several *in vitro* models that have been developed. A simple monolayer of brain endothelial cells on inserts could be sufficient for high-throughput screening. Incorporating astrocytes, pericytes, and neurons as well as brain endothelial cells adds further complexity to the developed model, which more closely resembles the *in vivo* environment.

REFERENCES

1. Islam Y, Leach AG, Smith J, Pluchino S, Coxon CR, Sivakumaran M et al. Physiological and pathological factors affecting drug delivery to the brain by nanoparticles. Adv Sci (Weinh). 2021;8(11):e2002085.
2. Williams-Medina A, Deblock M, Janigro D. In vitro models of the blood-brain barrier: Tools in translational medicine. Front Med Technol. 2020;2:623950.
3. da Silva Meirelles L, Caplan AI, Nardi NB. Chapter 12 - pericytes as the source of mesenchymal stem cells. In: Goldenberg RCdS, Campos de Carvalho AC, editors. Resident Stem Cells and Regenerative Therapy: Academic Press; 2013. p. 233–50.
4. Khoury N, Pizzo ME, Discenza CB, Joy D, Tatarakis D, Todorov MI et al. Fc-engineered large molecules targeting blood-brain barrier transferrin receptor and CD98hc have distinct central nervous system and peripheral biodistribution. Nature Communications. 2025;16(1):1822.
5. Wiley DT, Webster P, Gale A, Davis ME. Transcytosis and brain uptake of transferrin-containing nanoparticles by tuning avidity to transferrin receptor. Proc Natl Acad Sci. U.S.A. 2013;110(21):8662–7.
6. González-Gallego J, Todorov-Völgyi K, Müller SA, Antesberger S, Todorov MI, Malik R, et al. A fully iPS-cell-derived 3D model of the human blood–brain barrier for exploring neurovascular disease mechanisms and therapeutic interventions. Nat Neurosci. 2025;29(2):479–92.
7. Nair AL, Groenendijk L, Overdevest R, Fowke TM, Annida R, Mocellin O et al. Human BBB-on-a-chip reveals barrier disruption, endothelial inflammation, and T cell migration under neuroinflammatory conditions. Front Mol Neurosci. 2023;16:1250123.
8. Blanchard JW, Bula M, Davila-Velderrain J, Akay LA, Zhu L, Frank A et al. Reconstruction of the human blood–brain barrier in vitro reveals a pathogenic mechanism of APOE4 in pericytes. Nat Med. 2020;26(6):952–63.
9. Islam Y, Ehtezazi P, Cashmore A, Marinsalda E, Leach AG, Coxon CR et al. The inclusion of a matrix metalloproteinase-9 responsive sequence in self-assembled peptide-based brain-targeting nanoparticles improves the efficiency of nanoparticles crossing the blood-brain barrier at elevated MMP-9 levels. J Pharm Sci. 2021;110(3):1349–64.
10. Omidi Y, Campbell L, Barar J, Connell D, Akhtar S, Gumbleton M. Evaluation of the immortalised mouse brain capillary endothelial cell line, b.End3, as an in vitro blood–brain barrier model for drug uptake and transport studies. Brain Res. 2003;990(1):95–112.
11. Ohshima M, Kamei S, Fushimi H, Mima S, Yamada T, Yamamoto T. Prediction of drug permeability using in vitro blood-brain barrier models with human induced pluripotent stem cell-derived brain microvascular endothelial cells. BioRes Open Access. 2019;8(1):200–9.
12. Thomsen MS, Humle N, Hede E, Moos T, Burkhart A, Thomsen LB. The blood-brain barrier studied in vitro across species. PLoS One. 2021;16(3):e0236770.
13. Stone NL, England TJ, O'Sullivan SE. A novel transwell blood brain barrier model using primary human cells. Front Cell Neurosci. 2019;13:230.

14. Nakagawa S, Deli MA, Kawaguchi H, Shimizudani T, Shimono T, Kittel Á et al. A new blood–brain barrier model using primary rat brain endothelial cells, pericytes and astrocytes. Neurochem Int. 2009;54(3):253–63.
15. Perrière N, Yousif S, Cazaubon S, Chaverot N, Bourasset F, Cisternino S et al. A functional in vitro model of rat blood–brain barrier for molecular analysis of efflux transporters. Brain Res. 2007;1150:1–13.
16. Bagchi S, Chhibber T, Lahooti B, Verma A, Borse V, Jayant RD. In-vitro blood-brain barrier models for drug screening and permeation studies: An overview. Drug Des Devel Ther. 2019;13:3591–605.

9 *In Vivo* Evaluations of Drug Delivery to the Brain

Touraj Ehtezazi

9.1 INTRODUCTION

Back in 2007 or even 2011, the blood-brain barrier (BBB) was considered to be a major challenge in the development of effective therapeutics for the central nervous system (CNS) disorders (1, 2). The development of nanoparticle (NP) formulations and antibodies since then has shown targeting the brain both *in vivo* and clinical trials, such as aducanumab for the treatment of Alzheimer's disease (AD) (3). However, there are still no effective treatments for most CNS disorders. As part of developing an effective therapeutic approach for the CNS treatment, successful transversing of the BBB could be essential. *In vitro* models provide initial screening tests (4) while *in vivo* evaluations being the gold standard for assessing the efficacy, safety, and mechanisms of brain-targeted drug delivery systems. This chapter examines the methodologies, considerations, and recent advances in the *in vivo* evaluation of drug delivery to the brain. Different animal models, imaging techniques, and quantitative assessments will be explored to provide crucial insights into BBB penetration, drug distribution within the CNS, and therapeutic efficacy. Although important and essential but not promising, these evaluation methods may help researchers better design and optimise delivery systems for treating devastating neurological conditions ranging from neurodegenerative diseases to brain malignancies.

9.2 THE IMPORTANCE OF *IN VIVO* TESTING FOR BRAIN-TARGETED THERAPEUTICS

In most cases, it is crucial to demonstrate targeting the brain for the treatment of brain-disorders. Therefore, a model should be selected that closely simulates human physiological barriers to the brain, which are the neurovascular barrier (normally referred to as the BBB), the arachnoid barrier at the meninges, the blood cerebrospinal fluid (CSF) barrier at the choroid plexus, the circumventricular organs barrier, and the glia limitans (5). These are physiologically and anatomically complex barriers, and are difficult to build *in vitro*. Currently, animal models such as mice and rats are used to demonstrate that active ingredients of a drug delivery system reach the brain. Therefore, the clinical outcomes of the treatment in the animal models may be correlated with targeting the brain. In certain neurological disorders such as brain tumours, it would be important to demonstrate that the active ingredients/NPs cross the CNS physiological barriers, diffuse through the brain parenchyma, target, and accumulate in the tumour cells (6). In addition, the *in vivo* models can be employed to determine brain permeation parameters for compounds such as blood-to-brain influx constant, the permeability-surface area product, and the brain uptake index (2). *In vivo* models also allow researchers to determine pharmacokinetic parameters of the brain-targeting drug formulation such as peak concentration (C_{max}), time to reach C_{max} (T_{max}), area under the curve (AUC) of plasma/CSF drug concentration versus time, mean drug residence time (MRT) indicating the average amount of time the drug molecule spends in the body before it is eliminated, and plasma/CSF drug concentration half-life ($t_{1/2}$) referring to the time required for plasma/CSF concentration of the drug to decrease by 50% (7). Animal models (transgenic mice) have also been recruited to evaluate reduction of amyloid burden following dosing with aducanumab (3).

DOI: 10.1201/9781003482406-9

9.3 VARIOUS ANIMAL MODELS USED

Mice and rats remain the most popular animal models for testing drug delivery to the brain and development of neurological disorders *in vivo* (8–11). They are low-cost and require relatively low maintenance. This allows conducting studies with a larger number of animals to determine dose response as well as to demonstrate statistically significantly better therapeutic outcomes compared to control/untreated animal models. However, mouse models are limited by gross anatomical differences with humans (12). Due to the greater similarity of the brains of non-human primates to humans (12), non-human primates are better positioned than rodents for the translation of preclinical data related to brain-targeting drug delivery systems (13). However, due to the need for bigger spaces and dedicated facilities, non-human primates are not commonly used as animal models of neurological disorders (14). Canines (dogs) have been used as models of human brain tumours as they develop tumours that better mimic those of humans, as well as having larger brain sizes that allow the positioning of several probes in the brain in a research related to brain-targeting drug delivery systems (15). Zebrafish have been employed as a model of the human BBB for assessing of the permeability of drug delivery systems (16). Zebrafish models offer cost-effective maintenance and simple housing requirements, which make them suitable for high-throughput testing (17), such as identifying active components of *Gastrodia elata* with therapeutic effects on cerebral ischemic reperfusion injury (18). Zebrafish embryos not only are small enough to fit inside into wells of 24-well plates with 10 embryos per well, but also because the embryos are transparent, they provide the opportunity for live monitoring of thrombolysis in the treated embryos with cerebral ischemic reperfusion injury (18). In this study, the rat middle cerebral artery occlusion (MCAO) model of ischemic stroke was employed to demonstrate the therapeutic efficacy of *Gastrodia elata* (in general), while the zebrafish embryo model allowed researchers to test three active components of *Gastrodia elata* separately. Fruit flies have been employed to trace the distribution of trypan-blue-dyed Ringer buffer in the body including the head (19). Again, the transparent body of the fruit fly allowed to detect the penetration of the dye into the brain 15 s post-injection (19).

9.4 WILD TYPE AND DISEASED ANIMAL MODELS

9.4.1 Wild-type and Sham Animal Models

Wild-type animal models provide baseline assessments of BBB permeability and brain drug distribution in healthy animals. These are typically employed during early stages of the drug development process to demonstrate penetration of target drug molecules into the brain (20, 21). While sham animal models may undergo a particular process such as surgery but with the intention of causing minimal damage to the CNS. For example, in the development of MCAO, the sham animal will undergo surgery and will recover, but without blocking the middle cerebral artery. This is done to evaluate the extent of the BBB damage that may occur due to the surgery and not the occlusion of the middle cerebral artery (22).

9.4.2 Transgenic and Knockout Rodent Models

Knockout animal models are defined as species that genetically lack a particular biomolecule, mostly a protein. These animals provide valuable information about the efficacy of a drug by studying its interaction with the target protein. Knockout mice have been employed to study the penetration of several drugs such as digoxin and cyclosporin A, showing that P-glycoprotein (P-gp) in the BBB prevents penetration of digoxin and cyclosporin to protect the brain from toxins. The brain-penetration of these drugs increased 20- to 50-fold in *mdr1a (-/-)* knockout mice (which lack P-gp) (23). Chemically induced P-gp knockout mice were developed by pre-treatment of wild-type mice with the P-gp inhibitor GF120918, exhibiting increased SB-487946 brain-penetration, SB-487946 is a P-gp substrate (24). Knockout mice lacking P-gp and breast cancer resistance protein (BCRP) were employed to study efflux of erlotinib (a targeted chemotherapy drug) at the BBB (25). It was found that the brain-to-plasma concentration ratio of erlotinib significantly increased in the transgenic mice compared to the wild-type mice, suggesting that failure of clinical trials of new drugs for the treatment of glioma could be due to the active efflux at the BBB (25). In another study, peptide-based brain-targeting ligands were developed, which targeted classical BBB transmembrane receptors such as the insulin receptor(26). This was designated as an allosteric targeting strategy. Liposome formulations, designated as ITP-LIP, contained an allosteric peptide (ELGILIFLYLFSLILGIIYWKK) targeting transmembrane domain of the insulin-receptor. ITP-LIP did not penetrate into the brains of brain-specific Insulin-receptor-knockout mice (IR CKO), while it penetrated into the brains of normal mice (non-fasting wild-type) following i.v. injection (26).

Transgenic animal models have been developed for several neurodegenerative diseases such as AD, amyotrophic lateral sclerosis (ALS), Huntington's disease, Parkinson's disease (PD) (27), and glioblastoma (28). These animal models are genetically modified to recapitulate most (ideally all) of the pathological features of the target neurodegenerative disease. Transgenic models of AD have been far more investigated than other transgenic animal models. A mutation (V717F) in the *Amyloid beta A4 Precursor Protein (APP)* gene has been linked to familial AD (FAD) (29). This discovery prompted the development of the first transgenic mouse model of AD back in 1995 (30). Combined transgenic rodent models have been developed to capture more pathological features of neurodegenerative diseases such as AD. For example, the TAPP transgenic mouse model of AD was developed, which combined the P301L tau protein mutation ($Pro^{301} \rightarrow$ Leu, responsible for generation of neurofibrillary tangles) and the amyloid precursor protein mutant Tg2576 (31). To further simulate amyloid beta pathology in the brain of FAD, 5xFAD mouse models were developed (32), which express five mutations in *APP* and *presenilin-1* genes to increase Aβ production in the brain (33). 5xFAD transgenic mice (APP K670N/M671L (Swedish) +I716V (Florida) + V717I (London) and PS1 M146L + L286V) produced massive cerebral Aβ levels from the age of 2 months (33). The 5xFAD transgenic mouse model was employed to validate whether hybrid liposomes coated with cell membranes from platelets and chemokine C-C motif receptor 2 cells could cross the BBB and migrate to inflammatory lesions in the brain (34). The liposomes were loaded with rapamycin and 1-trifluoromethoxyphenyl-3-(1-propionylpiperidin-4-yl) urea (TPPU), a compound that alleviates cognitive deficits in AD. The formulation was administered i.v. and reduced Aβ plaques, and alleviated neuroinflammation by reducing the population of microglia in the brains of 5xFAD mouse models. Untreated mouse models showed significantly higher population of microglia in the brain compared to wild-type mice (34).

APP/PS1 is another type of AD mouse model with two combined genetic mutations: the Swedish mutation (KM670/671NL mutated APP), which overexpresses APP, and the presenilin-1 L166p mutation (35). Interestingly, the transgenic mouse model was designated as APPPS1 (not APP/PS1). It should be noted that other variants of APP/PS1 have been reported, such as mutant $APP_{K670N,M671L}$-mutant $PS1_{M146L}$ (36), mutant $APP_{K595N,M596L}$-mutant $PS1_{A246E}$ (37), and mutant $APP_{K670N,M671L}$-mutant $PS1_{M146L}$ (38). It was shown that the average plaque number was significantly higher in the prefrontal cortex of 5xFAD transgenic mice compared to APP/PS1 transgenic mice (39). The APP/PS1 mouse model of AD presented higher expression of activated microglia compared to wild-type mice, which was reduced by employing polydomapine-based NPs loaded with immune-activating compounds (to reduce brain Aβ load) such as hydroxychloroquine [targeting microglia] and all-trans retinoic acid [targeting astrocytes] (40). The reduction of brain Aβ load led to the reduction of activated microglia. Furthermore, the APP/PS1 mouse model of AD presented higher levels of IL-1β (pro-inflammatory cytokine) in the brain tissues, which were reduced to wild-type levels by intranasal administration (IN) of liposomes loaded with donepezil, memantine, and beta-site amyloid precursor protein-cleaving enzyme (BACE-1) siRNA (41).

The J20 mouse model of AD is a transgenic model that overexpresses human APP with FAD mutations (Swedish and Indiana) (42). The J20 mice exhibited impairment in retaining information over a long period, even before Aβ deposition and glia activation in the brain, compared to wild-type mice (43). Treatment of J20 mice with IN administration of docosahexaenoic acid (DHA) nano-emulsions significantly reduced cognitive impairment compared to the untreated mice (44). It was shown that nasal absorption and brain uptake of dulaglutide (an incretin receptor agonist that improves cognition in AD) were significantly lower in the hippocampus regions of female APP/PS1 mice, and in the striatum and thalamus of male APP/PS1 mice compared to wild-type mice (45).

Mutations in Cu/Zn-binding superoxide dismutase 1 (SOD1) have been linked to familial ALS (46). Therefore, a transgenic mouse model was developed that expressed human SOD1 in the brain without signs of overt motor neuron disease (47), although pathological signs were observed at neuromuscular junctions (48). To develop transgenic mice that represent familial ALS in humans, transgenic mice were developed that had only one mutation in the human SOD1, and that was the substitution of glycine with alanine at position 93 (49). This human SOD1-G93A mutation captured signs of motor neuron disease seen in humans such as paralysis in the limbs (49). Human SOD1-G93A transgenic mice were employed to evaluate calcium phosphate-based lipid NPs loaded with antisense oligonucleotide (ASO) targeting SOD1(SOD1-ASO-CaP NPs). The SOD1-ASO-CaP NPs in combination with microbubbles were administered intravenously to human SOD1-G93A transgenic mice. Focused ultrasound (FUS) was applied to increase the BBB permeability to the NPs with the help of microbubbles. The treatment significantly reduced the expression of SOD1 in the brain tissues of treated animals compared to the control group, which was treated with FUS only (50). Similarly, human SOD1-G93A transgenic mice were employed to evaluate brain delivery of arctiin (a natural compound with anti-inflammatory activities) following i.v. injections with employing FUS (51). It was found that microglia morphology was different in the brains of the ALS mouse models compared with wild-type mice. ALS microglia were enlarged and had irregular protuberant branching processes. Following the treatment with FUS plus arctiin, the average body size of microglia significantly reduced ($40.1 \pm 7.5\ \mu m^2$) compared to the untreated ALS animal models ($51.0 \pm 6.6\ \mu m^2$) (51). Wild-type mice had microglia with average body sizes less than $40\ \mu m^2$ (51).

Spinocerebellar ataxia type 3 (SCA3), or Machado-Joseph disease, is a hereditary disease which is caused by the accumulation of mutant ATXN3 protein in the brain, leading to ataxia. To investigate suitable therapeutic methods for this disease, SCA3 transgenic mice have been developed. The development of seven SCA3 transgenic mouse models have been reported, which differ in the corresponding DNA constructs with differences in the encoded human ataxin-3 isoforms, number of polyglutamine repeats, and the promoter driving transgenic expression (52). SCA3 transgenic mice were employed to evaluate the therapeutic efficacy of RVG-peptide-decorated exosomes loaded with miR-25 or miR-181a to inhibit mutant *ATXN3* expression. The exosomes were administered intravenously and they reached the brain with significantly increased expressions of miR-25 and miR-181a in the SCA3 transgenic mice. The treatment with both exosome formulations carrying miR-25 or miR-181a significantly improved motor functions compared to untreated SCA3 transgenic mice, and these became comparable with wild-type mice (53).

Transgenic animal models of PD have been developed. This has been done to identify responsible genes for the development of PD, and ultimately to develop therapies that are shown to be neuroprotective and neuroregenerative in humans with PD (54). Transgenic mice overexpressing human α-synuclein have been developed which showed an accumulation of α-synuclein in neuronal inclusion bodies found in brain regions such as the substantia nigra (55).

9.4.3 Induced Disease Animal Models for CNS Disorders

Induced disease animal models are derived from healthy animal models following certain manipulations, such as surgery or administration of chemicals. These models are described in the following sections.

9.4.3.1 Chemically Induced Models

Healthy animal models are exposed to toxins or chemicals to create models of brain diseases. PD, AD, and multiple sclerosis (MS) animal models are examples of this category.

9.4.3.1.1 PD

PD mouse models were developed by oral gavage administration of rotenone suspended in 0.5% carboxymethyl cellulose sodium salt once daily at a dose of 30 mg/kg body weight for 28 days (56). The animal models developed key symptomatic features of PD, namely impaired motor balance and coordination. These PD animal models were treated with curcumin-loaded NPs administered orally at a dose of 200 mg/kg, 30 min before the administration of rotenone. The treatment significantly improved behavioural deficit (measured by the rotarod task) in the animal models and made them comparable to wild-type mice (56).

In another approach, 1-methyl-4-phenyl-1,2,3,6-tetrahydropyridine (MPTP), a neurotoxin compound, was used to induce a PD animal model (57). MPTP was injected intraperitoneally into mice at a dose of 30 mg/kg over five consecutive days to develop PD mouse model. The animal model demonstrated significant deficits in motor functions such as, the time required to climb a pole (58). It was shown that oral administration of human albumin NPs loaded with selenium at a dose of 0.4 mg/kg for 3 days significantly improved motor functions of the diseased animal models to levels comparable to control mice (injected with saline and administered saline orally) (58).

A PD animal model can also be generated by the administration of 6-hydroxydopamine (6-OHDA) into discrete brain regions to destroy dopaminergic neurons and produce several non-motor comorbidities commonly associated with PD, including cognitive deficits, depression, and anxiety (39). Wistar rats were administered 2 μL of 6-OHDA solution (a dose of 2 μg) via stereotaxic injection into the right striatum (60). It has been shown that intracerebral administration of 6-OHDA led to the development of motor function deficits in rats (60, 61). Intranasal administration of vitamin E-loaded naringenin nonemotions into 6-OHDA PD rat models significantly improved motor functions in behavioural tests such as forced swimming time compared to untreated PD animal models, with results almost comparable to those of wild-type animals (60).

9.4.3.1.2 AD

AD animal models have been developed by intraperitoneal (i.p.) injection of scopolamine for 21 consecutive days at a dose of 4.5 mg/kg daily into Sprague-Dawley rats (62). The scopolamine-induced memory impairment is based on the cholinergic damage hypothesis. The pathology is based on the fact that scopolamine blocks muscarinic acetylcholine receptors, leading to impaired cognitive functions via decreased neuronal signalling. The animal models developed cognitive function impairment, as measured by the Morris water maze test, as well as pathological changes in the brain such as increased expression of acetylcholinesterase (62). In this study, AD rat models were also developed by co-administration of D-galactose and $AlCl_3$. The animals were administered $AlCl_3$ at a dose of 24 mg/kg by oral gavage for 90 consecutive days, as well as 150 mg/kg of D-galactose via i.p. injections. The addition of D-galactose was intended to accelerate the development of memory impairment. These AD animal models also developed cognitive function impairment, as measured by the Morris water maze test, as well as pathological changes in the brain such as increased BACE1 expression and Aβ levels. The D-galactose and $AlCl_3$ animal models were treated with intranasal microneedles loaded with cyclodextrin and potassium-based NPs loaded with huperzine A and stigmasterol with lactoferrin surface decoration, which significantly reduced brain Aβ and BACE1 levels (62). The treatment of scopolamine-induced AD rat models with the formulation also improved the memory function of the animal models and reduced brain acetylcholinesterase levels.

An AD animal model was also developed by intracerebroventricular injection of human Aβ oligomers into wild-type rats (63). This approach was intended to elucidate the initial steps of AD pathology. The Aβ oligomers were extracted from human brain samples of patients with severe AD dementia. The animals began to develop behavioural deficits 2.5 weeks post-surgery and significant differences were observed later such as a decline in social recognition memory on Day 21 post-surgery. Streptozotocin is another compound that was administered via bilateral intracerebroventricular injections into Wistar rats at a dose of 3 mg/kg (1.5 mg/kg per side) (64). These animals developed spatial memory and learning deficits 21 days post streptozotocin intracerebral administration. It was found that intravenous (i.v.) injections of ipriflavone loaded PLGA NPs with surface decoration of tet-1 peptide (a CNS targeting peptide: HLNILSTLWKYR), also known as G23, given three times a week at a dose of 5 mg/kg [equivalent to 0.4 mg/kg of ipriflavone] significantly improved spatial learning and memory function measured by the Morris water maze test (64). Ipriflavone is a naturally occurring estrogen mimic that promotes neuronal growth and cognitive function (64).

9.4.3.1.3 Stroke

This model is developed by injecting a photosensitive chemical such as Rose Bengal or erythrosine B. When the photosensitive compound is circulating in the bloodstream, a desired region of the brain is irradiated with a laser beam of a specific wavelength, which induces focal cerebral ischemia (65). For example, Rose Bengal was injected intravenously at a dose of 30 mg/kg to C57BL/6 mice (wild-type). The scalp of the animal was shaved, followed by removal of the skull, and the right sensorimotor cortex of the brain was irradiated (66), typically with blue/green laser light at 560 nm (67). It should be noted that the removal of skull may not be necessary; simply removing the scalp and exposing the skull could be sufficient (68). The advantages of this approach include easy manipulation, reproducibility, and less trauma to the animal compared to surgical procedures to produce stroke models (65). It has been stated that the photochromic stroke model would exhibit poor response to recombinant tissue-plasminogen activator (65), a typical medicine used for the treatment of ischemic stroke (69). It was found that intrathecal administration of PLGA NPs loaded with perampanel in a 20 μL solution decreased the infarct volume from 3 mm^3 to 1 mm^3 (67).

9.4.3.1.4 Epilepsy and Depression

Epilepsy is a chronic, noncommunicable brain disease that affects about 50 million people worldwide, according to the WHO reports (70). Temporal lobe epilepsy is a partial epileptic disorder in which seizures occur on the sides of the brain. Drug resistance in epilepsy has been reported, and it is estimated to affect about one-third of individuals (71). Inadequate BBB penetration, enhanced expression and activity of efflux transporters, and drug-metabolising enzymes have been considered as factors contributing to antiepileptic drug resistance (72). To evaluate novel formulations for the treatment of epilepsy, kainic acid has been employed to develop temporal lobe epilepsy, administered either intracerebrally into the hippocampus or via systemic administration (73). In one approach, hollow mesoporous silica NPs were loaded with the death-associated protein kinase I inhibitor (TC-DAPK6) and surface-coated with macrophage membrane to facilitate crossing the BBB in the epileptic inflammatory regions (74). To evaluate the therapeutic effects of the NP formulation (MA@RT-HMSN), a mouse model of acute seizures was developed by i.p. injection of kainic acid at a dose of 30 mg/kg. I.v. administration of MA@RT-HMSN NPs significantly improved behavioural assessments in kainic acid-treated mice compared to the untreated mouse models. Pilocarpine (75) and pentylenetetrazol (76) are other compounds have been used to induce epileptic animal models.

9.4.3.1.5 Depression

Depression, or major depressive disorder (MDD) is a mental health condition, which is characterised by depressed mood, loss of interest or pleasure, disturbed sleep, poor concentration, and self-harm or suicidal thoughts (77). According to the WHO report, an estimated 280 million people experienced depression in 2019. MDD may be classified as a brain disease (78). Recent clinical and preclinical evidence suggests that neuroinflammation is a major factor in MDD (79), observed by the association of depression and CSF proinflammatory cytokine levels such as IL-1β (80). A cumulative meta-analysis also revealed higher mean levels of IL-6 and C-reactive protein in patients with MDD compared to non-depressed controls (81). Although MDD and neurodegenerative diseases are two entirely different clinical entities, their increased concomitant presentation has been reported (82). A bidirectional relationship has been suggested between neurodegeneration and MDD (83). A reduction in neurogenesis has been reported in patients with depression, indicated by reduced serum BDNF levels (84). Therefore, animal models of depression would be useful for screening and assessing antidepressant drugs (85). Depressed animal models could be produced by compounds such as reserpine at a low dose of 0.5 mg/kg once daily injected intraperitoneally for 14 days (86) or by restraining animals for 14 days (87). To target neuroinflammation associated with depression, microglia-membrane-coated polydopamine NPs loaded with memantine (PDA-Mem@M) were developed (88). PDA-Mem@M NPs were administered intravenously to chronically restrained, stressed depressive mice, which significantly reduced depressive symptoms such as decreased time in the forced swimming test (88).

9.4.3.1.6 MS

MS is a chronic-autoimmune neurological disorder, which is characterised by the loss of myelin. A demyelination model is achieved by cuprizone diet, leading to toxic demyelination. In this approach, 8-week-old C57BL/6 mice are fed with 0.2% cuprizone for 5 (89) or 6 weeks (90). The diet causes death of oligodendrocytes; however, spontaneous remyelination is observed as early as 4 days after the diet withdrawal (90), and myelin proteolipid protein levels in the cerebellum reach control-animal (unfed with cuprizone) levels (89). Mitochondrial dysfunction in oligodendrocytes causes the death of these cells by the cuprizone diet (91). Cuprizone diet leads to neuroinflammation, which was determined by increased expression of Iba1+ cells in the corpus callosum of mice (92). Cuprizone diet has been applied to develop rat demyelinating models (93). It was found that oral administration of iron-oxide NPs (10 mg/mL) for 2 weeks to cuprizone-diet-rat models (developed over 5 weeks) reduced brain TNF-α mRNA levels compared to the untreated animals (93). Therefore, care should be taken when studying cuprizone-diet MS animal models, as part of treatment benefits may be masked by the spontaneous recovery after cuprizone cessation over the two weeks.

9.4.4 Surgically Induced Animal Models

9.4.4.1 Ischemic Stroke

MCAO is the most commonly used ischemic stroke animal model with pathological outcomes similar to human ischemic stroke. MCAO is induced by surgery through the insertion of a customised "T"-shaped PE-10 catheter (94) /filament coated with silicon resin initially into the external carotid artery and advancing it to the internal carotid artery and then middle carotid artery (95) (Figure 9.1).

Typical wild-type animal models are employed for the development of MCAO. Animals are anesthetised typically with 1.5% isoflurane in 68.5% N_2O and 30% O_2. The regional blood flow may be monitored using a laser-Doppler flowmeter (95). This is classified as focal ischemic stroke where there is reduced blood flow into a specific region of the brain (65). As there is a reperfusion following the ischemic stroke, the animal models undergo another surgery typically 1 h post-occlusion, and the filament is removed (96). When the animal is recovered, and it is assessed for the neurological improvements by applying treatments. The MCAO animal model develops hypoxic brain regions, which are visualised by TTC (2,3,5-triphenyltetrazolium chloride) staining (97). This technique is used to measure the infarct volume and its reduction by applying treatments (98). To measure the infarct volume by the TTC staining, the brain tissue is harvested and sliced at approximately 2 mm thickness (98). The brain hypoxic regions appear pale pink compared to the normoxic regions, which appear as dark-red areas (98).

9.4.4.2 Traumatic Brain Injury

The direct result of an external force to the head or severe shaking of the skull may lead to a mechanical deformation of the brain tissue, which is categorised as traumatic brain injury (TBI). TBI occurs in civilian life, on the battlefield, and in sports. The mechanical tissue deformation is known as the

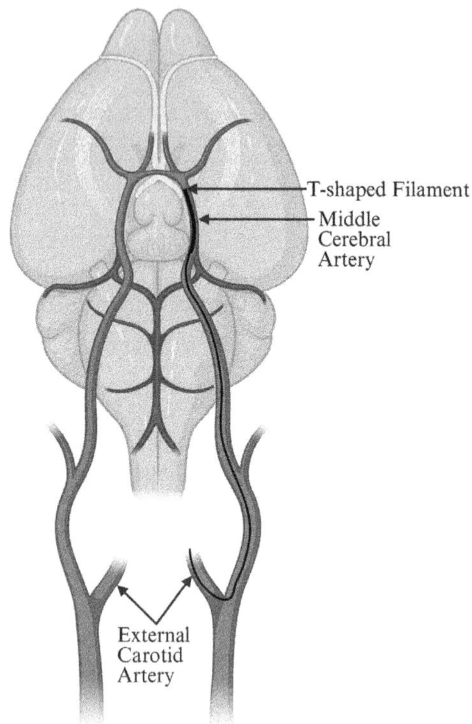

FIGURE 9.1 Schematic presentation of middle cerebral artery occlusion in mouse brain by a "T"-shaped filament.

primary injury. This involves the formation of necrotic cells (neurons, astrocytes, microglia), tearing of the brain blood vessels, and formation of hematomas, causing ischaemia, hypoxia, oedema, and increased intracranial pressure. A secondary injury follows the primary injury, which involves neuroinflammation and other metabolic cascades, ultimately leading to brain cell death. The secondary injury evolves over minutes to several months following the primary injury. Therefore, the animal model of TBI may recapitulate either primary or secondary injuries of TBI, although a model that covers both injuries would be preferred. Several animal models of TBI have been developed, such as the fluid percussion injury (FPI), the controlled cortical impact (CCI), the penetrating ballistic-like brain injury (PBBI), Feeney's weight-drop model, Marmarou's weight-drop model, Shohami's model, and blast TBI (99). Table 9.1 compares these different animal models.

9.4.5 Other Induction Methods

A recent work showed that hippocampal injection of adeno-associated viruses encoding human tau protein (4R isoform) into adult rhesus monkeys produced high levels of tau expression throughout the hippocampus (111). The diseased monkeys developed neuroinflammation (overexpression of $Iba1^+$ cells) as well as learning and cognitive function decline (111).

Experimental autoimmune encephalomyelitis (EAE) is a model of autoimmune disease that has been widely used for the human inflammatory demyelinating disease, namely MS (112). Typically, EAE is induced by subcutaneous injections of $MOG_{35\text{-}55}$, incomplete Freund's Adjuvant, and 8 mg/ml *Mycobacterium tuberculosis* into the flanks and the base of the tails of wild-type mice. Alternatively, 100 µL of Pertussis Toxin (5 ng/ml) is injected intravenously and again after 48 h (113).

The lysolecithin injection model is a pre-clinical laboratory technique to produce a focal area of demyelination (114). Lysolecithin (multiple CAS numbers: 17364-18-0, 14863-27-5) is a detergent

TABLE 9.1
Different Animal Models of TBI

Model	Description	Ref
FPI	Initially, a plastic cannula is implanted in the brain, with one end inside the brain and the other end outside of the skull with a plastic cap on it. FPI is developed by the application of a brief, transient fluid pressure pulse through the cannula directly to the exposed dura mater of an anesthetised animal.	(100)
CCI	A bore of 4 mm is drilled in the skull of an anesthetised mouse, and an actuator with a velocity of 3 m/s creates deformation in the brain at depths of 0.2–2 mm to simulate TBI with different severities. The injury site is closed for recovery.	(101)
PBBI	The scalp of an anesthetised rat is incised, and a 1 mm burr hole is created. The PBBI probe is inserted into the bore, and it travels at a velocity of 2 mm/s to a depth of 1.2 cm from the brain surface. The cavity in the brain is created by the rapid inflation/deflation of the PBBI probe using a water-filled balloon.	(102, 103)
Feeney's Model	The scalp of an anesthetised rat is opened, and a section of skull is removed. Contusions are made in the hindpaw area by the falling of a weight with impacts in the range of 600–1000 g/cm. There is a high rate of mortality in this model.	(104, 105)
Marmarou's Model	This model replicates diffuse TBI. The skull of an anesthetised rat is shaved and, two canulae are inserted into the brain to measure mean arterial blood pressure and blood gas analysis during the experiment. Also, a fibre optic probe is inserted through a burr-hole to measure the intracranial pressure before and after trauma. The animal is moved to a foam bed and placed prone under a hollow Plexiglass tube, and the skull is covered by a metallic disc to serve as a helmet. Weights are dropped from a certain height through the Plexiglass tube onto the helmet.	(106)
Shohami's Model	A 95 g weight is dropped from a 7 cm height onto the exposed skull (by a midline longitudinal incision) of an anesthetised wild-type mouse.	(107)
Blast TBI	To simulate TBI in warfare, an anesthetised mouse is located in a chamber that receives pneumatic shockwaves of up to 330 kPa, which represents a severe shock.	(108–110)

derived from lecithin. Lysolecithin is injected at a concentration of 1% and a volume of about 0.5 μL into the white matter of the spinal cord of a C57BL/6 mouse (114). Demyelination of axons occurs by Day 7 post-operation in the injected area. Remyelination starts to be detected at Day 14 of post-operation (114). A zebrafish lysolecithin-demyelinating model was developed to visualise the dynamics of demyelination process (115). The i.p. injection of curcumin-loaded chitosan-alginate NPs into lysolecithin-induced focal demyelination rat models on Day -10, Day 3, and Day 7 post-operation significantly reduced the extent of demyelination on Day 7 of post-operation compared to control animal models that received saline (116). Interestingly, the average demyelination levels did not change by Day 14 of post-operation for control animals (treated with saline), while for curcumin-treated animals, the demyelination levels further reduced (116).

9.4.6 Spontaneous Disease Models

There are species such as dogs, cats, and non-human primates that spontaneously develop AD; however, these are not routinely investigated due to their long lifespan and costs (117). These are animals that naturally develop neurological diseases similar to humans without manipulation. For example, examining the entorhinal cortex regions of the brains of 14–17-year old beagle dogs revealed β-amyloid immunopositive plaques comparable to those plaques observed in the brains of patients with AD (118). It should be noted that the dogs did not die due to the neurological disorders. Therefore, the aged dog brains were considered for understanding the mechanisms involved in initial β-amyloid deposition (118). Examining the brains of 16–21-year-old cats revealed the presence of abnormally phosphorylated tau in the CA1 regions, which showed similarity to phosphorylated tau in the AD human brain by Western blotting (119). It should be noted that the aged cats developed cognitive dysfunction syndrome prior to death by natural causes as disclosed by the owners (119). Non-human primates, in particular vervet monkeys, are considered valuable models for age-related neurological disorders (120) due to their highly conserved genomic sequence (121). It was found that aged (20–24 years old) female vervet/African green monkeys developed Aβ plaques in the brain cortex similar to those in humans with AD (120). Furthermore, statistically significant cognitive function decline was associated with age in the vervet monkeys (120).

9.5 TECHNIQUES FOR *IN VIVO* EVALUATION

Several *in vivo* techniques have been developed to evaluate targeting the brain by drug delivery systems. These include measuring the amounts of the drug in the CNS, and imaging techniques. These are briefly described below.

9.5.1 Brain Microdialysis

Intracerebral microdialysis requires surgery on the CNS to implant a microdialysis probe in the brain (122). This technique directly samples molecules from the brain's extracellular fluid. A schematic presentation of a microdialysis probe is given in Figure 9.2.

When a microdialysis probe is implanted in the brain tissue, a perfusion solution (typically an aqueous solution of 147 mM NaCl, 2.7 mM KCl, 1.2 mM $CaCl_2$, 0.85 mM $MgCl_2$) is circulated (typically at a flow rate of 1 μL/min) through the probe. The perfusion solution equilibrates with the interstitial fluid in the brain, which allows small molecules such as drug molecules to diffuse through the microdialysis membrane into the perfusion solution. Then the collected perfusion samples are analysed by applying techniques such as high-performance liquid chromatography

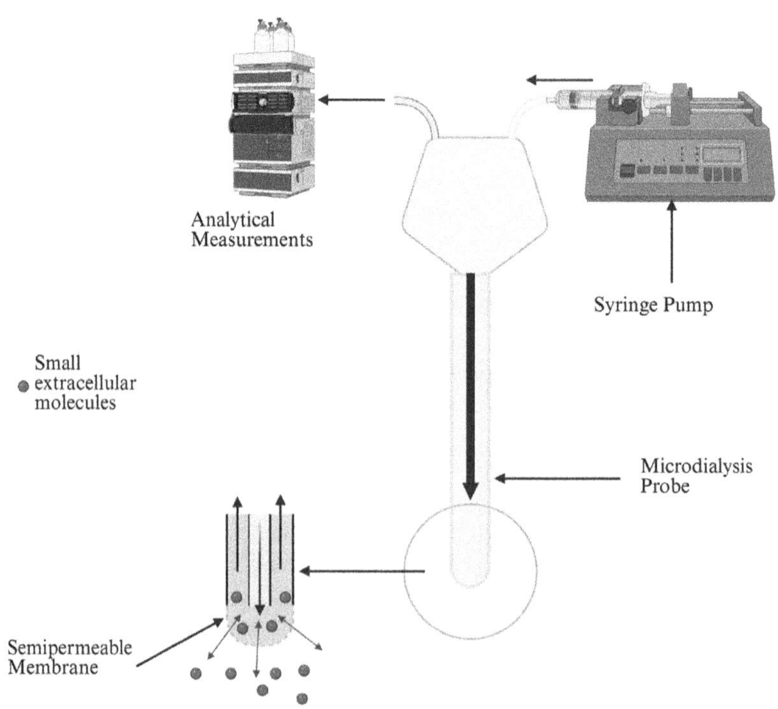

FIGURE 9.2 Schematic presentation of a microdialysis process.

equipped with fluorescence detection for quantification of drug molecules such as camptothecin (123). The limit of quantification was 1 ng/mL. It should be noted that a recovery correction (also termed extraction efficiency) may be applied for measurements using microdialysis (123), as the drug molecules may adhere to the microdialysis probe or drug molecules may not have sufficient time to fully diffuse through the membrane into the probe solution (122). Microdialysis also allows direct measurements of free drug (protein unbound) concentrations in brain parenchyma of freely moving animals (124), over extended periods of time (125), as well as drug metabolites separately (126) or simultaneously (127). Microdialysis cannot provide the spatial resolution and coverage as imaging techniques do (122). Microdialysis technique was employed to measure the release of FITC-dextran Mw 4000 (FD4) from self-assembled polymeric NPs (made of methoxy-polyethylene glycol-polylactic acid (Met-PEG-PLA) & maleimide-polyethylene glycol-polylactic acid (Mal-PEG-PLA) copolymers) surface-coated with CD-71/OX-26 antibodies for brain targeting (125). The NPs achieved a FITC-dextran brain concentration (in the ventral hippocampus) of 360 ng/mL 1 h after i.v. administration of NPs (0.4 mg/kg FD4 equivalent) and maintained an FD4 concentration of about 100 ng/mL 50 h post-drug administration (125). Recently, multiplexed microdialysis probes have been developed, which permit simultaneously measuring 21 amino acids in the brain using high-resolution tandem mass spectroscopy (128) or as many as 70 neurochemicals using BzCl with LC-MS/MS (129), or sampling multiple brain regions (130).

9.5.2 Imaging Techniques

Imaging techniques evaluate brain targeting by drug delivery systems, mostly by NPs, typically at the whole organ level. These include magnetic resonance imaging (MRI), X-ray computed tomography (CT), positron emission tomography (PET), single-photon emission computed tomography (SPECT), fluorescence *in vivo* imaging, and fluorescence molecular tomography (FMT). These techniques are briefly described below.

9.5.2.1 Fluorescence *In Vivo* Imaging

In fluorescence imaging technique, generally animals are anesthetised and then euthanised (131), or euthanised directly. This is needed to keep the animal motionless under the imaging camera. To visualise accumulation of NPs in the brain, the NPs should be fluorescently labelled with a dye such as IR780 (132), DiD (133) also known as DilC 18, or DiR (134). An imaging system such as IVIS (Revvity, Massachusetts, USA) (135) or another small imaging system is used (134).

9.5.2.2 MRI Imaging

MRI typically is used to visualise lesions in the brain such as tumours or cerebral ischemia. Animals should be anesthetised for conducting MRI sessions. The animals may be kept alive during MRI sessions to follow up the therapeutic effects of the intervention such as using miR-125a-5p-loaded extracellular vesicles (136). MRI provides 2D images of the brain at different layers, and assembling these images allows 3D reconstruction of the entire brain organ only (136). Typically, T2W MRI images are obtained, which do not require the use of MRI contrast agents.

9.5.2.3 PET Imaging

PET imaging allows quantitative measurements with high resolution compared to MRI (137). PET is sufficiently sensitive to quantify the BBB permeability following the application of FUS (138). PET imaging requires radiolabelling of either the drug molecule or the NP by radiotracers such as [^{89}Zr]Zr(oxinate) (139). For example, Talidox® liposome formulation (loaded with doxorubicin) was radiolabelled with ^{89}Zr and administered intravenously to wild-type mice, which received FUS treatment prior to drug administration (139). The whole-body distribution and pharmacokinetics were then determined by PET imaging at different time points. Preliminary experiments demonstrated similar particle size distribution of unlabelled and labelled Talidox liposome formulations. The organ uptake was calculated as the percentage of the injected activity per gram of tissue (%IA/g). The PET image analysis revealed that about 2%IA/g was present in the brain at the time (0–1 h) of administration of the ^{89}Zr-Talidox liposome formulation, which decreased to 1% at 72 h post-administration. Harvesting the mouse brains at 72 h post-administration (the body was perfused with PBS to remove the blood) and measuring the radioactivity revealed that only 0.05 ± 0.02%IA/g was delivered to the brain with no FUS application. There was a difference between PET imaging data and direct measurements of brain ^{89}Zr-Talidox liposome load, which was explained by the presence of ^{89}Zr-Talidox liposome in the blood of the brain (139). The radiolabelling of the liposome formulations does not appear to be a complex process. The liposomes were incubated with [^{89}Zr]Zr(oxinate) for 30 min at 50°C, and then the formulation was purified to remove unbound ^{89}Zr (139).

9.5.2.4 X-ray Computed Tomography

X-ray CT provides high-resolution 2D slices of anatomical parts, which can be computed to prepare 3D reconstructions. However, CT requires administration of contrast agents such as barium sulphate suspension or iodinated aromatic molecules (140). Barium sulphate suspension is used for gastrointestinal imaging. While iodinated aromatic compounds such as iohexol (brand name Omnipaque) are injectables. These compounds exhibit a short blood circulation half-life, such as 121 min for iohexol following i.v. administration with glomerular filtration being the main clearance route (141). Therefore, multiple injections of iodinated compounds may be required, which has been associated with the development of hyperthyroidism incident (142). Therefore, in the search of alternative X-ray contrast agents, it was found that gold-NPs can be used as X-ray contrast agents with superior properties over iodine-based agents (143). Therefore, gold-NPs (144) or NPs encapsulated with iohexol (145) were employed to trace NPs reaching the brain following IN administration. Gold-NPs were loaded into exosomes derived from bone marrow mesenchymal stem cells (144). CT revealed that exosomes specifically targeted the diseased regions of the brain in animal models of PD (striatum), AD (hippocampus), stroke (ischemic region), and autism (cerebellum) (144). In this published work, it appears that CT allowed real-time monitoring of NP migration in the brain.

9.5.2.5 Single-Photon Emission Computed Tomography

SPECT is an imaging technique to evaluate physiological and pathological functions such as presence of an infarction zone in the brain (146) or brain perfusion in dementia (147). Both technetium-99m (^{99m}Tc) and iodine-123 (^{123}I) are commonly used radionuclides (148, 149). To trace targeting of glioma by NPs, polyethylenimine (PEI)-based NPs were conjugated with diethylenetriaminepentaacetic acid (DTPA), and glioma-targeting peptide chlorotoxin. The choice of DTPA allows conjugation with ^{99m}Tc for SPECT (150). The PEI NPs were loaded with doxorubicin and administered intravenously to orthotopic glioma rat models and SPECT images were obtained at 0.5, 2, 4, 6, 8, and 12 h, followed by extraction of the tumour and other major organs. SPECT images showed targeting of the glioma tumour in the brain compared to PEI NPs that lacked the glioma targeting peptide (150). It should be noted that a typical SPECT acquires 2D images of different brain layers, which are then computationally processed for a 3D reconstruction and analysis of human brains (147). To analyse data from small animals such as mice or rats, small animal SPECT and SPECT/CT systems have been developed (151). SPECT/CT appears to be more interesting, which provides 3D reconstruction of anatomical parts with high resolution (CT part) overlaid with functional imaging (SPECT part) (151).

It should be noted that imaging small animals generally requires anaesthesia (152). This is to ensure constant restraining of the animals. However, anaesthesia may affect physiological functions of the brain, which has been demonstrated for dopamine receptor D1 ligands (153). In addition, different anesthetising agents may alter brain physiology differently (153). Therefore, it may be recommended that for the purpose of drug delivery, the imaging techniques such as PET may be performed in the conscious state small animals (152) by employing devices such as the RatCAP (SynchroPET, Brookhaven, New York, USA).

9.6 TRANSLATION TO CLINICAL EVALUATIONS

Currently, conducting animal studies is unavoidable in the development of novel drug delivery systems for targeting the brain. Certainly, it is important to choose an animal model that simulates human neurological disorders as closely as possible. Mice and Rats have been the main animal models (14). When rodents are used, the personnel may be blinded about the treatments (154). In addition, a suitable number of animals should be employed in each treatment group. This is to detect a difference such as 30% between the treatment group and control group based on an acceptable statistical power such as 80% (154). Animal models have been successfully used in developing m266, a murine version of Solanezumab (155). Preclinical studies confirmed the reduction of brain amyloid load by m266 (156), and this was also observed in clinical trials of Solanezumab (157). Preclinical studies demonstrated reversing memory deficits in the PDAPP mouse model of AD (158); However, improvements in cognitive functions were not observed in the clinical trials (157). Observing these discrepancies, further emphasis was placed on developing animal models with high translational validity (159). Given the inability of animal models to predict efficacy in clinical trials, the Alzheimer's Association Business Consortium arranged a think tank in November 2018 (160). It was proposed to develop next-generation animal models that will more accurately predict clinical efficacy, which may enable the translation of drugs from bench to bedside (160).

9.7 CONCLUSIONS

In vivo evaluation of brain drug delivery systems remains an indispensable and unavoidable component of CNS therapeutic development. Animal models have been developed for almost all CNS neurological disorders. These animal models recapitulate part of human CNS disorders. In addition, animal models tend to recover from CNS disorders, which may not occur in humans. Sophisticated imaging and analytical techniques have been developed to measure the extent to which drug molecules reach the brain. Animal models have been successful in predicting part of clinical outcomes such as reduction of Aβ load in the brain but generally overestimate cognitive function improvements. As another example, the lack of full representation of the secondary injury in animal models of TBI could be one of the reasons for drugs that showed promise in preclinical studies but failed in clinical studies of TBI (99). Therefore, next-generation animal models are needed for better prediction of clinical trial outcomes.

REFERENCES

1. Pardridge WM. Blood-brain barrier delivery. Drug Discov Today. 2007;12(1–2):54–61.
2. van Rooy I, Cakir-Tascioglu S, Hennink WE, Storm G, Schiffelers RM, Mastrobattista E. In vivo methods to study uptake of nanoparticles into the brain. Pharmaceutical Research. 2011;28(3):456–71.
3. Sevigny J, Chiao P, Bussière T, Weinreb PH, Williams L, Maier M et al. The antibody aducanumab reduces aβ plaques in Alzheimer's disease. Nature. 2016;537(7618):50–6.
4. Islam Y, Ehtezazi P, Cashmore A, Marinsalda E, Leach AG, Coxon CR et al. The inclusion of a matrix metalloproteinase-9 responsive sequence in self-assembled peptide-based brain-targeting nanoparticles improves the efficiency of nanoparticles crossing the blood-brain barrier at elevated MMP-9 levels. J Pharm Sci. 2021;110(3):1349–64.
5. Islam Y, Leach AG, Smith J, Pluchino S, Coxon CR, Sivakumaran M, et al. Physiological and pathological factors affecting drug delivery to the brain by nanoparticles. Adv Sci. 2021;8(11:2002085.
6. You H, Zhang S, Zhang Y, Chen Q, Wu Y, Zhou Z et al. Engineered bacterial outer membrane vesicles-based doxorubicin and CD47-siRNA co-delivery nanoplatform overcomes immune resistance to potentiate the immunotherapy of glioblastoma. Adv Mater. (Deerfield Beach, Fla). 2025;37(15):e2418053.

7. Trivedi S, Agade R, Belgamwar V. Engineering of dual-functionalized intranasal nanovesicles embeded with thymoquinone for targeted modulation of the PI3K/AKT pathway in glioblastoma therapy. Mol Pharm. 2025;22(8):5075–93.
8. Haney MJ, Zhao Y, Fay J, Duhyeong H, Wang M, Wang H et al. Genetically modified macrophages accomplish targeted gene delivery to the inflamed brain in transgenic parkin Q311X(A) mice: Importance of administration routes. Sci Rep. 2020;10(1):11818.
9. Kumar P, Sharma G, Kumar R, Malik R, Singh B, Katare OP et al. Enhanced brain delivery of dimethyl fumarate employing tocopherol-acetate-based nanolipidic carriers: Evidence from pharmacokinetic, biodistribution, and cellular uptake studies. ACS Chem Neurosci. 2017;8(4):860–5.
10. Kopec BM, Kiptoo P, Zhao L, Rosa-Molinar E, Siahaan TJ. Noninvasive brain delivery and efficacy of BDNF to stimulate neuroregeneration and suppression of disease relapse in EAE mice. Mol Pharm. 2020;17(2):404–16.
11. Tuladhar A, Morshead CM, Shoichet MS. Circumventing the blood–brain barrier: Local delivery of cyclosporin A stimulates stem cells in stroke-injured rat brain. J Control Release. 2015;215:1–11.
12. Wise SP. Forward frontal fields: Phylogeny and fundamental function. Trends in Neurosci. 2008; 31(12):599–608.
13. Wilson MG, Webb TD, Odéen H, Kubanek J. Remotely controlled drug release in deep brain regions of non-human primates. J Control Release. 2024;369:775–85.
14. Azkona G, Sanchez-Pernaute R. Mice in translational neuroscience: What r we doing? Prog Neurobiol. 2022;217:102330.
15. Bredlau AL, Motamarry A, Chen C, McCrackin MA, Helke K, Armeson KE et al. Localized delivery of therapeutic doxorubicin dose across the canine blood-brain barrier with hyperthermia and temperature sensitive liposomes. Drug Delivery. 2018;25(1):973–84.
16. Saleem S, Kannan RR. Zebrafish: A promising real-time model system for nanotechnology-mediated neurospecific drug delivery. Nanoscale Res Lett. 2021;16(1):135.
17. Nayaka NMDMW, Adnyana IK, Anggadiredja K, Wibowo I. Drug screening for ischemic stroke using larvae and adult zebrafish model: A review. Lab Anim Res. 2025;41(1):1.
18. Liu Z, Wang M, Ding X, Tian J, Sun D, Gao X et al. Exploration the effective components of Gastrodia elata in improving cerebral ischemia reperfusion injury based on "Spectrum-effect" correlation and zebrafish verification experiment. Phytomedicine. 2024;135:156211.
19. Manev H, Dimitrijevic N, Dzitoyeva S. Techniques: Fruit flies as models for neuropharmacological research. Trends Pharmacol Sci. 2003;24(1):41–3.
20. Shen H, Aggarwal N, Cui B, Foo GW, He Y, Srivastava SK et al. Engineered commensals for targeted nose-to-brain drug delivery. Cell. 2025;188(6):1545–62.e16.
21. Tolomeo AM, Zuccolotto G, Malvicini R, De Lazzari G, Penna A, Franco C et al. Biodistribution of intratracheal, intranasal, and intravenous injections of human mesenchymal stromal cell-derived extracellular vesicles in a mouse model for drug delivery studies. Pharmaceutics. 2023;15(2):548.
22. Han D, Wang M, Dong N, Zhang J, Li D, Ma X et al. Selective homing of brain-derived reconstituted lipid nanoparticles to cerebral ischemic area enables improved ischemic stroke treatment. J Control Release. 2024;365:957–68.
23. Schinkel AH, Wagenaar E, van Deemter L, Mol CA, Borst P. Absence of the mdr1a p-glycoprotein in mice affects tissue distribution and pharmacokinetics of dexamethasone, digoxin, and cyclosporin A. J Clin Invest. 1995;96(4):1698–705.
24. Cutler L, Howes C, Deeks NJ, Buck TL, Jeffrey P. Development of a p-glycoprotein knockout model in rodents to define species differences in its functional effect at the Blood–brain barrier. J Pharm Sci. 2006;95(9):1944–53.
25. Agarwal S, Manchanda P, Vogelbaum MA, Ohlfest JR, Elmquist WF. Function of the blood-brain barrier and restriction of drug delivery to invasive glioma cells: Findings in an orthotopic rat xenograft model of glioma. Drug Metab Dispos. 2013;41(1):33–9.
26. Tang K, Tang Z, Niu M, Kuang Z, Xue W, Wang X et al. Allosteric targeted drug delivery for enhanced blood-brain barrier penetration via mimicking transmembrane domain interactions. Nat Commun. 2025;16(1):3410.
27. Harvey BK, Richie CT, Hoffer BJ, Airavaara M. Transgenic animal models of neurodegeneration based on human genetic studies. Journal of Neural Transmission (Vienna, Austria: 1996). 2011;118(1):27–45.
28. Pasupuleti V, Vora L, Prasad R, Nandakumar DN, Khatri DK. Glioblastoma preclinical models: Strengths and weaknesses. Biochim Biophys Acta Rev Cancer. 2024;1879(1):189059.
29. Murrell J, Farlow M, Ghetti B, Benson MD. A mutation in the amyloid precursor protein associated with hereditary Alzheimer's disease. Science. 1991;254(5028):97–9.

30. Games D, Adams D, Alessandrini R, Barbour R, Berthelette P, Blackwell C et al. Alzheimer-type neuropathology in transgenic mice overexpressing V717F beta-amyloid precursor protein. Nature. 1995;373(6514):523–7.
31. Lewis J, Dickson DW, Lin WL, Chisholm L, Corral A, Jones G et al. Enhanced neurofibrillary degeneration in transgenic mice expressing mutant tau and APP. Science. 2001;293(5534):1487–91.
32. Soundara Pandi SP, Ratnayaka JA, Lotery AJ, Teeling JL. Progress in developing rodent models of age-related macular degeneration (AMD). Exp Eye Res. 2021;203:108404.
33. Oakley H, Cole SL, Logan S, Maus E, Shao P, Craft J et al. Intraneuronal β-amyloid aggregates, neurodegeneration, and neuron loss in transgenic mice with five familial Alzheimer's disease mutations: Potential factors in amyloid plaque formation. J Neurosci. 2006;26(40):10129–40.
34. Lin RR, Jin LL, Xue YY, Zhang ZS, Huang HF, Chen DF et al. Hybrid membrane-coated nanoparticles for precise targeting and synergistic therapy in Alzheimer's disease. Adv Sci. 2024;11(24):e2306675.
35. Radde R, Bolmont T, Kaeser SA, Coomaraswamy J, Lindau D, Stoltze L et al. Abeta42-driven cerebral amyloidosis in transgenic mice reveals early and robust pathology. EMBO Reports. 2006;7(9): 940–6.
36. Kurt MA, Davies DC, Kidd M, Duff K, Rolph SC, Jennings KH et al. Neurodegenerative changes associated with β-amyloid deposition in the brains of mice carrying mutant amyloid precursor protein and mutant presenilin-1 transgenes. Exp Neurol. 2001;171(1):59–71.
37. Borchelt DR, Ratovitski T, van Lare J, Lee MK, Gonzales V, Jenkins NA et al. Accelerated amyloid deposition in the brains of transgenic mice coexpressing mutant presenilin 1 and amyloid precursor proteins. Neuron. 1997;19(4):939–45.
38. Holcomb L, Gordon MN, McGowan E, Yu X, Benkovic S, Jantzen P et al. Accelerated Alzheimer-type phenotype in transgenic mice carrying both mutant amyloid precursor protein and presenilin 1 transgenes. Nat Med. 1998;4(1):97–100.
39. Locci A, Orellana H, Rodriguez G, Gottliebson M, McClarty B, Dominguez S et al. Comparison of memory, affective behavior, and neuropathology in APPNLGF knock-in mice to 5xFAD and APP/PS1 mice. Behav Brain Res. 2021;404:113192.
40. Liu P, Zhang T, Wu Y, Chen Q, Sun T, Jiang C. A peptide-drug conjugate-based nanoplatform for immunometabolic activation and in situ nerve regeneration in advanced-stage Alzheimer's disease. Adv Mater. (Deerfield Beach, Fla). 2024;36(46):e2408729.
41. Lee D, Shen AM, Shah M, Garbuzenko OB, Minko T. In vivo evaluation of nose-to-brain delivery of liposomal donepezil, memantine, and BACE-1 siRNA for Alzheimer's disease therapy. Int J Mol Sci. 2024;25(19):10357.
42. Tosh JL, Rickman M, Rhymes E, Norona FE, Clayton E, Mucke L et al. The integration site of the APP transgene in the J20 mouse model of Alzheimer's disease. Wellcome Open Res. 2017;2:84.
43. Ameen-Ali KE, Simpson JE, Wharton SB, Heath PR, Sharp PS, Brezzo G et al. The time course of recognition memory impairment and glial pathology in the hAPP-J20 mouse model of Alzheimer's disease. J Alzheimers Dis. 2019;68(2):609–24.
44. Otaegui L, Urgin T, Zaiter T, Zussy C, Vitalis M, Pellequer Y et al. Nose-to-brain delivery of DHA-loaded nanoemulsions: A promising approach against Alzheimer's disease. Int J Pharma. 2025;670:125125.
45. Abdulhameed N, Babin A, Hansen K, Weaver R, Banks WA, Talbot K et al. Comparing regional brain uptake of incretin receptor agonists after intranasal delivery in CD-1 mice and the APP/PS1 mouse model of Alzheimer's disease. Alzheimers Res Ther. 2024;16(1):173.
46. Rosen DR, Siddique T, Patterson D, Figlewicz DA, Sapp P, Hentati A et al. Mutations in Cu/Zn superoxide dismutase gene are associated with familial amyotrophic lateral sclerosis. Nature. 1993;362(6415):59–62.
47. Epstein CJ, Avraham KB, Lovett M, Smith S, Elroy-Stein O, Rotman G et al. Transgenic mice with increased Cu/Zn-superoxide dismutase activity: Animal model of dosage effects in down syndrome. Proc Natl Acad Sci U.S.A. 1987;84(22):8044–8.
48. Yarom R, Sapoznikov D, Havivi Y, Avraham KB, Schickler M, Groner Y. Premature aging changes in neuromuscular junctions of transgenic mice with an extra human CuZnSOD gene: A model for tongue pathology in Down's syndrome. J Neurol Sci. 1988;88(1):41–53.
49. Gurney ME, Pu H, Chiu AY, Dal Canto MC, Polchow CY, Alexander DD et al. Motor neuron degeneration in mice that express a human cu, Zn superoxide dismutase mutation. Science. 1994; 264(5166):1772–5.
50. Ediriweera GR, Sivaram AJ, Cowin G, Brown ML, McAlary L, Lum JS et al. Lipid nanoparticles and transcranial focused ultrasound enhance the delivery of SOD1 antisense oligonucleotides to the murine brain for ALS therapy. J Control Release. 2025;378:221–35.

51. Zhang J, Chen K, Chen Y, Hua L, Chen S, Chen X et al. Pathology reduction and motor behavior improvement associated with ultrasound-mediated delivery of arctiin to the motor cortex in a mutant SOD1 mouse model of amyotrophic lateral sclerosis. Ultrasonics. 2024;144:107449.
52. Colomer Gould VF. Mouse models of spinocerebellar ataxia type 3 (Machado-Joseph disease). Neurotherapeutics. 2012;9(2):285–96.
53. Tang Z, Hu S, Wu Z, He M. Therapeutic effects of engineered exosome-based miR-25 and miR-181a treatment in spinocerebellar ataxia type 3 mice by silencing ATXN3. Mol Med. (Cambridge, Mass). 2023;29(1):96.
54. Dawson TM, Ko HS, Dawson VL. Genetic animal models of Parkinson's disease. Neuron. 2010;66(5): 646–61.
55. Masliah E, Rockenstein E, Veinbergs I, Mallory M, Hashimoto M, Takeda A et al. Dopaminergic loss and inclusion body formation in alpha-synuclein mice: Implications for neurodegenerative disorders. Science. 2000;287(5456):1265–9.
56. Kundu P, Das M, Tripathy K, Sahoo SK. Delivery of dual drug loaded lipid based nanoparticles across the blood-brain barrier impart enhanced neuroprotection in a rotenone induced mouse model of Parkinson's disease. ACS Chem Neurosci. 2016;7(12):1658–70.
57. Mustapha M, Mat Taib CN. MPTP-induced mouse model of Parkinson's disease: A promising direction of therapeutic strategies. Bosn J Basic Med Sci. 2021;21(4):422–33.
58. Xu K, Huang P, Wu Y, Liu T, Shao N, Zhao L et al. Engineered Selenium/Human serum albumin nanoparticles for efficient targeted treatment of Parkinson's disease via Oral gavage. ACS Nano. 2023;17(20):19961–80.
59. Masini D, Plewnia C, Bertho M, Scalbert N, Caggiano V, Fisone G. A guide to the generation of a 6-hydroxydopamine mouse model of Parkinson's disease for the study of non-motor symptoms. Biomedicines. 2021;9(6).
60. Gaba B, Khan T, Haider MF, Alam T, Baboota S, Parvez S et al. Vitamin E-loaded naringenin nanoemulsion via intranasal delivery for the management of oxidative stress in a 6-OHDA Parkinson's disease model. BioMed Res Int. 2019;2019:2382563.
61. Lu CT, Jin RR, Jiang YN, Lin Q, Yu WZ, Mao KL et al. Gelatin nanoparticle-mediated intranasal delivery of substance P protects against 6-hydroxydopamine-induced apoptosis: An in vitro and in vivo study. Drug Des Devel Ther. 2015;9:1955–62.
62. Ruan S, Li J, Ruan H, Xia Q, Hou X, Wang Z et al. Microneedle-mediated nose-to-brain drug delivery for improved Alzheimer's disease treatment. J Control Release. 2024;366:712–31.
63. Baerends E, Soud K, Folke J, Pedersen A-K, Henmar S, Konrad L et al. Modeling the early stages of Alzheimer's disease by administering intracerebroventricular injections of human native aβ oligomers to rats. Acta Neuropathol Commun. 2022;10(1):113.
64. Saleh SR, Khamiss SE, Aly Madhy S, Khattab SN, Sheta E, Elnozahy FY et al. Biochemical investigation and in silico analysis of the therapeutic efficacy of ipriflavone through tet-1 surface-modified-PLGA nanoparticles in streptozotocin-induced Alzheimer's like disease: Reduced oxidative damage and etiological descriptors. Int J Pharm. 2025;669:125021.
65. Li Y, Zhang J. Animal models of stroke. Animal Model Exp Med. 2021;4(3):204–19.
66. Zhu Z, Jin L, Wang Q, Shi H, Cheng K, Mao Z. Inhalable ce nanozyme-backpacked phage aims at ischemic cerebral injury by M1-microglia hitchhiking. Adv Mater (Deerfield Beach, Fla). 2025; 37(27):e2419903.
67. Shin HJ, Lee KY, Kang JW, Choi SG, Kim DW, Yi YY. Perampanel reduces brain damage via induction of M2 microglia in a neonatal rat stroke model. Int J Nanomed. 2022;17:2791–804.
68. Watson BD, Dietrich WD, Busto R, Wachtel MS, Ginsberg MD. Induction of reproducible brain infarction by photochemically initiated thrombosis. Ann Neurol. 1985;17(5):497–504.
69. Roth JM. Recombinant tissue plasminogen activator for the treatment of acute ischemic stroke. Proc (Bayl Univ Med Cent). 2011;24(3):257–9.
70. Kestal DW, S; Brodie; M EPILEPSY A Public Health Imperative. WHO; 2019.
71. Perucca E, Perucca P, White HS, Wirrell EC. Drug resistance in epilepsy. Lancet Neurol. 2023;22(8):723–34.
72. Han H, Mann A, Ekstein D, Eyal S. Breaking bad: The structure and function of The blood-brain barrier in epilepsy. AAPS J. 2017;19(4):973–88.
73. Lévesque M, Avoli M. The kainic acid model of temporal lobe epilepsy. Neurosci Biobehav Rev. 2013;37(10, Part 2):2887–99.
74. Geng C, Ren X, Cao P, Chu X, Wei P, Liu Q et al. Macrophage membrane–biomimetic nanoparticles target inflammatory microenvironment for epilepsy treatment. Theranostics. 2024;14(17):6652–70.

75. Cavalheiro EA, Santos NF, Priel MR. The pilocarpine model of epilepsy in mice. Epilepsia. 1996;37(10):1015–9.
76. Shimada T, Yamagata K. Pentylenetetrazole-induced kindling mouse model. J Vis Exp. 2018(136):56573.
77. Fava M, Kendler KS. Major depressive disorder. Neuron. 2000;28(2):335–41.
78. Gibson QH. Interventionism and intelligibility: Why depression is not (Always) a brain disease. J Med Philos. 2024;49(2):160–77.
79. Troubat R, Barone P, Leman S, Desmidt T, Cressant A, Atanasova B et al. Neuroinflammation and depression: A review. Eur J Neurosci. 2021;53(1):151–71.
80. Miller ES, Sakowicz A, Roy A, Yang A, Sullivan JT, Grobman WA et al. Plasma and cerebrospinal fluid inflammatory cytokines in perinatal depression. Am J Obstet Gynecol. 2019;220(3):271.e1–.e10.
81. Haapakoski R, Mathieu J, Ebmeier KP, Alenius H, Kivimäki M. Cumulative meta-analysis of interleukins 6 and 1β, tumour necrosis factor α and C-reactive protein in patients with major depressive disorder. Brain, Behav, Immun. 2015;49:206–15.
82. Hussain M, Kumar P, Khan S, Gordon DK, Khan S. Similarities between depression and neurodegenerative diseases: Pathophysiology, challenges in diagnosis and treatment options. Cureus. 2020;12(11):e11613.
83. Papa D, Ingenito A, von Gal A, De Pandis MF, Piccardi L. Relationship between depression and neurodegeneration: Risk factor, prodrome, consequence, or something else? A scoping review. Biomedicines. 2025;13(5):1023.
84. Xie X-h, Xu S-x, Yao L, Chen M-m, Zhang H, Wang C et al. Altered in vivo early neurogenesis traits in patients with depression: Evidence from neuron-derived extracellular vesicles and electroconvulsive therapy. Brain Stimul. 2024;17(1):19–28.
85. Deussing JM. Animal models of depression. Drug Discov Today Dis Models. 2006;3(4):375–83.
86. Qian X, Zhong Z, Lu S, Zhang Y. Repeated reserpine treatment induces depressive-like behaviors accompanied with hippocampal impairment and synapse deficit in mice. Brain Res. 2023;1819: 148541.
87. Ma S, Chen M, Jiang Y, Xiang X, Wang S, Wu Z et al. Sustained antidepressant effect of ketamine through NMDAR trapping in the LHb. Nature. 2023;622(7984):802–9.
88. Jiang C, Yang X, Huang Q, Lei T, Luo H, Wu D et al. Microglial-biomimetic memantine-loaded polydopamine nanomedicines for alleviating depression. Adv Mater (Deerfield Beach, Fla). 2025; 37(9):e2417869.
89. Palavra F, Viana SD, Henriques S, Dinis J, Martins J, Madeira MH et al. Defining milestones for the study of remyelination using the cuprizone mouse model: How early is early? Mult Scler and Relat Disord. 2022;63:103886.
90. Torkildsen O, Brunborg LA, Myhr KM, Bø L. The cuprizone model for demyelination. Acta Neurol Scand Suppl. 2008;188:72–6.
91. Zirngibl M, Assinck P, Sizov A, Caprariello AV, Plemel JR. Oligodendrocyte death and myelin loss in the cuprizone model: An updated overview of the intrinsic and extrinsic causes of cuprizone demyelination. Mol Neurodegener. 2022;17(1):34.
92. Toomey LM, Papini M, Lins B, Wright AJ, Warnock A, McGonigle T et al. Cuprizone feed formulation influences the extent of demyelinating disease pathology. Sci Rep. 2021;11(1):22594.
93. El-Sayed SAM, Fouad GI, Rizk MZ, Beherei HH, Mabrouk M. Comparative neuroprotective potential of nanoformulated and free resveratrol against cuprizone-induced demyelination in rats. Mol Neurobiol. 2025;62(3):2710–25.
94. Guan Y, Wang Y, Yuan F, Lu H, Ren Y, Xiao T et al. Effect of suture properties on stability of middle cerebral artery occlusion evaluated by synchrotron radiation angiography. Stroke. 2012;43(3):888–91.
95. Steele EC, Guo Q, Namura S. Filamentous middle cerebral artery occlusion causes ischemic damage to the retina in mice. Stroke. 2008;39(7):2099–104.
96. Gao Z, Wang X, Zhang W, Wang J, Liu C. Inflammatory oriented nanospheres-reconstructed extracellular matrix in ischemic stroke. ACS Nano. 2025;19(24):22065–78.
97. Kramer M, Dang J, Baertling F, Denecke B, Clarner T, Kirsch C et al. TTC staining of damaged brain areas after MCA occlusion in the rat does not constrict quantitative gene and protein analyses. J Neurosci Methods. 2010;187(1):84–9.
98. Zhu N, Wang F, Han Z, Ling S, Wong DW, Ye S et al. Noninvasive method for achieving the regeneration of damaged nerves via ultrasonic nasal drops. Bioact Mater. 2025;49:342–61.
99. Xiong Y, Mahmood A, Chopp M. Animal models of traumatic brain injury. Nature Reviews Neuroscience. 2013;14(2):128–42.

100. da Silva Fiorin F, Cunha do Espírito Santo C, Santos do Nascimento R, França AP, Freire Royes LF. Behavioral deficits after mild traumatic brain injury by fluid percussion in rats. Neurosci Lett. 2024;818:137550.
101. Romine J, Gao X, Chen J Controlled cortical impact model for traumatic brain injury. J Vis Exp. 2014(90):e51781.
102. Williams AJ, Hartings JA, Lu XC, Rolli ML, Tortella FC. Penetrating ballistic-like brain injury in the rat: Differential time courses of hemorrhage, cell death, inflammation, and remote degeneration. J Neurotrauma. 2006;23(12):1828–46.
103. Thangavelu B, Wilfred BS, Johnson D, Gilsdorf JS, Shear DA, Boutté AM. Penetrating ballistic-like brain injury leads to MicroRNA dysregulation, BACE1 upregulation, and amyloid precursor protein loss in lesioned rat brain tissues. Front Neurosci. 2020;14 - 2020.
104. Feeney DM, Boyeson MG, Linn RT, Murray HM, Dail WG. Responses to cortical injury: I. Methodology and local effects of contusions in the rat. Brain Res. 1981;211(1):67–77.
105. Albert-Weissenberger C, Sirén AL. Experimental traumatic brain injury. Exp Transl Stroke Med. 2010;2(1):16.
106. Marmarou CR, Prieto R, Taya K, Young HF, Marmarou A Marmarou weight drop injury model. In: Chen J, Xu ZC, Xu X-M, Zhang JH, editors. Animal Models of Acute Neurological Injuries. Totowa, NJ: Humana Press; 2009. p. 393–407.
107. Friedman-Levi Y, Liraz-Zaltsman S, Shemesh C, Rosenblatt K, Kesner EL, Gincberg G et al. Pharmacological blockers of CCR5 and CXCR4 improve recovery after traumatic brain injury. Exp Neurol. 2021;338:113604.
108. Liang R, Wang H, Gao J, Wang J, Zhang W, Qian A. Assessment model of blast injury: A narrative review. iScience. 2025;28(7):112830.
109. Aravind A, Ravula AR, Chandra N, Pfister BJ. Behavioral deficits in animal models of blast traumatic brain injury. Front Neurol. 2020;11 - 2020.
110. Cernak I, Merkle AC, Koliatsos VE, Bilik JM, Luong QT, Mahota TM et al. The pathobiology of blast injuries and blast-induced neurotrauma as identified using a new experimental model of injury in mice. Neurobiol Dis. 2011;41(2):538–51.
111. Jiang Z, Wang J, Qin Y, Liu S, Luo B, Bai F et al. A nonhuman primate model with Alzheimer's disease-like pathology induced by hippocampal overexpression of human tau. Alzheimer's Res Ther. 2024;16(1):22.
112. Constantinescu CS, Farooqi N, O'Brien K, Gran B. Experimental autoimmune encephalomyelitis (EAE) as a model for multiple sclerosis (MS). Br J Pharmacol. 2011;164(4):1079–106.
113. Peruzzotti-Jametti L, Bernstock JD, Vicario N, Costa ASH, Kwok CK, Leonardi T et al. Macrophage-derived extracellular succinate licenses neural stem cells to suppress chronic neuroinflammation. Cell Stem Cell. 2018;22(3):355–68.e13.
114. Keough MB, Jensen SK, Yong VW. Experimental demyelination and remyelination of murine spinal cord by focal injection of lysolecithin. J Vis Exp. 2015(97).
115. Morris AD, Kucenas S. A novel lysolecithin model for visualizing damage in vivo in the larval zebrafish spinal cord. Front Cell Dev Biol. 2021;9:654583.
116. Naeimi R, Safarpour F, Hashemian M, Tashakorian H, Ahmadian SR, Ashrafpour M et al. Curcumin-loaded nanoparticles ameliorate glial activation and improve myelin repair in lyolecithin-induced focal demyelination model of rat corpus callosum. Neurosci Lett. 2018;674:1–10.
117. Van Dam D, De Deyn PP. Animal models in the drug discovery pipeline for Alzheimer's disease. Br J Pharmacol. 2011;164(4):1285–300.
118. Cummings BJ, Su JH, Cotman CW, White R, Russell MJ. β-amyloid accumulation in aged canine brain: A model of early plaque formation in Alzheimer's disease. Neurobiol Aging. 1993;14(6):547–60.
119. Head E, Moffat K, Das P, Sarsoza F, Poon WW, Landsberg G et al. β-amyloid deposition and tau phosphorylation in clinically characterized aged cats. Neurobiol Aging. 2005;26(5):749–63.
120. Latimer CS, Shively CA, Keene CD, Jorgensen MJ, Andrews RN, Register TC et al. A nonhuman primate model of early Alzheimer's disease pathologic change: Implications for disease pathogenesis. Alzheimer's & Dementia: J Alzheimer's Association. 2019;15(1):93–105.
121. Jasinska AJ, Schmitt CA, Service SK, Cantor RM, Dewar K, Jentsch JD et al. Systems biology of the vervet monkey. ILAR J. 2013;54(2):122–43.
122. Shannon RJ, Carpenter KL, Guilfoyle MR, Helmy A, Hutchinson PJ. Cerebral microdialysis in clinical studies of drugs: Pharmacokinetic applications. J Pharmacokinet Pharmacodyn. 2013;40(3):343–58.

123. Tsai TH, Chen YF, Chou CJ, Chen CF. Measurement and pharmacokinetics of unbound 20(s)-camptothecin in rat blood and brain by microdialysis coupled to microbore liquid chromatography with fluorescence detection. J Chromatogr A. 2000;870(1–2):221–6.
124. Alavijeh MS, Palmer AM. Measurement of the pharmacokinetics and pharmacodynamics of neuroactive compounds. Neurobiol Dis. 2010;37(1):38–47.
125. Bommana MM, Kirthivasan B, Squillante E. In vivo brain microdialysis to evaluate FITC-dextran encapsulated immunopegylated nanoparticles. Drug Delivery. 2012;19(6):298–306.
126. Clinckers R, Smolders I, Michotte Y, Ebinger G, Danhof M, Voskuyl RA et al. Impact of efflux transporters and of seizures on the pharmacokinetics of oxcarbazepine metabolite in the rat brain. Br J Pharmacol. 2008;155(7):1127–38.
127. Chang Y-L, Tsai P-L, Chou Y-C, Tien J-H, Tsai T-H. Simultaneous determination of nicotine and its metabolite, cotinine, in rat blood and brain tissue using microdialysis coupled with liquid chromatography: Pharmacokinetic application. J Chromatogr A. 2005;1088(1):152–7.
128. Heininen J, Julku U, Myöhänen T, Kotiaho T, Kostiainen R. Multiplexed analysis of amino acids in mice brain microdialysis samples using isobaric labeling and liquid chromatography-high resolution tandem mass spectrometry. J Chromatogr A. 2021;1656:462537.
129. Zestos AG, Kennedy RT. Microdialysis coupled with LC-MS/MS for in vivo neurochemical monitoring. AAPS J. 2017;19(5):1284–93.
130. Akbar L, Castillo VCG, Olorocisimo JP, Ohta Y, Kawahara M, Takehara H et al. Multi-region microdialysis imaging platform revealed dorsal raphe nucleus calcium signaling and serotonin dynamics during nociceptive pain. Int J Mol Sci. 2023;24(7).
131. Song Y, Wang L, Peng T, Shan L, Wan B, Tang M et al. Brain-targeting biomimetic nanozyme enhances neuroprotection in ischemic stroke by remodeling the neurovascular unit. J Control Release: Official J Control Release Soc. 2025;382:113750.
132. Han D, Zhang J, Li D, Wang C. Celery seed derived reconstituted lipid nanoparticles as an innate neuron-targeted neuroprotective nanomedicine for ischemic stroke treatment. J Nanobiotechnol. 2025;23(1):298.
133. Zhang T, Liu P, Shen W, Li C, Zhao Z, Wu Y, et al. DNase I-mediated chemotactic nanoparticles for NETs targeting and microenvironment remodeling treatment of acute ischemic stroke. Adv Sci (Weinh). 2025;12(34):e03689.
134. Bai L, Li N, Li H, Dai Y, Bai Y, Ma F et al. Macrophage-mimetic liposomes co-delivering ceria and Ac2-26 peptide for penumbra protection in ischemic stroke. J Nanobiotechnol. 2025;23(1):488.
135. Zhao M, Li Q, Chai Y, Rong R, He L, Zhang Y et al. An anti-CD19-exosome delivery system navigates the blood–brain barrier for targeting of central nervous system lymphoma. J Nanobiotechnol. 2025;23(1):173.
136. Liu Q, Ma H, Liao J, Zhu Z, Chen H, Sun D et al. MiR-125a-5p in extracellular vesicles of neural stem cells acts as a crosstalk signal modulating neuroinflammatory microenvironment to alleviate cerebral ischemia-reperfusion injury. Theranostics. 2025;15(14):7064–89.
137. Ding H, Wu F, Nair MP. Image-guided drug delivery to the brain using nanotechnology. Drug Discovery Today. 2013;18(21):1074–80.
138. Okada M, Kikuchi T, Okamura T, Ikoma Y, Tsuji AB, Wakizaka H et al. In-vivo imaging of blood-brain barrier permeability using positron emission tomography with 2-amino-[3-11C]isobutyric acid. Nucl Med Commun. 2015;36(12):1239–48.
139. Mishra A, Payne C, Carrascal-Miniño A, Sunassee K, Halbherr S, Pouliopoulos AN et al. PET imaging for non-invasive monitoring of 89Zr-talidox delivery to the brain following focused ultrasound-mediated blood-brain barrier opening. J Control Release. 2025;387:114183.
140. Cole LE, Ross RD, Tilley JMR, Vargo-Gogola T, Roeder RK. Gold nanoparticles as contrast agents in X-ray imaging and computed tomography. Nanomedicine. 2015;10(2):321–41.
141. Olsson B, Aulie A, Sveen K, Andrew E. Human pharmacokinetics of iohexol. A new nonionic contrast medium. Invest Radiol. 1983;18(2):177–82.
142. Rhee CM, Bhan I, Alexander EK, Brunelli SM. Association between iodinated contrast media exposure and incident hyperthyroidism and hypothyroidism. Arch Intern Med. 2012;172(2):153–9.
143. Hainfeld JF, Slatkin DN, Focella TM, Smilowitz HM. Gold nanoparticles: A new x-ray contrast agent. Br J Radiol. 2006;79(939):248–53.
144. Perets N, Betzer O, Shapira R, Brenstein S, Angel A, Sadan T et al. Golden exosomes selectively target brain pathologies in neurodegenerative and neurodevelopmental disorders. Nano Lett. 2019;19(6):3422–31.

145. Law LH, Huang J, Xiao P, Liu Y, Chen Z, Lai JHC et al. Multiple CEST contrast imaging of nose-to-brain drug delivery using iohexol liposomes at 3T MRI. J Control Release. 2023;354:208–20.
146. Alexandrov AV, Black SE, Ehrlich LE, Bladin CF, Smurawska LT, Pirisi A et al. Simple visual analysis of brain perfusion on HMPAO SPECT predicts early outcome in acute stroke. Stroke. 1996;27(9):1537–42.
147. Imokawa T, Yokoyama K, Takahashi K, Oyama J, Tsuchiya J, Sanjo N et al. Brain perfusion SPECT in dementia: What radiologists should know. Jpn J Radiol. 2024;42(11):1215–30.
148. Goffin K, van Laere K. Chapter 13 - single-photon emission tomography. In: Masdeu JC, González RG, editors. Handb Clin Neurol. 135: Elsevier; 2016. p. 241–50.
149. Mizutani A, Kobayashi M, Ohuchi M, Sasaki K, Muranaka Y, Torikai Y et al. Indirect SPECT imaging evaluation for possible nose-to-brain drug delivery using a compound with poor Blood–Brain barrier permeability in mice. Pharmaceutics. 2022;14(5):1026.
150. Zhao L, Zhu J, Gong J, Song N, Wu S, Qiao W et al. Polyethylenimine-based theranostic nanoplatform for glioma-targeting single-photon emission computed tomography imaging and anticancer drug delivery. J Nanobiotechnol. 2020;18(1):143.
151. Franc BL, Acton PD, Mari C, Hasegawa BH. Small-animal SPECT and SPECT/CT: Important tools for preclinical investigation. J Nucl Med. 2008;49(10):1651–63.
152. Hildebrandt IJ, Su H, Weber WA. Anesthesia and other considerations for in vivo imaging of small animals. ILAR J. 2008;49(1):17–26.
153. Momosaki S, Hatano K, Kawasumi Y, Kato T, Hosoi R, Kobayashi K et al. Rat-PET study without anesthesia: Anesthetics modify the dopamine D1 receptor binding in rat brain. Synapse. 2004;54(4):207–13.
154. Hersch SM, Ferrante RJ. Translating therapies for Huntington's disease from genetic animal models to clinical trials. NeuroRx. 2004;1(3):298–306.
155. Mably AJ, Liu W, Mc Donald JM, Dodart J-C, Bard F, Lemere CA et al. Anti-aβ antibodies incapable of reducing cerebral aβ oligomers fail to attenuate spatial reference memory deficits in J20 mice. Neurobiol Disease. 2015;82:372–84.
156. DeMattos RB, Bales KR, Cummins DJ, Dodart JC, Paul SM, Holtzman DM. Peripheral anti-a beta antibody alters CNS and plasma a beta clearance and decreases brain a beta burden in a mouse model of Alzheimer's disease. Proc Natl Acad Sci U.S.A. 2001;98(15):8850–5.
157. Doody RS, Thomas RG, Farlow M, Iwatsubo T, Vellas B, Joffe S et al. Phase 3 trials of solanezumab for mild-to-moderate Alzheimer's disease. N Engl J Med. 2014;370(4):311–21.
158. Dodart JC, Bales KR, Gannon KS, Greene SJ, DeMattos RB, Mathis C et al. Immunization reverses memory deficits without reducing brain abeta burden in Alzheimer's disease model. Nat Neurosci. 2002;5(5):452–7.
159. Homberg JR, Adan RAH, Alenina N, Asiminas A, Bader M, Beckers T et al. The continued need for animals to advance brain research. Neuron. 2021;109(15):2374–9.
160. Vitek MP, Araujo JA, Fossel M, Greenberg BD, Howell GR, Rizzo SJS et al. Translational animal models for Alzheimer's disease: An Alzheimer's association business consortium think tank. Alzheimer's & Dementia: Translational Res Clinic Interventions. 2020;6(1):e12114.

10 Introduction to Hematological and Immunological Safety in Brain Drug Delivery

Riya Patel, Bindu Kumari Yadav, Shreeraj Shah, and Bhupendra Prajapati

10.1 INTRODUCTION TO BRAIN DRUG DELIVERY

An estimated 1.5 billion people worldwide suffer from disorders of the central nervous system (CNS). The aging population is expected to make the situation worse in the twenty-first century, increasing the prevalence of CNS illnesses (1). Brain-targeted drug delivery has assumed a vital role in modern neurotherapeutics since the brain has some of the strict biological barriers in the body and notably the blood–brain barrier (BBB) and the blood cerebrospinal fluid barrier (BCSFB). These specialized barriers protect the neural environment from fluctuations in the systemic circulation, but they also severely restrict the passage of pharmacological agents, with more than 98% of small molecules and nearly all large therapeutic biologics failing to enter the CNS in therapeutically meaningful concentrations (2). The BBB has endothelial cells that are closely joined, proactive efflux transporters, and metabolic enzymes, which guarantee that only a few small lipophilic molecules enter the CNS. Consequently, a very large percentage of currently available therapeutics, such as large molecules, biologics, and many hydrophilic drugs, fail to reach the effective concentrations in the brain, making it extremely difficult to treat CNS diseases, particularly neurodegenerative diseases, brain tumors, and infections (3). To address these challenges, CNS-targeted delivery uses advanced methods of nanoparticles (NPs), liposomes, nanostructured lipid carriers (NLCs), extracellular vesicles, and receptor/transporter-mediated delivery systems to deliver therapeutic agents across or around the BBB (4). These approaches can boost drug concentrations in the brain, lower systemic exposures, and be able to deliver otherwise impermeable molecules such as peptides, proteins, or even gene-based therapies (5,6). Furthermore, more modern non-invasive delivery methods, such as intranasal nose-to-brain delivery, may have direct access to the CNS with no transit via the BBB and hence improved patient adherence and fewer side effects (7). In addition to treatment, delivery systems can also be utilized in diagnostics since imaging agents or biomarkers can be delivered directly to the CNS tissue (8). Overall, brain-targeted drug delivery is of critical importance for the effective treatment of brain and spinal cord disorders, as it represents the sole method by which many promising therapeutics can be utilized for CNS diseases. This approach enables enhanced efficacy, reduced systemic toxicity, and the potential to address conditions that were previously beyond reach.

Despite substantial breakthroughs, CNS drug delivery approaches, especially those employing nanomedicines, immune cell-based carriers, or immunotherapies, introduce complex hematological and immunological risks that must be thoroughly evaluated. In brain drug delivery techniques, especially those using nanomedicines, immune cell-based carriers, immunotherapies, and similar methods, there are distinct hematological and immunological risks. Most CNS-targeted therapies, including NPs, liposomes, viral vectors, biologics, and gene-delivery systems, are administered systemically and therefore inevitably interact with blood components, such as erythrocytes, leukocytes, platelets, plasma proteins, and endothelial cells. These interactions can cause hemolysis, changes in red blood cell shape, platelet activation, or disturbances in coagulation pathways, potentially

DOI: 10.1201/9781003482406-10

leading to thrombosis or bleeding. Systemic administration of drugs or NPs may also result in off-target effects, such as cytopenias, immune stimulation or suppression, and systemic toxicities like liver or heart damage if localization is poor (9). For instance, immune cell-mediated delivery systems, like macrophage-based carriers, are designed to reduce systemic toxicity and prevent sensitive tissues from exposure to cytotoxic drugs. However, improper encapsulation or release can still lead to hematological toxicity or immune-related adverse events (10). Immunotherapies, including CAR-T cells used to treat brain tumors, carry risks of neurotoxicity, cytokine release syndrome, systemic inflammation, and complications depending on the route of administration (intravenous versus intrathecal), resulting in mild or severe (11). Furthermore, the transferrin receptors or other targeting ligands, along with permeating enhancers that cross the BBB, may unintentionally affect immunity or cause leakage of blood components into the brain, which could trigger local or systemic immune responses (12). These immunological and hematological reactions also impact pharmacokinetics, biodistribution, and bioavailability, resulting in a complicated interaction between the therapeutic goal of brain targeting and immune system recognition. While advanced delivery systems aim to minimize these issues by improving targeting and reducing systemic exposure, hematological and immunological complications related to brain drug delivery remain a concern, and further research is necessary.

Besides immunological issues, barrier-related and systemic exposure factors are also critical in deciding the efficacy and safety of the CNS-targeted delivery systems. The BBB is a necessity in the neuroprotection process, but it significantly heterogeneously penetrates the endothelial tight junctions, transporters, and metabolic enzymes regionally. Few attempts to open or penetrate the BBB by chemical permeability enhancers, hyperosmotic perturbation, ultrasound, or nanocarriers should therefore be critically considered in terms of unwanted physiological effects, including the persistent disruption of barrier functions, a larger influx of toxins or pathogens, and deregulation of cerebral homeostasis (13). Systemic exposure is also problematic because several drug carriers can be easily cleared by the mononuclear phagocyte system, liver, and spleen, leading to accumulation in non-target tissues and unintended side effects. Plasma protein opsonization further alters the biological identity of NPs, impacting immune recognition, biodistribution, and circulation time (14). Finding a balance between limited systemic accumulation and good CNS penetration is crucial since too much off-target presence can lead to hematological and immunological overload in addition to decreasing efficacy. Furthermore, any unintentional immune activation can result in neuroinflammation, microglial activation, or long-term alterations in the brain's functional skills, according to CNS immunology, which was formerly believed to be immune-privileged but is now recognized to be immunologically specialized (15).

This chapter thoroughly examines the hematological and immunological safety variables that are critical for medication transport to the brain. The consequences of distribution vectors, BBB-modulation methods, immunological activation issues, and systemic blood compatibility are all discussed. Highlighted are coagulation pathways, hypersensitivity reactions, interactions between nanocarriers, and regulatory expectations. Future strategies for creating safer, tailored CNS-targeted medications while combining preclinical and clinical evaluation techniques are highlighted in this chapter.

10.2 HEMATOLOGICAL SAFETY IN BRAIN DRUG DELIVERY

Delivering drugs to the brain faces significant challenges due to the BBB, which regulates interactions between systemic circulation and the CNS. While overcoming the BBB is crucial, the hematological safety of drug delivery systems, particularly those given intravenously, is equally important. This involves understanding how delivery vehicles and drug molecules interact with blood components, impacting blood rheology, coagulation, microcirculation, and oxygen transport. Neglecting these interactions can lead to serious adverse effects, including hemolysis, thrombosis, and impaired oxygen delivery. This section discusses hematological considerations related to blood composition, RBC integrity, platelet/coagulation dynamics, and effects on microcirculation.

10.2.1 Blood Composition, Rheology, and Compatibility

Blood is a complex mixture of leukocytes, platelets, and red blood cells (RBCs) suspended in plasma, which is made up of various components such as proteins, ions, and metabolites. Leukocytes and platelets only make up a small portion of the cellular volume, but they carry out vital physiological tasks. RBCs make up the majority of the cellular volume (16). Hemodynamic parameters such as hematocrit, plasma viscosity, RBC deformability, and RBC aggregation behavior affect the rheology of blood. RBCs undergo deformation and alignment under high shear to allow smooth flow, and may aggregate under low shear to form structures, which makes the blood viscous (17). In the design of brain-drug delivery systems, specifically NPs or colloidal carriers that are delivered by intravenous injection, such devices will necessarily face the blood milieu. Much attention is therefore necessary to ensure that hemocompatibility, the carriers have no significant effect on plasma viscosity, no effect on red cell behavior, and no excessive stimulation of platelets, coagulation pathways, or complement system (18). The bad hemocompatibility can lead to poor flow and disturbed cell and platelet distribution, as well as elevated chances of clotting or microvascular blockage, which is an extremely unfavorable event should the aim be to deliver drugs to the brain. In fact, stringent hemocompatibility testing is currently seen as part of the required process of preclinical evaluation of nanoparticle-based drug delivery systems (19).

10.2.2 Hemolysis, RBC Interactions, and Membrane Stability

The interaction of foreign particles or drug carriers with RBCs is one of the main concerns during their introduction into the bloodstream. The result of these interactions may be a compromise of the RBC membrane integrity, resulting in hemolysis (rupture of RBCs) or sublethal damage, thereby impairing RBC deformability, or inducing exposure of previously concealed components of the membrane, both able to cause dire consequences downstream. Examinations of the influence of NPs on RBCs have indicated that under various conditions, several negative effects can take place, like aggregation of the RBCs, reduction in deformability, and destruction of the plasma membrane, resulting in the exposure of phosphatidylserine (PS), a procoagulant molecule (20). The deformability of RBCs is especially crucial because RBCs need to deform to squeeze through small capillaries and microvessels to perfuse tissues efficiently. When drug carriers or other foreign agents make RBCs more rigid or destroy their membrane, it becomes possible that microcirculatory flow is impaired, local stasis or obstruction occurs, and more haemolysis takes place. Indeed, delivery systems based on the coating of red-cell membranes by RBCs have been investigated precisely in order to exploit the compatibility of the red-cell membranes. Further, in an experiment that loaded RBCs with polymeric nanoparticle carriers, high loading ratios enhanced the sensitivity of RBCs to mechanical, osmotic, and oxidative stresses in in vitro harsh conditions; some RBC damage and agglutination were seen. Although these negative effects were not so extensive with modest loading, the results highlight the delicacy of such a combination, as adequate drug payload/storage and preservation of RBC viability and functionality (21). When applied to brain drug delivery, particularly to repeated or high-dose regimes, RBC integrity concerns might inhibit oxygen supply, deform blood circulation, cause inflammatory responses due to hemolysis, or even initiate immunologic clearance of aberrant RBCs. Therefore, any possible brain-targeting delivery system should also include careful consideration of RBC interactions, hemolysis possibility, membrane integrity during physiological shear, osmotic, oxidative, and complement stress.

10.2.3 Platelet Activation, Thrombosis, and Coagulation Pathways

In addition to RBCs, platelets, and the plasma coagulation system, the central role of platelets in hemostasis is particularly vulnerable to disruptions caused by foreign materials in blood circulation. Platelets are small, discoid cells derived from megakaryocytes that form platelet plugs and clots (22). The entry of NPs or drug carriers into circulation can lead to adsorption of plasma proteins, platelets, blood plasma proteins such as fibrinogen, coagulation factors, or even the vascular endothelium, resulting in a cascade of events including platelet adhesion, activation, and aggregation; coagulation factor activation (intrinsic/extrinsic), thrombin generation, and fibrin formation, all of which can potentially cause thrombosis (23). Specifically, NPs contacting RBCs can indirectly stimulate thrombosis, which makes platelets more prone to collisions with the vessel wall and platelet activation. In addition, the damage or stress of the RBC membranes exposes pro-coagulant phospholipids, which act as a catalytic surface to coagulation complexes (e.g., prothrombinase), which further accelerate the formation of a thrombus (24). Some nanocarriers can also trigger another possible pathway, which is complement activation resulting in inflammatory and pro-coagulant signaling. Thus, to achieve brain-targeted drug delivery, particularly through systemic delivery, carrier surface properties, i.e., charge, hydrophobicity/hydrophilicity, functionalization should be designed to reduce unwanted platelet or coagulation activation. Lack of attention can result in some of the most dangerous, although less frequent, adverse effects, including thrombosis, embolism, or systemic coagulation disorders (19). Therefore, platelet-coagulation safety cannot be compromised in preclinical assessment of brain-drug delivery platforms.

10.2.4 Effects on Microcirculation and Oxygen Transport

Although the RBC membranes and platelets may not be severely damaged, changes in the blood rheology caused by modifications in the deformability, aggregation, or changes in plasma viscosity of RBCs can severely disrupt microcirculatory perfusion and oxygen delivery. Microcirculation (capillaries, microvessels) is particularly susceptible since the flow is slow, the size of vessels is small, and RBCs tend to undergo significant deformation to squeeze through. To have good efficacy of therapeutic effect in the brain, the drug must cross the BBB, but once in the brain vasculature, there must be proper perfusion to deliver oxygen and nutrients and ensure uniform distribution, particularly in the sensitive microenvironment of the brain (24). When a drug-delivery system raises blood viscosity accidentally, it enhances RBC aggregation (e.g, forms rouleaux), impairs RBC deformability, or even causes some minimal hemolysis, this may slow capillary perfusion, lead to local stasis, and reduce oxygen supply, and may even cause ischemic damage or microinfarcts. In fact, experiments have demonstrated that rouleaux can be stabilized at physiological concentrations by plasma proteins such as fibrinogen; such aggregates can be maintained in microcapillaries and can have a dramatic impact on perfusion and flow distribution. In addition, in computational and hydrodynamic models, viscous variations in RBC cytosolic viscosity, membrane characteristics, and cell deformability have been demonstrated to have an effect on flow resistance and flow behavior in microvessels, which may compromise oxygen delivery without apparent indicators of hemolysis (25). In addition to RBCs, in case carriers lead to subclinical platelet/coagulation protein activation, microthrombi might arise in capillaries, which impacts local perfusion again. Also, non-thrombotic but disturbed flow patterns can act on shear forces on endothelial cells that can affect BBB integrity, endothelial functionality, and even encourage inflammation (26). In conclusion, these considerations point out that not only is hematological safety concerned with the prevention of overt hemolysis or clotting, but also with the maintenance of normal microcirculatory flow, perfusion, and oxygen delivery, and in particular within the most delicate microenvironment, the cerebral.

10.3 IMMUNOLOGICAL RESPONSES TO CNS-DIRECTED THERAPIES

Therapies targeting the CNS, particularly those designed to cross the BBB, encounter physical, biochemical, and immunological challenges. The immune system can respond to the therapy itself or to secondary effects such as cell damage. These immunological reactions can significantly affect the safety and efficacy of CNS-directed therapies. The key immunological mechanisms discussed include innate immune activation, adaptive immune responses, hypersensitivity/systemic inflammation, and immune surveillance at the BBB.

10.3.1 Innate Immune Activation (Macrophages, Neutrophils, Complement)

The innate immune system is the first line of body's defense, and it involves phagocytic cells (e.g., macrophages, monocytes), granulocytes (e.g., neutrophils), and molecular systems (e.g., complement cascade (27)). Once CNS-directed therapy is introduced into the bloodstream, particularly when it is using a synthetic carrier, i.e., NP, liposome, extracellular vesicle (EV), or other so-called non-self-construct to deliver drugs/genes, which can be recognized by innate immunity components. In fact, communication between engineered nanomedicines and immune cells has been widely reported (28). The complement activation is especially significant. The complement system may be activated in either the classical, lectin, or alternative pathway, depending on the foreign surface nature. Upon activation, complement proteins (e.g., C3a, C5a) may produce inflammatory mediators (anaphylatoxins), coat the particles to be ingested, and facilitate the clearance or destruction of the foreign substance (29). Nanocarriers or other therapeutic constructs may be phagocytosed by phagocytic cells such as macrophages, monocytes, and neutrophils. These cells may polarize to assume a pro-inflammatory phenotype (e.g., M1-like macrophages), produce pro-inflammatory cytokines (e.g., IL-1β-1 2, IL-6, TNF-22) and chemokines, and trigger either a systemic or local inflammatory response depending on their state of activation. The ensuing inflammation may have several adverse results, like disturbed pharmacokinetics of the therapy, unwanted tissue damage, or even aggravated CNS pathology in case inflammatory mediators gain access to the brain (30). Therefore, therapies need to be engineered and experimented with to reduce innate immune activation that is not intended by careful regulation of surface chemistry, size, stealth coating, and other physicochemical characteristics that determine immunogenicity.

10.3.2 Adaptive Immune Responses (T/B Cells, Antibody Formation)

In addition to the first innate response, CNS-directed repeat or long-term therapies can activate adaptive immunity mediated by T and B lymphocytes. The adaptive system provides specificity and memory, and the responses generated by it (e.g., the formation of antibodies) can impair treatment efficacy and safety. Once a carrier (e.g., non-viral or viral vectors, biologic, etc.) containing a therapeutic is considered a foreign antigen, antigen-presenting cells like dendritic cells or macrophages can process and present fragments of the therapeutic (or part of a therapeutic) to T cells. This may cause the activation of helper T cells and cytotoxic T cells, B-cell activation, and eventually, the development of antibodies against the therapeutic or carrier. These types of immune reactions can decrease the effectiveness by evacuating the therapeutic agent through the blood vessels, by destroying its effect, or by obstructing its reaching the brain (31). This is especially true in the case of treatments that need to be repeated frequently, as the body builds higher and higher levels of immune response; this measure will become counterproductive to the treatment process, or will result in progressively greater adverse autoimmune toxicity. Further, antibody generation could also have cross-reactivity with self-proteins, particularly when the therapeutic or carrier protein is similar or reacts with host molecules, which may cause autoimmune or off-target responses. Although the CNS is commonly referred to as immune-privileged, adaptive immune processes do take place in this location in some situations (e.g., autoimmunity, neuroinflammation) (32). Hence, immunogenicity should be carefully considered in the design of CNS-directed therapies.

10.3.3 Hypersensitivity and Systemic Inflammation

Immunological reactions against CNS-directed therapies are sometimes not only classical adaptive immune activation, but also acute hypersensitivity or systemic inflammatory reactions. Indicatively, some nanomedicines, such as liposomes, lipid NPs, or iron-based NPs, have been linked to infusion reactions in human beings. These reactions may be non-IgE-mediated, quick, and might be severe. Complement activation related pseudoallergy (CARPA), activation of complement after initial exposure to the nanocarrier results in the release of anaphylatoxins (ex. C3a, C5a), which causes vasodilation, increased vascular permeability, recruitment of leukocytes, and release of inflammatory mediators, which ultimately cause systemic effects, such as hypotension, bronchospasm, or other adverse effects, is one of the hypothesized mechanisms (33). Additionally, inflammation can be further enhanced by the activation of innate immune cells (e.g., macrophages, neutrophils) by the therapeutic (e.g., IL-6, TNF-alpha), which recruits more leukocytes and may also promote coagulation cascades. Systemic inflammation or over-sensitivity of such a system could impair patient safety, particularly among at-risk groups of individuals, or where repeat dosing is necessary (34). In the case of CNS-directed therapies that can be high-dose or repetitive usage, the risk is enhanced. Therefore, pharmacokinetics and efficacy of delivery to the brain should not be considered the final goal in preclinical testing, but also immunotoxicity, complement activation tests, cytokine profiling, assessing immune cell reactions (monocytes/macrophages, neutrophils, dendritic cells), and testing possible hypersensitivity reactions under real-life dose conditions (35).

10.3.4 Immune Surveillance at the BBB

Although the CNS is regarded as an immune-privileged site, it is also subject to a type of immune surveillance of both resident innate immune cells (e.g., microglia, perivascular macrophages), specific circumstances, and peripheral immune cells (T/B cells, monocytes) capable of crossing the BBB (36). The BBB is a tight network consisting of specialized endothelial cells, closely attached by tight junctions, and anchored by pericytes and astrocyte end-feet, which restricts the passage of most cells and molecules into the CNS from the blood. Nevertheless, some physiological or pathological cues (e.g., inflammatory cytokines, chemokines, and cell adhesion molecules) may cause a change in BBB permeability and allow permeability-based infiltration of immune cells. When it comes to CNS-directed therapy, two opposite immunological phenomena play a critical role (37). First, immune surveillance may restrict the effectiveness of therapeutic agents such as nanocarriers because these compounds may be neutralized by complement/immune molecules or destroyed by peripheral immune cells at the endothelium interface. In order to improve delivery, recent strategies seek to alter leukocyte trafficking across brain barriers, which calls for a careful balancing act between immune activation and safety. Second, especially with long-term treatments, inadvertent immune activation at the BBB might weaken its integrity and result in leaks, peripheral immune cell infiltration, neuroinflammation, or autoimmunity. Inappropriate immune infiltration leads to neuroinflammatory and neurodegenerative disorders, and the immunological milieu of the CNS is strictly regulated (38). Furthermore, recent studies show that the CNS can create complement proteins or immune mediators locally, suggesting that immunological responses to treatments should be taken into account both peripherally and in the brain microenvironment. Therefore, techniques that limit immune surveillance-mediated clearance while maintaining or restoring BBB integrity and minimizing neuroinflammation are essential for developing safe and efficacious CNS-directed treatments. Stealthy carrier design, the utilization of autologous or biomimetic systems, immune response modification, or cautious dosage techniques are some possible approaches. Analysis of BBB integrity, neuroinflammation, immune cell infiltration, and local complement/immune mediator activation in CNS tissue following delivery should all be part of preclinical research (32).

10.4 SAFETY CONSIDERATIONS OF NANOCARRIERS AND ADVANCED DELIVERY SYSTEMS

10.4.1 Nanoparticle–blood Interactions and Opsonization

The differential morphology and electrostatic interactions between erythrocytes are mostly responsible for erythrocyte aggregation. Rouleaux are linear face-to-face cell stacks or arrays that erythrocytes often generate at low shear rates. These linear arrays are easily dispersed by normal shear pressures and contribute to the non-Newtonian behavior of blood under physiological conditions. However, these aggregates are more difficult to disperse and can block tiny blood arteries and lower tissue perfusion when erythrocyte is excessively increased, for instance, through interactions with NPs. It is well

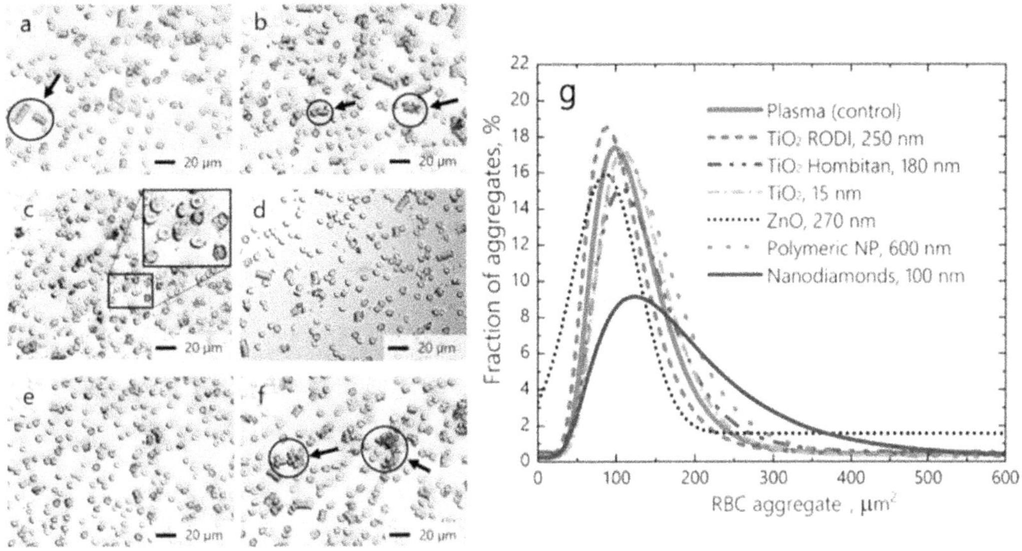

FIGURE 10.1 Relative size distribution of erythrocyte aggregates observed by conventional optical microscopy under the influence. (a) Polymeric NPs 600 nm, (b) TiO_2 NPs 250 nm, (c) TiO_2 NPs 180 nm, (d) TiO_2 NPs 15 nm, (e) ZnO NPs 270 nm, and (f) nanodiamonds 100 nm. (g) Distribution of erythrocyte aggregates by occupied area based on the quantitative assessment of images in (a–f) (adapted with permission (40)).

known that erythrocyte aggregates promote platelet margination, which in tiny arteries causes blood viscosity to rise in the vessel's center and fall close to the vascular wall. Increased endothelial activation and platelet aggregation may result from this phenomenon, known as the Fahraeus effect.

Recent research has demonstrated that cell-NP interactions can generate considerable erythrocyte aggregation with negative effects, despite the NPs' low hemolytic activity. Erythrocyte aggregation is being used in an increasing number of studies to show the hemocompatibility of NPs in addition to cell viability and hemolysis tests. According to a recent study by Avsievich et al., when erythrocytes were incubated with various inorganic and polymeric NPs, they produced distinct aggregates with variable forms and sizes Figure 10.1 (39). The largest increase in the attraction forces between erythrocyte membranes among the NPs investigated was induced by nanodiamonds, which resulted in the creation of huge, abnormal aggregates. However, once erythrocytes were incubated with polymeric NPs, only standard erythrocyte rouleaux were seen, and these NPs had no discernible effect on erythrocyte interaction.

10.4.2 Viral Vectors, Exosomes, Peptides Immunotoxicity Risks

For nearly 40 years, recombinant viral vectors have been employed to convey antigens from some illnesses. There have been a lot of changes since the SV40 virus was utilized to construct the first viral vector in 1972. Adenoviruses and lentiviruses are superior to conventional vaccines because they create robust antibody and cellular responses that are needed to kill cells that are infected. They can give long-lasting immunity without any outside help, and they can be made to send vaccine antigens to specific cells. Some viral vectors are made so that they can't create copies of themselves, which makes things safer (40). This means that doses need to be carefully watched. Replicating vectors act like genuine illnesses and make the immune system stronger by generating cytokines. There are many techniques to give individuals vaccines, such as intradermal and intranasal injections, that work for both systemic and mucosal immunity. Researchers are utilizing clinical studies to see if viral vectors can work against diseases, including HIV, malaria, and SARS-CoV-2. Each one has its own perks and cons when it comes to safety and immunogenicity (41,42).

Exosomes resemble the cells from where they derive and the environment that produced them. They are less effective due to their short duration, as macrophages rapidly eliminate them. They may accumulate in organs at increased dosages (43). Obtaining regulatory approval and ensuring safety is more challenging due to the absence of established methods for their separation and analysis. Researchers are continuing to investigate the long-term implications of exosome treatment (44). The short-term benefits, such as fixing the heart, may come with bad side effects, such as more scarring. Peptides used as drugs and nanocarriers have a lot of problems and dangers. They usually don't make the immune system react as strongly as bigger viral vectors, but immune responses can still make them less effective. They need to be put in nanocarriers, like modified exosomes or liposomes, to make them easier to find because they don't stay stable for long in biological settings. Peptides and their nanocarrier systems could also hurt other cells, especially if they aren't very picky. The changes to the surface could change how the peptides move around the body (45,46).

10.4.3 Strategies to Improve Biocompatibility

Biocompatibility strategies that can help medicines work better and have fewer side effects include using better targeting mechanisms, changing the surface, and choosing the right materials. Nanomaterials are more dangerous than regular materials because they are very small and have a lot of surface area for their volume (46,47). Metal NPs, like silver and zinc oxide, and carbon-based nanoparticles, like carbon nanotubes, can make reactive oxygen species (ROS). These can hurt cells, make oxidative stress worse, and make inflammation inferior. Opsonization helps the immune system get rid of NPs that don't belong there quickly. This could make treatments work less well and cause bad side effects. Nanocarriers can cross biological barriers like the BBB, which makes it easier to deliver drugs to the right places (48,49). NPs may or may not be safe, depending on shape, size, and surface charge, and the ingredients used during the development process. Cationic NPs are more readily absorbed by cells; nevertheless, they also pose more risks. Utilize biocompatible materials that are biodegradable, such as some polymers like PLGA, employ PEGylation and additional techniques to modify the surface for enhanced mobility and to evade detection by the immune system, as well as incorporating ligands that selectively adhere to specific cells to enhance the drug's efficacy and minimize the likelihood of adverse effects (50,51). The stability of nanocarriers can be enhanced through their morphology and charge. It may be suggested to develop systems capable of adjusting the quantity of medication dispensed based on received inputs. This will reduce the likelihood of adverse effects and facilitate the administration of the appropriate dosage. Encapsulating NPs within biological membranes utilizing biomimetic techniques will render them more selective and inhibit the immune system from targeting them (52,53).

10.5 BBB MODULATION APPROACHES AND THEIR HEMATO-IMMUNOLOGICAL IMPACT

BBB modulation approaches are broadly categorized into chemical, biological, and physical strategies, each impacting the hemato-immunological system differently, primarily by altering BBB permeability and potentially causing inflammation or immune responses.

10.5.1 Chemical, Biological, and Physical BBB-opening Strategies

Strategies for regulating the BBB concentrate on temporarily enhancing its permeability to facilitate the entry of therapeutic drugs into the CNS. Chemical methods use things that modify how cells work or tear up strong connections. Hyperosmolar medicines, such as mannitol, generate osmotic pressure that dehydrates endothelial cells and facilitates the opening of tight junctions. But this procedure that

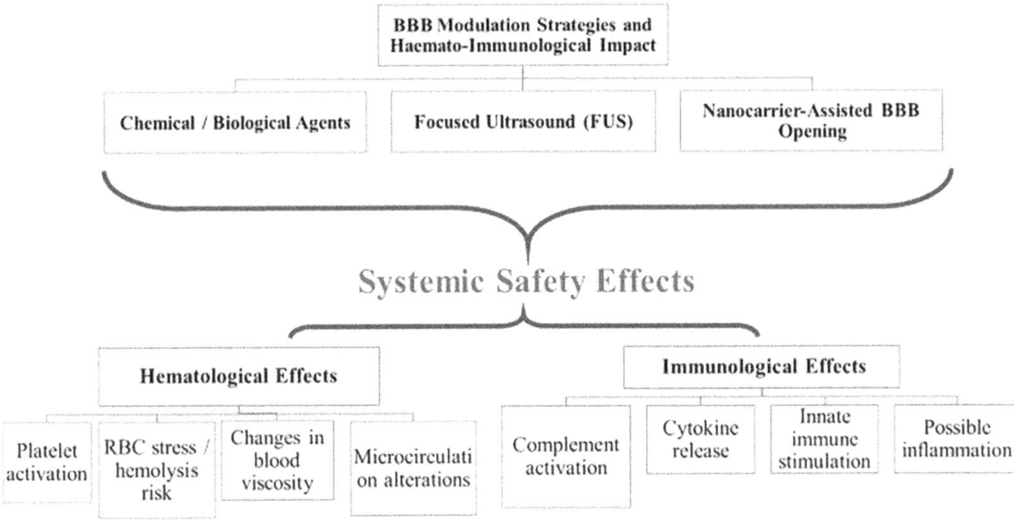

FIGURE 10.2 BBB modulation strategies and consequent hemato-immunological impacts.

isn't specific can induce cerebral edema, convulsions, and harm to the nervous system since plasma components flow into the brain (54,55). Vasoactive chemicals, like bradykinin, temporarily make blood vessels more permeable, but they may also have effects on other parts of the body. Efflux pump inhibitors limit medications from leaving the brain, but they often make the body sick and mess with enzymes that break down food. Biological techniques utilize existing transport mechanisms within the body. Receptor-mediated transcytosis (RMT) changes drugs by adding ligands that stick to receptors on brain endothelial cells. This makes it possible to target transport and slows down immune clearance. Cell-penetrating peptides (CPPs) let medications get into cells; however, they aren't highly selective. NP-based systems offer protection and enhanced drug delivery; nevertheless, poorly designed nanoparticles may compromise BBB integrity and induce inflammation. Physical methods use energy from outside sources to affect how the BBB is built. Microbubbles and focused ultrasound (FUS) make materials more permeable by making them vibrate. However, if not handled correctly, this can cause microbleeds. Lasers are employed in optical treatment to break apart tight bonds. If done wrong, this can hurt the tissue. Electrical stimulation can make things more permeable for a short time, but it can also turn on things that aren't specific (56,57). Figure 10.2 shows BBB modulation strategies and hemato-immunological impact.

10.5.2 Hematological Alterations after BBB Modulation

Alterations in hematology resulting from changes to the BBB present considerable difficulties, despite the improved drug delivery linked to this modulation. Major alterations consist of the leakage of neurotoxic blood elements, increased immune cell entry into the body leading to inflammation, the likelihood of microhemorrhages due to methods like FUS, alterations in platelet functions that influence hemostasis, detrimental hematopoiesis resulting from elevated erythropoietin (EPO) levels, and modifications in plasma biomarkers associated with inflammation (58,59). The impact of BBB modulation primarily relies on the technique employed. Osmotic disruption consistently raises permeability, potentially resulting in swelling and seizures. Concentrated ultrasound enables precise interruption but may harm blood vessels at elevated energy levels. Improper administration of drugs might lead to transient inflammatory responses or severe adverse consequences. Targeted techniques, such as nanoparticles, aim to enhance safety by reducing systemic impacts. The effectiveness of BBB alteration in treatment is hindered by the necessity to balance improved medication delivery with the reduction of hematological and immunological side effects caused by non-specific environmental disturbances (60).

10.5.3 Immune Activation and Inflammatory Consequences

Damaging openings in the BBB permit the entrance of immune cells, such as leukocytes and macrophages, as well as inflammatory mediators, including cytokines and plasma proteins, into the CNS. This technique identifies endothelial cell adhesion molecules and interferes with tight junctions (61,62). This compels immune cells to exert greater effort, exacerbating inflammation. Upon activation, glial cells secrete pro-inflammatory cytokines such as IL-1β, TNF-α, and IL-6. FUS can induce a sterile inflammatory response by altering the expression of genes that regulate inflammation and apoptosis, influenced by the permeability of the BBB. This could hurt the nervous system and tissues, changing how neurons work and making them swollen and confused. Systemic inflammation can harm the BBB. However, targeted treatments restrict immune responses to certain brain areas, which is good for the body (63,64). After modifications to the BBB, the central and peripheral immune systems work together. This shows how important it is to keep a close eye on things to protect CNS immune privilege and adjust how the body reacts to inflammation. To safeguard brain tissue, alterations to the BBB must achieve a balance between permitting the entry of drugs, activating the immune system, and inducing inflammation (65,66).

10.5.4 Recovery, Reversibility, and Long-Term Effects

Most cutting-edge methods for changing the BBB are meant to be used just once and can be undone. FUS can break up the BBB in a small area by using microbubbles. This injury usually heals on its own in 6 to 24 hours. Enhanced sonic energy can extend the suppression of cerebral vasomotion for a duration of up to 72 hours. Regulated applications, on the other hand, lessen this risk to protect tissues. Pharmacological drugs, such as hyperosmotic mannitol, angubindin-1, and claudin-5 siRNA, induce transient elevations in BBB permeability, distinguished by fluctuations in both duration and magnitude of these effects (67,68). Nanotechnology and laser stimulation temporarily increase paracellular permeability, but only for a short time. Function is restored within 24 hours. The modulation has hazards, including the possibility of neurotoxic blood components penetrating the brain, leading to inflammation, edema, and neurological issues. Techniques may cause temporary symptoms of neuroinflammation, such as gliosis, which can help keep the brain safe. When the BBB is broken, immune cells can get in, which makes systemic inflammation worse. Some drugs can affect several organs because the proteins they target are found all over the body. Longitudinal studies show that when done within certain limits, regulated, temporary changes to the BBB usually don't cause damage to tissues, neurons, or blood vessels (69,70). The long-term effects of repeated modulation are yet unknown, especially for people who already have health problems that make them more sensitive to neurotoxic chemicals. The ongoing and uncontrolled violation of the BBB has been linked to chronic inflammation and structural alterations in the brain, particularly in those with protracted COVID-19, who exhibit symptoms such as cognitive impairment (71).

10.6 PRECLINICAL AND CLINICAL EVALUATION OF SAFETY

A comprehensive preclinical and clinical assessment is essential to ascertain the safety of cerebral medicine administration for the hematologic and immune systems. Preclinical research employs diverse animal models, encompassing both rodent and non-rodent species, to identify potential toxicities and establish safe initial doses for human trials. These studies are conducted in accordance with the drug delivery strategy (72,73). Immunological safety studies investigate how administration of medicines influence the immune system. They look for signs of immunological activity and potential unforeseen immune responses, particularly when the BBB is only temporarily broken.

Preclinical investigations are conducted before clinical trials to ensure the safety of the participants (74). This necessitates stringent supervision of clinical trials and post-market operations. This means that safety markers from preclinical studies should be used to plan clinical trials. Immunological tests, on the other hand, look at how the body responds to infusions, hypersensitivity, and the production of anti-drug antibodies (ADAs). The FDA and EMA are two agencies that set regulations for testing biopharmaceuticals' safety in areas other than hospitals. They underline how crucial it is to undertake extensive safety assessments so that it is easier to use in the clinic. Sponsors are also told to take care of issues like immunogenicity and long-term safety straight away (75).

The following is a structured research summary as shown in Table 10.1 about Hematological and Immunological Safety in brain drug delivery, emphasizing nanoparticles, biologics, intranasal systems, BBB-modifying approaches, and immune-blood compatibility results.

TABLE 10.1
Work Done in the Field of Hematological & Immunological Safety in Brain Drug Delivery

Sr. No.	Formulation	Brain Targeting Strategy	Animal Model Used	Key Findings	Outcome	Refs
1	PLGA–PEG Nanoparticles	Receptor-mediated (TfR) BBB targeting	Rat plasma & *in vivo*	Complement activation (C3a, C5a) reduced vs uncoated PLGA	Hemocompatible with <5% hemolysis; mild cytokine elevation	(75)
2	Liposome-based Doxorubicin	Passive + ligand targeting	Mouse *in vivo*	↓ RBC aggregation; reduced splenic macrophage uptake	No immunotoxicity; safe chronic dosing profile	(76)
3	Solid Lipid Nanoparticles (SLNs)	Intranasal pathway to olfactory bulb	Sprague–Dawley rats	No coagulation pathway disruption (PT/aPTT normal)	No platelet activation; immunologically silent	(77)
4	Gold Nanoparticles (AuNPs) PEGylated	Enhanced neuronal uptake via PEG stealth	Human plasma *ex vivo*	PEG reduced opsonization & complement fixation	High immune tolerance vs un-PEGylated AuNPs	(78)
5	Exosome-mediated Delivery (Neural isolate)	Natural immune-evading vesicles	Mouse model	No cytokine storm; low IL-6, TNF-α	Highly biocompatible; no RBC damage	(79)
6	Chitosan-coated NLCs	Mucoadhesive intranasal BBB route	Rat model	Slight increase in neutrophils; no hemolysis	Safe, transient immune modulation only	(80)
7	Polysorbate-80 Coated Nanocarriers	BBB uptake via ApoE adsorption	Human serum *in vitro*	Initial complement activation, normalized by 24h	Acceptable safety with PEG stabilizer co-coating	(81)
8	Cell-penetrating Peptide Nanocarriers	Tight junction passage	Rabbit blood *ex vivo*	No platelet aggregation; normal fibrinogen	No hematotoxic effects	(82)
9	Viral Vector AAV-based Brain Delivery	BBB crossing via capsid	Non-human primates	Triggered adaptive response (IgG increase)	Safe with immune pre-conditioning	(83)
10	Intranasal Insulin Delivery	Nose-to-brain transport	Clinical volunteers	No immune cell elevation; stable leukocyte ratio	Hematologically inert, safe for chronic use	(84)

(Continued)

TABLE 10.1 (*Continued*)
Work Done in the Field of Hematological & Immunological Safety in Brain Drug Delivery

Sr. No.	Formulation	Brain Targeting Strategy	Animal Model Used	Key Findings	Outcome	Refs
11	Polymeric Micelles (PEG-PCL)	LRP-1 mediated transcytosis	Human plasma	Decrease complement factors vs non-PEG micelles	PEGylation crucial for hemocompatibility	(85)
12	Magnetite Fe3O4 Nanocarriers	Magnetic-guided BBB targeting	Rats	Mild ROS-linked neutrophil rise	Requires antioxidant functional surface	(86)
13	LDL-mimetic Nanoparticles	Endogenous transport mimicry	Mouse *in vivo*	Negligible macrophage uptake; no coagulation impact	Immune stealth confirmed	(87)
14	Stem Cell-loaded Nanocarriers	Cell-assisted targeting	Rat brain tumor model	No hemolysis, transient immune fluctuation	Safe for controlled dosing	(88)
15	PAMAM Dendrimers (surface neutralized)	Enhanced intracellular release	*In vitro* blood model	Cationic versions toxic, neutralized safe	Neutral charge critical to avoid RBC lysis	(89)
16	Polypeptide Nanogels	Synaptic cleft targeting	Rodent CNS delivery	IL-1β and TNF-α not significantly elevated	Non-inflammatory	(90)
17	Albumin Nanoparticles	gp60 receptor brain uptake	Rat blood	No RBC morphology changes	Clinically promising biocompatibility	(91)
18	CRISPR-Cas9 Brain Delivery Liposomes	Gene editing payload	Mouse CNS	Transient innate immune activation only	Safety manageable via steroid pre-dosing	(92)
19	Nanoemulsion Intranasal Delivery	Rapid nose-to-brain	Rat & ex vivo	No platelet activation; normal ESR	Hematologically compatible	(93)
20	Hyaluronic Acid Brain Nanocarriers	CD44 targeting	Human blood assay	No complement activation	Highly safe immune-inert carrier	(94)

10.7 FUTURE DIRECTIONS & CONCLUSION

Further investigations into the hematological and immunological safety of brain-targeted pharmaceutical administration must emphasize the incorporation of modern materials science with predictive, human-relevant safety assessments to enhance the secure implementation of future technologies. The efficacy of brain-targeted delivery systems, increasingly reliant on nanoparticles, biologics, cell-derived carriers, and mechanisms that alter BBB, is significantly influenced by their interactions with blood constituents and immune pathways. Future research should concentrate on creating materials that are biocompatible and immunologically inactive. These materials will block the complement system from working, the platelets from clumping together, and the cytokines from being released by using sophisticated surface engineering, biomimetic coatings, and enhanced physicochemical qualities. The field needs testing frameworks that are the same and have been used before. The tests consist of whole-blood assays, complement activation panels, coagulation profiles, and cytokine evaluations. Ex vivo human blood models and BBB-on-chip technologies are essential for the precise replication of immunological and neurovascular microenvironments. Single-cell transcriptomics and high-dimensional proteomics are multi-omic techniques that facilitate the investigation of immune cell motility and the temporal dynamics of the protein corona. This helps us learn more about how living things work. To assess immunotoxicity, computer modeling and machine learning prediction techniques are employed to examine both individual patient characteristics and formulation variables. They can do a number of things.

After that, translational research needs to keep a careful eye on blood and immune system cells, sentinel dosage delivery strategies, and biomarker-guided patient stratification to reduce the consequences of neuroinflammation, infusion-related reactions, and thrombosis. Comprehensive post-marketing surveillance is essential to ascertain the long-term or delayed effects on the immunological and circulatory systems. This necessitates meticulous documentation of unfavorable events and the utilization of real-world data. To make sure that brain drug delivery systems are safe and effective, they need to be evaluated for safety and accuracy on a regular basis, and tests that can predict their effectiveness need to be done.

ACKNOWLEDGMENT

Dr. Prajapati, extends his sincere appreciation to the Faculty of Pharmacy, Silpakorn University, Thailand, for their generous support that enabled the completion of this work.

FUNDING

Not Applicable

DECLARATION OF GENERATIVE AI AND AI-ASSISTED TECHNOLOGIES IN THE WRITING PROCESS

During the preparation of this work, the author(s) used Quill Bot and Grammarly SERVICE to improve scientific writing and the English language. After using this tool/service, the author(s) reviewed and edited the content as needed and take(s) full responsibility for the content of the publication.

REFERENCES

1. Doke RR, Naik TS, Lamkhade DL, Bhise TS, Khokrale VN, Gosavi YB. Novel therapeutic delivery for neurodegenerative diseases: Strategies to overcome CNS barriers. J Pharm Biol Sci. 2023;11(1):1–8.
2. Goyal K, Koul V, Singh Y, Anand A. Targeted drug delivery to central nervous system (CNS) for the treatment of neurodegenerative disorders: Trends and advances. Cent Nerv Syst Agents Med Chem [Internet]. 2014 Nov 2 [cited 2025 Dec 4];14(1):43–59. Available from: https://pubmed.ncbi.nlm.nih.gov/25360772/
3. Khambhla E, Shah V, Baviskar K. Drug delivery to CNS: Challenges and opportunities with emphasis on biomaterials based drug delivery strategies. Curr Pharm Des [Internet]. 2016 Apr 6 [cited 2025 Dec 4];22(19):2913–22. Available from: https://pubmed.ncbi.nlm.nih.gov/27033697/
4. Das M, Sarma A, Baruah H, Basak D. Insight into central nervous system targeted nanostructured lipid carriers via the nose to brain pathway. RSC Pharm [Internet]. 2024 Dec 10 [cited 2025 Dec 4];1(5):904–27. Available from: https://pubs.rsc.org/en/content/articlehtml/2024/pm/d4pm00057a
5. Chahar RK, Tiwari C, Malik P, Jaiswal PK. Brain-targeted drug delivery system: A novel approach. J Drug Deliv Ther [Internet]. 2022 Nov 15 [cited 2025 Dec 4];12(6):171–8. Available from: https://jddtonline.info/index.php/jddt/article/view/5776/5084

6. Trivedi S, Shah S, Patel R. Review on novel oral iron formulations with enhanced bioavailability for the treatment of iron deficiency. J Drug Deliv Sci Technol. 2023;90:105181.
7. Liu S, Jin X, Ge Y, Dong J, Liu X, Pei X, et al. Advances in brain-targeted delivery strategies and natural product-mediated enhancement of blood–brain barrier permeability. J Nanobiotechnol 2025 23:1 [Internet]. 2025 May 26 [cited 2025 Dec 4];23(1):382. Available from: https://link.springer.com/article/10.1186/s12951-025-03415-w
8. Getting drugs to the brain: Advances and prospects of organic nanoparticle delivery systems for assisting drugs to cross the blood–brain barrier. J Mater Chem B (RSC Publishing) https://doi.org/10.1039/D2TB01440H[Internet]. [cited 2025 Dec 4]. Available from: https://pubs.rsc.org/en/content/articlehtml/2022/tb/d2tb01440h?utm_source=chatgpt.com
9. Liaw K, Sharma R, Sharma A, Salazar S, Appiani La Rosa S, Kannan RM. Systemic dendrimer delivery of triptolide to tumor-associated macrophages improves anti-tumor efficacy and reduces systemic toxicity in glioblastoma. J Control Release [Internet]. 2021 Jan 10 [cited 2025 Dec 4];329:434–44. Available from: https://pubmed.ncbi.nlm.nih.gov/33290796/
10. Wang X, Song B, Wu M, Qin L, Liang W. Immune cell targeting-mediated cytomimetic drug delivery system for BBB-penetrating and precise therapy of in situ glioma. Mater Today Bio [Internet]. 2025 Jun 1 [cited 2025 Dec 4];32:101694. Available from: https://www.sciencedirect.com/science/article/pii/S2590006425002534
11. Del Baldo G, Del Bufalo F, Pinacchio C, Carai A, Quintarelli C, De Angelis B et al. The peculiar challenge of bringing CAR-t cells into the brain: Perspectives in the clinical application to the treatment of pediatric Central nervous system tumors. Front Immunol. 2023;14.
12. Johnsen KB, Burkhart A, Thomsen LB, Andresen TL, Moos T. Targeting the transferrin receptor for brain drug delivery. Prog Neurobiol [Internet]. 2019 Oct 1 [cited 2025 Dec 5];181:101665. Available from: https://www.sciencedirect.com/science/article/abs/pii/S0301008219301340
13. Wu D, Chen Q, Chen X, Han F, Chen Z, Wang Y. The blood–brain barrier: Structure, regulation, and drug delivery. Signal Transduct Target Ther [Internet]. 2023 Dec 1 [cited 2025 Dec 5];8(1):217. Available from: https://www.nature.com/articles/s41392-023-01481-w
14. Parhiz H, Khoshnejad M, Myerson JW, Hood E, Patel PN, Brenner JS, et al. Unintended effects of drug carriers: Big issues of small particles. Adv Drug Deliv Rev [Internet]. 2018 May 1 [cited 2025 Dec 5];130:90. Available from: https://pmc.ncbi.nlm.nih.gov/articles/PMC6588191/
15. Xu J, Ma C, Hua M, Li J, Xiang Z, Wu J CNS and CNS diseases in relation to their immune system. Front Immunol [Internet]. 2022 Nov 16 [cited 2025 Dec 5];13:1063928. Available from: https://pmc.ncbi.nlm.nih.gov/articles/PMC9708890/
16. Trejo-Soto C, Lázaro GR, Pagonabarraga I, Hernández-Machado A. Microfluidics approach to the mechanical properties of red blood cell membrane and their effect on blood rheology. Membranes 2022, Vol 12, [Internet]. 2022 Feb 13 [cited 2025 Dec 5];12(2). Available from: https://www.mdpi.com/2077-0375/12/2/217
17. Baskurt OK, Meiselman HJ. Blood rheology and hemodynamics. Semin Thromb Hemost [Internet]. 2003 Oct [cited 2025 Dec 5];29(5):435–50. Available from: https://pubmed.ncbi.nlm.nih.gov/14631543/
18. Guo S, Shi Y, Liang Y, Liu L, Sun K, Li Y Relationship and improvement strategies between drug nanocarrier characteristics and hemocompatibility: What can we learn from the literature. Asian J Pharm Sci [Internet]. 2021 Sep 1 [cited 2025 Dec 5];16(5):551. Available from: https://pmc.ncbi.nlm.nih.gov/articles/PMC8609445/
19. Malehmir S, Esmaili MA, Khaksary Mahabady M, Sobhani-Nasab A, Atapour A, Ganjali MR, et al. A review: Hemocompatibility of magnetic nanoparticles and their regenerative medicine, cancer therapy, drug delivery, and bioimaging applications. Front Chem [Internet]. 2023 [cited 2025 Dec 5];11. Available from: https://pubmed.ncbi.nlm.nih.gov/37711315/
20. Tran HDN, Moonshi SS, Xu ZP, Ta HT. Influence of nanoparticles on the haemostatic balance: Between thrombosis and haemorrhage. Biomater Sci [Internet]. 2021 Dec 21 [cited 2025 Dec 6];10(1):10–50. Available from: https://pubs.rsc.org/en/content/articlehtml/2022/bm/d1bm01351c
21. Pan D, Vargas-Morales O, Zern B, Anselmo AC, Gupta V, Zakrewsky M, et al. The effect of polymeric nanoparticles on biocompatibility of carrier red blood cells. PLoS One [Internet]. 2016 Mar 1 [cited 2025 Dec 6];11(3). Available from: https://pubmed.ncbi.nlm.nih.gov/27003833/
22. Mehrizi TZ, Kafiabad SA. Evaluation of the effects of nanoparticles on the therapeutic function of platelet: A review. J Pharm Pharmacol [Internet]. 2022 Feb 1 [cited 2025 Dec 6];74(2):179–90. Available from: https://dx.doi.org/10.1093/jpp/rgab089

23. Tian Y, Tian Z, Dong Y, Wang X, Zhan L Current advances in nanomaterials affecting morphology, structure, and function of erythrocytes. RSC Adv [Internet]. 2021 Feb 10 [cited 2025 Dec 6];11(12):6958–71. Available from: https://pubs.rsc.org/en/content/articlehtml/2021/ra/d0ra10124a
24. De La Harpe KM, Kondiah PPD, Choonara YE, Marimuthu T, Du Toit LC, Pillay V. The hemocompatibility of nanoparticles: A review of Cell–Nanoparticle interactions and hemostasis. Cells 2019, Vol 8, [Internet]. 2019 Oct 7 [cited 2025 Dec 6];8(10). Available from: https://www.mdpi.com/2073-4409/8/10/1209
25. Hemodynamic effects of red blood cell aggregation [Internet]. [cited 2025 Dec 6]. Available from: https://www.researchgate.net/publication/6554565_Hemodynamic_effects_of_red_blood_cell_aggregation
26. Zhao R, Kameneva MV, Antaki JF. Investigation of platelet margination phenomena at elevated shear stress. Biorheology. 2007;44(3):161–77.
27. Advances in engineered nanosystems: immunomodulatory interactions for therapeutic applications - Nanoscale (RSC Publishing) DOI:10.1039/D4NR00680A [Internet]. [cited 2025 Dec 6]. Available from: https://pubs.rsc.org/en/content/articlehtml/2024/nr/d4nr00680a?utm_source=chatgpt.com
28. Viana IM de O, Roussel S, Defrêne J, Lima EM, Barabé F, Bertrand N Innate and adaptive immune responses toward nanomedicines. Acta Pharm Sin B [Internet]. 2021 Apr 1 [cited 2025 Dec 6];11(4):852–70. Available from: https://pubmed.ncbi.nlm.nih.gov/33747756/
29. Moghimi SM, Haroon HB, Yaghmur A, Hunter AC, Papini E, Farhangrazi ZS, et al. Perspectives on complement and phagocytic cell responses to nanoparticles: From fundamentals to adverse reactions. Journal of Controlled Release [Internet]. 2023 Apr 1 [cited 2025 Dec 6];356:115–29. Available from: https://www.sciencedirect.com/science/article/pii/S0168365923001335?utm_source=chatgpt.com
30. Liu J, Liu Z, Pang Y, Zhou H. The interaction between nanoparticles and immune system: Application in the treatment of inflammatory diseases. J Nanobiotechnol 2022 20:1 [Internet]. 2022 Mar 12 [cited 2025 Dec 6];20(1):127. Available from: https://link.springer.com/article/10.1186/s12951-022-01343-7
31. Chen J, Wang C, Chen S, Cai H, Wang M, Chang J, et al. Trained immunity modulators: A new frontier in immunoregulation and disease intervention. J Adv Res [Internet]. 2025 Sep 18 [cited 2025 Dec 6]; Available from: https://www.sciencedirect.com/science/article/pii/S2090123225007349
32. Iwasaki A. Immune regulation of antibody access to neuronal tissues. Trends Mol Med [Internet]. 2017 Mar 1 [cited 2025 Dec 6];23(3):227–45. Available from: https://pubmed.ncbi.nlm.nih.gov/28185790/
33. Szebeni J Complement activation-related pseudoallergy: A stress reaction in blood triggered by nanomedicines and biologicals. Mol Immunol [Internet]. 2014 Oct 1 [cited 2025 Dec 6];61(2):163–73. Available from: https://www.sciencedirect.com/science/article/abs/pii/S0161589014001692
34. Saxena S, Sharma S, Kumar G, Thakur S. Unravelling the complexity of CARPA: A review of emerging advancements in therapeutic strategies. Arch Dermatol Res [Internet]. 2025 Dec 1 [cited 2025 Dec 6];317(1). Available from: https://pubmed.ncbi.nlm.nih.gov/39971823/
35. Chen S, Saeed AFUH, Liu Q, Jiang Q, Xu H, Xiao GG, et al. Macrophages in immunoregulation and therapeutics. Signal Transduct Targeted Ther 2023 8:1 [Internet]. 2023 May 22 [cited 2025 Dec 6];8(1):207. Available from: https://www.nature.com/articles/s41392-023-01452-1
36. Moreira R, Nóbrega C, de Almeida LP, Mendonça L. Brain-targeted drug delivery - nanovesicles directed to specific brain cells by brain-targeting ligands. J Nanobiotechnol 2024 22:1 [Internet]. 2024 May 17 [cited 2025 Dec 6];22(1):260-. Available from: https://link.springer.com/article/10.1186/s12951-024-02511-7
37. Yuan S, Hu D, Gao D, Butch CJ, Wang Y, Zheng H, et al. Recent advances of engineering cell membranes for nanomedicine delivery across the blood–brain barrier. J Nanobiotechnol 2025 23:1 [Internet]. 2025 Jul 8 [cited 2025 Dec 6];23(1):493. Available from: https://link.springer.com/article/10.1186/s12951-025-03572-y
38. Charabati M, Rabanel JM, Ramassamy C, Prat A. Overcoming the brain barriers: From immune cells to nanoparticles. Trends Pharmacol Sci [Internet]. 2020 Jan 1 [cited 2025 Dec 6];41(1):42–54. Available from: https://pubmed.ncbi.nlm.nih.gov/31839374/
39. Avsievich T, Popov A, Bykov A, Meglinski I. Mutual interaction of red blood cells influenced by nanoparticles. Scientific Reports. 2019;9(1).
40. Travieso T, Li J, Mahesh S, Mello JDFRE, Blasi M. The use of viral vectors in vaccine development. Npj Vaccines 2022 7:1 [Internet]. 2022 Jul 4 [cited 2025 Dec 3];7(1):75. Available from: https://www.nature.com/articles/s41541-022-00503-y
41. Lu B, Lim JM, Yu B, Song S, Neeli P, Sobhani N et al. The next-generation DNA vaccine platforms and delivery systems: Advances, challenges and prospects. Front Immunol. 2024;15:1332939.
42. Sugandhi VV, Gadhave DG, Khot SV, Jadhav K, Patel D, Thapa R, et al. Exploring natural components in vaccine delivery systems: Types, routes of administration, in-vitro and in-vivo quantification techniques. J Drug Deliv Sci Technol [Internet]. 2025 Jun 1 [cited 2025 Dec 3];108:106960. Available from: https://www.sciencedirect.com/science/article/pii/S1773224725003636

43. Kumar MA, Baba SK, Sadida HQ, Marzooqi SA, Jerobin J, Altemani FH, et al. Extracellular vesicles as tools and targets in therapy for diseases. Signal Transduction and Targeted Therapy 2024 9:1 [Internet]. 2024 Feb 5 [cited 2025 Dec 3];9(1):27. Available from: https://www.nature.com/articles/s41392-024-01735-1
44. Kalluri R, LeBleu VS. The biology, function, and biomedical applications of exosomes. Science [Internet]. 2020 Feb 7 [cited 2025 Dec 3];367(6478):eaau6977. Available from: https://pmc.ncbi.nlm.nih.gov/articles/PMC7717626/
45. Yadav D, Malviya R. Exploring potential of exosomes drug delivery system in the treatment of cancer: Advances and prospective. Med Drug Discov [Internet]. 2023 Dec 1 [cited 2025 Dec 3];20:100163. Available from: https://www.sciencedirect.com/science/article/pii/S2590098623000131
46. Dhall S, Nigam A, Harshavardhan M, Mukherjee A, Srivastava P. A comprehensive overview of methods involved in nanomaterial production and waste disposal from research labs and industries and existing regulatory guidelines for handling engineered nanomaterials. Environmental Chemistry and Ecotoxicology [Internet]. 2024 Jan 1 [cited 2025 Dec 3];6:269–82. Available from: https://www.sciencedirect.com/science/article/pii/S2590182624000195
47. Zhang L, Gu FX, Chan JM, Wang AZ, Langer RS, Farokhzad OC. Nanoparticles in medicine: Therapeutic applications and developments. Clin Pharmacol Ther. 2008 May 30;83(5):761–9.
48. Xuan L, Ju Z, Skonieczna M, Zhou PK, Huang R Nanoparticles-induced potential toxicity on human health: Applications, toxicity mechanisms, and evaluation models. MedComm (Beijing) [Internet]. 2023 Aug 1 [cited 2025 Dec 3];4(4):e327. Available from: https://pmc.ncbi.nlm.nih.gov/articles/PMC10349198/
49. Burgess DJ, Duffy E, Etzler F, Hickey AJ. Particle size analysis: AAPS workshop report, cosponsored by the food and drug administration and the United States pharmacopeia. AAPS J. 2004;6(3).
50. Kawada K, T M. Significance and mechanism of lymph node metastasis in cancer ProgressionLymph. Node Metastasis in Cancer Progression. 2011;71(4):1214–8.
51. Porter CJH, Trevaskis NL. Targeting immune cells within lymph nodes. Nat Nanotechnol. 2020;15(6):423–5.
52. Thorek DLJ, Ulmert D, Diop NFM, Lupu ME, Doran MG, Huang R et al. Non-invasive mapping of deep-tissue lymph nodes in live animals using a multimodal PET/MRI nanoparticle. Nat Commun. 2014 Jan 20;5(1):3097.
53. Lan HR, Zhang YN, Han YJ, Yao SY, Yang MX, Xu XG, et al. Multifunctional nanocarriers for targeted drug delivery and diagnostic applications of lymph nodes metastasis: A review of recent trends and future perspectives. J Nanobiotechnol 2023 21:1 [Internet]. 2023 Aug 2 [cited 2025 Dec 3];21(1):247. Available from: https://link.springer.com/article/10.1186/s12951-023-01990-4
54. Lingineni K, Belekar V, Tangadpalliwar SR, Garg P. The role of multidrug resistance protein (MRP-1) as an active efflux transporter on blood–brain barrier (BBB) permeability. Mol Divers. 2017;21(2):355–65.
55. Asaduzzaman S, Chakma R, Rehana H, Raihan M. Regulatory gene network pathway among brain cancer and associated disease: A computational analysis. Biointerface Res Appl Chem. 2021;11(5):12973–84.
56. Teleanu RI, Preda MD, Niculescu AG, Vladâcenco O, Radu CI, Grumezescu AM, et al. Current strategies to enhance delivery of drugs across the Blood–Brain barrier. Pharmaceutics [Internet]. 2022 May 1 [cited 2025 Dec 3];14(5):987. Available from: https://pmc.ncbi.nlm.nih.gov/articles/PMC9145636/
57. Matsuhisa K, Kondoh M, Takahashi A, Yagi K. Tight junction modulator and drug delivery. Expert Opin Drug Deliv. 2009;6(5):509–15.
58. Ronaldson PT, Davis TP. Targeting blood-brain barrier changes during inflammatory pain: An opportunity for optimizing CNS drug delivery. Ther Deliv. 2011;2(8):1015–41.
59. Zeiadeh I, Najjar A, Karaman R. Strategies for enhancing the permeation of CNS-active drugs through the blood-brain barrier: A review. Molecules : A Journal of Synthetic Chemistry and Natural Product Chemistry [Internet]. 2018 [cited 2025 Dec 3];23(6):1289. Available from: https://pmc.ncbi.nlm.nih.gov/articles/PMC6100436/
60. González-Mariscal L, Nava P, Hernández S. Critical role of tight junctions in drug delivery across epithelial and endothelial cell layers. Journal of Membrane Biology. 2005;207(2):55–68.
61. Yang C, Hawkins KE, Doré S, Candelario-Jalil E. Neuroinflammatory mechanisms of blood-brain barrier damage in ischemic stroke. Am J Physiol Cell Physiol. 2019;316(2):C135–53.
62. Kim JH, Byun HM, Chung EC, Chung HY, Bae ON. Loss of integrity: Impairment of the blood-brain barrier in heavy metal-associated ischemic stroke. Toxicol Res. 2013;29(3):157–64.
63. Benz F, Liebner S. Structure and function of the Blood–Brain barrier (BBB). Handb Exp Pharmacol. 2022;273:3–31.

64. Khan E. An examination of the blood-brain barrier in health and disease. Br J Nurs. 2005;14(9): 509–13.
65. Dyrna F, Hanske S, Krueger M, Bechmann I. The blood-brain barrier. J Neuroimmune Pharmacol. 2013;8(4):763–73.
66. Gryka-Marton M, Grabowska AD, Szukiewicz D Breaking the barrier: The role of proinflammatory cytokines in BBB dysfunction. Int J Mol Sci 2025, Vol 26, Page 3532 [Internet]. 2025 Apr 9 [cited 2025 Dec 3];26(8):3532. Available from: https://www.mdpi.com/1422-0067/26/8/3532/htm
67. Hawkins BT, Davis TP. The blood-brain barrier/neurovascular unit in health and disease. Pharmacol Rev. 2005;57(2):173–85.
68. Zlokovic BV. Cerebrovascular permeability to peptides: Manipulations of transport systems at the blood-brain barrier. Pharma Res: An Official Journal of the American Association of Pharmaceutical Scientists. 1995;12(10):1395–406.
69. Zloković BV, Lipovac MN, Begley DJ, Davson H, Rakić L. Transport of leucine-enkephalin across the blood-brain barrier in the perfused Guinea pig brain. J Neurochem. 1987;49(1):310–5.
70. Zlokovic BV, Hyman S, McComb JG, Lipovac MN, Tang G, Davson H. Kinetics of arginine-vasopressin uptake at the blood-brain barrier. BBA – Biomembranes. 1990 Jun 27;1025(2):191–8.
71. Zlokovic BV, Begley DJ. Chain-eliash DG. Blood-brain barrier permeability to leucine-enkephalin, d-Alanine2-d-leucine5-enkephalin and their n-terminal amino acid (tyrosine). Brain Res. 1985 Jun 10;336(1):125–32.
72. Kilkenny C, Parsons N, Kadyszewski E, Festing MFW, Cuthill IC, Fry D et al. Survey of the quality of experimental design, statistical analysis and reporting of research using animals. PLoS One. 2013 Mar 27;4(11).
73. Arabi F, Mansouri V, Ahmadbeigi N. Gene Therapy clinical trials, where do we go? An overview. Biomed Pharmacother. 2022;153.
74. Shchaslyvyi AY, Antonenko SV, Tesliuk MG, Telegeev GD. Current state of human Gene therapy: Approved products and vectors. Pharmaceuticals. 2023;16(10).
75. Li J, Feng L, Fan L, Zha Y, Guo L, Zhang Q, et al. Targeting the brain with PEG-PLGA nanoparticles modified with phage-displayed peptides. Biomaterials [Internet]. 2011 Jul [cited 2025 Dec 8];32(21):4943–50. Available from: https://pubmed.ncbi.nlm.nih.gov/21470674/
76. Samad A, Sultana Y, Aqil M. Liposomal drug delivery systems: An update review. Curr Drug Deliv. 2007;4(4):297–305.
77. Djupesland PG, Messina JC, Mahmoud RA. The nasal approach to delivering treatment for brain diseases: An anatomic, physiologic, and delivery technology overview. Ther Deliv [Internet]. 2014 [cited 2025 Dec 8];5(6):709–33. Available from: https://pubmed.ncbi.nlm.nih.gov/25090283/
78. Quach QH, Kong RLX, Kah JCY. Complement activation by PEGylated gold nanoparticles. Bioconjug Chem [Internet]. 2018 Apr 18 [cited 2025 Dec 8];29(4):976–81. Available from: https://pubmed.ncbi.nlm.nih.gov/29431995/
79. Kamerkar S, Lebleu VS, Sugimoto H, Yang S, Ruivo CF, Melo SA, et al. Exosomes facilitate therapeutic targeting of oncogenic kras in pancreatic cancer. Nature [Internet]. 2017 Jun 22 [cited 2025 Dec 8];546(7659):498. Available from: https://pmc.ncbi.nlm.nih.gov/articles/PMC5538883/
80. Gong X, Gao Y, Shu J, Zhang C, Zhao K Chitosan-based nanomaterial as immune adjuvant and delivery carrier for vaccines. Vaccines (Basel) [Internet]. 2022 Nov 1 [cited 2025 Dec 8];10(11). Available from: https://www.researchgate.net/publication/365381751_Chitosan-Based_Nanomaterial_as_Immune_Adjuvant_and_Delivery_Carrier_for_Vaccines
81. Hartl N, Adams F, Merkel OM. From adsorption to covalent bonding: Apolipoprotein e functionalization of polymeric nanoparticles for drug delivery across the blood-brain barrier. Adv Ther (Weinh) [Internet]. 2020 Jan 1 [cited 2025 Dec 8];4(1):2000092. Available from: https://pmc.ncbi.nlm.nih.gov/articles/PMC7116687/
82. Milletti F Cell-penetrating peptides: Classes, origin, and current landscape. Drug Discov Today [Internet]. 2012 Aug [cited 2025 Dec 8];17(15–16):850–60. Available from: https://pubmed.ncbi.nlm.nih.gov/22465171/
83. Mingozzi F, High KA. Overcoming the Host Immune Response to Adeno-Associated Virus Gene Delivery Vectors: The Race Between Clearance, Tolerance, Neutralization, and Escape. 2025 [cited 2025 Dec 8];8. Available from: https://doi.org/10.1146/annurev-virology-101416-041936
84. Craft S, Raman R, Chow TW, Rafii MS, Sun CK, Rissman RA, et al. Safety, efficacy, and feasibility of intranasal insulin for the treatment of mild cognitive impairment and alzheimer disease dementia: A randomized clinical trial. JAMA Neurol [Internet]. 2020 Sep 1 [cited 2025 Dec 8];77(9):1099–109. Available from: https://pubmed.ncbi.nlm.nih.gov/32568367/

85. Tenchov R, Sasso JM, Zhou QA. PEGylated lipid nanoparticle formulations: Immunological safety and efficiency perspective. Bioconjug Chem. 2023 Jun 21;34(6):941–60.
86. Mahmoudi M, Sant S, Wang B, Laurent S, Sen T. Superparamagnetic iron oxide nanoparticles (SPIONs): Development, surface modification and applications in chemotherapy. Adv Drug Deliv Rev [Internet]. 2011 Jan [cited 2025 Dec 8];63(1–2):24–46. Available from: https://pubmed.ncbi.nlm.nih.gov/20685224/
87. Di L, Maiseyeu A. Low-density lipoprotein nanomedicines: Mechanisms of targeting, biology, and theranostic potential. Drug Deliv [Internet]. 2021 [cited 2025 Dec 8];28(1):408. Available from: https://pmc.ncbi.nlm.nih.gov/articles/PMC7894439/
88. Qiu Z, Yu Z, Xu T, Wang L, Meng N, Jin H, et al. Novel nano-drug delivery system for brain tumor treatment. Cells 2022, Vol 11, Page 3761 [Internet]. 2022 Nov 24 [cited 2025 Dec 8];11(23):3761. Available from: https://www.mdpi.com/2073-4409/11/23/3761/htm
89. Duncan R, Izzo L. Dendrimer biocompatibility and toxicity. Adv Drug Deliv Rev [Internet]. 2005 Dec 14 [cited 2025 Dec 8];57(15):2215–37. Available from: https://pubmed.ncbi.nlm.nih.gov/16297497/
90. Blackburn WH, Dickerson EB, Smith MH, McDonald JF, Lyon LA. Peptide-functionalized nanogels for targeted siRNA delivery. Bioconjug Chem [Internet]. 2009 May 20 [cited 2025 Dec 8];20(5):960–8. Available from: https://pubmed.ncbi.nlm.nih.gov/19341276/
91. Elsadek B, Kratz F. Impact of albumin on drug delivery–new applications on the horizon. J Control Release [Internet]. 2012 Jan 10 [cited 2025 Dec 8];157(1):4–28. Available from: https://pubmed.ncbi.nlm.nih.gov/21959118/
92. Kazemian P, Yu SY, Thomson SB, Birkenshaw A, Leavitt BR, Ross CJD. Lipid-nanoparticle-based delivery of CRISPR/Cas9 genome-editing components. Mol Pharm [Internet]. 2022 Jun 6 [cited 2025 Dec 8];19(6):1669–86. Available from: https://pubmed.ncbi.nlm.nih.gov/35594500/
93. Bonferoni MC, Rossi S, Sandri G, Ferrari F, Gavini E, Rassu G, et al. Nanoemulsions for "Nose-to-brain" drug delivery. Pharmaceutics [Internet]. 2019 Feb 1 [cited 2025 Dec 8];11(2):84. Available from: https://pmc.ncbi.nlm.nih.gov/articles/PMC6409749/
94. Yan J, Siwakoti P, Shaw S, Bose S, Kokil G, Kumeria T. Porous silicon and silica carriers for delivery of peptide therapeutics. Drug Deliv Transl Res [Internet]. 2024 Dec 1 [cited 2025 Dec 8];14(12):3549–67. Available from: https://www.researchgate.net/publication/381042953_Porous_silicon_and_silica_carriers_for_delivery_of_peptide_therapeutics

11 Cytotoxic Evaluations of the Drug Delivery Systems to the Brain

Humaira Liaqat, Huma Hameed,
Mahtab Ahmad Khan, Ana Cláudia Paiva-Santos,
Muhammad Zaman, and Anam Hameed

11.1 CYTOTOXICITY

Cytotoxicity is a term used to describe how harmful a substance can be to living cells whether it's chemical, biological, or physical. When a substance is cytotoxic, it can damage or destroy cells by interfering with their normal functions. This damage can happen in different ways and often leads to cell death. The death might be controlled (like in apoptosis, where the cell essentially shuts itself down) or chaotic (as in necrosis, where the cell dies suddenly and messily). Reviewing harmful effects to cells is indeed imperative in advance drug progression, checking if they are safe to use and scheming resources which can damage body. It supports investigators in identifying that these substances are dangerous for the cells that are healthy and it is decisive for warranting the therapies are harmless and operative (Çelik 2018).

In the meantime, DDS commonly formed to navigate the blood-brain barrier (BBB) so it is indispensable to methodically appraise that they are safe or not and predominantly their capability to harm cells of the brain which is stated as cytotoxicity. Even though successfully distributing medicines in brain is critical and it is fair and imperative to assure that these delivery systems don't perpetrate unplanned destruction that might aggravate prevailing neural disorders or else conceivably start novel diseases. This subdivision scrutinizes diverse methods implemented to estimate the cytotoxic influences of drug delivery systems (DDS) targeting the brain and the obstinate struggle of accomplishing robust medication effect, although diminishing noxiousness (Zeng et al. 2024).

11.1.1 Fundamental Concepts in Neurotoxicity Assessment

CNS holds moderately sporadic extraordinary potentials that protect it chiefly against susceptible and toxic elements. All these features are influenced by means used to estimate possible damage

11.1.1.1 Diverse Cell Types

The brain isn't made up of just one kind of cell it contains a wide mix, including neurons, astrocytes, microglia, oligodendrocytes, and endothelial cells. Each plays a different role and reacts differently to harmful substances, so toxicity testing must take into account how each type is affected and how they interact with one another (Langen et al. 2019)

11.1.1.2 Limited Ability to Heal

Unlike other tissues in the body, the adult brain can't easily repair itself. Once damage occurs especially to neurons it's often irreversible, which makes preventing toxicity all the more important.

DOI: 10.1201/9781003482406-11

11.1.1.3 Delicate Neural Networks

Neurons are part of highly complex communication systems that rely on precise electrical activity and connections. Even small disruptions from toxic exposure can throw off these networks, potentially impacting brain function without necessarily killing the cells (Li et al. 2022).

11.1.1.4 High Metabolic Demands

The brain is an energy-intensive organ, using about 20% of the body's oxygen even though it makes up only 2% of the body's weight. If this delicate metabolic balance is disturbed, neurons can be harmed quickly.

11.1.1.5 BBB Effects

While the BBB protects the brain from many toxins, it also complicates things. It not only blocks the entry of many substances but can also slow down the removal of harmful compounds that do get in or are introduced directly into the CNS (Schurhoff and Toborek 2023).

11.1.1.6 Glial Cell Responses

Astrocytes and microglia, two types of supportive brain cells, respond to toxins in complex ways. Sometimes they help by reducing damage, but they can also make things worse by triggering inflammation (Yao et al. 2023).

11.2 COMMON MECHANISMS OF NEUROTOXICITY INDUCED BY DDS

Several mechanisms underlie the potential neurotoxicity of brain-targeted DDS, as shown in Figure 11.1.

11.2.1 Oxidative Stress

Many components used in DDS can either produce reactive oxygen species (ROS) or weaken the body's natural antioxidant defenses. When that happens, it leads to oxidative stress essentially, a chemical imbalance that damages cells. The brain is particularly at risk because it uses a lot of oxygen, contains many easily oxidized fatty acids, and doesn't have strong antioxidant defenses.

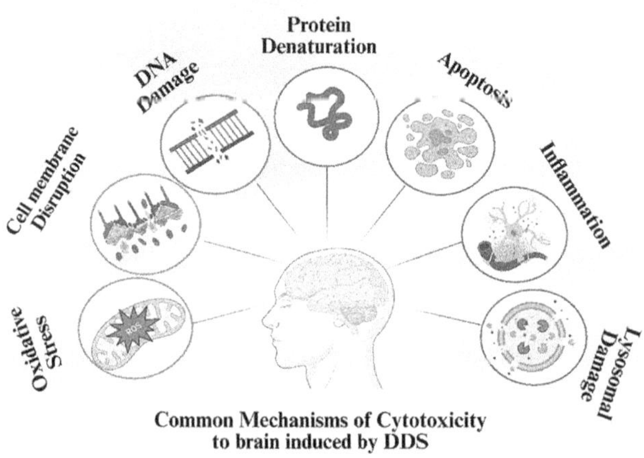

FIGURE 11.1 Common mechanisms of neurotoxicity caused by different DDS are described in this figure. These mechanisms are oxidative stress, disruption of the cell membrane, DNA damage. Induction of apoptosis, inflammation, and damage to lysosomes. (Image created by Bio render.com)

When ROS levels rise too high, they can harm vital parts of the cell like lipids, proteins, and DNA. This kind of damage can interfere with how cells function and even lead to their death. Considering the brain's vulnerability, controlling oxidative stress is an essential aspect of assessing the safety of DDS (Jelinek et al. 2021).

11.2.2 Cell Membrane Disruption

Cell membrane which is made up of lipid bilayer is indispensable for preserving viability of cell in addition regulating flow of materials into and out of cells. However, some DDS might disturb this barrier for instance predominantly which contains cationic nanoparticles, dendrimers or specific surfactants. By interacting with cell membrane at molecular level they can interrupt, increase penetrability and, in some cases, cause estrangement of cell also known as lysis (Alotaibi et al. 2021).

Some substances for instance surfactant and lipids have the ability to substantially integrate in the cell membrane. The membrane becomes additional porous as a result of its framework disruption.

Specific peptides and some nanoparticles can create gaps in the membrane that allow substances in the cells to discharge frequently leading to death of the cell (Li et al. 2023).

11.2.3 DNA Damage

Cell viability and its role is essential for veracity of DNA. Certain DDS and their constituents have the potential to either directly or indirectly damage DNA leading to chromosomal abnormalities, mutations and even cell death. These mechanisms involve:

Direct DNA interaction: Certain agents insert into the double helix of DNA, creating adducts or strand breaks.

ROS-mediated damage: ROS generated by the DDS or cellular responses can oxidize DNA bases, leading to DNA strand breaks and mutations.

Interference with DNA repair process: Some agents interfere with the enzymes of the DNA repair process, thereby increasing the vulnerability of cells to DNA damage (Pongrac et al. 2016).

11.2.4 Protein Denaturation

Proteins play a role in nearly all cellular functions. Proteins can be induced to unfold and lose their native conformation by exposure to a certain environment or materials, disrupting vital cellular functions and causing cell death.

11.2.5 Induction of Apoptosis

Induction of programmed cell death, also known as apoptosis, is a strictly controlled process that removes damaged or unwanted cells. Inappropriate triggering of apoptotic pathways by some DDS can cause the death of neurons and tissue damage in the CNS. Some of the important apoptotic pathways are: Extrinsic pathway and Intrinsic pathway. In Extrinsic Pathway Death receptor activation on the surface of the cell, including Fas or TNF receptors, initiates a caspase activation cascade to induce apoptosis while in Intrinsic pathway Damage to the mitochondrion and release of cytochrome c may induce activation of caspase-9, triggering the apoptotic process.

11.2.6 Inflammation

The CNS possesses a specialized immune environment in which the resident microglia are important for immune surveillance and inflammation. Certain DDSs induce inflammation in the brain, promoting the release of cytokines, chemokines, and other inflammatory mediators that may cause damage to neurons and activation of glial cells.

11.2.7 Lysosomal Damage

Lysosomes are cellular organelle comprising of hydrolytic enzymes that break down cellular debris. Destruction of the lysosomes may allow these enzymes to be released into the cytoplasm, leading to cellular damage as well as death of the cells.

11.3 FACTORS INFLUENCING CYTOTOXICITY OF BRAIN-TARGETED DDS

The cytotoxicity of a brain-targeted DDS is a function of many factors:

11.3.1 Physicochemical Properties

The size of DDS influences cytotoxicity. Nanoparticles smaller than 100 nm can penetrate deeper into brain tissue but may also have enhanced cellular uptake and potential toxicity (Albanese et al. 2012; Lakkadwala and Singh 2018).

Another factor is surface charge, cationic surfaces usually exhibit greater cytotoxicity as a result of strong interactions with negatively charged cellular membranes (Menichetti et al. 2024). Surface chemistry also influence cytotoxicity. Functional groups, ligands, and coatings play a major role in cellular interactions, uptake mechanisms, and intracellular fate (Menichetti et al. 2024). The aspect ratio and shape of nanoparticles play a critical role in determining their biological fate. Non-spherical particles (e.g., rods, discs, filaments) exhibit distinct cellular internalization pathways compared to spherical particles, often showing enhanced adhesion, prolonged circulation, or differential uptake depending on orientation and cell type. These particles can also demonstrate unique intracellular trafficking patterns, influencing endosomal escape and drug release efficiency.

11.3.2 Material Composition

Materials are Biodegradable or Non-biodegradable. Non-biodegradable materials may accumulate in the brain, causing chronic toxicity (Martano et al. 2022). Degradation Products such as Toxic metabolites may be produced even by biodegradeable materials during degradation process. Certain materials (such as some metals) present in DDS possess an inherent toxicity to neural cells (Sawicki et al. 2019).

11.3.3 Drug-Related Factors

The Interactions between drug and carrier can modify drug release kinetics or produce reactive complexes which cause cytotoxicity (Sawicki et al. 2019). Another factor is Drug Loading Capacity. High drug loading can decrease the quantity of carrier required, but can also elevate local drug concentrations to toxic levels (Sawicki et al. 2019).

The Release Kinetics mechanism of delivery system influence cytotoxicity. Burst release can cause acute toxicity, whereas incomplete release can lead to the accumulation of the carrier (Was et al. 2022).

11.3.4 Administration Route and Dosing Schedule

The most important factor that can influence cytotoxicity is Systemic vs. Local Administration of drug. Direct CNS administration circumvents the BBB but poses an increased potential for local toxicity (Li et al. 2019). Repetitive Dosing also Can cause cumulative toxicity, especially in non-biodegradable systems. Exposure Duration also influence cytotoxicity. Different patterns of acute and chronic exposure can induce differing toxicity levels (Li et al. 2019).

11.3.5 Biological

Adsorption of proteins to the surface of the nanoparticles in biological fluids forms a "corona" that mediates cellular interactions results in toxic effects (Liu et al. 2020). Brain Region Specificity also important to consider because variable vulnerability to toxin insults in different brain regions. Age and Disease State of the patient should also considered because pathologic and developing brains can be more susceptible to the toxicity of DDS (Li et al. 2019).

11.4 METHODOLOGIES FOR THE EVALUATION OF CYTOTOXICITY

Assessment of toxicity of cells is crucial for analyzing the effectiveness and biological compatibility of different DDS used as targeted system for brain particularly when it comes to delicate neuronic matters and BBB. In both laboratory settings also known as *in vitro* techniques and in vivo or animal models these sort of analysis is commonly performed by the use of neural cells, glial cells and brain slices.

11.4.1 In-Vitro Cytotoxicity Test

Prior moving on to in vivo research in vitro cytotoxicity evaluations are indispensable initial tools for scrutinizing the biocompatibility, effectiveness and probable noxiousness of DDS. These assessments create precise settings to learn how cells react to various doses and formulations of different DDS. They are predominantly critical in preparing treatments aiming for brain where it is vital to vigilantly assessing the sensitivity of glial and neural cells (Ilyas et al. 2021).

11.4.1.1 Assays for Metabolic Activity

One of the most widely employed in vitro method for evaluating sustainability of cells particularly in preliminary phases of developing a DDS is metabolic activity method These assessments are predominantly advantageous for assessing the potential noxiousness of DDS in neural cells since they count on the perception that exclusively metabolically active (living) cells are capable of processing specific dyes and substrates. Metabolic activity tests encompass MTT assay, XTT assay, Alamar blue assay, and assay for ATP content. The MTT assay is most commonly utilized metabolic activity-based tests for evaluating cellular viability and cytotoxicity, predominantly in the initial assessment of brain-targeted DDS. Neurons depend heavily on oxidative metabolism, which makes the MTT assay a superior marker for detecting early mitochondrial stress induced by DDS formulations. Frequently utilized for cell lines pertinent to the brain, including: SH-SY5Y (human neuroblastoma cell line), PC12 (rata feocromocitoma), C6 glioma cells, and Main cortical neurons or astrocytes. The principle of the MTT assay is 3-(4,5-dimethylthiazol-2-yl)-2,5-diphenyltetrazolium bromide (MTT) is reduced to purple formazan crystals by metabolically active cells, primarily through NAD(P)H-dependent cellular oxidoreductase enzymes. Its limitations are Susceptible to nanomaterial interference by direct MTT decrease or by overlapping absorbance. Nanomaterial-induced mitochondrial dysfunction can change the outcome irrespective of cell death. Lower densities of cells can be employed for the culture of neurons; longer incubation periods can be required for some of the less metabolically active neural cells (de Medeiros et al. 2019; Jo et al. 2015; Sánchez-Martínez et al. 2022).

MTS, WST-1, and XTT assays are colorimetric metabolic activity tests that act as improved substitutes for the classic MTT assay. These tests are especially beneficial for evaluating the cytotoxicity of DDS designed for the central nervous system (CNS) because their water-soluble formazan byproducts streamline the process and lessen cell stress. In these tests, water-soluble tetrazolium salts that yield soluble formazan derivatives bypass the solubilization step found in MTT. Beneficial because less interference with nanomaterials by soluble end products; it can be used in kinetic measurements. Remain prone to possible redox interference by nanomaterials (Gu et al. 2019).

The Resazurin assay, also referred to as Alamar Blue in commercial terms, is a sensitive, non-toxic assay based on fluorescence/colorimetric techniques employed to assess cellular metabolic activity. It holds significant importance in evaluating cell viability, proliferation, and cytotoxicity of DDS, particularly in neuronal and glial cells, where non-invasive, extended monitoring is frequently required. In this assay, non-fluorescent resazurin is converted to highly fluorescent resorufin in living cells. Resazurin is a blue, non-fluorescent substance that gets converted to resorufin, a pink and very fluorescent substance, by mitochondrial, cytosolic, and microsomal enzymes in cells that are metabolically active. Colorimetric Detection can be performed by observing a shift in absorbance from blue (~600 nm) to pink (~570 nm), and by Fluorometric Detection in which excitation approximately 530–560 nm/Emission around 590 nm (Longhin et al. 2022; Petiti et al. 2024).

ATP content assays are extremely sensitive, luminescence-driven techniques utilized to evaluate cellular viability by measuring intracellular adenosine triphosphate (ATP) concentrations a direct marker of metabolic activity. Because brain cells, especially neurons, rely heavily on energy, ATP-based assays are regarded as the gold standard for identifying early cytotoxic effects of DDS aimed at the CNS. The assay generally employs the luciferase-luciferin reaction, in which ATP from live cells interacts with luciferin alongside luciferase to generate light. The released luminescence is directly related to the ATP levels and, consequently, to the quantity of metabolically active cells (Greiner and Glonek 2021).

11.4.1.2 Membrane Integrity Assays

Membrane integrity tests evaluate the structural and functional stability of the cell's plasma membrane, essential for the survival of the cell. The degradation of membrane integrity is a key indicator of irreversible cytotoxicity and is frequently employed to assess necrosis or advanced-stage apoptosis triggered by DDS, particularly in vulnerable brain cells like neurons and astrocytes. These assays include

Lactate Dehydrogenase (LDH) Release Assay, Trypan Blue/Propidium Iodide (PI) Exclusion, and Calcein-AM/Ethidium Homodimer (Live/Dead) Assay. LDH Release Assay is a commonly utilized and dependable technique to evaluate cell membrane integrity and cytotoxic effects, especially in neuropharmacology and brain-targeted DDS. Because neurons and glial cells are very responsive to agents that disrupt membranes, this assay is essential for assessing potential neurotoxic impacts of new DDS. This assay quantifies the discharge of cytoplasmic enzyme LDH into culture media following damage to the cell membrane. It evaluates actual cell death rather than mere metabolic alterations, which is helpful in detecting the lytic action of DDS. Limitations are interference by nanomaterial binding to LDH or by the enzyme's adsorption onto the surface of the particles (Chavarria et al. 2022).

Trypan Blue/PI Exclusion is a traditional but commonly utilized method for evaluating cell viability by measuring the integrity of membranes. These tests are particularly significant in the area of neurotoxicity evaluation, as membrane damage could indicate lasting impacts of delivery systems aimed at the CNS. In this test, membrane-impermeable dyes penetrate into cells with damaged membranes, identifying dead cells as opposed to living cells. Beneficial for direct observation of dead cells; little disturbance by nanomaterials (Teixeira et al. 2021). Calcein-AM/Ethidium Homodimer (Live/Dead) Assay is a dual-staining fluorescence method employed to evaluate cell viability and membrane integrity. It is particularly beneficial for assessing brain-targeted DDS, due to its capacity to offer clear visual and quantitative contrasts between living and deceased cells in fragile and intricate neural cultures. In this test, non-fluorescent calcein-AM is converted to fluorescent calcein in living cells (green), whereas ethidium homodimer enters cells with compromised membranes and intercalates into nucleic acids (red fluorescence). It is beneficial for visualization of live and dead cells in a single experiment; spatial resolution of toxicity patterns (Amereh et al. 2022).

11.4.1.3 Apoptosis Detection Methods

Apoptosis, known as programmed cell death, is a highly controlled cellular mechanism crucial for brain development and balance. In the realm of brain-targeted DDS, it is essential to differentiate between apoptotic and necrotic cell death, particularly when assessing possible neurotoxicity or therapeutic effectiveness (e.g., in brain tumors). Different techniques are employed, such as Annexin V/PI Staining, Caspase Activity Assays, and TUNEL Assay.

The Annexin V/PI assay is among the most dependable and commonly employed methods for identifying and quantifying apoptotic and necrotic cells, particularly in neuronal and glial cell models utilized for evaluating brain-targeted DDS. Annexin V is a protein that strongly binds to phosphatidylserine (PS), typically found on the inner layer of the cell membrane. During early apoptosis, PS moves to the outer leaflet, where it becomes accessible to Annexin V. PI is a dye that binds to DNA and cannot pass through intact membranes, allowing it to only enter cells with damaged membranes, usually late apoptotic or necrotic cells. By merging the two stains, cells may be categorized as follows: Viable: Annexin V^- / PI^-, Initial Apoptotic: Annexin V+ / PI^-, Late Apoptotic or Necrotic: Annexin V+ / PI+, and Necrotic: Annexin V^- / PI+ (uncommon). Widely utilized to assess the neurotoxic impacts of nanoparticles, liposomes, and polymeric carriers in neuronal (e.g., SH-SY5Y) and astrocyte cell lines. It is usable in BBB co-culture systems to observe endothelial and neuronal cell death following DDS treatment (Kari et al. 2022).

Cysteine-aspartate proteases (Caspases) are a set of enzymes essential for the execution phase of apoptosis. Caspase activity assays are often utilized in both in vitro and in vivo studies to assess whether a brain-targeted DDS induces programmed cell death through the intrinsic (mitochondrial) or extrinsic (death receptor) apoptotic pathways (Flores et al. 2022). The TUNEL assay is a primary technique for detecting DNA fragmentation, an essential characteristic of advanced apoptosis. It is especially useful for assessing neuronal apoptosis in both in vitro cultures and in vivo brain tissue specimens following DDS treatment. The TUNEL assay detects both single- and double-stranded DNA breaks by enzymatically labeling the unpaired 3'-OH ends with modified dUTP nucleotides via the enzyme terminal deoxynucleotidyl transferase (Mitsui et al. 2023).

11.4.1.4 Mitochondrial Membrane Potential Assays

Mitochondria which are the energy-producing hubs in brain cells, preserve a distinct electrical gradient across their inner membrane referred to as the mitochondrial membrane potential. An alteration of this potential may serve as an early sign of cellular stress and the apoptotic pathway, frequently occurring before obvious indications of cell death after exposure to a brain-targeted DDS. Particular assays enable the assessment of this vital mitochondrial function. The technique utilizes cationic fluorescent dyes like JC-1 or TMRE, which are taken up by healthy mitochondria due to their significant negative internal charge. JC-1 shows potential-dependent characteristics; in mitochondria with elevated membrane potential, it creates aggregates that emit red fluorescence. In contrast, a decrease in membrane potential results in JC-1 forming as a monomer, which emits green fluorescence. Mitochondria with higher membrane potential have greater concentration of TMRE. To ascertain the state of mitochondrial membranes within the cells the method measures the intensity of total fluorescence in case of TMRE and the change from red to green fluorescence in case of JC-1. These assessments serve as an early warning system. In neurons exposed to DDS, they are able to detect minute alterations in function of mitochondria which may occur prior to discernible death of cell using other techniques. Early diagnosis is crucial for identifying potential delivery systems issues and allow for quick adjustments to lessen negative effects on brain cell's vital energy centers (Xu et al. 2022).

11.4.2 Cell-Type-Specific Assessments

Understanding how brain-targeted DDS intermingle with diverse brain cell types, including endothelial cells, neurons, and glial cells. Because different cell types respond variably to therapeutic agents, evaluating the specificity and efficiency of a DDS across multiple cell types is essential to confirm its therapeutic effectiveness and safety.. Evaluations specific to neuronal cells offer a broader picture that the brain is a complex environment with various kinds of cells, each of which carries out a specific function. In order to truly understand the impact of different DDS targeted to the brain, there should be specific methods which allow to precisely evaluate different ways in which various cell types, particularly priceless neurons, are affected.

Analyzing neurite outgrowth and morphology is crucial for assessing neurotoxicity, protective qualities against neurodegeneration, and the efficiency of DDS aimed at the brain. These evaluations help determine if DDS adversely affect neuron development or promote neuronal repair, regeneration, and growth in relation to neurological disorders. Neurites are the thin extensions of neurons, comprising axons (that send signals) and dendrites (that receive signals). The length, complexity of branching patterns, and form of the growth cone (the active tip of a developing neurite) play vital roles in forming accurate neural circuits (Sharma and Schwarzbauer 2022).

Electrophysiological methods assess the electrical characteristics of neurons, such as their membrane potential, ion channel functionality, and action potential discharge. These methods can show how DDS influence neuronal function at the single neuron level, within neural networks, or across overall brain activity. They assist in evaluating both acute (immediate impacts on neuronal activity) and chronic effects (long-term alterations in neuronal function) (Xu et al. 2024).

Synaptic Function Assays Synaptic function tests assess different components of synaptic activity, such as neurotransmitter release, reactions to synaptic stimulation, and enduring changes in synaptic strength (including long-term potentiation and long-term depression). These tests assist in understanding how DDS affect synaptic transmission, communication between neurons, and plasticity. Synapses are the vital connections at this place neurons exchange information. The efficiency of this communication depends on the recycling of synaptic vesicles (tiny sacs that hold neurotransmitters), the release of these neurotransmitters, and the distribution of receptors on the receiving neuron. Problems with any of these steps can disrupt neural signaling. Various tools are used, like FM dyes that light up when synaptic vesicles are recycled and track the movement of synaptophysin (a protein on synaptic vesicles), and fluorescent tags used to see how postsynaptic receptors are clustered on the receiving neuron (Appelbaum et al. 2023).

Calcium imaging ions play a crucial role in many neuronal processes, from neurotransmitter release to gene expression. Abnormal fluctuations in intracellular calcium levels can indicate problems like excitotoxicity (overstimulation that can damage neurons), altered signaling pathways, or a general disruption of the neuron's internal balance. Fluorescent indicators like Fluo-4 or Fura-2 are used that change their brightness or color when they bind to calcium ions inside the neuron. Via observing these alterations in fluorescence over time with microscopy, calcium dynamics were monitored in real time. This is a non-invasive method that enables observation of neuronal responses to a DDS during its application, offering important insights into the immediate impacts on neuronal signaling (Grienberger et al. 2022; Zong et al. 2022).

Evaluations of Glial Cells, such as astrocytes, microglia, and oligodendrocytes, are non-neuronal cells in the CNS crucial for sustaining homeostasis, overseeing immune responses, controlling synapses, and aiding in repair mechanisms. In the field of DDS for the brain, analyzing the reactions of glial cells is vital for determining biocompatibility, efficacy, and neurotoxicity. Immunostaining (GFAP, Iba1, MBP) to assess assembly and activation, qRT-PCR/Western Blot to measure glial marker and cytokine concentrations, ELISA/Cytokine Arrays for detecting inflammatory mediators (such as IL-6, TNF-α) and Phagocytosis Tests to scrutinize DDS uptake by microglia and immune response.

11.4.3 BBB Cell Evaluations

By controlling the flow of chemicals between the CNS and the brain, the BBB is a unique and strategically leaky fence that shields the brain. Assessing the collaboration between DDS and BBB cellular machineries is decisive for determining the effectiveness, permeability safety, and potential barrier disruption in order to ensure successful DDS targeted to brain. The function of tight junctions is gauged by Trans endothelial electrical resistance. This quantitative method is frequently used to evaluate the permeability and integrity of cell monolayers predominantly brain microvascular endothelial cells (BMECs) which impersonate BBB. It is one of the most accurate *in vitro* marker of barrier function and tight junction integrity (Thomsen et al. 2021). Transport

Protein Function also employed. Transport proteins are essential in controlling the movement of substances into and out of the BBB. This test evaluates whether a nanocarrier or formulation can evade P-gp efflux (e.g., through size alteration or surface enhancement), assess ligand-targeted approaches that utilize particular influx transporters (e.g., glucose-coated nanoparticles aimed at GLUT1) and examine possible interactions between drugs caused by the inhibition of transporters (Bors and Erdő 2019).

Inflammatory responses serve as vital markers of neurotoxicity and immune activation after the introduction of DDS into the brain. Given that the brain is an immune-privileged organ, any DDS, particularly those utilizing nanoparticles, viral vectors, or chemical enhancers, needs to be thoroughly assessed for pro-inflammatory effects to guarantee safety and therapeutic effectiveness.

Adhesion Molecule Expression, such as ICAM-1, VCAM-1 upregulation, and Production of cytokines: IL-6, IL-1β, TNF-α secretion, is monitored. Endothelial inflammation can allow the entry of immune cells into the CNS (Al Mamun et al. 2022).

11.4.4 Advanced In Vitro Models for Neurotoxicity Assessment

Sophisticated in vitro models are crucial for assessing the safety and possible neurotoxicity of DDS prior to moving on to animal or clinical trials. These models create well-regulated settings that replicate the intricate microenvironment of the human brain, enhancing the predictive precision for neurotoxicity results. A few models are listed in Table 11.1.

TABLE 11.1
The Advanced In Vitro Models for Neurotoxicity Assessment

Model Type	Description	Advantages	Applications	Limitations	References
3D Neurol Spheroids	Spherical aggregates of neural stem/ progenitor cells mimicking early CNS development	Simple to generate, 3D structures, scalable	Developmental neurotoxicity, DDS penetration, viability assays	Limited structural complexity, lacks vasculature	(Lee et al. 2016)
Brain Organoids	Self-organizing 3D brain-like tissue mimicking regions like the cortex or the midbrain	High complexity, human-specific, mimics brain regions	Long-term DDS exposure, developmental toxicity, synaptic activity	High variability, expensive, lacks full BBB	(Mayhew and Singhania 2023)
Microfluidic Brain-on-a-Chip	Miniaturized device integrating brain cells under fluid flow with multiple compartments	Dynamic flow, tight control, reproducible, mimics BBB	Real-time drug transport, BBB modeling, DDS crossing	Technical complexity, cost, lower throughput	(Wang et al. 2023)
Co-culture Systems	2D or 3D systems combining multiple brain-relevant cell types	Cell-cell interaction, customizable, more physiological	Neuroinflammation, cell-type-specific DDS effects	May not fully mimic in vivo spatial organization	(Yang et al. 2023)
High-Throughput Screening (HTS) Platforms	Automated 2D/3D assays with fluorescent markers and robotic imaging	Fast, multi-parametric, scalable	Large-scale screening of DDS libraries	Requires expensive equipment, 2D bias in some setups	(Curtaz et al. 2024)

11.5 IN VIVO METHODS FOR CYTOTOXIC EVALUATION OF BRAIN DDS

Aimed at assessing the cytotoxicity of brain-targeted DDS *in vivo* techniques are essential. These techniques offer a thorough comprehension of the connection between therapeutic medicines and the intricate environment of living organisms, especially with regard to the CNS. One of the main benefits of *in vivo* research is that it can accurately mimic the human body's complex physiological conditions. Drug distribution to the CNS is significantly hampered by the BBB that specifically regulates leakage to protect the brain from dangerous compounds. Researchers can evaluate DDS' capacity to efficiently traverse the BBB and arrive at its target location using *in vivo* models. Drug levels in brain tissues are assessed using methods including microdialysis and intravenous injections which provide information about absorption, distribution, metabolism, and elimination of the drugs (Chesnut et al. 2021).

Additionally, DDS' possible neurotoxicity and therapeutic effectiveness can be assessed using in vivo approaches. For example, behavioral evaluations can expose elusive neural variations that biochemical testing might miss (Fitzgerald et al. 2021). Table 11.2 describes many in vivo techniques utilized to evaluate the cytotoxic behavior of brain-targeted DDS.

TABLE 11.2
In Vivo Methods for Neurotoxicity Assessment

Method	Description	Purpose	Reference
Histopathological Analysis	Microscopic examination of brain tissue using H&E, Nissl, or immunohistochemistry to detect neuronal damage, inflammation, or necrosis.	Identify structural changes, neurodegeneration, gliosis, apoptosis.	(Ocak et al. 2025)
Behavioral Testing	Observation of animal behavior using tests like Morris Water Maze, Open Field Test, and Rotarod.	Assess functional impairments such as memory, motor coordination, or anxiety.	(Yan et al. 2021)
Biochemical Assays	Quantification of biomarkers such as ROS, lipid peroxidation, and cytokines from brain homogenates.	Evaluate oxidative stress, inflammation, and apoptotic markers.	(Wang et al. 2022)
In Vivo Imaging (MRI/PET)	Non-invasive visualization of brain drug distribution and pathology.	Track biodistribution, BBB crossing, edema, and lesion formation.	(Qiao et al. 2023)
Electrophysiological Recording	Measurement of neural activity using EEG or implanted electrodes.	Assess neural function, seizure activity, or synaptic disruption.	(Wei et al. 2022)
Micro dialysis	Sampling of extracellular fluid in the brain to measure drug levels.	Determine drug pharmacokinetics and brain penetration.	(Kondo 2024)
Cerebrospinal Fluid (CSF) Analysis	Collection and biochemical examination of CSF to detect inflammatory markers, protein levels, and drug presence.	Monitor neuroinflammation, barrier integrity, and drug diffusion into CSF.	(Dabbagh et al. 2022)
In Vivo Bioluminescence Imaging	Use of luciferase-tagged cells or probes to visualize biological processes in real-time.	Track tumor progression, apoptosis, or inflammation in brain tissue non-invasively.	(Yang et al. 2023)
Fluorescent Tracer Studies	Injection of dyes or fluorescently labeled compounds to assess BBB permeability and drug transport.	Evaluate real-time penetration and distribution of DDS across the BBB.	(Wu et al. 2021)
Magnetic Resonance Spectroscopy (MRS)	Non-invasive technique to assess metabolic changes in the brain following DDS treatment.	Evaluate neuronal viability, energy metabolism, and neurotransmitter levels.	(Weinberg et al. 2021)
In Vivo Two-Photon Microscopy	Real-time imaging technique to visualize cellular processes in live brain tissue.	Study cellular uptake, neuroinflammation, and microvascular function with high resolution.	(Yang et al. 2021)

11.6 SMART STRATEGIES TO MINIMIZE THE CYTOTOXICITY OF DDS TO THE BRAIN

Because some brain DDSs might have the potential to cause harm to brain cells, researchers are actively developing clever ways to make them safer and more effective. The main goals are to ensure the DDS is well-tolerated by the body, releases its cargo in a controlled manner, and precisely targets the intended cells or brain regions.

Here are some key approaches adapted to minimize the risk of cytotoxicity:

11.6.1 Use of Biocompatible and Biodegradable Nanocarriers

Biocompatibility and biodegradability are among the most crucial considerations in creating DDS intended for the brain. The brain is an organ sensitive to the immune system, safeguarded by the BBB, and any foreign material must be harmless and non-toxic to prevent triggering unwanted immune reactions or harming neural tissues. Utilizing biocompatible and biodegradable nanocarriers greatly minimizes the likelihood of cytotoxicity while enhancing therapeutic effectiveness. Examples are delivery systems based on PLGA extensively researched for therapies aimed at the brain because of their regulated release characteristics and low cytotoxicity (Lagreca et al. 2020), and nanoparticles composed of chitosan specifically transport the drug to the site of action, reducing unwanted adverse effects (Ahmad et al. 2022).

11.6.2 Surface Modification

Changing the outer layer of a DDS can significantly affect how it interacts with the body and reduce potential harm. It's like putting a protective coating on something to make it less reactive. This can be achieved through PEGylation, Zwitter ionic coating.

PEGylation is attaching a water-loving polymer called polyethylene glycol (PEG) to the surface, which can make nanoparticles "invisible" to the immune system, prevent proteins from sticking to them (which can trigger unwanted reactions), and help them circulate longer in the bloodstream. This longer circulation can indirectly reduce toxicity by allowing more targeted delivery. PEG is non-reactive, harmless, and does not elicit an immune response. Its presence inhibits direct contact between the nanocarrier surface and biological membranes, thereby decreasing cytotoxicity (Zhang et al. 2022).

Zwitterionic coatings, zwitterionic substances are becoming an effective approach to lower cytotoxicity and improve the biocompatibility of nanocarriers utilized in brain DDS. These materials have both positive and negative charges that balance each other out, resulting in an overall neutral charge. This neutrality helps them resist protein buildup and reduces their sticking to cells nonspecifically, which minimizes potential harm. Their distinct molecular composition resembles natural bio membranes, allowing them to be highly efficient in resisting protein adsorption and immune detection (Harijan and Singh 2022).

11.6.3 Targeted Delivery through Ligand Attachment

Ligand conjugation is a specific and efficient approach to improve drug delivery to the brain while reducing cytotoxic effects on non-target tissues. This method entails binding particular ligands like peptides, antibodies, or small molecules to the exterior of nanocarriers. These ligands identify and attach to receptors that are overexpressed on the BBB or particular brain cells (such as neurons, astrocytes, and tumor cells), promoting selective uptake and transcytosis. For example, nanoparticles conjugated with Angiopep-2 have demonstrated notable penetration of the BBB and targeted accumulation in glioblastoma models (Sahoo et al. 2023), Transferrin-coated liposomes improve the delivery of doxorubicin to brain tumors and decrease systemic toxicity (Mojarad-Jabali et al. 2022) and RVG peptide-altered nanocarriers aim at neuronal cells through nicotinic acetylcholine re,ceptors and are being investigated for gene delivery in neurological conditions (Wang et al. 2021).

11.6.4 DDS Receptive to Stimuli

Stimuli-sensitive DDS, commonly known as smart DDS, are designed to release their therapeutic agents when activated by specific internal or external triggers. These systems provide accurate spatiotemporal regulation of drug release, thus improving treatment effectiveness and reducing cytotoxicity to healthy brain tissues. This method is particularly effective for penetrating the BBB and for addressing affected brain areas like tumors, ischemic regions, or sites of neuroinflammation. pH-Sensitive medication release in acidic tumor environments or inflamed brain tissue (Wang et al. 2022). Some examples of stimuli responsive DDD are Redox-Responsive Drug release activated by elevated glutathione (GSH) concentrations in cancer cells (Peng et al. 2021), Release responsive to enzymes through cleavage by overexpressed enzymes, for instance MMPs, cathepsins (Yang et al. 2022), Temperature-Responsive Activated by localized hyperthermia or external heat treatment (Gai et al. 2024) and External activation by means of magnetic fields or NIR/UV light through magnetic or light-triggered methods (Shivanna et al. 2022).

11.6.5 Biomimetic Nanoparticles

Biomimetic nanoparticles signify an innovative method in nanomedicine, especially for delivering drugs aimed at the brain. These are designed nanocarriers covered with natural cell membranes or created to replicate biological systems. Their design improves biocompatibility, lowers immunogenicity, and greatly boosts transport through the BBB, tackling the typical issue of cytotoxicity linked to synthetic systems. For instance, biomimetic nanoparticles covered with glioma-cell membranes exhibit homotypic targeting, enhancing absorption by tumor cells (Ferreira et al. 2024). Particles coated with neutrophils specifically target inflamed endothelium to effectively deliver anti-inflammatory agents (Liao et al. 2024).

11.6.6 Exosome Based Delivery Methods

Exosomes are tiny, extracellular vesicles (30–150 nm) released by almost all types of cells. Involved in intercellular communication, these vesicles transport proteins, lipids, and nucleic acids, and have recently been recognized as effective, biocompatible nanocarriers for targeted drug delivery to the brain. Thanks to their natural capacity to traverse the BBB and interact with neural cells, exosomes are being thoroughly investigated for the treatment of neurodegenerative diseases, brain tumors, and neuroinflammation, exhibiting markedly lower cytotoxicity than synthetic alternatives (Wang et al. 2021).

11.7 CONCLUSION

Creating secure and efficient methods for transporting medications to the brain poses a significant challenge, yet the opportunity to address severe neurological disorders is substantial. Carefully assessing cytotoxicity is an essential aspect of this process. The goal is to make sure that the delicate and invaluable components of CNS are not subject to harm by state-of-the-art delivery methods. The doors to truly innovative treatments for brain ailments can be opened by carefully regulating the standards and placing strong emphasis on protection and safety.

GLOSSARY

Apoptosis: A type of programmed cell death characterized by membrane architectural alterations, DNA disintegration and contraction of cell, usually non-inflammatory.

Blood-Brain Barrier: A unique barrier designed by endothelial cells that prevents substances from entering the brain from the circulatory system.

Biocompatibility: The capacity of a sustenance to react with the body without producing adverse impacts.

Caspase: A class of protease enzymes necessary for apoptosis or programmed cell death.
Drug Delivery System: A method or device that precisely delivers drugs into the body in a precise and controlled way.
Exosome: Tiny extracellular vesicle that cells released naturally. Employed in medicine distribution owing to its capability of crossing the BBB.
Fas Receptor: A member of tumor necrosis factor receptor (TNFR) family, which controls programmed cell death.
Glial Cells: Non-neuronal cells including astrocytes and microglia that assist and shield neurons in CNS
Histopathological Analysis: examining tissues under microscope to investigate the consequences of disease.
Lysis: The breakdown or demolition of a cell due to cracking of the cell membrane consequential in spilling out its components into adjacent settings
MTT Assay: A colorimetric method that evaluates mitochondrial activity as a measure of cell viability.
Neurotoxicity: The extent to which a substance or system can damage nerve tissue.
PEGylation: The method of attaching PEG chains to nanoparticles to improve stability and reduce immunogenicity.
Reactive Oxygen Species (ROS): Chemically reactive molecules containing oxygen that can potentially damage DNA, RNA, and proteins in cells.
Trans Endothelial Electrical Resistance: A measure for assessing the integrity of tight junctions in cellular monolayers, often employed to study BBB models.
Zwitter ionic Coating: A technique for modifying surfaces that enables materials to carry both positive and negative charges, improving biocompatibility and decreasing protein adhesion.

REFERENCES

Ahmad, M. Z., M. Rizwanullah, J. Ahmad, M. Y. Alasmary, M. H. Akhter, B. A. Abdel-Wahab, M. H. Warsi, and A. Haque. 2022. "Progress in nanomedicine-based drug delivery in designing of chitosan nanoparticles for cancer therapy" *International Journal of Polymeric Materials and Polymeric Biomaterials* **71**(8): 602–623.

Albanese, A., P. S. Tang, and W. C. Chan. 2012. "The effect of nanoparticle size, shape, and surface chemistry on biological systems" *Annual Review of Biomedical Engineering* **14**(1): 1–16.

Al Mamun, A., M. S. Uddin, A. Perveen, N. K. Jha, B. S. Alghamdi, P. Jeandet, H.-J. Zhang, and G. M. Ashraf. 2022. "Inflammation-targeted nanomedicine against brain cancer: From design strategies to future developments" *Seminars in Cancer Biology* **86**(2): 101–116.

Alotaibi, B. S., Buabeid, M., Ibrahim, N. A., Kharaba, Z. J., Ijaz, M., Noreen, S. and **Murtaza, G.** (2021). "Potential of nanocarrier-based drug delivery systems for brain targeting: A current review of literature" *International Journal of Nanomedicine* **16**, 7517–7533.

Amereh, M., A. Seyfoori, and M. Akbari. 2022. "In vitro brain organoids and computational models to study cell death in brain diseases" *Neuronal Cell Death: Methods and Protocols, Springer* 281–296.

Appelbaum, L. G., M. A. Shenasa, L. Stolz, and Z. J. N. Daskalakis. 2023. "Synaptic plasticity and mental health: Methods, challenges and opportunities" *Neuropsychopharmacology* **48**(1): 113–120.

Bors, L. A., and F. J. S. P. Erdő. 2019. "Overcoming the blood–brain barrier. Challenges and tricks for CNS drug delivery" *Scientia Pharmaceutica* **87**(1): 6.

Çelik, T. A. (2018). Introductory chapter: Cytotoxicity. In: Çelik, T. A. (ed.), *Cytotoxicity*. InTechOpen, London, UK, pp. 1–5.

Chavarria, V., E. Ortiz-Islas, A. Salazar, V. Pérez-de la Cruz, A. Espinosa-Bonilla, R. Figueroa, A. Ortíz-Plata, J. Sotelo, F. J. Sánchez-García, and B. J. P. Pineda. 2022. "Lactate-loaded nanoparticles induce glioma cytotoxicity and increase the survival of rats bearing malignant glioma brain tumor" *Pharmaceutics* **14**(2): 327.

Chesnut, M., T. Hartung, H. Hogberg, and D. J. Pamies. 2021. "Human oligodendrocytes and myelin in vitro to evaluate developmental neurotoxicity" *International Journal of Molecular Sciences* **22**(15): 7929.

Curtaz, C. J., S. Wucherpfennig, E. Al-Masnaea, S.-L. Herbert, A. Wöckel, P. Meybohm, and M. Burek. 2024. "High-throughput drug screening to investigate blood-brain barrier permeability in vitro with a focus on breast cancer chemotherapeutic agents" *Frontiers in Drug Delivery* **4**: 1331126.

Dabbagh, F., H. Schroten, and C. Schwerk. 2022. "In vitro models of the blood–cerebrospinal fluid barrier and their applications in the development and research of (neuro) pharmaceuticals" *Pharmaceutics* **14**(8): 1729.

de Medeiros, L. M., M. A. De Bastiani, E. P. Rico, P. Schonhofen, B. Pfaffenseller, B. Wollenhaupt-Aguiar, L. Grun, F. Barbé-Tuana, E. R. Zimmer, and M. A. Castro. 2019. "Cholinergic differentiation of human neuroblastoma SH-SY5Y cell line and its potential use as an in vitro model for Alzheimer's disease studies" *Molecular Neurobiology* **56**: 7355–7367.

Ferreira, N. N., C. M. Leite, N. S. Moreno, R. R. Miranda, P. M. Pincela Lins, C. F. Rodero, E. de Oliveira Junior, E. M. Lima, R. M. Reis, and V. Zucolotto. 2024. "Nose-to-brain delivery of biomimetic nanoparticles for glioblastoma targeted therapy" *ACS Applied Materials & Interfaces* **17**(1): 484–499.

Fitzgerald, J. A., S. Könemann, L. Krümpelmann, A. Županič, and C. Vom Berg. 2021. "Approaches to test the neurotoxicity of environmental contaminants in the zebrafish model: From behavior to molecular mechanisms" *Environmental Toxicology and Chemistry* **40**(4): 989–1006.

Flores, J., A. Noël, M.-L. Fillion, and A. C. LeBlanc. 2022. "Therapeutic potential of Nlrp1 inflammasome, caspase-1, or caspase-6 against Alzheimer disease cognitive impairment" *Cell Death & Differentiation* **29**(3): 657–669.

Gai, Y., H. Zhou, Y. Yang, J. Chen, B. Chi, P. Li, Y. Yin, Y. Wang, and J. Li. 2024. "Injectable body temperature responsive hydrogel for encephalitis treatment via sustained release of nano-anti-inflammatory agents" *Biomaterials Translational* **5**(3): 300.

Greiner, J. V., and T. Glonek. 2021. "Intracellular ATP concentration and implication for cellular evolution" *Biology* **10**(11): 1166.

Grienberger, C., A. Giovannucci, W. Zeiger, and C. P. Portera-Cailliau. 2022. "Two-photon calcium imaging of neuronal activity" *Nature Reviews Methods Primers* **2**(1): 67.

Gu, Q., E. Cuevas, S. F. Ali, M. G. Paule, V. Krauthamer, Y. Jones, and Y. Zhang. 2019. "An alternative in vitro method for examining nanoparticle-induced cytotoxicity" *International Journal of Toxicology* **38**(5): 385–394.

Harijan, M. and Singh, M. 2022. "Zwitterionic polymers in drug delivery: A review" *Journal of Molecular Recognition* **35**(1), e2944.

Ilyas, S., R. Tabasum, A. Iftikhar, M. Nazir, A. Hussain, A. Hussain, M. S. Ali, F. Saleem, U. Saleem, and M. Froeyen. 2021. "Effect of Berberis vulgaris L. root extract on ifosfamide-induced in vivo toxicity and in vitro cytotoxicity" *Scientific Reports* **11**(1): 1708.

Jelinek, M., M. Jurajda, and K. Duris. 2021. "Oxidative stress in the brain: Basic concepts and treatment strategies in stroke" *Antioxidants* **10**(12): 1886.

Jo, H. Y., Y. Kim, H. W. Park, H. E. Moon, S. Bae, J. Kim, D. G. Kim, and S. H. Paek. 2015. "The unreliability of MTT assay in the cytotoxic test of primary cultured glioblastoma cells" *Experimental Neurobiology* **24**(3): 235.

Kari, S., K. Subramanian, I. A. Altomonte, A. Murugesan, O. Yli-Harja, and M. Kandhavelu. 2022. "Programmed cell death detection methods: A systematic review and A categorical comparison" *Apoptosis* **27**(7): 482–508.

Kondo, M. 2024. "In vivo brain microdialysis" *Cerebral Cortex Development: Methods and Protocols, Springer*: 259–269.

Lagreca, E., V. Onesto, C. Di Natale, S. La Manna, P. A. Netti, and R. Vecchione. 2020. "Recent advances in the formulation of PLGA microparticles for controlled drug delivery" *Progress in Biomaterials* **9**: 153–174.

Lakkadwala, S., and J. Singh. 2018. "Dual functionalized 5-fluorouracil liposomes as highly efficient nanomedicine for glioblastoma treatment as assessed in an in vitro brain tumor model" *Journal of Pharmaceutical Sciences* **107**(11): 2902–2913.

Langen, U. H., S. Ayloo, and C. Gu. 2019. "Development and cell biology of the blood-brain barrier" *Annual Review of Cell and Developmental Biology* **35**(1): 591–613.

Lee, H.-K., C. Velazquez Sanchez, M. Chen, P. J. Morin, J. M. Wells, E. B. Hanlon, and W. Xia. 2016. "Three dimensional human neuro-spheroid model of Alzheimer's disease based on differentiated induced pluripotent stem cells" *PLOS One* **11**(9): e0163072.

Li, J., Y. Wei, C. Zhang, R. Bi, Y. Qiu, Y. Li, and B. Hu. 2023. "Cell-membrane-coated nanoparticles for targeted drug delivery to the brain for the treatment of neurological diseases" *Pharmaceutics* **15**(2): 621.

Li, X., Z. Han, T. Wang, C. Ma, H. Li, H. Lei, Y. Yang, Y. Wang, Z. Pei, and Z. Liu. 2022. "Cerium oxide nanoparticles with antioxidative neurorestoration for ischemic stroke" *Biomaterials* **291**: 121904.

Li, Y., J. Fan, and D. Ju. 2019. "Neurotoxicity concern about the brain targeting delivery systems" *Brain Targeted Drug Delivery System, Elsevier* 377–408.

Liao, J., L. Gong, Q. Xu, J. Wang, Y. Yang, S. Zhang, J. Dong, K. Lin, Z. Liang, and Y. Sun. 2024. "Revolutionizing neurocare: Biomimetic nanodelivery via cell membranes" *Advanced Materials* **36**(26): 2402445.

Liu, N., M. Tang, and J. Ding. 2020. "The interaction between nanoparticles-protein corona complex and cells and its toxic effect on cells" *Chemosphere* **245**: 125624.

Liu, P., and C. Jiang. 2022. "Brain-targeting drug delivery systems" *WIREs Nanomedicine and Nanobiotechnology* **14**(5): e1818.

Longhin, E. M., N. El Yamani, E. Rundén-Pran, and M. Dusinska. 2022. "The alamar blue assay in the context of safety testing of nanomaterials" *Frontiers in Toxicology* **4**: 981701.

Martano, S., V. De Matteis, M. Cascione, and R. Rinaldi. 2022. "Inorganic nanomaterials versus polymer-based nanoparticles for overcoming neurodegeneration" *Nanomaterials* **12**(14): 2337.

Mayhew, C. N., and R. Singhania. 2023. "A review of protocols for brain organoids and applications for disease modeling" *STAR Protocols* **4**(1): 101860.

Menichetti, A., D. Mordini, and M. J. Montalti. 2024. "Penetration of microplastics and nanoparticles through skin: Effects of size, shape, and surface chemistry" *Journal of Xenobiotics* **15**(1): 6.

Mitsui, S., J. Yamaguchi, C. Suzuki, Y. Uchiyama, and I. Tanida. 2023. "TUNEL-positive structures in activated microglia and SQSTM1/p62-positive structures in activated astrocytes in the neurodegenerative brain of a CLN10 mouse model" *GLIA* **71**(12): 2753–2769.

Mojarad-Jabali, S., S. Mahdinloo, M. Farshbaf, M. Sarfraz, Y. Fatahi, F. Atyabi, and H. Valizadeh. 2022. "Transferrin receptor-mediated liposomal drug delivery: Recent trends in targeted therapy of cancer" *Expert Opinion on Drug Delivery* **19**(6): 685–705.

Ocak, M., Ş. Ateş, S. Kahveci, A. Okan, Z. Doğanyiğit, S. Uçar, and S. Yılmaz. 2025. "Evaluation of the anticarcinogenic effects of Rutin on brain tissue in mice with Ehrlich ascites carcinoma by micro-computed tomography and histological methods" *Asia-Pacific Journal of Clinical Oncology* **21**(2): 174–179.

Peng, Y., J. Lu, R. Li, Y. Zhao, L. Hai, L. Guo, and Y. Wu. 2021. "Glucose and triphenylphosphonium co-modified redox-sensitive liposomes to synergistically treat glioma with doxorubicin and lonidamine" *ACS Applied Materials & Interfaces* **13**(23): 26682–26693.

Petiti, J., L. Revel, and C. Divieto. 2024. "Standard operating procedure to optimize resazurin-based viability assays" *Biosensors* **14**(4): 156.

Pongrac, I. M., I. Pavičić, M. Milić, L. Brkić Ahmed, M. Babič, D. Horák, I. Vinković Vrček, and S. Gajović 2016. "Oxidative stress response in neural stem cells exposed to different superparamagnetic iron oxide nanoparticles" *International Journal of Nanomedicine* **11**: 1701–1715.

Qiao, R., C. Fu, H. Forgham, I. Javed, X. Huang, J. Zhu, A. K. Whittaker, and T. P. Davis. 2023. "Magnetic iron oxide nanoparticles for brain imaging and drug delivery" *Advanced Drug Delivery Reviews* **197**: 114822.

Sahoo, R. K., H. Kumar,, V. Jain, S. Sinha, Ajazuddin, and U. Gupta. 2023. "Angiopep-2 grafted PAMAM dendrimers for the targeted delivery of temozolomide: In vitro and in vivo effects of PEGylation in the management of glioblastoma multiforme" *ACS Biomaterials Science & Engineering* **9**(7): 4288–4301.

Sánchez-Martínez, J. D., A. Valdés, R. Gallego, Z. J. Suárez-Montenegro, M. Alarcón, E. Ibañez, G. Alvarez-Rivera, and A. Cifuentes. 2022. "Blood–brain barrier permeability study of potential neuroprotective compounds recovered from plants and agri-food by-products" *Frontiers in Nutrition* **9**: 924596.

Sawicki, K., M. Czajka, M. Matysiak-Kucharek, B. Fal, B. Drop, S. Męczyńska-Wielgosz, K. Sikorska, M. Kruszewski, and L. Kapka-Skrzypczak. 2019. "Toxicity of metallic nanoparticles in the central nervous system" *Nanotechnology Reviews* **8**(1): 175–200.

Schurhoff, N., and M. Toborek. 2023. "Circadian rhythms in the blood–brain barrier: Impact on neurological disorders and stress responses" *Molecular Brain* **16**(1): 5.

Sharma, A., and J. E. Schwarzbauer. 2022. "Differential regulation of neurite outgrowth and growth cone morphology by 3D fibronectin and fibronectin-collagen extracellular matrices" *Molecular Neurobiology* **59**(2): 1112–1123.

Shivanna, A. T., B. S. Dash, and J.-P. Chen. 2022. "Functionalized magnetic nanoparticles for alternating magnetic field-or near infrared light-induced cancer therapies" *Micromachines* **13**(8): 1279.

Singh, A., S. Maheshwari, J. P. Yadav, R. Kumar, A. Verma, S. Singh, and B. G. Prajapati. 2025. "Bioactive compound-fortified nanocarriers in the management of neurodegenerative disease: A review" *Journal of Chemistry & Biodiversity* **22**(6): 202402018.

Teixeira, M. C., C. G. Rodrigues, A. Zielińska, J. R. Campos, A. R. Fernandes, A. A. Shimojo, and E. B. Souto. 2021. "In vitro methodologies for toxicological assessment of drug delivery nanocarriers" *Nanotoxicology & Nanoecotoxicology* **.** : 203–227..

Thomsen, M. S., N. Humle, E. Hede, T. Moos, A. Burkhart, and L. B. Thomsen. 2021. "The blood-brain barrier studied in vitro across species" *PLOS One* **16**(3): e0236770.

Wang, H., Y. Chao, H. Zhao, X. Zhou, F. Zhang, Z. Zhang, Z. Li, J. Pan, J. Wang, and Q. Chen. 2022. "Smart nanomedicine to enable crossing blood–brain barrier delivery of checkpoint blockade antibody for immunotherapy of glioma" *ACS Nano* **16**(1): 664–674.

Wang, J., W. Tang, M. Yang, Y. Yin, H. Li, F. Hu, L. Tang, X. Ma, Y. Zhang, and Y. Wang. 2021. "Inflammatory tumor microenvironment responsive neutrophil exosomes-based drug delivery system for targeted glioma therapy" *Biomaterials* **273**: 120784.

Wang, P., L. Yu, J. Gong, J. Xiong, S. Zi, H. Xie, F. Zhang, Z. Mao, Z. Liu, and J. S. Kim. 2022. "An activity-based fluorescent probe for imaging fluctuations of peroxynitrite (ONOO−) in the Alzheimer's disease brain" *Angewandte Chemie* **134**(36): e202206894.

Wang, Q., S. Cheng, F. Qin, A. Fu, and C. Fu. 2021. "Application progress of RVG peptides to facilitate the delivery of therapeutic agents into the central nervous system" *RSC Advances* **11**(15): 8505–8515.

Wang, Z., Y. Zhang, Z. Li, H. Wang, N. Li, and Y. Deng. 2023. "Microfluidic brain-on-a-chip: From key technology to system integration and application" *Small* **19**(52): 2304427.

Was, H., A. Borkowska, A. Bagues, L. Tu, J. Y. Liu, Z. Lu, J. A. Rudd, K. Nurgali, and R. Abalo. 2022. "Mechanisms of chemotherapy-induced neurotoxicity" *Frontiers in Pharmacology* **13**: 750507.

Wei, W., M. Hao, K. Zhou, Y. Wang, Q. Lu, H. Zhang, Y. Wu, T. Zhang, and Y. Liu. 2022. "In situ multimodal transparent electrophysiological hydrogel for in vivo miniature two-photon neuroimaging and electrocorticogram analysis" *Acta Biomaterialia* **152**: 86–99.

Weinberg, B. D., M. Kuruva, H. Shim, and M. E. Mullins. 2021. "Clinical applications of magnetic resonance spectroscopy (MRS) in of brain tumors: From diagnosis to treatment" *Radiologic Clinics of North America* **59**(3): 349.

Wu, J., C. Shao, X. Ye, X. Di, D. Li, H. Zhao, B. Zhang, G. Chen, H.-K. Liu, and Y. Qian. 2021. "In vivo brain imaging of amyloid-β aggregates in Alzheimer's disease with a near-infrared fluorescent probe" *ACS Sensors* **6**(3): 863–870.

Xu, J., W. Du, Y. Zhao, K. Lim, L. Lu, C. Zhang, and L. Li. 2022. "Mitochondria targeting drugs for neurodegenerative disease: Design, mechanism and application" *Acta Pharmaceutica Sinica B* **12**(6): 2778–2789.

Xu, X., Y. Zuo, S. Chen, A. Hatami, and H. Gu. 2024. "Advancements in brain research: The in vivo/in vitro electrochemical detection of neurochemicals" *Biosensors* **14**(3): 125.

Yan, Y., Y. Chen, Z. Liu, F. Cai, W. Niu, L. Song, H. Liang, Z. Su, B. Yu, and F. Yan. 2021. "Brain delivery of curcumin through low-intensity ultrasound-induced blood–brain barrier opening via lipid-plga nanobubbles" *International Journal of Nanomedicine* **16**: 7433–7447.

Yang, G., Y. Liu, J. Chen, J. Ding, and X. Chen. 2022. "Self-adaptive nanomaterials for rational drug delivery in cancer therapy" *Accounts of Materials Research* **3**(12): 1232–1247.

Yang, J., W. Ding, B. Zhu, S. Zhen, S. Kuang, J. Yang, C. Zhang, P. Wang, F. Yang, and L. Yang. 2023. "Bioluminescence imaging with functional amyloid reservoirs in Alzheimer's disease models" *Analytical Chemistry* **95**(38): 14261–14270.

Yang, J., Y. Li, A. Bhalla, M. Maienschein-Cline, and K. Fukuchi. 2023. "A novel co-culture model for investigation of the effects of LPS-induced macrophage-derived cytokines on brain endothelial cells" *Journals PLOS One* **18**(7): e0288497.

Yang, Q., A. L. Vazquez, and X. T. Cui. 2021. "Long-term in vivo two-photon imaging of the neuroinflammatory response to intracortical implants and micro-vessel disruptions in awake mice" *Biomaterials* **276**: 121060.

Yao, S., M.-D. Xu, Y. Wang, S.-T. Zhao, J. Wang, G.-F. Chen, W.-B. Chen, J. Liu, G.-B. Huang, and W. Sun. 2023. "Astrocytic lactate dehydrogenase A regulates neuronal excitability and depressive-like behaviors through lactate homeostasis in mice" *Nature Communications* **14**(1): 729.

Zeng, H., H. Lu, J. Yang, and P. Hu. 2024. "An update on recent drug delivery systems targeting brain diseases via the transnasal pathway" *Pharmaceutical Research* **41**: 1–21.

Zhang, X., Z. Wei, X. Wang, Y. Wang, Q. Tang, Q. Huang, and C. Xue. 2022. "Fabrication and characterization of core-shell gliadin/tremella polysaccharide nanoparticles for curcumin delivery: Encapsulation efficiency, physicochemical stability and bioaccessibility" *Current Research in Food Science* **5**: 288–297.

Zong, W., H. A. Obenhaus, E. R. Skytøen, H. Eneqvist, N. L. de Jong, R. Vale, M. R. Jorge, M.-B. Moser, and E. Moser. 2022. "Large-scale two-photon calcium imaging in freely moving mice" *International Journal of Cell* **185**(7): 1240–1256.e1230.

12 Personalised Therapies for Rare Neurological Disorders

Bhupendra Prajapati, Touraj Ehtezazi, Yashwant Pathak, Riya Patel, and Disha Joshi

12.1 INTRODUCTION

The National Institute of Neurological Disorders and Stroke states that our delicate bodies can suffer greatly from harm to the brain and neurological system. There are more than 600 different kinds of neurological disorders, and the rarest ones can be severe (or worse), regardless of whether the damage is brought on by disease, trauma, or infection. A total of 3% of people suffer from rare neurogenetic conditions, which frequently present with complicated multiorgan comorbidity (1). The advancement of disease-modifying and other customised drugs, which have immense potential to improve health and quality of life, depends significantly on early genetic diagnosis. Since this affect 3% of the population directly and another 5% indirectly, that is, as family members, coping with care and potential recurrence risks, these strategies may benefit a large number of people (2). Patients frequently experience severe and complicated somatic and neuropsychiatric comorbidities, such as epilepsy, intellectual impairment (ID), behavioural and cognitive abnormalities, sensory deficiencies, and other organ malfunctions. These patient populations offer a difficult problem for health-care providers and systems to provide optimal tailored treatment, taking into account both the features of the condition and the individual care needs, due to their lifelong care demands on various life domains(3). For rare neurogenetic diseases, the possibilities for tailored treatment interventions, both symptomatic and disease-modifying, are expanding significantly. These include vitamins and diets, (repurposed) drugs, organ and stem cell transplants, and gene and RNA treatment. However, there are particular difficulties with methodology and outcome metrics when implementing therapeutic trials for these frequently minor patient populations (4). Financial, organisational, and regulatory obstacles can make it difficult to implement individualised treatment. At both the local and global levels, translating novel drugs into patient care with assured access and funding is difficult and frequently fails. This leads to an undesirable situation where patients do not receive novel medicines and evidence-based therapy is unavailable (5).

In this chapter, we examine a number of success stories to discuss the state of personalised medicine for rare neurogenetic diseases and highlight areas for improvement in trial design, outcome measurements, and collaboration.

12.2 AMYOTROPHIC LATERAL SCLEROSIS

A neurodegenerative condition called amyotrophic lateral sclerosis, or amyotrophic lateral sclerosis (ALS), affects the brain and spinal cord's neurons. Your ability to control your muscles is impacted. Over time, symptoms get worse (6). Muscle weakness and twitching that impair your ability to walk on your own, reach for objects, chew meals, and speak may be noticed by the patient. Muscles progressively deteriorate due to ALS (atrophy). Lou Gehrig's disease was the previous name for amyotrophic lateral sclerosis. Gehrig was a well-known baseball player who suffered from the condition in the 1920s and 1930s. Every year, an estimated 5,000 Americans are diagnosed with ALS (7). Treatments are always getting better, even while there is no cure. The correct mix could enhance your quality of life and reduce the disease's progression. According to their cause, there are two

DOI: 10.1201/9781003482406-12

kinds. ALS that is both familial and sporadic. Roughly 90% of instances are of the most prevalent kind. It is not inherited and happens at random. Ten percent or so of ALS cases are familial. It is brought on by genetic alterations acquired from one's parents. Muscle weakness in the arms, legs, or neck, cramping, twitching in the hands, feet, shoulders, or tongue, stiff or spastic muscles, slurred or difficult-to-form words, drooling, inadvertent emotional outbursts like laughing or crying, exhaustion, and difficulty swallowing (dysphagia) are all possible symptoms of ALS (8). Early symptoms typically include difficulty speaking and swallowing, as well as muscle weakness or stiffness in the patient's arms and legs. Daily activities like writing or eating may become more difficult as a result. Symptoms usually spread throughout the patient's body over time. Each person experiences symptoms at a different rate (9).

The patient may experience difficulty breathing, standing, or walking as symptoms worsen. The patient may lose a substantial amount of weight. If your symptoms get worse, get in touch with your doctor; if you're having breathing difficulties, get emergency assistance. There are hereditary forms of amyotrophic lateral sclerosis. The patient's biological parents may pass on genetic alterations that cause ALS (10). However, inherited ALS is uncommon. Genetic alterations can occasionally occur at random and have no biological family history. As symptoms worsen, the patient's life expectancy will be reduced. Patients' mental health may suffer as a result of learning about and managing this condition on a daily basis. The patient may experience stress, hopelessness, overwhelm, or loss. Because of this, many ALS patients also experience anxiety and sadness (11). As you collaborate with various professionals to address the physical health of your patients, be sure to attend to their emotional well-being as well. Seek assistance from a mental health professional or the patient care team. ALS can be identified through a physical examination, a neurological examination, and testing to understand more about the symptoms and how they affect the body. Different tests and examinations are required for an appropriate diagnosis. Damage from ALS cannot be undone. However, it might help make things more comfortable and slow the progression of symptoms (12).

Medication like Riluzole (Rilutek®, Exservan®, and Tiglutik®) could reduce motor neuron damage. It might increase survival by several months. Therapies or rehabilitation, nutritional assistance, and breathing support are among the available treatments. As the disease worsens, the patient may require more or different treatment options. Supportive care is also offered to address requirements. There are more drugs available for symptom management. This could include drugs for pain, mental health issues, stiffness, cramping in the muscles, and excessive urination (13). The main federal funding source for studies on the brain and nervous system, including conditions like ALS, is the National Institute of Neurological Disorders and Stroke (NINDS), a division of the National Institutes of Health (NIH). In order to accelerate the discovery of successful therapies for the diagnosis, treatment, management, prevention, and cure of ALS, NINDS released strategic priorities for ALS research in 2023. These strategic priorities were developed with assistance from scientists, doctors, advocates, ALS patients, and the general public. In accordance with the Accelerating Access to Critical Therapies for ALS Act, NINDS also provides funding for studies aimed at providing individuals with ALS who are ineligible for clinical trials with increased access to investigational new medications (14). Adult skin and blood cells are being transformed by scientists into stem cells that can develop into any kind of cell, including motor neurons and other cells that might be involved in ALS. To better understand the function of neurons involved in ALS, researchers sponsored by NINDS are growing human spinal cord sections on tissue chips using stem cells. NINDS-funded clinical research projects are examining the evolution of ALS symptoms in individuals with C9orf72 mutations. Additional genes that may cause or increase a person's susceptibility for either familial or sporadic ALS are being investigated (15).

Development of biomarkers, which are measurable biological indicators of disease. Biomarkers can be found by brain scans, blood tests, and other methods. They can demonstrate whether a person has a disease or condition and how it is developing over time. The rate of progression and the efficacy of present and upcoming treatments can be determined with the use of ALS biomarkers (16).

12.3 ACUTE DISSEMINATED ENCEPHALOMYELITIS

The central nervous system (CNS) is affected by acute, severe, polyfocal, inflammatory autoimmune disease known as acute disseminated encephalomyelitis (ADEM) or post-infectious encephalomyelitis. Although adult cases are not unusual, children are the most commonly affected. Due to inflammation brought on by an earlier illness or vaccination, ADEM is characterised by demyelination in the brain, spinal cord, and possibly the optic nerve (17). With typical imaging results on brain magnetic resonance imaging (MRI), the diagnosis is primarily clinical. The current diagnostic criteria were created by the International Pediatric Multiple Sclerosis Study Group, which defined ADEM as an acute demyelinating syndrome that necessitates the presence of encephalopathy, polyfocal CNS involvement, and demyelinating lesions on brain MRI (18). ADEM is primarily regarded as a CNS post-infectious disease. In between 50% and 85% of instances, an identifiable infection or vaccination has been observed. Cytomegalovirus, Epstein-Barr virus, herpes simplex virus, human herpes-virus-6, influenza virus, hepatitis A, human immunodeficiency virus, and mycoplasma pneumonia are the most often linked organisms; nevertheless, the causative pathogen is typically unknown. Leptospira, beta-hemolytic streptococci, and Borrelia burgdorferi are other related bacterial diseases (19). ADEM was most frequently linked to measles prior to the creation of immunisation programs, and it was also more frequently linked to rubeola, rubella, mumps, varicella, and smallpox. The probability of getting ADEM is determined by a variety of factors, including an individual's genetics, exposure to pathogenic organisms, immunisation history, and lighter skin pigmentation. ADEM, which occurs around the world, affects all ethnic groups. ADEM is linked to a previous infection or vaccination in 50%–85% of patients, with the majority of cases occurring after a viral or bacterial infection. However, the underlying pathogen is not usually recognised (20).

Differential diagnosis of ADEM includes Aseptic meningitis and acute viral encephalitis are characterised by constitutional symptoms indicating an underlying infection, such as fever, neck rigidity, and raised acute phase reactants such as C-reactive protein and erythrocyte sedimentation rate(21). The first clinical episode of multiple sclerosis: clinical presentation of multifocal neurological impairments and MRI appearance may be similar to ADEM, with some modest MRI alterations; multiple sclerosis relapse often lacks acute encephalopathy characteristics. Diseases related with myelin oligodendrocyte glycoprotein antibody disease, MOGAD: Check myelin oligodendrocyte glycoprotein (MOG) antibody as part of the examination, especially in children (22).

Patients with meningeal symptoms, fever, acute encephalopathy, and evidence of inflammation in the blood or CSF may require supportive care and empiric treatment with intravenous (IV) acyclovir and antibiotics. The primary acute treatment for ADEM is immunosuppression with high-dose IV glucocorticoids. The typical protocol is to administer 30 mg/kg/day of methylprednisolone intravenously for 3–5 days, followed by a 4- to 6-week decrease of oral prednisone (23). Early withdrawal of steroids may increase the probability of recurrence. These can be begun alongside acyclovir or antibiotics at the time of the patient's initial presentation. If a patient does not improve or responds badly to glucocorticoid drugs, IV immunoglobulin is an effective second-line treatment. The dosage is 2 g/kg every 2–5 days. Refractory patients are treated with plasma exchange (3–7 exchanges) or cyclophosphamide (24). ADEM consequences are heavily influenced by the severity and extent of CNS involvement. Involvement of the long fibre tracts of the white matter can lead to long-term consequences of physical disability, such as spasticity and varied degrees of extremity weakness. Post-ADEM epilepsy is a common complication in children. Children with MOGAD-associated ADEM are more likely to experience ADEM relapses as well as seizures. Impaired cognitive function may be the most serious long-term side effect of ADEM in children. According to studies, children aged 6–15 exhibited modest cognitive abnormalities and lower intelligence quotient scores 3.5–4 years after ADEM. Patients who developed ADEM later in life showed delayed verbal processing(25).

Aggressive treatment strategies, such as surgical decompression, must be considered and implemented in patients with fulminant variants of ADEM who show evidence of ongoing clinical deterioration due to increased intracranial pressure and are unresponsive to conventional medical treatment and critical care. With current treatment regimens, the long-term prognosis for ADEM is usually excellent. Full recovery occurs in approximately 57%–89% of patients, with 20%–30% showing mild residual impairments. Following the start of treatment, a rapid improvement is commonly noted, often within hours, though recovery typically takes days (26). The goal of personalised therapeutic approaches for ADEM is to customise treatments based on the patient's immunological profile, response patterns, and severity. Precision treatment is guided by early biomarkers such as cytokine profiles, myelin-specific antibodies, and neuroimaging signs. Although they can be tailored according to steroid response, high-dose corticosteroids continue to be the first choice. Depending on the antibody-mediated disease, refractory cases are treated with either plasma exchange or intravenous immunoglobulin (IVIG) (27). Personalised decision-making is further supported by genetic predispositions and past infection history. New strategies, such as neuroprotective drugs and tailored immunomodulators, seek to minimise treatment-related side effects while maximising neurological recovery and lowering the chance of relapse (28).

12.4 AGENESIS OF THE CORPUS CALLOSUM

One of the most prevalent brain abnormalities in humans is Agenesis Of The Corpus Callosum (ACC). Its frequency in children with developmental disabilities is around 230 in 10,000 (2.3%), and its incidence ranges from 0.5 to 70 in 10,000. It appears in one out of every 19,000 autopsies. ACC is a congenital condition in which the corpus callosum is absent or partially absent at birth. The brain's left and right hemispheres are connected by this dense bundle of nerve fibres (29). It enables communication between the two sides of the brain. The following are possible indicators and symptoms of ACC includes delays in development (e.g., rolling over, sitting up, walking, talking)(30): intellectual disability or cognitive impairment; difficulties with speech, hearing, and vision and, moreover, seizures; and feeding challenges (infants) and tight or loose muscular tone, as well as an unusually big or tiny head size. During an ultrasound performed after the 16th week of pregnancy, a doctor may suspect agenesis of the ACC. Typically, a formal diagnosis is made after birth (31).

A doctor or nurse will conduct a physical examination and testing to do this. The provider will learn more about the patient's medical history and symptoms throughout the examination. When it comes to children, a provider will ask parents or other caregivers if their child is reaching age-appropriate developmental milestones, such as walking or sitting up straight on their own. To confirm a diagnosis, a provider could arrange imaging testing. MRI or computed tomography (CT) can reveal the presence or absence of the corpus callosum (32). Consider the corpus callosum to be a wooden bridge that spans a river. It makes it possible for people to go freely across the river. The bridge is either nonexistent or has missing planks due to agenesis of the corpus callosum. In certain situations, the patient could still be able to cross the bridge, but it becomes considerably more difficult to get there (33).

The patient with agenesis of the corpus callosum may benefit from an interprofessional team approach that includes the obstetrician for the mother's prenatal care, neonatologist, pediatric neurologist, pediatric neurosurgeon, geneticist, maxillofacial surgeon, ophthalmologist, otorhinolaryngologist, plastic surgeon, pediatric psychiatrist and psychologist, dietician, and pediatric neuro nurse (34). Agenesis of the corpus callosum has no known treatment, although advances in medical technology and research continue to give hope for better results. For instance, improvements in neuroimaging methods have made it possible to diagnose corpus callosum agenesis earlier and more accurately, which can result in earlier intervention and better outcomes for those who are impacted (35).

New customised therapies are being developed to improve neurodevelopmental outcomes. Precision diagnostics employing genetic sequencing offer tailored management based on individual gene abnormalities. Early-intervention programs speech therapy, occupational therapy, and motor training, are increasingly personalised utilising neuropsychological profiles (36). Novel neuromodulation technologies such as transcranial direct current stimulation and neurofeedback show promise for increasing cognitive integration. Personalised medication targets related symptoms such as epilepsy, ADHD, or anxiety. Customised educational planning, AI-guided cognitive training, and digital therapies all help children with ACC improve their functional abilities (37).

Access to resources and assistance is crucial for people with corpus callosum agenesis and their families. To help people deal with the difficulties brought on by the disease, this might involve counseling and support groups. Advocacy groups and online forums can also offer impacted people and their families helpful information and support (38).

12.5 AICARDI–GOUTIÈRES SYNDROME DISORDER

Aicardi–Goutières syndrome (AGS) is an extremely rare genetic disorder classified as a type I interferonopathy. The AGS primarily affects the brain, body, and skin; individuals diagnosed with AGS typically exhibit an early-onset encephalopathy, with the severity of the condition varying significantly from one individual to another. With fewer than a thousand cases recorded worldwide, recent registry and population-based estimates have provided an approximate birth incidence of 1 in every 100,000 live births, reflecting the sporadic nature of this disorder. AGS is caused by mutations in genes involved in nucleic acid metabolism or sensing, which drive chronic activation of type I interferon pathways and sterile neuroinflammation (39). The identified genetic mutations involve the *TREX1, RNASEH2A/B/C, SAMHD1, ADAR, IFIH1, LSM11*, and *RNU7-1* genes; inheritance can be both autosomal recessive and autosomal dominant. This explains the genetic and clinical heterogeneity observed in families (40).

In clinical settings, the AGS will usually present as an encephalopathy appearing in the first two years of life, characterised by somewhat acquired microcephaly, rigidity, dystonia, seizures, and developmental delay or impairment on a global scale (39). However, more recently, AGS has been identified with less severe variants that maintain cognitive ability and even asymptomatic individuals who carry the AGS genetic mutation. From a neuroimaging perspective, the most common indicators of AGS are calcifications present within the brain, especially in the basal ganglia, white matter changes consistent with Leukoencephalopathy, and cerebral atrophy, suggesting a major disruption with the normal development of myelin and other structures, resulting in the disruption of normal function within the brain. An infant's encephalitic symptoms, such as increased irritability, feeding difficulty, abnormal movements, seizures, and diminished head growth, result in microcephalic features (41). Most infants with this form will develop some sort of periodic spiking fever, elevated hepatic enzymes, and hepatosplenomegaly, which is very similar to the symptoms of prenatal infection and makes it down to the case of AGS very difficult to evaluate precisely. In regard to neurological symptomatology, each of the genetically defined subtypes has different patterns of neurological involvement (42). Examples of neurologic involvement might include the combination of spastic paresis but intact cognitive function versus severe quadriparesis, with a huge degree of accompanying intellectual impairment (43).

Similarly, all genetic subtypes may present with similar extraneural symptoms, including chilblains on fingers or toes, or complications associated with the liver, kidneys, or eyes associated with certain genetic subtypes. Uncommon or broadened phenotypes encompass conditions such as stroke-like episodes, glaucoma, opsoclonus–myoclonus syndrome, and pigmentary skin disorders such as dyschromatosis symmetrica hereditaria, which can lead to misdiagnosis as other inflammatory or vasculopathic diseases. In some cases of childhood-onset AGS, the clinical symptoms stabilise over time, and an adult diagnosis is frequently made only when chilblains are found in combination with basal ganglia calcifications that have been incidentally discovered during neuroimaging (44).

AGS is an exceptionally rare condition, with an estimated patient count of 500 or fewer, placing it among the rarest diseases as per Orphanet's classification system (45). A recent Danish study based on population data indicates that the birth incidence of AGS is approximately 1.01 per 100,000 live births. This suggests that AGS may have been considerably underdiagnosed, given its resemblance to congenital infections like congenital cytomegalovirus infection, cerebral palsy, and various autoimmune diseases (46). Furthermore, AGS affects individuals regardless of gender or race and is present across numerous racial groups, implying that the genetic factors contributing to AGS have been inherited by human populations worldwide over generations.

To diagnose a patient, it is important to integrate clinical findings, neuroimaging, laboratory results, and molecular testing. The brain MRI and CT will typically show characteristic findings such as basal ganglia and deep white matter calcifications, leukoencephalopathy or delayed myelination, and cerebral atrophy. However, in some patients, particularly those who are early in the course of the disease, the calcifications may be absent or subtle. Biochemically, a patient with AGS will usually have a persistent type I interferon "signature", which includes elevated levels of interferon-α in their cerebrospinal fluid and serum and increased expression of interferon-stimulated genes in their peripheral blood cells. These biochemical markers help to distinguish AGS from congenital infections (47). The definitive diagnosis of AGS is made through genetic testing of the AGS-associated genes, using either a targeted gene panel or whole-exome or whole-genome sequencing, to identify pathogenic variants in the AGS genes. This also provides genotype-phenotype correlations and family counselling with regard to developing personalised therapeutic approaches for rare neurological disorders (48).

There is no cure for AGS, so treatment focuses on early inflammation control, preventing tissue damage, and providing supportive care. New disease-modifying methods are being developed, particularly those reducing type I interferon signalling or stopping immunostimulatory nucleic acid buildup, enabling personalised treatment options (49). Daily management primarily involves addressing symptoms and providing support. This generally encompasses respiratory and nutritional assistance, controlling seizures with antiseizure drugs, and engaging in intensive physiotherapy, occupational therapy, and speech therapy to enhance motor abilities, communication, and feeding. Endocrine issues like diabetes insipidus or hypothyroidism are managed with standard hormone replacement therapies, while hematological issues such as thrombocytopenia might necessitate transfusions. Traditional immunosuppression using corticosteroids, immunoglobulin, or thiopurines has shown limited effectiveness, particularly for neurological outcomes, and is considered empirical rather than targeted treatment. Monitoring interferon activity has become crucial in modern management. The reduction of interferon-stimulated gene expression in blood serves as a pharmacodynamic marker for targeted therapies, as clinical improvement can be difficult to measure in severely impaired children. Additionally, family counselling, palliative care, and psychosocial support for caregivers are essential for improving quality of life (49,50).

Recognising that AGS is a form of Type I Interferonopathy has resulted in the development of targeted treatments (51). As all AGS genotypes ultimately result in Chronic Type 1 Interferon activation through different triggers, personalising treatments based upon the specific type of genetic mutation that causes AGS is a new approach to personalised medicine development (52). Janus kinase (JAK) inhibitors represent the most advanced targeted therapies currently available in clinical practice. These agents theoretically reduce Interferon-induced inflammation by inhibiting the JAK1/2-dependent signalling pathways that occur following the binding of Interferons to their receptors in the downstream pathway of activation (53,54). The therapeutic agents currently being studied include Ruxolitinib, Baricitinib, etc., and several case series or open-label studies have demonstrated improvement in both systemic and cutaneous manifestations of the disorder (e.g., improvement of motor function, reduction in Dystonia, resolution or improvement in Chilblain Lesion, and improvement in clinical symptoms such as Pericardial Effusion) (55). However, these therapeutic agents have limitations as they do not seem to improve existing neurological disabilities, and there is a risk of severe infection and cytopenias associated with these agents; not all agents appear to cross the blood–brain barrier effectively (43). This highlights the importance of genotype-guided RTI selection for a personalised approach.

Reverse transcriptase inhibitors (RTIs) such as zidovudine, lamivudine, and abacavir are designed to inhibit endogenous retroelements from reverse transcribing and activating host cell recognition systems through interferon signalling pathways, due to the fact that endogenous retroelements can either inhibit or activate interferon signalling based on the pathway that is being activated. In a small clinical trial, using an RTI regimen was able to lower serum levels of interferon α and increase cerebral perfusion, indicating a possible biological effect; however, the clinical benefits are still unclear. RTIs are not likely to be beneficial for those forms of AGS that do not utilise retroelement reverse transcription (e.g., certain forms of ADAR1- or IFIH1-related disease), and additionally, changes in nucleotide pool metabolism by SAMHD1 may also affect RTI efficacy. Establishing a genotype-guided selection process for RTI use is critical to provide a more individualised approach to treatment.

Several of these targeted immunomodulatory strategies are still experimental. Another work has suggested the use of monoclonal antibodies against IFN-α or its receptor, such as sifalimumab and anifrolumab. This suggestion is based on data derived from studies on systemic lupus erythematosus and lupus mouse models, where anti-IFN receptor antibodies have been shown to ameliorate the disease (56). Preclinical investigations by Li and their colleagues on the cGAS-STING pathway, including the genetic rescue of TREX1-/- mice through cGAS deletion, as well as the development of small-molecule or nucleic acid-based cGAS/STING inhibitors, suggest that targeting this axis could be pertinent to TREX1- and related AGS forms (57). In addition, Henrickson and colleagues have observed that tocilizumab, an IL-6 inhibitor, exhibited a positive response in a case of a cerebral vasculopathy caused by the SAMHD1 mutation; therefore, blockade of IL-6 and other downstream cytokines may confer therapeutic benefits in certain phenotypes (58).

Using induced pluripotent stem cell (iPSC)-derived neural stem cells from patients with AGS, Braidotti et al. conducted studies that enhanced understanding of protective mechanisms against toxicity. By studying glucocorticoids, this team was able to demonstrate no loss in viability when AGS-derived neural stem cells were treated with glucocorticoids. JAK inhibitors, including ruxolitinib, baricitinib, and tofacitinib, were found to be safe and, in most cases, promoted the growth of AGS-derived cells. Conversely, some thiopurines and reverse transcriptase inhibitors, such as zidovudine, in AGS2 cells were shown to be selectively toxic. This study establishes the tolerability of JAK inhibitors and reverse transcriptase inhibitors with neural stem cells while identifying a risk of toxicity from particular agents based on genetic background (40).

In addition, personalised approaches also extend to psychiatric and behavioural features. One such illustrative case report by Gil et al. described an 18-year-old with SAMHD1-related AGS who presented with severe psychiatric manifestations unresponsive to classic psychotropic medications but demonstrated dramatic improvement following electroconvulsive therapy (59). This case broadens treatment perspectives for significant psychiatric symptoms in adolescents and adults and serves to illustrate how innovations in individual cases can further develop mechanism-based pharmacological approaches in this developing area of personalised treatment of AGS.

12.6 ATAXIAS AND CEREBELLAR OR SPINOCEREBELLAR DEGENERATION

Degenerative cerebellar ataxias and cerebellar/spinocerebellar degenerations (ACSD) encompass a wide-ranging group of disorders characterised by genetic and clinical diversity, where the gradual impairment of cerebellar function results in debilitating motor and non-motor symptoms. While many forms of spinocerebellar ataxias (SCAs) do not yet have a definitive cure available, rapidly developing gene and cell therapies personalised for specific patient needs are providing increasingly individualised options for treating SCAs (60). ACSD & SCAs refer to both hereditary and sporadic neurodegenerative ataxias, which involve the progressive degeneration of the cerebellum and related connections (e.g., spinocerebellar pathways). Included within this group are autosomal dominant SCAs, autosomal recessive cerebellar ataxias (ARCA), X-linked & mitochondrial ataxias, and degenerative disorders (e.g., multiple system atrophy, cerebellar type) (61).

SCAs are one of the best-understood parts of the ACSD genetic family, with at least 40 types caused by various genetic changes, including CAG/polyglutamine repeat expansion and non-CAG/polyglutamine mutations. While clinical characteristics can be similar, each SCA genetic mutation affects cellular pathways differently. Due to this genetic diversity, next-generation sequencing is essential for SCA diagnosis and provides the foundation for developing genetically specific therapeutic interventions (62).

ACSD primarily manifests as a worsening cerebellar syndrome characterised by difficulties with balance and walking, uncoordinated limb movements, speech disturbances, eye movement issues such as nystagmus and saccadic intrusions, and tremors during intentional movements. These symptoms significantly hinder mobility and fine motor abilities. As the condition progresses, patients often experience swallowing problems, frequent falls, and a loss of independence, necessitating the use of mobility aids like walkers and wheelchairs, as well as assistance with daily activities (63). Most hereditary neuropathies, especially ARCAs, have numerous neurological and systemic features, including peripheral neuropathy, pyramidal/extrapyramidal signs, seizures, retinopathy, endocrine disorders, and cardiomyopathy. Recently, there has been growing recognition of ACSD's impact on the "cerebellar cognitive-affective" network, with executive dysfunctions, visuospatial deficits, language changes, depressive symptoms, anxiety, apathy, and psychosis associated with ACSD. These findings indicate a clear link between cerebello-cortical circuit impaired function and cerebellar neuroimaging alterations. Non-motor symptoms can severely reduce quality of life and add to care complexities for patients and families (64).

Although each ACSD entity is rare on its own, together they constitute a major contributor to chronic neurological disability. Studies on populations reveal that hereditary ataxia impacts about 2–5 individuals per 100,000, with ARCA comprising nearly half of these genetic cases and generally manifesting before the age of 30. The prevalence of these disorders varies significantly across different regions and due to specific founder mutations; Friedreich ataxia is the most prevalent ARCA in numerous cohorts, while certain SCAs are more common in particular populations (65).

Diagnosis requires a stepwise clinical approach. Detailed history and examination help distinguish hereditary from acquired ataxias, identify red flags for ARCAs (early onset, multiple affected siblings, consanguinity), and document features that may indicate specific subtypes. MRI shows cerebellar atrophy, often with brainstem or spinal involvement, and helps exclude structural or inflammatory mimics. Baseline laboratory work-up screens for treatable metabolic, nutritional, autoimmune, or toxic causes (e.g., vitamin E deficiency, Wilson disease). When hereditary ACSD is suspected, multigene ataxia panels, exome, or genome sequencing are now standard to identify SCA and ARCA genes, guide prognosis and reproductive counselling, and determine eligibility for emerging gene-targeted trials (66).

The conventional method for addressing ACSD focuses on symptom management through a team of specialists from various disciplines. Consensus based on evidence endorses the use of physiotherapy, training for balance and walking, occupational therapy, and speech and swallowing therapy to help maintain functionality, decrease the risk of falls, and enhance communication. Patients receive support with assistive devices, modifications to their homes, management of spasticity, and treatment for complications. Social and psychological support is also provided to help both patients and caregivers adjust to the progression of the disability (66–70).

Pharmacological interventions are available for motor ataxia. Riluzole has shown effectiveness in patients with mixed SCA and Friedreich ataxia through SARA score assessments. However, evidence for its effectiveness in SCA2 is inconsistent, affecting its efficacy for specific subtypes. Other agents like amantadine, buspirone, varenicline, and rovatirelin have shown mixed or negative outcomes in trials, lacking a scientific foundation for routine use. Various symptomatic treatments exist for ARCA patients: high-dose vitamin E helps those with vitamin E deficiency ataxia, dopamine agonists and SSRIs may help ataxia-telangiectasia patients, baclofen can treat early ARSACS spasticity, and Friedreich ataxia patients might benefit from idebenone and antioxidant formulations. However, phase III trials of idebenone in Friedreich's ataxia have not shown conclusive neurological outcomes (60,71).

Personalised therapeutic strategies for adult-onset cerebellar atrophy and spinal cerebellar atrophy diseases are contingent upon genetic diagnosis and a mechanistic understanding, particularly concerning dominantly inherited SCAs and ARCAs. Gene editing techniques, such as RNA interference (RNAi) and antisense oligonucleotides, represent a prominent approach for addressing SCAs (72). A CRISPR/Cas9-based methodology, as employed by He, Pappadà, and collaborators, has demonstrated efficacy in normalising neural differentiation and reducing pathogenic protein levels in iPSCs for SCA1 and SCA3. Nevertheless, challenges related to the in vivo delivery of CRISPR/Cas9 and potential off-target effects are anticipated to limit the applicability of these therapies (73–75).

Studies by Santos, Kotowska-Zimmer, Carrell, and Conceição have shown that RNAi strategies utilising miRNA, siRNA, or shRNA constructs can decrease mutant proteins and enhance histopathology and motor function in preclinical models of SCA1, SCA3, SCA7, and similar disorders (76,77). Efforts are underway to develop advanced delivery mechanisms, including AAV vectors, nanoparticles, cell-penetrating peptides, and liposomal carriers, to enhance targeting of the cerebellum and decrease the need for invasive procedures.

Stem cell-based strategies present a personalised approach for treating SCA. Mesenchymal stem cell (MSC) transplantation has shown efficacy in SCA models. Tsai and colleagues reported that human umbilical MSC administration to the cerebellar cortex of SCA1 mice resulted in improved motor behaviour, reduced cerebellar atrophy, and Purkinje cell rescue (78,79). Similarly, Correia and co-workers found that MSC secretome administration in SCA3 mice led to mild, transient improvements. Preliminary clinical trials with MSCs in SCA1, SCA2, SCA3, and SCA6 have indicated symptomatic benefits (80).

Downstream pathway targeted therapies are emerging. Calpain inhibition (e.g., BDA410 in SCA3 mice, BLD2736 in SCA3 zebrafish) reduced toxic ATXN3 fragments, decreased aggregates, enhanced autophagy, and improved motor function, suggesting effects on proteolysis and mitophagy, though broad protease inhibition raises off-target concerns (81). Autophagy induction is active: agents like trehalose, lithium, temsirolimus, carbamazepine, cordycepin, and n-butylidenephthalide lowered mutant protein levels and aggregates while improving neuropathology in SCA models, and trehalose showed stabilisation or modest improvement in small SCA3 clinical series, especially when started early.

Ion channel modulators represent a personalised direction, as Purkinje cell excitability changes are shared across SCAs. Riluzole and chlorzoxazone-baclofen combinations that enhance BK channel function have normalised cerebellar firing and improved motor deficits in SCA1 models, positioning them for subtype-specific trials. Other strategies targeting transcriptional dysregulation (valproic acid), mitochondrial dysfunction (intranasal NeuroEPO in SCA2), and metabolic deficits like docosahexaenoic acid in SCA38 show how interventions can be tailored to each genetic context (60).

ACSD is evolving from a uniform symptomatic model to a tailored framework that integrates genetic diagnosis, disease stage, and both motor and non-motor profiles. This shift aims to identify the most suitable interventions, including gene-silencing tools, stem cell therapies, pathway-modulating drugs, and targeted rehabilitation. The advancement of sensitive biomarkers and preataxic cohorts will be vital for implementing these therapies early, at a stage when neuronal networks are most responsive to treatment (82–84). Figure 12.1 shows Pathophysiological clusters underlying major rare neurological disorders.

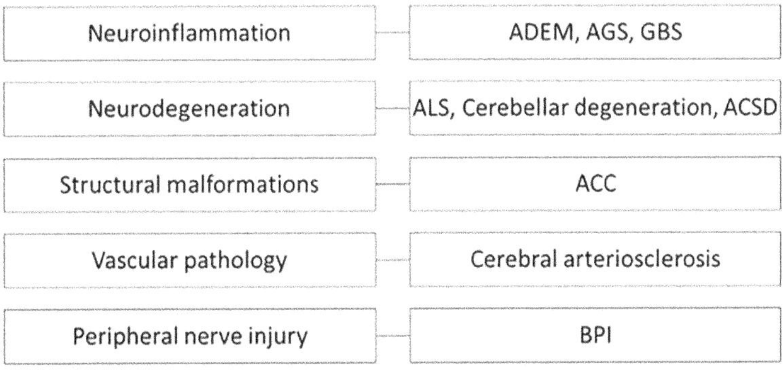

FIGURE 12.1 Pathophysiological clusters underlying major rare neurological disorders

12.7 BRACHIAL PLEXUS INJURIES

The brachial plexus is a network of spinal nerve roots (C5–T1) that come together in the neck and then branch out to supply the shoulder girdle, arm, forearm, and hand with nerves. An injury to the brachial plexus happens when these nerves are stretched, compressed, torn, or pulled away from the spinal cord, often due to traction between the neck and shoulder, high-speed car accidents, falls, penetrating injuries, or, in newborns, challenging deliveries (85).

Common indicators include weakness or paralysis in the arm and shoulder, reduced mobility, numbness or altered sensations, and neuropathic pain in the affected limb. Damage to the upper plexus (C5–C6) typically leads to difficulties with shoulder abduction and elbow flexion. Conversely, injuries to the lower plexus (C8–T1) mainly affect hand and wrist functions, although the intermediate segments might also be impacted (86,87).

Patients often report experiencing intense burning or electric shock-like sensations that follow specific nerve pathways, such as the lateral shoulder in C5 or the radial forearm and thumb in C6. They may also present with muscle atrophy, reduced reflexes, and abnormal arm postures. In cases of preganglionic root avulsions, additional symptoms like Horner's syndrome, a winged scapula, or persistent deafferentation pain indicate a more proximal injury and are associated with a worse prognosis. Besides physical limitations, brachial plexus injuries (BPI) can significantly impede the ability to return to work and perform daily activities. It is also linked to chronic pain, depression, and a reduced quality of life (87,88).

BPI is rare among the general populace, with current estimates of traumatic BPI ranging from about 0.2 to 3.9 cases per 100,000 people each year, depending on the country and the methods used. In major trauma groups, BPI is found in roughly 1%–2% of patients with multiple injuries, with most research showing a higher occurrence among young male motorcyclists or those involved in road traffic accidents (89).

A precise diagnosis is contingent upon a detailed patient history, an exhaustive neurological assessment, and specific diagnostic tests. Evaluating the mechanism, speed, and timing of the injury, alongside any related fractures or vascular and thoracic injuries, as well as the initial presentation of weakness and pain, is essential for determining the injury's severity. A comprehensive physical examination assesses motor strength, sensory function, and reflexes across each myotome and dermatome, and identifies signs of preganglionic involvement, which would rule out the root as a donor for reconstruction. MRI or CT myelography are utilsed to detect root avulsions and examine the plexus structure. Electrodiagnostic tests, including EMG and nerve conduction studies, verify denervation, locate the lesion, evaluate axonal loss, and distinguish between pre- and postganglionic injuries (90–92).

Effectively managing BPI requires a multidisciplinary team, including peripheral nerve surgeons, neurologists, rehabilitation physicians, therapists, pain specialists, and psychological support services. The initial focus is on addressing any life-threatening injuries and stabilising fractures. Subsequently, early rehabilitation is crucial to ensure joint mobility, prevent contractures, improve posture, and guide individuals in adapting their daily activities to suit their needs (87,93).

The type and severity of injury influence the decision for conservative or surgical treatment. Low-energy stretch injuries without signs of root avulsion may resolve spontaneously and require close monitoring with electrodiagnostic evaluations over months. Confirmed root avulsions or high-grade postganglionic ruptures require prompt surgical intervention within 3–6 months, due to irreversible degeneration of motor end plates from prolonged inactivity. Surgical interventions include neurolysis, nerve grafting, and nerve transfers. In nerve transfers, expendable donor nerves, such as the spinal accessory, intercostal, or fascicles of the ulnar/median nerves, are redirected to reinnervate critical targets, including elbow flexors or shoulder abductors. In chronic cases, tendon transfers, free functional muscle transfers, and orthopaedic procedures like shoulder fusion may restore limited functional movement (87,88,92,93).

Enhancing long-term outcomes necessitates the optimisation of pain management through the use of neuropathic agents and rehabilitation support, particularly in light of the substantial economic burden demonstrated in national studies. Personalised care in BPI is contingent upon early lesion characterisation and the customisation of reconstruction according to injury patterns and functional priorities. Advances in microsurgery now enable surgeons to utilise donor nerves with favourable axon counts and reduced distances to restore essential functions, such as elbow flexion. Recent investigations into nerve transfer configurations in neonatal brachial plexus birth palsy have indicated improved shoulder function, facilitating the tailoring of procedures based on age, lesion severity, and functionality (94–96).

Regenerative therapies are a swiftly evolving area of research. Initial preclinical and clinical studies indicate that MSCs, neural stem/progenitor cells, and other cell types can promote axonal regeneration, remyelination, and the survival of motor neurons in models of peripheral nerve and BPI (97). Systematic reviews have documented an increase in axon numbers and the density of myelinated fibres following the transplantation of stem cells derived from bone marrow or umbilical cords. In human cases, autologous cell injections have led to enhanced strength in individuals with traumatic BPI. A clinical pilot study that combined intercostal nerve transfers with umbilical cord-derived MSCs showed improvements in quality of life and function, although histological regeneration was limited, and the long-term outcomes remain unclear (98,99). Emerging strategies are focused on enhancing the repair environment and axon–target connections through the use of bioengineered conduits, growth factor delivery, extracellular vesicle therapies, and gene-based approaches. Additionally, researchers are investigating neuroprosthetic and brain–computer interface technologies to enable patients with complete BPI to control external orthoses using myoelectric or cortical signals (100–102).

12.8 CEREBRAL ARTERIOSCLEROSIS

Cerebral arteriosclerosis in small brain vessels, known as brain arteriolosclerosis or arteriosclerotic cerebral small vessel disease (aCSVD), is a common age-related vascular pathology causing lacunar stroke, cognitive impairment and gait disturbance (103). Brain arteriolosclerosis describes thickening and narrowing of small penetrating arterioles in deep white matter and subcortical grey nuclei, with loss of smooth muscle cells and hyaline deposition. It belongs to cerebral small vessel disease (CSVD), which includes changes in intracranial vessels that damage brain parenchyma (104).

Cerebral arteriosclerosis remains clinically silent for years, but progressive vessel damage leads to characteristic manifestations. Major consequences include lacunar ischemic stroke, intracerebral haemorrhage, cognitive decline, and gait disturbances. Patients may present with lacunar syndromes (such as pure motor or sensory stroke), subcortical vascular dementia, depression, urinary urgency or incontinence, and Parkinsonian gait (105). Cognitive symptoms involve slowed processing speed, impaired executive function, and reduced attention rather than aphasia or amnesia, typical of cortical disease. Psychiatric changes (depression, emotional lability), abnormal gait, and falls are common in advanced CSVD and correlate with white matter lesions and lacunes. Because these features overlap with neurodegenerative conditions like Alzheimer's or Parkinson's disease, careful clinical and imaging evaluation is essential (104).

CSVD imaging markers such as white matter hyperintensities (WMH), lacunes, enlarged perivascular spaces, and cerebral microbleeds are common in older adults, with WMH in 40%–90% and lacunes in 8%–30% of elderly populations. Brain arteriolosclerosis is prevalent in aging cohorts and strongly associates with vascular cognitive impairment and dementia. Intracranial atherosclerotic disease (ICAD) accounts for cerebral infarction rates of 20–40 per 100,000 individuals globally and is a leading cause of ischemic stroke (106).

Diagnosis is primarily radiological and risk-factor-based. MRI with T2/FLAIR sequences detects WMH, small subcortical infarcts and lacunes, while T2* sequences identify microbleeds and siderosis. WMH appear bilaterally in periventricular and deep white matter, correlating with reduced blood flow and demyelination; brain arteriolosclerosis is a major underlying cause. Lacunes (≈3–15 mm infarcts) must be distinguished from enlarged perivascular spaces, and deep microbleeds often reflect hypertensive arteriolosclerosis rather than CAA (107).

Advanced structural and functional imaging (high-resolution 3T/7T vessel wall MRI, arterial spin labelling perfusion, diffusion tensor imaging) allows in vivo assessment of microvascular structure, perfusion, and white matter integrity, though these remain research tools. Vascular risk assessment (blood pressure, glycaemia, lipids, renal function) and carotid/intracranial ultrasound help characterise systemic involvement, while blood-based biomarkers (inflammatory markers, homocysteine, LpPLA2, cystatin C) and gut microbiota signatures show promise for early detection and risk stratification (108).

Current management focuses on controlling vascular risk factors and preventing stroke and cognitive decline. Hypertension is the strongest modifiable driver; intensive blood pressure lowering in SPRINTMIND reduced WMH progression and lowered mild cognitive impairment, supporting safe BP control. Glycaemic control in type 2 diabetes, treating dyslipidaemia, managing obesity, smoking cessation, physical activity, and sleep optimisation contribute to slowing CSVD progression (109).

Antiplatelet therapy (typically low-dose aspirin) and statins are used per secondary prevention guidelines for ischaemic stroke, although evidence for CSVD subtypes remains limited against microbleed risk. For arteriosclerotic cerebral infarction, vascular imaging and plasma markers like LpPLA2 and cystatin C help guide preventive therapy. Multidisciplinary management, including cognitive rehabilitation, physiotherapy, and mood disorder treatment, is crucial for maintaining function in symptomatic CSVD patients (110).

Personalised therapy for cerebral arteriosclerosis integrates vascular risk profiles, imaging markers, and biological signatures to tailor prevention and intervention. Multiomics studies of the gut–brain axis have identified microbiota alterations in arteriosclerotic CSVD, including depletion of butyrate-producing species like *Faecalibacterium prausnitzii* and *Roseburia intestinalis* and enrichment of Bacteroides and *Proteobacteria taxa*, with shifts in lipopolysaccharide biosynthesis and phenylalanine-tyrosine metabolism. These changes correlate with cognitive performance, inflammation, and MRI markers of CSVD, suggesting that microbiota-based biomarkers and gut-directed interventions could support risk prediction and disease activity modulation (111,112). Novel pharmacological approaches include vasculoprotective and neuroprotective agents evaluated in arteriosclerotic cerebrovascular disease. A multicentre randomised trial in China showed that Ginkgo biloba extract improved traditional Chinese medicine "blood stasis" symptoms and dizziness handicap measures in patients with cerebral arteriosclerosis, showing favourable safety versus controls. Preclinical work with next-generation Ginkgo biloba extracts (nGBE) demonstrates reduced infarct volumes, improved white matter integrity and recovery after cerebral ischaemia, mediated by dampening microglial inflammation and promoting remyelination, supporting evaluation in small-vessel ischaemic disease (113,114).

Drug delivery innovations targeting the brain microvasculature are being explored. A thermoresponsive intranasal gel system for tolperisone has been developed to bypass the blood-brain barrier, achieving rapid gelation at nasal temperatures, mucoadhesion, and sustained drug release with favourable safety in preclinical testing, proposed as a strategy to enhance CNS exposure while limiting systemic toxicity in cerebral arteriosclerosis (115). High-resolution vessel wall imaging, quantitative perfusion MRI, and blood-based biomarkers of CSVD burden may enable risk-stratified treatment targets and monitoring of response in future precision-medicine protocols (116,117).

Integrating vascular genomics, microbiome profiles, imaging and digital cognitive metrics could support machine-learning models that classify CSVD subtypes and predict progression. These tools may transform cerebral arteriosclerosis from a radiological finding to a modifiable target within personalised cerebrovascular prevention (106).

12.9 GUILLAIN–BARRÉ SYNDROME

Guillain–Barré syndrome (GBS) is an acute, typically monophasic, immune-mediated disorder that affects peripheral nerves and spinal nerve roots. Pathologically, it is characterised by demyelination and/or axonal injury, while clinically, it presents as rapidly progressing, symmetrical limb weakness accompanied by diminished or absent deep tendon reflexes (118). In approximately 2/3rd of cases, symptoms emerge 1 to 6 weeks following a preceding infection, most commonly of the respiratory or gastrointestinal tract. Pathogens such as *Campylobacter jejuni, Zika virus,* and *SARS-CoV-2* are strongly implicated, likely due to molecular mimicry between microbial antigens and neuronal gangliosides (119).

GBS typically commences with nonspecific prodromal symptoms, such as fever, upper respiratory or gastrointestinal infection, and malaise. These are followed over several days by symmetrical ascending weakness, initially manifesting in the legs and subsequently progressing to the arms, facial, and bulbar muscles. Patients frequently report paraesthesias, numbness, or neuropathic pain in the distal limbs. Clinical examination often reveals diminished or absent tendon reflexes, which serve to differentiate GBS from central causes of acute weakness (120,121).

Cranial nerve involvement commonly manifests as bilateral facial weakness, ophthalmoplegia, and dysphagia. About 25–30% of patients experience respiratory muscle weakness requiring ventilatory assistance. Many patients exhibit autonomic dysfunction, including blood pressure fluctuations, cardiac arrhythmias, sweating irregularities, and gastrointestinal dysmotility (122). In pediatric patients, the condition may present aggressively, with rapid progression to quadriplegia, cranial nerve palsies, and an elevated risk of respiratory failure and dysautonomia. The clinical trajectory involves an initial progressive phase lasting 2–4 weeks, followed by a plateau, and then a recovery phase extending over months to years, with variability in residual deficits (123).

GBS is an infrequent yet globally occurring condition, with an estimated annual incidence ranging from approximately 0.8–2 cases per 100,000 person-years, equating to around 100,000 new cases worldwide each year. The incidence of GBS increases with age, from about 0.5 per 100,000 in children to roughly 2 per 100,000 in older adults. Additionally, the incidence is approximately 1.5–1.7 times higher in males compared to females, with slight regional and seasonal variations associated with circulating infectious triggers (124).

The diagnosis is predominantly clinical, supported by specific laboratory and neurophysiological findings, and involves the exclusion of similar conditions. The primary criteria encompass rapidly progressing, relatively symmetrical limb weakness with diminished or absent reflexes, reaching its peak within four weeks, often following a recent infection, without any alternative explanation such as spinal cord compression. Cerebrospinal fluid typically demonstrates albuminocytologic dissociation, elevated protein levels with normal or slightly increased cell count, from the second week onwards. Nerve conduction studies and electromyography reveal demyelination (slowed conduction, conduction block, prolonged F-waves) and/or axonal damage, which assist in disease subtyping. MRI of the spine may show gadolinium enhancement of nerve roots, particularly in children, but it is primarily utilised to exclude other causes (125–127). The International GBS Outcome Study (IGOS) has standardised data and biosample collection in over 1800 patients, facilitating the refinement of diagnostic criteria, prognostic models, and biomarker discovery to enhance the understanding of individual disease courses (128,129).

Managing GBS involves the use of immunotherapy, supportive care, and rehabilitation. Patients with suspected cases should be admitted to the hospital for monitoring of respiratory function, autonomic signs, and bulbar involvement, with a transfer to intensive care if necessary. Essential supportive measures include thromboprophylaxis, pressure care, nutrition, bladder and bowel management, and physiotherapy to prevent complications and maintain functionality (129).

Two immunomodulatory therapies, IVIG (0.4 g/kg/day for 5 days) and plasma exchange, are first-line treatments of equal efficacy when initiated within two weeks of onset. Both reduce time to independent ambulation and decrease mechanical ventilation duration. Corticosteroids, alone or with IVIG

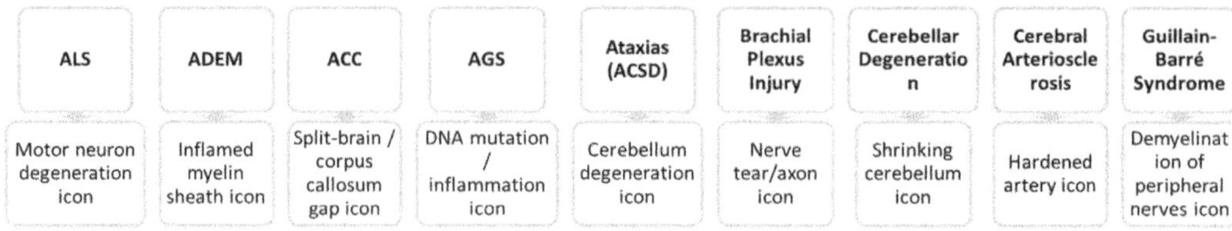

FIGURE 12.2 Overview of rare neurological disorders and personalised therapeutic approaches

or plasma exchange, have not shown consistent benefits. About 40–50% of patients show no significant improvement within four weeks after IVIG or plasma exchange and need extended supportive care. A second IVIG course has not improved outcomes and increases adverse events (126,127).

Long-term recovery requires comprehensive rehabilitation. Programs encompassing physical, occupational, and speech therapy are essential for addressing muscle strength, preventing contractures, and improving gait, balance, upper limb function, swallowing, and communication (130). Psychological and social support are vital for helping patients and families manage pain, fatigue, mood disorders, and vocational challenges. Studies show most patients achieve independent ambulation within 6–12 months; however, 10–20% continue to experience significant disability, and approximately 5% succumb, particularly among older patients with severe initial weakness, rapid disease progression, ventilation needs, or marked autonomic instability (128,131,132).

Emerging personalised therapies under development include FcRn blockers (efgartigimod) (133) and IgG-degrading enzymes (imlifidase) aimed at expediting the clearance of pathogenic IgG, complement inhibitors (anti-C1q antibody ANX005) designed to mitigate complement-mediated nerve damage, engineered Fc multimers and glyco-optimised IVIG analogues targeting Fcγ receptors, and T-cell-directed agents tailored to specific immunopathological conditions (134). Clinical data from studies involving efgartigimod and ANX005 indicate accelerated strength recovery and reduced ventilation durations in small cohorts, thereby justifying the initiation of larger phase II and III trials (135). Concurrently, the IGOS biobank is correlating serological and electrophysiological subtypes with treatment responses to facilitate future stratified trials. Preliminary case series and a systematic review of robot-assisted therapy in GBS suggest that exoskeleton-based limb training may enhance motor recovery, with a double-blind randomised controlled trial currently underway to ascertain its efficacy within personalised rehabilitation programs (136–138). Figure 12.2 shows the overview of rare neurological disorders and personalised therapeutic approaches.

12.10 CONCLUSION

Rare neurological disorders constitute a severe clinical challenge due to their variability, low prevalence, and generally restricted therapy options. The common thread is the complexity of underlying mechanisms and the necessity for therapies tailored to individual patients. Advances in genetics, neuroimaging, biomarker discovery, and machine-learning–based analysis are altering the landscape by enabling faster diagnosis and richer phenotypic characterisation. These resources serve as the basis for individualised treatment plans that go beyond treating symptoms to address the underlying biological processes causing each illness.

Approaches like precision medicine range from gene-specific innovation in AGS, neuromodulation for ACC-associated cognitive dysfunction, cerebellar degeneration, and stem cell-based strategies for BPI. highlight the evolving and growing potential of customised treatments. The evolution of AI in healthcare also plays a crucial role in it, like AI-guided cognitive training, digital therapies, remote patients, and rehabilitation, which also help for patient monitoring and hence a more comprehensive and patient-centered approach to healthcare. Involvement of pharmacogenomics also has great potential to enhance the effectiveness and reduce the side effects, especially metabolic profiles like ALS and cerebellar ataxia.

While significant advancements have been made, the sector continues to encounter challenges including limited clinical trial populations, high costs of innovative medicines, and the need for longitudinal outcome data. Nonetheless, the trajectory is clear: personalised medicine is redefining how rare neurological disorders are understood and treated. Continued interdisciplinary research, collaborative worldwide networks, and patient-centred innovation will be vital for converting emerging technologies into real clinical improvements and increasing quality of life for impacted individuals.

FUNDING

Not Applicable

DECLARATION OF GENERATIVE AI AND AI-ASSISTED TECHNOLOGIES IN THE WRITING PROCESS

During the preparation of this work, the author(s) used Quill Bot and Grammarly SERVICE to improve scientific writing and the English language. After using this tool/service, the author(s) reviewed and edited the content as needed and take(s) full responsibility for the content of the publication.

REFERENCES

1. 10 Rarest Brain (Neurological) Diseases in the World - Rarest.org. [accessed 20 Nov 2025] Available from: https://rarest.org/people/brain-diseases
2. Liu F, Li R, Zhu Z, Yang Y, Lu F. Current developments of gene therapy in human diseases. MedComm (Beijing). 2024;5(9):e645. doi: 10.1002/mco2.645
3. Rayman G, Akpan A, Cowie M, Evans R, Patel M, Posporelis S, et al. Managing patients with comorbidities: Future models of care. Future Healthc J. 2022;9(2):101. doi: 10.7861/FHJ.2022-0029
4. van Eeghen AM, Bruining H, Wolf NI, Bergen AA, Houtkooper RH, van Haelst MM, et al. Personalized medicine for rare neurogenetic disorders: Can we make it happen? Cold Spring Harb Mol Case Stud. 2022;8(2):a006200. doi: 10.1101/mcs.a006200.
5. Assadourian AA, Martinez-Agosto JA. Precision diagnostic and therapeutic interventions in rare genetic neurodevelopmental disorders. Pediat Res. 2025. 2025;1–12. doi: 10.1038/s41390-025-04611-y
6. Amyotrophic Lateral Sclerosis (ALS): What It Is & Symptoms. [accessed 1 Dec 2025] Available from: https://my.clevelandclinic.org/health/diseases/16729-amyotrophic-lateral-sclerosis-als
7. Feldman EL, Goutman SA, Petri S, Mazzini L, Savelieff MG, Shaw PJ, et al. Amyotrophic lateral sclerosis. The Lancet. 2022;400(10360):1363–80. doi: 10.1016/S0140-6736(22)01272-7
8. Masrori P, Van Damme P. Amyotrophic lateral sclerosis: a clinical review. Eur J Neurol. 2020;27(10):1918. doi: 10.1111/ENE.14393
9. Rizea RE, Corlatescu AD, Costin HP, Dumitru A, Ciurea AV. Understanding amyotrophic lateral sclerosis: Pathophysiology, diagnosis, and therapeutic advances. Int J Mol Sci. 2024, Vol 25, Page 9966. 2024;25(18):9966. doi: 10.3390/IJMS25189966
10. Rizuan A, Shenoy J, Mohanty P, dos Passos PM, Mercado Ortiz JF, Bai L, et al. Structural details of helix-mediated multimerization of the conserved region of TDP-43 C-terminal domain. Nat Commun. 2025;16(1):10528. doi: 10.1038/s41467-025-65546-w
11. Al-Khayri JM, Ravindran M, Banadka A, Vandana CD, Priya K, Nagella P, et al. Amyotrophic lateral sclerosis: Insights and new prospects in disease pathophysiology, biomarkers and therapies. Pharmaceuticals. 2024;17(10):1391. doi: 10.3390/PH17101391
12. Wei Y, Zhong S, Yang H, Wang X, Lv B, Bian Y, et al. Current therapy in amyotrophic lateral sclerosis (ALS): A review on past and future therapeutic strategies. Eur J Med Chem. 2024;272:116496. doi: 10.1016/J.EJMECH.2024.116496
13. Miller RG, Mitchell JD, Moore DH. Riluzole for amyotrophic lateral sclerosis (ALS)/motor neuron disease (MND). Cochrane Database Syst Rev. 2012;2012(3):CD001447. doi: 10.1002/14651858.CD001447.PUB3
14. Meyer T, Funke A, Münch C, Kettemann D, Maier A, Walter B, et al. Real world experience of patients with amyotrophic lateral sclerosis (ALS) in the treatment of spasticity using tetrahydrocannabinol:cannabidiol (THC:CBD). BMC Neurol. 2019;19(1). doi: 10.1186/S12883-019-1443-Y

15. An Organ-on-a-Chip Model of ALS Will Help Uncover Early Disease Characteristics and Find Potential Therapies | National Center for Advancing Translational Sciences. [accessed 1 Dec 2025] Available from: https://ncats.nih.gov/news-events/news/an-organ-on-a-chip-model-of-als-will-help-uncover-early-disease-characteristics-and-find-potential-therapies
16. Yadav A, Matson KJE, Li L, Hua I, Petrescu J, Kang K, et al. A cellular taxonomy of the adult human spinal cord. Neuron. 2023;111(3):328–344.e7. doi: 10.1016/J.NEURON.2023.01.007
17. Anilkumar AC, Foris LA, Tadi P. Acute Disseminated Encephalomyelitis. StatPearls. 2024 Jan 26.
18. Rare Neuroimmune Association S. Acute disseminated encephalomyelitis (ADEM) Factsheet.
19. Pohl D, Alper G, Van Haren K, Kornberg AJ, Lucchinetti CF, Tenembaum S, et al. Acute disseminated encephalomyelitis: Updates on an inflammatory CNS syndrome. Neurology. 2016;87(9):S38–45. doi: 10.1212/WNL.0000000000002825;REQUESTEDJOURNAL:JOURNAL:WNL;JOURNAL:JOURNAL:WNL;WGROUP:STRING:PUBLICATION
20. Paolilo RB, Deiva K, Neuteboom R, Rostásy K, Lim M. Acute disseminated encephalomyelitis: Current perspectives. Children. 2020;7(11):210. doi: 10.3390/CHILDREN7110210
21. Young NP, Weinshenker BG, Lucchinetti CF. Acute Disseminated Encephalomyelitis: Current Understanding and Controversies. doi: 10.1055/s-2007-1019130
22. Li K, Li M, Wen L, Wang Q, Ding X, Wang J. Clinical presentation and outcomes of acute disseminated encephalomyelitis in adults worldwide: Systematic review and meta-analysis. Front Immunol. 2022;13:870867. https://doi.org/10.3389/fimmu.2022.870867
23. Bunyan RF, Tang J, Weinshenker B. Acute demyelinating disorders: Emergencies and management. Neurol Clin. 2012;30(1):285–307. doi: 10.1016/J.NCL.2011.09.013
24. Thakur KT, Legouy C, Sonneville R. Treating acute encephalitis. Intens Care Med. 2024 50:11. 2024;50(11):1916–9. doi: 10.1007/S00134-024-07569-3
25. Anilkumar AC, Foris LA, Tadi P. Acute Disseminated Encephalomyelitis. StatPearls. 2024 Jan 26;
26. Zolkefli Y. Managing aggressive behaviour in healthcare: Balancing of patients and staff interests. Malays J Med Sci. 2024;31(3):252. doi: 10.21315/mjms2024.31.3.20.
27. Goetz LH, Schork NJ. Personalized medicine: Motivation, challenges and progress. Fertil Steril. 2018;109(6):952. doi: 10.1016/J.FERTNSTERT.2018.05.006
28. Stumpp NE, Sauer-Zavala S. Evidence-based strategies for treatment personalization: A review. Cogn Behav Pract. 2022;29(4):902–13. doi: 10.1016/J.CBPRA.2021.10.004
29. Agenesis of the Corpus Callosum: Symptoms & Treatment. [accessed 1 Dec 2025] Available from: https://my.clevelandclinic.org/health/diseases/6029-agenesis-of-the-corpus-callosum-acc
30. Rashid R, Bach A, Gebb J, Soni S, Patel V, Saha K, et al. Agenesis of corpus callosum: A clinical study of complete versus partial agenesis in a 20-year retrospective cohort. J Child Neurol. 2025 Jan 1; doi: 10.1177/08830738251345652;WGROUP:STRING:PUBLICATION
31. Hofman J, Hutny M, Sztuba K, Paprocka J. Corpus callosum agenesis: An insight into the etiology and spectrum of symptoms. Brain Sci. 2020;10(9):625. doi: 10.3390/BRAINSCI10090625
32. Rotmensch S, Monteagudo A. Agenesis of the corpus callosum. Am J Obstet Gynecol. 2020;223(6): B17–22. doi: 10.1016/j.ajog.2020.08.179
33. Chiappedi M, Bejor M. Corpus callosum agenesis and rehabilitative treatment. Ital J Pediatr. 2010;36(1):64. doi: 10.1186/1824-7288-36-64
34. Rosewater JB, Zaydlin M, McLeod-Bryant SA. Agenesis of the corpus callosum in a patient with schizophrenia. Cureus. 2021;13(6):e16058. doi: 10.7759/CUREUS.16058
35. Lieb JM, Ahlhelm FJ. [Agenesis of the corpus callosum]. Radiologe. 2018;58(7):636–45. doi: 10.1007/S00117-018-0388-2
36. Singh S, Garge S. Agenesis of the corpus callosum. J Pediatr Neurosci. 2010;5(1):83. doi: 10.4103/1817-1745.66662
37. Bedeschi MF, Bonaglia MC, Grasso R, Pellegri A, Garghentino RR, Battaglia MA, et al. Agenesis of the corpus callosum: Clinical and genetic study in 63 young patients. Pediatr Neurol. 2006;34(3):186–93. doi: 10.1016/j.pediatrneurol.2005.08.008
38. Goldstein A, Covington BP, Mahabadi N, Mesfin FB. Neuroanatomy, Corpus Callosum. StatPearls. 2023 Apr 3; Available from: https://www.ncbi.nlm.nih.gov/books/NBK448209/
39. Aydin H, Bolat H. The rare syndrome Aicardi–Goutières 4: A case report and literature review. Dev Neurobiol [Internet]. 2025 Apr 1 [cited 2025 Dec 1];85(2):e22965. https://pubmed.ncbi.nlm.nih.gov/40183101/
40. Braidotti S, Ferraro RM, Franca R, Genova E, Giambuzzi F, Mancini A, et al. Pharmacological evaluation of drug therapies in Aicardi-Goutières syndrome: Insights from patient-derived neural stem cells. Front Pharmacol [Internet]. 2025 Mar 20 [cited 2025 Dec 1];16:1549183. Available from: https://www.frontiersin.org/journals/pharmacology/articles/10.3389/fphar.2025.1549183/full.

41. Adang L, Gavazzi F, De Simone M, Fazzi E, Galli J, Koh J, et al. Developmental outcomes of Aicardi Goutières syndrome. J Child Neurol [Internet]. 2020 Jan 1 [cited 2025 Dec 1];35(1):7–16. Available from: https://pubmed.ncbi.nlm.nih.gov/31559893/
42. Piccoli C, Bronner N, Gavazzi F, Dubbs H, De Simone M, De Giorgis V, et al. Late-onset Aicardi-Goutières syndrome: A characterization of presenting clinical features. Pediatr Neurol [Internet]. 2021 Feb 1 [cited 2025 Dec 1];115:1–6. Available from: https://pubmed.ncbi.nlm.nih.gov/33307271/
43. Liu A, Ying S. Aicardi–Goutières syndrome: A monogenic type I interferonopathy. Scand J Immunol [Internet]. 2023 Oct 1 [cited 2025 Dec 1];98(4):e13314. Available from: https://pubmed.ncbi.nlm.nih.gov/37515439/
44. Tasharrofi B, Karimzadeh P, Asadollahi M, Hasani S, Heidari M, Keramatipour M. Aicardi-Goutières syndrome type 1: A novel missense variant and review of the mutational spectrum. Iran J Child Neurol [Internet]. 2024 May 22 [cited 2025 Dec 1];18(3):117–29. Available from: https://journals.sbmu.ac.ir/ijcn/article/view/43274
45. Nguengang Wakap S, Lambert DM, Olry A, Rodwell C, Gueydan C, Lanneau V, et al. Estimating cumulative point prevalence of rare diseases: Analysis of the Orphanet database. Eur J Hum Genet 2019 28:2 [Internet]. 2019 Sep 16 [cited 2025 Dec 1];28(2):165–73. Available from: https://www.nature.com/articles/s41431-019-0508-0
46. Møller RS, Zhao L, Shoaff JR, Duno M, Andersen BN, Nguyen V, et al. Incidence of Aicardi-Goutières syndrome and KCNT1-related epilepsy in Denmark. Mol Genet Metab Rep [Internet]. 2022 Dec 1 [cited 2025 Dec 1];33:100924. Available from: https://pmc.ncbi.nlm.nih.gov/articles/PMC9574483/
47. Videira G, Malaquias MJ, Laranjinha I, Martins R, Taipa R, Magalhães M Diagnosis of Aicardi-Goutières syndrome in adults: A case series. Mov Disord Clin Pract [Internet]. 2020 Apr 1 [cited 2025 Dec 1];7(3):303–7. Available from: https://pubmed.ncbi.nlm.nih.gov/32258229/
48. Garau J, Cavallera V, Valente M, Tonduti D, Sproviero D, Zucca S, et al. Molecular genetics and interferon signature in the Italian Aicardi Goutières syndrome cohort: Report of 12 new cases and literature review. J Clin Med 2019;8:750 [Internet]. 2019 May 26 [cited 2025 Dec 1];8(5):750. Available from: https://www.mdpi.com/2077-0383/8/5/750/htm
49. Dell'Isola GB, Dini G, Culpepper KL, Portwood KE, Ferrara P, Di Cara G, et al. Clinical spectrum and currently available treatment of type I interferonopathy Aicardi–Goutières syndrome. World Journal of Pediatrics. 2023 Jul 1;19(7):635. doi:10.1007/S12519-022-00679-2 PubMed PMID: 36650407
50. Al-Agha AE, Bawahab NS, Nagadi SA, Alghamdi SA, Felemban DA, Milyani AA. Endocrinopathies complicating transfusion-dependent hemoglobinopathy. Saudi Med J [Internet]. 2020 Feb 1 [cited 2025 Dec 1];41(2):138. Available from: https://pmc.ncbi.nlm.nih.gov/articles/PMC7841625/
51. Tonduti D, Fazzi E, Badolato R, Orcesi S Novel and emerging treatments for Aicardi-Goutières syndrome. Expert Rev Clin Immunol [Internet]. 2020 Feb 1 [cited 2025 Dec 1];16(2):189–98. Available from: https://www.tandfonline.com/doi/abs/10.1080/1744666X.2019.1707663
52. Crow YJ, Stetson DB. The type I interferonopathies: 10 years on. Nature Reviews Immunology 2021 22:8 [Internet]. 2021 Oct 20 [cited 2025 Dec 1];22(8):471–83. Available from: https://www.nature.com/articles/s41577-021-00633-9
53. Politano D, Tonduti D, Battini R, Fazzi E, Orcesi S Exploring emerging JAK inhibitors in the treatment of Aicardi–Goutières syndrome. Expert Opin Emerg Drugs [Internet]. 2025 Jan 2 [cited 2025 Dec 1];30(1):21–39. Available from: https://www.tandfonline.com/doi/abs/10.1080/14728214.2024.2445508
54. Antonioli L, Armuzzi A, Fantini MC, Fornai M. JAK inhibitors: An evidence-based choice of the most appropriate molecule. Front Pharmacol. 2024 Oct 29;15:1494901.
55. Crow YJ, Shetty J, Livingston JH. Treatments in Aicardi–Goutières syndrome. Dev Med Child Neurol [Internet]. 2020 Jan 1 [cited 2025 Dec 1];62(1):42–7. Available from: https://pubmed.ncbi.nlm.nih.gov/31175662/
56. Dell'Isola GB, Dini G, Culpepper KL, Portwood KE, Ferrara P, Di Cara G, et al. Clinical spectrum and currently available treatment of type I interferonopathy Aicardi–Goutières syndrome. World J Pediatr 2023 19:7 [Internet]. 2023 Jan 17 [cited 2025 Dec 1];19(7):635–43. Available from: https://link.springer.com/article/10.1007/s12519-022-00679-2
57. Li T, Yum S, Wu J, Li M, Deng Y, Sun L, et al. cGAS activation in classical dendritic cells causes autoimmunity in TREX1-deficient mice. Proc Natl Acad Sci U S A [Internet]. 2024 Sep 17 [cited 2025 Dec 1];121(38):e2411747121. Available from: https://pubmed.ncbi.nlm.nih.gov/39254994/
58. Henrickson M, Wang H. Tocilizumab reverses cerebral vasculopathy in a patient with homozygous SAMHD1 mutation. Clin Rheumatol 2017 36:6 [Internet]. 2017 Mar 13 [cited 2025 Dec 1];36(6):1445–51. Available from: https://link.springer.com/article/10.1007/s10067-017-3600-2

59. Arévalo Gil ME, Caballero González M, Linero Ríos NA. Successful electroconvulsive therapy in Aicardi-Goutières syndrome presenting psychiatric symptoms: An unprecedented clinical case. J ECT [Internet]. 2025 Jun 1 [cited 2025 Dec 1];41(2):143–4. Available from: https://journals.lww.com/ectjournal/fulltext/2025/06000/successful_electroconvulsive_therapy_in.12.aspx
60. Sarva H, Shanker VL. Treatment options in degenerative cerebellar ataxia: A systematic review. Mov Disord Clin Pract [Internet]. 2014 Dec 1 [cited 2025 Dec 1];1(4):291. Available from: https://pmc.ncbi.nlm.nih.gov/articles/PMC6183008/
61. Poston KL, Rosenthal LS. Neurodegenerative cerebellar ataxia. Continuum (N Y) [Internet]. 2022 Oct 1 [cited 2025 Dec 1];28(5):1409–34. Available from: https://continuum.aan.com/doi/abs/10.1212/CON.0000000000001180
62. Cui ZT, Mao ZT, Yang R, Li JJ, Jia SS, Zhao JL et al. Spinocerebellar ataxias: From pathogenesis to recent therapeutic advances. Front Neurosci. 2024;18:1422442.
63. Ilg W, Bastian AJ, Boesch S, Burciu RG, Celnik P, Claaßen J, et al. Consensus paper: Management of degenerative cerebellar disorders. The Cerebellum 2013 13:2 [Internet]. 2013 Nov 13 [cited 2025 Dec 1];13(2):248–68. Available from: https://link.springer.com/article/10.1007/s12311-013-0531-6
64. Tamaš O, Kostić M, Marić G, Milovanović A, Janković M, Salak Đokić B, et al. Neuropsychiatric manifestations of degenerative cerebellar ataxia. Brain Sci [Internet]. 2024 Oct 1 [cited 2025 Dec 1];14(10):1003. Available from: https://pmc.ncbi.nlm.nih.gov/articles/PMC11506162/
65. Roberts LJ, McVeigh M, Seiderer L, Harding IH, Corben LA, Delatycki M, et al. Overview of the clinical approach to individuals with cerebellar ataxia and neuropathy. Neurol Genet [Internet]. 2022 Oct 28 [cited 2025 Dec 1];8(5):e200021. Available from: https://doi.org/10.1212/NXG.0000000000200021
66. Salem IH, Beaudin M, Klein CJ, Dupré N. Treatment and management of autosomal recessive cerebellar ataxias: Current advances and future perspectives. CNS Neurol Disord Drug Targets. 2022 Apr 20;22(5):678–97.
67. Coarelli G, Wirth T, Tranchant C, Koenig M, Durr A, Anheim M. The inherited cerebellar ataxias: An update. J Neurol. 2023;270(1):208–22.
68. Takao H, Iijima H, Odagiri R, Hayakawa I, Ogimi C. Acute cerebellar ataxia during acute COVID-19: A case series and review of the literature. J Infect Chemother. 2023;29(8):792–5.
69. Lee A. Omaveloxolone: First approval. Drugs. 2023;83(8):725–9.
70. Radmard S, Zesiewicz TA, Kuo SH. Evaluation of cerebellar ataxic patients. Neurol Clin. 2023;41(1):21–44.
71. Lynch DR, Perlman SL, Meier T. A phase 3, double-blind, placebo-controlled trial of idebenone in friedreich ataxia. Arch Neurol [Internet]. 2010 Aug [cited 2025 Dec 1];67(8):941–7. Available from: https://pubmed.ncbi.nlm.nih.gov/20697044/
72. Bushart DD, Murphy GG, Shakkottai VG. Precision medicine in spinocerebellar ataxias: Treatment based on common mechanisms of disease. Ann Transl Med [Internet]. 2016 Jan 1 [cited 2025 Dec 1];4(2):25. Available from: https://pmc.ncbi.nlm.nih.gov/articles/PMC4731605/
73. Singh A, Babu S, Phan M, Yuan SH CRISPR/Cas9-based protocol for precise genome editing in induced pluripotent stem cells. Bio Protoc [Internet]. 2024 Dec 20 [cited 2025 Dec 1];14(24). Available from: https://pubmed.ncbi.nlm.nih.gov/39735291/
74. Mout R, Ray M, Lee YW, Scaletti F, Rotello VM. In vivo delivery of CRISPR/Cas9 for therapeutic Gene editing: Progress and challenges. Bioconjug Chem [Internet]. 2017 Apr 19 [cited 2025 Dec 1];28(4):880. Available from: https://pmc.ncbi.nlm.nih.gov/articles/PMC5846329/
75. De Masi C, Spitalieri P, Murdocca M, Novelli G, Sangiuolo F Application of CRISPR/Cas9 to human-induced pluripotent stem cells: From Gene editing to drug discovery. Hum Genomics [Internet]. 2020 Jun 26 [cited 2025 Dec 1];14(1):25. Available from: https://pmc.ncbi.nlm.nih.gov/articles/PMC7318728/
76. Scoles DR, Meera P, Schneider MD, Paul S, Dansithong W, Figueroa KP, et al. Antisense oligonucleotide therapy for spinocerebellar ataxia type 2. Nature [Internet]. 2017 Apr 20 [cited 2025 Dec 1];544(7650):362. Available from: https://pmc.ncbi.nlm.nih.gov/articles/PMC6625650/
77. Ding Y, Zhang Y, Liu X. Combinational treatments of RNA interference and extracellular vesicles in the spinocerebellar ataxia. Front Mol Neurosci. 2022 Oct 13;15:1043947.
78. Lee GB, Park SM, Jung UJ, Kim SR. The potential of mesenchymal stem cells in treating spinocerebellar ataxia: Advances and future directions. Biomedicines [Internet]. 2024 Nov 1 [cited 2025 Dec 1];12(11):2507. Available from: https://pmc.ncbi.nlm.nih.gov/articles/PMC11591855/
79. Tsai PJ, Yeh CC, Huang WJ, Min MY, Huang TH, Ko TL, et al. Xenografting of human umbilical mesenchymal stem cells from Wharton's jelly ameliorates mouse spinocerebellar ataxia type 1. Transl Neurodegener [Internet]. 2019 Sep 5 [cited 2025 Dec 1];8(1):29. Available from: https://pmc.ncbi.nlm.nih.gov/articles/PMC6727337/

80. Correia JS, Duarte-Silva S, Salgado AJ, Maciel P Cell-based therapeutic strategies for treatment of spinocerebellar ataxias: An update. Neural Regen Res [Internet]. 2022 Jun 1 [cited 2025 Dec 1];18(6):1203. Available from: https://pmc.ncbi.nlm.nih.gov/articles/PMC9838137/
81. Robinson KJ, Yuan K, Plenderleith SK, Watchon M, Laird AS. A novel calpain inhibitor compound has protective effects on a zebrafish model of spinocerebellar ataxia type 3. Cells [Internet]. 2021 Oct 1 [cited 2025 Dec 1];10(10):2592. Available from: https://pmc.ncbi.nlm.nih.gov/articles/PMC8533844/
82. V Azquez-Mojena Y, Le On-Arcia K, Gonz Alez-Zaldivar Y, Rodríguez-Labrada R, Vel Azquez-Pérez L. Gene Therapy for Polyglutamine Spinocerebellar Ataxias: Advances, Challenges, and Perspectives. Mov Disord. 2021 Dec;36(12):2731–2744. doi: 10.1002/mds.28819
83. Brooker SM, Edamakanti CR, Akasha SM, Kuo SH, Opal P Spinocerebellar ataxia clinical trials: Opportunities and challenges. Ann Clin Transl Neurol [Internet]. 2021 Jul 1 [cited 2025 Dec 1];8(7):1543. Available from: https://pmc.ncbi.nlm.nih.gov/articles/PMC8283160/
84. Borbolla-Jiménez FV, García-Aguirre IA, Del Prado-Audelo ML, Hernández-Hernández O, Cisneros B, Leyva-Gómez G, et al. Development of a polymeric pharmacological nanocarrier system as a potential therapy for spinocerebellar ataxia type 7. Cells [Internet]. 2023 Dec 1 [cited 2025 Dec 1];12(23). Available from: https://pubmed.ncbi.nlm.nih.gov/38067163/
85. Luo TD, Levy ML, Li Z Brachial Plexus Injuries. Nerves and Nerve Injuries: Pain, Treatment, Injury, Disease and Future Directions: Vol 2 [Internet]. 2023 Feb 6 [cited 2025 Dec 1];2:681–706. Available from: https://www.ncbi.nlm.nih.gov/books/NBK482305/
86. Brachial Plexus Injuries - OrthoInfo - AAOS [Internet]. [cited 2025 Dec 1]. Available from: https://orthoinfo.aaos.org/en/diseases--conditions/brachial-plexus-injuries/
87. Brachial Plexus Injury: What It Is, Symptoms, Treatment & Types [Internet]. [cited 2025 Dec 1]. Available from: https://my.clevelandclinic.org/health/diseases/22822-brachial-plexus-injury
88. Brachial Plexus Injuries - Trauma - Orthobullets [Internet]. [cited 2025 Dec 1]. Available from: https://www.orthobullets.com/trauma/1008/brachial-plexus-injuries
89. Boyle A, Karia C, Wade RG, Lecky F, Yates D, Quick TJ, et al. The epidemiology of traumatic brachial plexus injuries in England and Wales—A 32-year review. JBJS Open Access [Internet]. 2025 Mar 18 [cited 2025 Dec 1];10(1):e24.00105. Available from: https://pmc.ncbi.nlm.nih.gov/articles/PMC11905974/
90. Cho ÁB, Guerreiro AC, Henrique C, Ferreira V, Kiyohara LY, Sorrenti L Epidemiological study of traumatic brachial plexus injuries estudo epidemiológico das lesões traumáticas do plexo braquial. Acta Ortop Bras. 2020;28(1):16–24.
91. Lin JS, Tabatabaei K, Bishop AT, Spinner RJ, Pulos NA, Shin AY. Demographics and etiologies of patients evaluated in a multidisciplinary brachial plexus clinic: A ten-year experience. World Neurosurg [Internet]. 2025 Jul 1 [cited 2025 Dec 1];199:124081. Available from: https://www.sciencedirect.com/science/article/pii/S1878875025004371
92. Bagheri A, Akbari H, Akbari P Evaluation and diagnostic-treatment approaches of brachial plexus injuries in adults. World J Plast Surg [Internet]. 2024 Nov 1 [cited 2025 Dec 1];13(3):41. Available from: https://pmc.ncbi.nlm.nih.gov/articles/PMC11629760/
93. Weekes MA, McGhee C, Miller C, Burahee A, Rouse I, Rajaratnam V, et al. Scoping review of the published guidelines for the management of traumatic brachial plexus injuries. Cureus [Internet]. 2025 Feb 19 [cited 2025 Dec 1];17(2). Available from: https://pubmed.ncbi.nlm.nih.gov/40125159/
94. Genedy MKA, Salama EY, Elsaadany MA, AbdelWahab MAF, Amin AF, Lashin AA, et al. Nerve transfer to musculocutaneous for elbow flexion restoration in brachial plexus injury (ulnar and/or median vs. intercostal): A systematic review and meta-analysis of comparative studies. Acta Neurochir (Wien) [Internet]. 2025 Dec 1 [cited 2025 Dec 1];167(1):229. Available from: https://pmc.ncbi.nlm.nih.gov/articles/PMC12380866/
95. Moisset X Neuropathic pain: Evidence based recommendations. Presse Med [Internet]. 2024 Jun 1 [cited 2025 Dec 1];53(2):104232. Available from: https://www.sciencedirect.com/science/article/pii/S0755498224000101
96. Emamhadi M, Behzadnia H, Zamanidoust M, Baghi I, Ebrahimian R, Emamhadi R, et al. Intercostal or ulnar nerve: Which donor nerve is to be used for reanimation of elbow flexion? Musculoskelet Surg 2020 105:2 [Internet]. 2020 Mar 2 [cited 2025 Dec 1];105(2):183–8. Available from: https://link.springer.com/article/10.1007/s12306-020-00653-z
97. Aldali F, Yang Y, Deng C, Li X, Cao X, Xu J, et al. Induced pluripotent stem cell-derived exosomes promote peripheral nerve regeneration in a rat sciatic nerve crush injury model: A safety and efficacy study. Cells [Internet]. 2025 Apr 1 [cited 2025 Dec 1];14(7):529. Available from: https://pmc.ncbi.nlm.nih.gov/articles/PMC11989054/
98. Myckatyn TM, Mackinnon SE, McDonald JW. Stem cell transplantation and other novel techniques for promoting recovery from spinal cord injury. Transpl Immunol [Internet]. 2004 Apr 1 [cited 2025 Dec 1];12(3–4):343–58. Available from: https://www.sciencedirect.com/science/article/abs/pii/S0966327403001151

99. Huang H, Chen L, Mao G, Sharma HS. Clinical neurorestorative cell therapies: Developmental process, current state and future prospective. J Neurorestoratol [Internet]. 2020 Jun 5 [cited 2025 Dec 1];8(2):61–82. Available from: https://www.sciencedirect.com/science/article/pii/S2324242622000730
100. Sousa JPM, Deus IA, Monteiro CF, Custódio CA, Gil J, Papadimitriou L, et al. Amniotic membrane-derived multichannel hydrogels for neural tissue repair. Adv Healthc Mater [Internet]. 2024 Oct 28 [cited 2025 Dec 1];13(27):2400522. Available from: https://pubmed.ncbi.nlm.nih.gov/38989725/
101. Khodve G, Banerjee S, Kumari M, Ravichandiran V, Banerjee S, Roy S Exploration of biomaterial and stem cell-based strategies for promoting neuronal regeneration and creating engineered 3D in-vitro disease models. J Transl Med 2025 23:1 [Internet]. 2025 Oct 30 [cited 2025 Dec 1];23(1):1197-. Available from: https://link.springer.com/article/10.1186/s12967-025-07266-9
102. Zhang M, An H, Zhang F, Jiang H, Wan T, Wen Y, et al. Prospects of using chitosan-based biopolymers in the treatment of peripheral nerve injuries. Int J Mol Sci [Internet]. 2023 Aug 1 [cited 2025 Dec 1];24(16):12956. Available from: https://pmc.ncbi.nlm.nih.gov/articles/PMC10454829/
103. Blevins BL, Vinters HV, Love S, Wilcock DM, Grinberg LT, Schneider JA, et al. Brain arteriolosclerosis. Acta Neuropathol [Internet]. 2020 Jan 1 [cited 2025 Dec 1];141(1):1. Available from: https://pmc.ncbi.nlm.nih.gov/articles/PMC8503820/
104. Ren B, Tan L, Song Y, Li D, Xue B, Lai X, et al. Cerebral small vessel disease: Neuroimaging features, biochemical markers, influencing factors, pathological mechanism and treatment. Front Neurol [Internet]. 2022 Jun 14 [cited 2025 Dec 1];13:843953. Available from: www.frontiersin.org
105. Cuadrado-Godia E, Dwivedi P, Sharma S, Santiago AO, Gonzalez JR, Balcells M et al. Cerebral small vessel disease: A review focusing on pathophysiology, biomarkers, and machine learning strategies. J Stroke. 2018;20(3):302–20.
106. Markus HS, Erik de Leeuw F Cerebral small vessel disease: Recent advances and future directions. Int J Stroke [Internet]. 2022 Jan 1 [cited 2025 Dec 1];18(1):4. Available from: https://pmc.ncbi.nlm.nih.gov/articles/PMC9806465/
107. Wu LY, Chai YL, Cheah IK, Chia RSL, Hilal S, Arumugam TV, et al. Blood-based biomarkers of cerebral small vessel disease. Ageing Res Rev [Internet]. 2024 Mar 1 [cited 2025 Dec 1];95:102247. Available from: https://www.sciencedirect.com/science/article/pii/S1568163724000655
108. Han F, Zhai FF, Wang Q, Zhou LX, Ni J, Yao M et al. Prevalence and risk factors of cerebral small vessel disease in a Chinese population-based sample. J Stroke. 2018;20(2):239–46.
109. Melgarejo JD, Maestre GE, Gutierrez J, Thijs L, Mena LJ, Gaona C, et al. Subclinical magnetic resonance imaging markers of cerebral small vessel disease in relation to office and ambulatory blood pressure measurements. Front Neurol [Internet]. 2022 Jul 14 [cited 2025 Dec 1];13:908260. Available from: www.frontiersin.org
110. What Is Cerebral Small Vessel Disease (CSVD)? | American Brain Foundation [Internet]. [cited 2025 Dec 1]. Available from: https://www.americanbrainfoundation.org/julies-story/
111. Shi Y, Zhao E, Li L, Zhao S, Mao H, Deng J, et al. Alteration and clinical potential in gut microbiota in patients with cerebral small vessel disease. Front Cell Infect Microbiol [Internet]. 2023 [cited 2025 Dec 1];13:1231541. Available from: https://pmc.ncbi.nlm.nih.gov/articles/PMC10366612/
112. Huang J, Liu S, Li P, Wei L, Lin G, Lin J, et al. Multi-omics analysis of gut-brain axis reveals novel microbial and neurotransmitter signatures in patients with arteriosclerotic cerebral small vessel disease. Pharmacol Res [Internet]. 2024 Oct 1 [cited 2025 Dec 1];208:107385. Available from: https://www.sciencedirect.com/science/article/pii/S104366182400330X?via%3Dihub
113. Sha R, Tang L, Du Y, Wu S, Shi H, Zou H, et al. Effectiveness and safety of Ginkgo biloba extract (GBE50) in the treatment of dizziness caused by cerebral arteriosclerosis: A multi-center, double-blind, randomized controlled trial. J Tradit Chin Med [Internet]. 2022 Feb 15 [cited 2025 Dec 1];42(1):83–9. Available from: https://pubmed.ncbi.nlm.nih.gov/35294126/
114. Dong W, Gong T, Zhao S, Wen S, Chen Q, Jiang M, et al. A novel extract from Ginkgo biloba inhibits neuroinflammation and maintains white matter integrity in experimental stroke. Neuroscience [Internet]. 2023 Jul 15 [cited 2025 Dec 1];523:7–19. Available from: https://pubmed.ncbi.nlm.nih.gov/37225050/
115. Gaikwad VB, Mahaparale PR, Shaikh MS, Pathan RR, Mahaparale SP. Design and characterization of a nasal thermogelling system loaded with tolperisone for the treatment of cerebral arteriosclerosis through the nose to brain delivery system. J Sol-Gel Sci Technol 2025 116:2 [Internet]. 2025 Sep 1 [cited 2025 Dec 1];116(2):1101–13. Available from: https://link.springer.com/article/10.1007/s10971-025-06906-2
116. Hakim AM. Small vessel disease. Front Neurol. 2019;10.

117. Singh A, Bonnell G, De Prey J, Buchwald N, Eskander K, Kincaid KJ et al. Small-vessel disease in the brain. Am Heart J Plus: Cardiol Res Pract. 2023;27.
118. Smith AG Guillain-Barré syndrome. Contin Lifelong Learn Neurol [Internet]. 2020 Oct 1 [cited 2025 Dec 1];26(5):1184–204. Available from: https://continuum.aan.com/doi/abs/10.1212/CON.0000000000000929
119. Gorson KC. Evolving understanding of Guillain-Barré syndrome pathophysiology and the central role of the classical complement pathway in axonal injury. Front Neurol. 2025 May 14;16:1572949.
120. Nguyen TP, Taylor RS. Guillain-Barre syndrome. StatPearls [Internet]. 2023 Feb 7 [cited 2025 Dec 1]; Available from: https://www.ncbi.nlm.nih.gov/books/NBK532254/
121. Shahrizaila N, Lehmann HC, Kuwabara S Guillain-Barré syndrome. The Lancet [Internet]. 2021 Mar 27 [cited 2025 Dec 1];397(10280):1214–28. Available from: https://www.thelancet.com/journals/lancet/article/PIIS0140-6736(21)00517-1/abstract
122. Bellanti R, Rinaldi S Guillain-Barré syndrome: A comprehensive review. Eur J Neurol [Internet]. 2024 Aug 1 [cited 2025 Dec 1];31(8):e16365. Available from: https://pubmed.ncbi.nlm.nih.gov/38813755/
123. Wachira VK, Peixoto HM, de Oliveira MRF. Systematic review of factors associated with the development of Guillain–Barré syndrome 2007–2017: What has changed? Trop Med Int Health [Internet]. 2019 Feb 1 [cited 2025 Dec 1];24(2):132–42. Available from: 10.1111/tmi.13181
124. Jacobs BC, van den Berg B, Verboon C, Chavada G, Cornblath DR, Gorson KC, et al. International Guillain-Barré syndrome outcome study: Protocol of a prospective observational cohort study on clinical and biological predictors of disease course and outcome in Guillain-Barré syndrome. J Peripher Nerv Syst [Internet]. 2017 Jun 1 [cited 2025 Dec 1];22(2):68–76. Available from: https://pubmed.ncbi.nlm.nih.gov/28406555/
125. Ariatti A, Mazzoli M, Bastia E, Bedin R, Benuzzi F, Agnoletto V, et al. Predictors of poor prognosis in Guillain-Barre' syndrome: A study of 22 years from A single Italian center. J Neurol Sci [Internet]. 2025 Aug 15 [cited 2025 Dec 1];475:123588. Available from: https://www.sciencedirect.com/science/article/abs/pii/S0022510X25002059
126. Shankar PS. Guillain-Barre syndrome. RGUHS J Med Sci [Internet]. 2025 [cited 2025 Dec 1];15(2). Available from: https://journalgrid.com/view/article/rjms/12434368
127. Hughes RAC. Guillain-Barré syndrome: History, pathogenesis, treatment, and future directions. Eur J Neurol [Internet]. 2024 Nov 1 [cited 2025 Dec 1];31(11):e16346. Available from: https://pubmed.ncbi.nlm.nih.gov/38752584/
128. Jacobs B, van den Berg B, Verboon C, Chavada G, Cornblath D, Gorson K, et al. International Guillain-Barré Syndrome Outcome Study (IGOS) Protocol of a prospective observational cohort study on clinical and biological predictors of disease course and outcome in Guillain-Barré syndrome. J Peripher Nerv Syst. 2017 Jun;22(2):68–76. doi: 10.1111/jns.12209.
129. Leonhard SE, Mandarakas MR, Gondim FAA, Bateman K, Ferreira MLB, Cornblath DR, et al. Diagnosis and management of Guillain–Barré syndrome in ten steps. Nat Rev Neurol [Internet]. 2019 Nov 1 [cited 2025 Dec 1];15(11):671–83. Available from: https://www.nature.com/articles/s41582-019-0250-9
130. Yabuki J, Yoshikawa K, Koseki K, Ishibashi K, Matsushita A, Kohno Y Improvement of Functional Mobility Using a Hip-Wearable Exoskeleton Robot in Guillain-Barré Syndrome With Residual Gait Disturbance: A Case Report. 2024 Jul 5;16(7):e63882. doi: 10.7759/cureus.63882.
131. Doets AY, Lingsma HF, Walgaard C, Islam B, Papri N, Davidson A et al. Predicting outcome in Guillain-Barré syndrome international validation of the modified erasmus GBS outcome score. Neurology. 2022;98(5):E518–32.
132. Ali BM, Alghamdi AA, Alanazi WR, Alsamti AA, Allahyani AF, Altalhi MM, et al. Guillain-Barré syndrome: A review of immunotherapy, rehabilitation, and long-term outcomes. Int J Community Med Public Health [Internet]. 2023 Nov 30 [cited 2025 Dec 1];10(12):5038–43. Available from: https://www.ijcmph.com/index.php/ijcmph/article/view/12016
133. Zhang H, Ma J, Feng Y, Ma H, Liu D, Pang X, et al. Efgartigimod in the treatment of Guillain–Barré syndrome. J Neurol 2024 271:6 [Internet]. 2024 Mar 26 [cited 2025 Dec 1];271(6):3506–11. Available from: https://link.springer.com/article/10.1007/s00415-024-12321-4
134. Sprenger-Svačina A, Svačina MKR, Gao T, Zhang G, Sheikh KA. Emerging treatment landscape for Guillain-Barré syndrome (GBS): What's new? Expert Opin Investig Drugs [Internet]. 2024 Sep 1 [cited 2025 Dec 1];33(9):881–6. Available from: https://www.tandfonline.com/doi/abs/10.1080/13543784.2024.2377323
135. Deng M, Kong Z, Wang Y, Wang X, Li T Efgartigimod in the treatment of Guillain-Barré syndrome: Case report. Front Immunol [Internet]. 2025 [cited 2025 Dec 1];16:1586663. Available from: https://pmc.ncbi.nlm.nih.gov/articles/PMC12443784/

136. Martino Cinnera A, D'Arienzo M, Piatti D, Casagrande Conti L, Deledda P, Tenore A, et al. Robot-assisted therapy in Guillain–Barrè syndrome: Systematic review of primary evidence and study protocol for a randomized clinical trial. J Clin Med [Internet]. 2024 Dec 1 [cited 2025 Dec 1];13(23):7153. Available from: https://www.mdpi.com/2077-0383/13/23/7153/htm
137. Study Details | NCT06620198 | Upper Limb Robot-Assisted Therapy in Patients with Guillain-Barré Syndrome | ClinicalTrials.gov [Internet]. [cited 2025 Dec 1]. Available from: https://clinicaltrials.gov/study/NCT06620198
138. Cardone D, Perpetuini D, Di Nicola M, Merla A, Morone G, Ciancarelli I, et al. Robot-assisted upper limb therapy for personalized rehabilitation in children with cerebral palsy: A systematic review. Front Neurol [Internet]. 2024 Jan 6 [cited 2025 Dec 1];15:1499249. Available from: https://osf.io/a78zb/

13 Industrial Aspects to Be Considered for Drug Delivery to the Brain

Selin Akpinar Adscheid, Akif E. Türeli, and Nazende Günday-Türeli

13.1 INTRODUCTION

Developing efficient therapeutic options for central nervous system (CNS) diseases remains challenging due to the blood-brain barrier, BBB. CNS consists of the spinal cord and the brain and is protected by the BBB [1]. While BBB is essential for protecting the brain from pathogens and toxins, it also does not allow almost all large molecules and most of the small molecules into the brain [2]. Therefore, researchers have focused on strategies to overcome BBB and provide efficient therapies for brain-related diseases [3]. One of the approaches is employing drug delivery systems (DDS) for overcoming BBB or employing DDS to target the brain by using alternative routes of administration [4]. DDS are considered as a formulation or device assisting in introducing the therapeutic substance into the body to improve its efficacy and safety while giving possibilities on controlling the release rate and promoting targeting capacities [5].

Regarding particulate formulations, polymeric nanoparticles (NPs), lipid NPs (e.g., liposomes, non-structured lipid carriers (NLCs), and solid-lipid NPs (SLNs)), inorganic NPs, and hybrid NPs stand out due to their various advantages to different drug delivery strategies to the body [5]. As an example of advantages, lipid NPs stand out due to their biocompatibility and biodegradability, ability to be a reservoir to both Hydrophilic and hydrophobic substances, and lipophilicity, an advantage to bypass BBB [6, 7]. Polymeric NPs, on the other hand, can further present extensive tunable characteristics and can be of biological or synthetic origin [8]. Among synthetic polymeric NPs, poly(lactic-co-glycolic) acid (PLGA) has been significantly studied due to its biocompatible and biodegradable structure, which has enabled the FDA approval of PLGA for therapeutic use [9, 10]. Moreover, polymeric NPs can be employed to modify the release profile of the drugs when administered to the body by carefully choosing the polymer to prepare the NPs [11]. Inorganic NPs, on the other hand, employ a wide range of materials such as gold, silver, and silica NPs. Thanks to these wide ranges of material options and their unique physicochemical, optical, and electrical properties, they can be used in various applications such as imaging and photothermal therapy [12]. Moreover, these unique properties allow researchers to tailor these NPs with high functionalization abilities for targeted drug delivery. In addition, when the surface of these DDS is decorated with proteins, peptides, polymers, and carbohydrates, it can exhibit additional targeting abilities [13] (Figure 13.1). Techniques for overcoming the BBB can be invasive (e.g., disruption of the BBB and surgical interventions to administrate the drug at the site of action focusing on the local delivery) or noninvasive (Figure 13.2) [14].

The use of DDS as an option to treat CNS diseases focusing on the noninvasive methods and routes of administration such as intranasal drug delivery, systemic administration of the DDS for targeting the BBB, ultrasound-mediated-drug delivery can be named among the noninvasive techniques that researchers have been using to overcome the BBB [16, 17]. The systemic route for employing DDS has interested researchers who want to cross the BBB as well [15]. While NPs hold great promise for drug delivery to the brain, their systemic administration faces significant challenges related to BBB

DOI: 10.1201/9781003482406-13

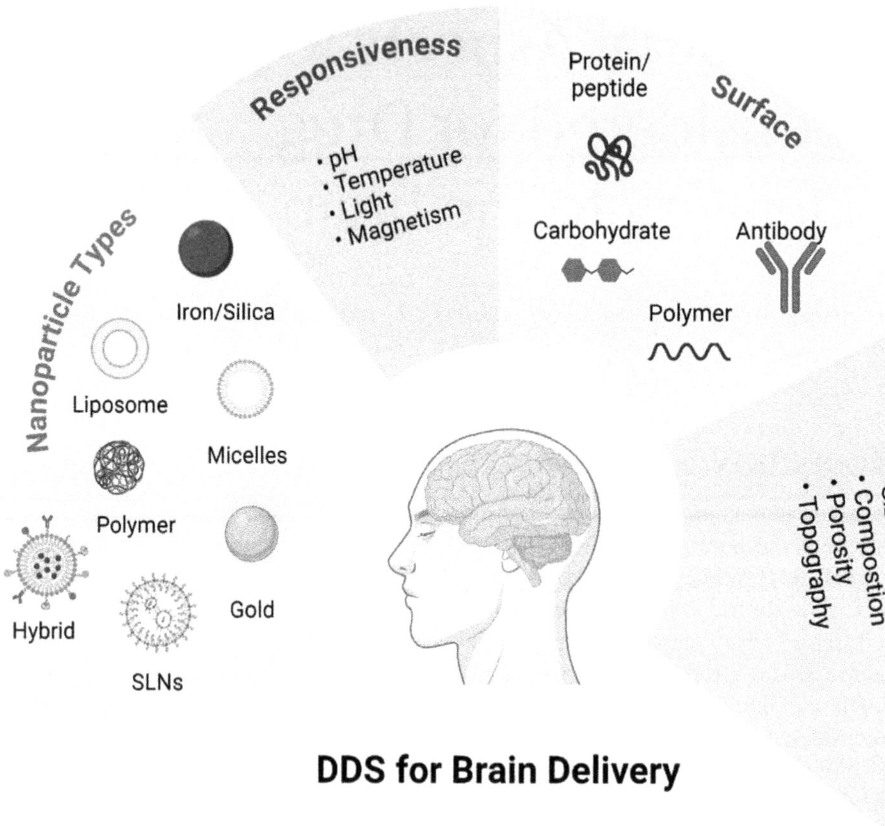

FIGURE 13.1 DDS for brain delivery. There are many different NP types, and they can have responsive properties. The surface of the NPs can be decorated with different materials to tailor them for therapeutic needs. Their physicochemical and morphological properties play a critical role in brain delivery. Created with Biorender.com

penetration, targeting specificity, and their particle size and surface properties that are crucial for crossing the BBB [3, 16]. In another perspective, the intranasal route of administration to administrate the drugs efficiently to the brain has been gaining interest. The nose-to-brain delivery (N2B delivery) exploits the direct anatomical interaction between the nasal cavity and the brain [18]. This approach highlights the advantages of bypassing the BBB, therefore enabling more efficient delivery of drug substances to the site of action, the brain. While N2B delivery has multiple benefits, such as enhanced drug delivery and bioavailability and reduced systemic side effects it has also been disadvantaged due to the high nasal clearance in the nasal cavity and the limited dosage due to the limited surface area that the intranasal route offers [19]. In another technique, ultrasound-mediated drug delivery for brain diseases, researchers employ ultrasound to open the BBB and simultaneously enhance the triggered release of the drugs from the loaded NPs [20]. While ultrasound mediation involves a noninvasive approach to successfully delivering the drug to the brain, one of the main challenges of obtaining clinically relevant results is the establishment of setups that are relevant in the transition from in vitro to in vivo that provide the efficient acoustic energy needed [21].

As mentioned, each route of administration and delivery approach has different advantages and disadvantages when successfully delivering the drug to the brain. The manufacturing techniques used to prepare these NPs determine the fate of the physicochemical properties (such as stability and surface properties) as well as the scalability of the NPs that have been prepared [22]. Therefore, in addition to the route of administration in focus, manufacturing techniques, including

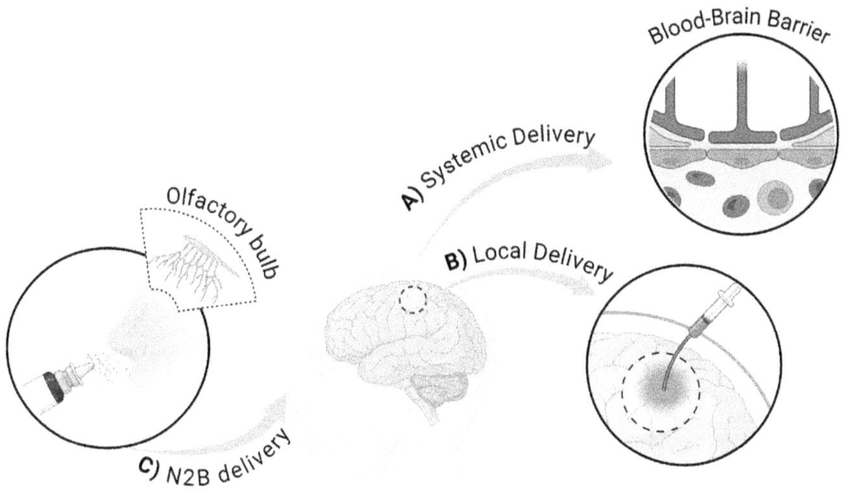

FIGURE 13.2 Schematic representation of different brain delivery routes. (A) The NPs meet the requirements to cross the BBB and can deliver the therapeutics through systemic delivery. (B) The local delivery focuses on the direct delivery/injection of the drug at the site of action. (C) The N2B delivery can exploit the direct anatomical interaction between the nasal cavity and the brain. The figure is modified from [15] and created with Biorender.com

both traditional and innovative methods, play a critical role in their clinical relevance and transition from bench to bedside [23]. This transition involves multiple steps from lab-scale development to good manufacturing practices (GMP) manufacturing (Figure 13.3). Thus, in the following sections, we introduce both traditional benchtop methods and innovative methods to prepare these DDS; scalability, different down-streaming techniques as well as GMP-compliance to be considered.

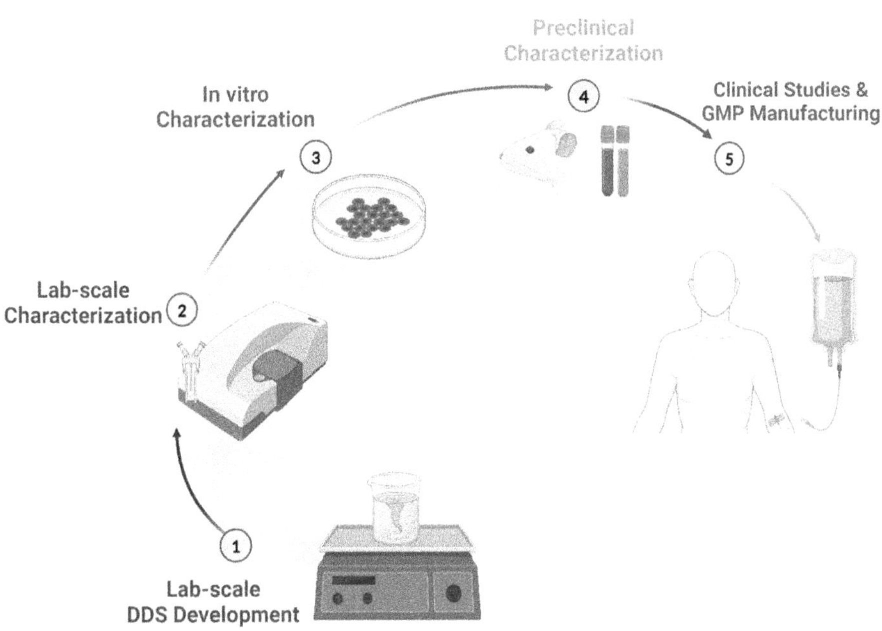

FIGURE 13.3 Schematic representation of essential steps for transition of nano-pharmaceutical DDS from bench to bedside. The figure is modified from [24] and created with the Biorender.com

13.2 MANUFACTURING OF DDS

13.2.1 Benchtop Methods for NP Production

NPs can be prepared with different techniques. These techniques are mainly divided into bottom-up and top-down methods. While the bottom-up methods focus on assembling molecules to form the nanomaterials, the top-down techniques are based on breaking the larger materials to form the nanomaterials (Figure 13.4) [25]. While there is no perfect technique to prepare NPs, and each NP type requires specific needs, understanding the pros and cons of the top-down and bottom-up methods enables researchers to develop innovative techniques suitable for continuous manufacturing and, thus, scalability in industrial applications [26].

13.2.2 Top-down Methods

Milling is one of the oldest, most efficient, and economical techniques for preparing NPs. Milling, or ball milling, focuses on using mechanical forces to grind the bulk material into fine particles [27].

High-pressure homogenization is an energy-intensive technique in which an emulsion, suspension, or dispersion undergoes a high pressure through a narrow gap of a homogenizing nozzle. Due to the high energy applied, the premix is subjected to a size reduction, breaking down, and subsequent homogenization [28]. The principles of high-pressure homogenization allow it to be a large-scale friendly technique to produce smaller droplet sizes and prepare DDS such as nanoemulsions, liposomes, and polymeric NPs. While it is considered a scale-up-friendly technique, several factors, such as flow rate, the volume of the fluids introduced, the geometry of the valve, the design

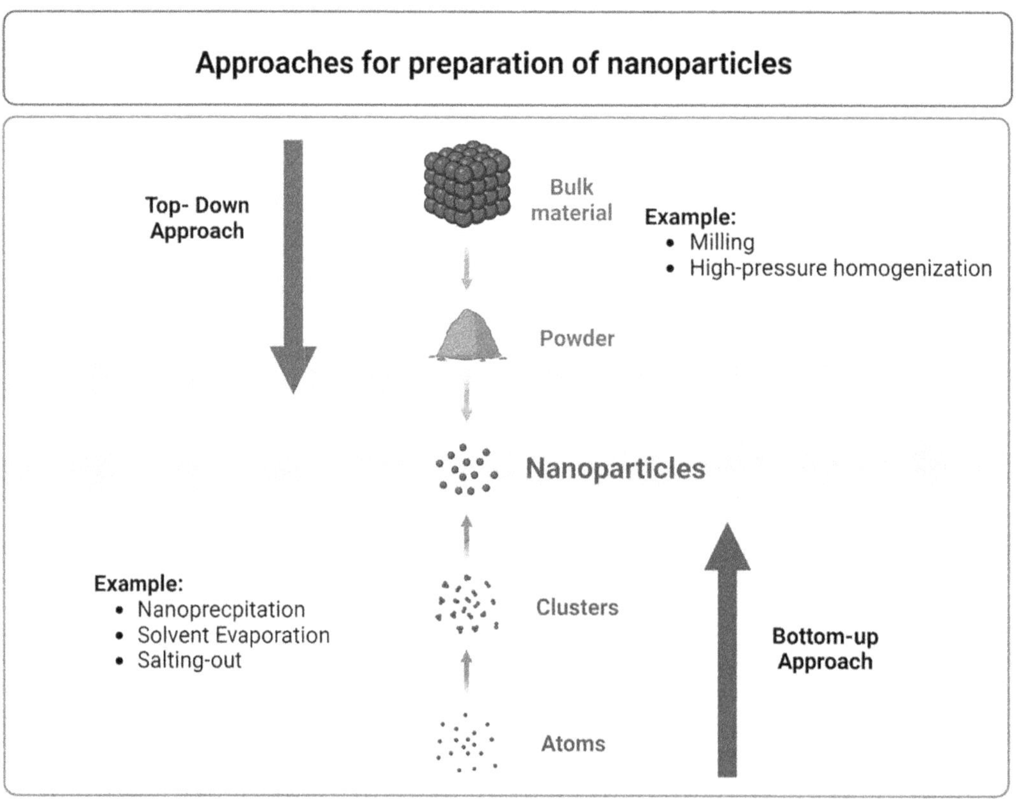

FIGURE 13.4 Top-down and bottom-up approaches to prepare NPs. Created with biorender.com

of the homogenizer, the number of cycles in the process and the pressure, should be regarded during scale-up processes [29, 30].

13.2.3 Bottom-up Methods

Nanoprecipitation is one of the common and straightforward methods to prepare NPs. The method requires two miscible solvents. In the benchtop nanoprecipitation method, the NP precursor solution containing the polymer dissolved in a water-miscible organic solvent phase is added into the non-solvent (aqueous phase) (Figure 13.5). The rapid mixture further supports the polymer assembly and formation of the NPs [31]. Unlike the high-pressure homogenization methods, this simple and cost-effective method does not require high energy input [32]. While simple, several factors were addressed in the literature as an essential step to prepare polymer NPs with this method; optimization of the NPs prepared by the nanoprecipitation method may require consideration of several factors. For instance, Draheim et al. conducted a design of experiment study to determine the critical factors in preparation of PLGA NPs by the nanoprecipitation method. They showed that the polymer concentration dominantly controls particle size and the polydispersity (PDI) of the prepared NPs [33]. Furthermore, et al. proposed a multi-criteria decision-making approach for developing camptothecin-loaded polymer NPs to identify the best possible method to prepare these particles according to several criteria based on instrument (e.g., ease of operation), process (e.g., yield, minimum number of excipients, minimum average particle size and PDI) and cost. The study showed that nanoprecipitation would be the most suitable method to prepare these NPs [34].

Solvent evaporation is a technique widely used for preparing polymeric NPs. It is based on where the polymer is dissolved in a volatile organic solvent and emulsified with an aqueous phase. The volatile organic solvent evaporates upon emulsification, and the polymer solidifies to form the NPs (Figure 13.6) [36]. While the single emulsification-solvent evaporation method is more suitable for hydrophobic substances, the double emulsion solvent evaporation method can be applied to encapsulate hydrophilic substances. The prepared primary emulsion is added into another aqueous phase in the double emulsion technique. These two immiscible phases are once more emulsified. The subsequent evaporation after the emulsification leads to the polymer's solidification and the NPs' formation [37]. In a study by Hernández-Giottonini et al., polymeric NPs were

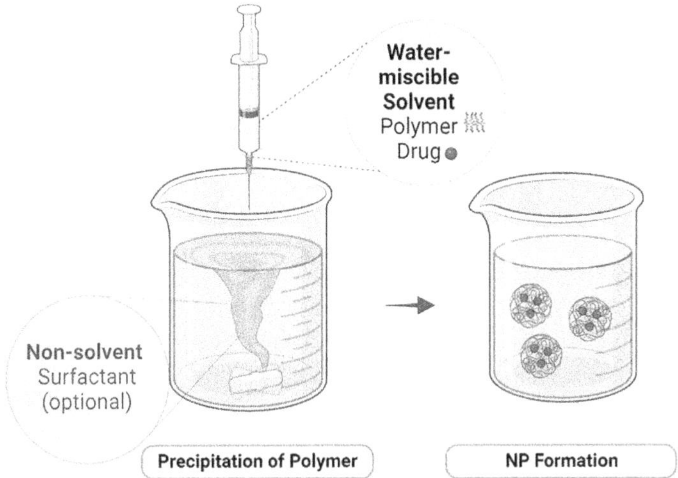

FIGURE 13.5 The formation of NPs by the nanoprecipitation method. The figure is modified from [35] and created with Biorender.com

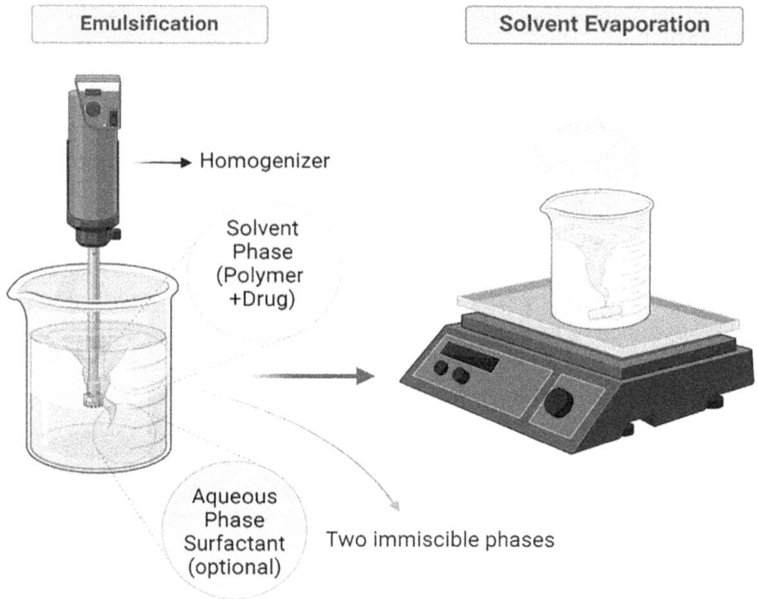

FIGURE 13.6 The formation of NPs by the emulsification solvent evaporation method. The figure is modified from [35] and created with Biorender.com

prepared to compare the factors affecting particle size and PDI values NPs by nanoprecipitation and emulsification-solvent evaporation methods. They showed that high velocities of agitation in solvent evaporation decreased the particle size in both techniques. While in the nanoprecipitation technique, particle size was controllable with polymer and surfactant concentration, in the emulsification-solvent evaporation technique, these could be controllable by organic solvent fractions [38].

Salting-out is also a variation of a solvent evaporation method widely used for developing polymeric NPs where the polymer is dissolved in a water-miscible organic solvent. The polymer-containing-organic solvent phase is added to the aqueous phase with a salting-out agent in the salting-out method. The process does not require high energy, and nucleation occurs when the addition of the excess water leads to diffusing the organic solvent. Subsequently, the polymer precipitates and forms the NPs [39]. Magnesium sulfate, sodium chloride, calcium chloride, and calcium sulfate are the most used salting-out agents [40]. While it is an easy and efficient way to prepare polymer NPs, extensive purification processes are required due to the presence of salting-out agents.

13.3 SCALABILITY AND CONSIDERATIONS

As presented, different traditional NP manufacturing methods represent different advantages and disadvantages in developing these particles. For instance, nanoprecipitation methods are simple and easy to operate, but the particle growth control is limited, leading to uneven particle size distribution. The salting-out method requires extensive purification cycles and is considered time-consuming, while high-pressure homogenization is not suitable for thermosensitive drugs due to the heat that the production process requires as well, as removing the solvent used could be problematic [41, 42]. While no single production technique fits in producing different types of NPs, developing nanomedicines relies on consideration of the material, process, and regulatory aspects. From a materials perspective, the materials that have been chosen could be biocompatible and regarded as safe to minimize any further biological intolerance. The material's particle size, shape, and degradability are essential parameters to consider [23]. In terms of processes, as mentioned, there is not a single process that fits the production of all kinds of NPs. Indeed, the top-down methods are considered more for large-scale industrial production due to ease of operation and control of the final product. Still, there is a need to develop continuous manufacturing technologies that can fill the gap from the lab-scale production to the industrial scale. These technologies offer control over the process parameters, and well-defined parameters enable easier transition in scale-up practices. In addition to process development, the cost of the process and the early employment of quality-by-design (QbD) practices are important steps toward the industrialization of these NPs. The EU defined QbD as an approach that aims to ensure the quality of medicines by employing statistical, analytical, and risk-management methodology in designing, developing, and manufacturing medicines. While the methodology can be applied to any product, it aims to shorten the gap between production and regulatory approvals. In the next part of the chapter, we focus on the different innovative manufacturing technologies and techniques to shorten this gap and see industrial-friendly practices such as DoE and QbD. We also give examples of applying these innovative approaches to CNS delivery.

13.4 INNOVATIVE MANUFACTURING METHODS NP PRODUCTION: ROAD TO SCALE-UP

Different innovative methods and technologies such as flash nanoprecipitation, micro reactors, and microfluidics are used to prepare NPs with fine physicochemical properties with continuous flow synthesis methods [43].

Flash nanoprecipitation is an innovative milli-fluidic industrial-friendly approach like T-junction mixing to produce polymeric NPs [44]. This method uses nanoprecipitation to stimulate the supersaturation through a jet mixer. The process takes place in milliseconds; therefore, this flash mixing induces instant precipitation [45]. One of the limitations of this technique is encapsulation of hydrophilic substances. To overcome this, Massella et al. developed caffeine-loaded poly-ε-caprolactone (PCL) NPs by a confined impinging jet mixer using acetone as the solvent and water as the antisolvent. Their study showed that while caffeine-loaded PCL NPs successfully prepared for entrapment of the substance as well as modifying the release profile, the solvent-drug interactions and the structure of the drug chosen played a vital role in NP production. In another study, Caggiano et al. presented a proof-of-concept study as an extension of the flash nanoprecipitation called the Sequential Flash Nanoprecipitation technique. They showed that using two connected FNP mixers, core-shell NPs could be prepared at higher core loading and total mass concentration than the traditional single-step procedure with up to 90% core loading [46]. In another study, researchers encapsulated ibuprofen, which showed a potential benefit in treating Alzheimer's disease. During the NP preparation, poly(ethylene glycol)-poly(lactic acid) diblock copolymer was used as a stabilizer, and NPs were prepared under optimized conditions using a four-stream multi-inlet vortex mixer. In vivo studies with mice showed that polysorbate-(PS)-80-coated NPs showed greater uptake of the drug into the brain compared to the drug solution itself [47].

Another technology, the microreactor technology, has also gained attention for the scale-up of NP production. The technology usually relies on a mixing chamber of defined size and configuration. [48]. For instance, Günday-Türeli developed ciprofloxacin-complex loaded PLGA NPs by the microreactor system for pulmonary treatment of cystic fibrosis infections and evaluated the effect of process parameters on NP quality by the DoE approach. Overall, the production of NPs by the continuous manufacturing technology yielded fine particle size and PDI values [49]. In another study, Lombardo and Moya et al. focused on preparing surfactant-coated (PS80 or P188) polymeric NPs by the same technology for delivery to the brain through the BBB. Their study demonstrated the interaction of the NPs with human brain-like endothelial cells by in vitro coculture human BBB models, and while both types of surfactant-coated NPs were internalized by the cells and showed biocompatibility, they differed in their uptake profiles [50].

There are also microfluidic approaches for NP production that can be suitable for large-scale production. Microfluidics offers advantages in NP productions such as being high-throughput, cost-effective, controllable, eco-friendly, time efficient, and due to their easy-to-operate characteristics [51]. In fact, several studies have been conducted to produce DDS for efficient CNS delivery using microfluidics [52]. As an example of these efforts, for instance, Wang et al. developed a microfluidic platform to enhance the delivery of miRNAs through intranasal delivery for glioblastoma therapy. Their microfluidic platform demonstrated that it could efficiently encapsulate therapeutic miRNAs and drugs into controllable-sized EVs using a pressure-based disruption and reconstitution procedure. Their neural stem cell-derived EVs loaded with miRNA and labeled with CXCR4 successfully target glioblastoma and enhance tumor cell apoptosis when combined with temozolomide in their in vivo mice studies [53]. In another study, Samaridou et al. investigated N2B delivery by focusing on delivering RNA via encapsulation into nanocomplexes with traditional and microfluidics methods. They reported that their microfluidic method yielded a fine particle size distribution with tunable mean particle size, protected the RNA from degradation, and prevented premature RNA release in the release media. Furthermore, the in vivo studies with mice models of Alzheimer's disease confirmed an increase in miRNA in the hippocampus after intranasal administration [54]. In another study also performed on Alzheimer's Disease, Mohebichamkhorami et al. prepared a novel microfluidic approach to prepare ultra-small and monodispersed chitosan/graphene quantum dot hybrid NPs to prepare optimal NPs for the intranasal delivery. Upon intranasal administration, the hybrid ultra-small NPs could penetrate the cytoplasm of C6 glioma cells and show cell viability depending on exposure and time [55]. Furthermore Martins et al., developed a microfluidic method to prepare PLGA NPs loaded with an anti-HIV drug to enhance its delivery to the brain. Microfluidic production yielded smaller particle sizes, higher drug association degrees, and encapsulation efficiency compared to the conventional method [56]. As a further example, Sommonte et al. worked on developing SLNs loaded with brain-derived neurotrophic factor (BDNF) through a microfluidic production method. The in vitro permeability study highlighted that BDNF-loaded-SLNS showed enhanced permeability compared to non-encapsulated BDNF [57].

QbD approach also applies to the preparation of NPs by the microfluidic approach [58]. For instance, Gurba-Bryśkiewicz et al. discussed in detail the possible applications of QbD and DoE applications on mRNA encapsulation into Lipid NPs [59], a topic that gained much interest in the commercialization of NPs with the COVID-19 vaccine development research. On the other hand, the application of QbD is not limited to innovative manufacturing approaches—for instance, Vitorino et al., lipid NPs for the treatment of depression via intranasal administration. The formulation was optimized through QbD principles by observing correlations of critical material variables on critical quality attributes (CQAs) for intranasal drug delivery. Overall, the lipid NPs administered intranasally decreased depressive behaviors in behavioral in vivo evaluations with mice [60].

13.5 GMP-COMPLIANT DOWNSTREAMING METHODS FOR NPs

Industrialization of the NPs would be possible in reasonable compliance with both production and downstream methods for purification, concentration, and storage. For instance, Operti et al. investigated industrial-friendly production and down-streaming of PLGA NPs suitable for continuous operation. They manufactured NPs using a scalable inline sonication method. To purify the particles continuously, they employed cross-flow filtration (CFF) and showed that the technique was suitable for sufficient purification of the NPs. Moreover, the choice of enclosed systems made sterilization possible for GMP-compliance manufacturing [61]. The next sections focus on these purification and sterilization techniques to better understand their application in GMP-compliant manufacturing strategies.

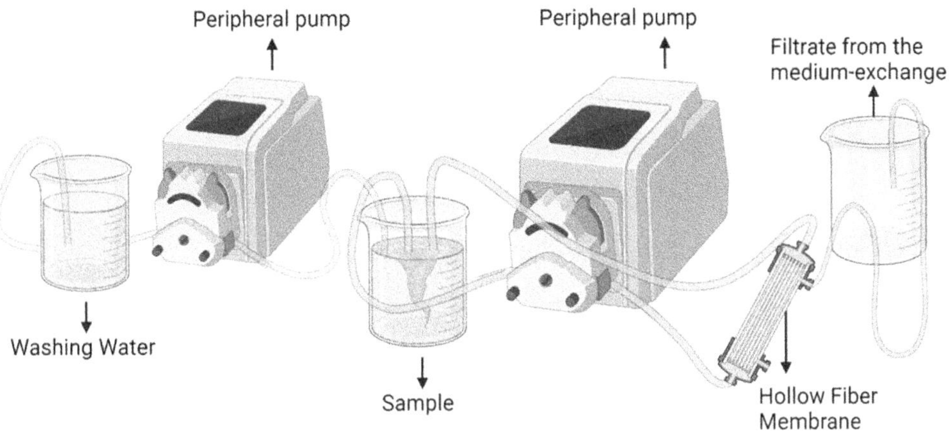

FIGURE 13.7 Illustration of a simple CFF setup. Created with Biorender.com

13.5.1 Purification

The purification of the NPs is an essential step during the particle development process. Sufficient cleaning is necessary to remove any excess substances and impurities that might lead to incompatibility and biological intolerance in future studies. The laboratory-scale purification processes generally involve centrifugation, extraction, dialysis, and filtration-based techniques [62]. While these techniques might suit small-scale laboratory-based production, large-scale NP production requires compliant and automatized methods to achieve sufficient cleaning procedures. One of the industrial-friendly purification techniques is known as CFF. CFF (Figure 13.7) is an industrial-friendly technique that purifies the NPs and removes excess substances. Unlike dead-end filtration methods where the solute goes through to the membrane perpendicularly, which could result in membrane caking and a decrease in the flux, in the CFF, the fluid passes through tangentially to a filter [63]. In the CFF process, the NPs are passed through a hollow fiber member with a specific membrane size cut-off. The hollow fiber membrane is chosen according to the particle size of the prepared NP, the molecular weight of the substance encapsulated, and the required time to perform the purification process. During the CFF, the NPs cannot cross the membrane, so they are collected in the retentate and circulated back to the sample reservoir. Due to their sizes, an excess amount of the encapsulated substance, the surfactant, is passed through the hollow fiber membrane and collected at the filtrate. The CFF system can concentrate the sample, or the initial sample concentration can be kept constant by adding an aqueous phase. For instance, in a study by Dalwadi et al., researchers compared diafiltration and the CFF procedure. The study showed that the CFF, with a concentration mode at trans membrane pressure (TMP) of 10 psi, could efficiently remove the excess polyvinyl alcohol (PVA) in 2.8h in large batches, unlike the diafiltration and dialysis methods. Moreover, this purification process had less impact on the physicochemical properties of NPs than the ultracentrifugation [64].

13.5.2 Sterilization

Many different methods have been used to sterilize NPs such as sterile filtration, autoclave, ionizing, and moist heat sterilization. Although these methods have been used, they also present drawbacks in the sterilization of NPs and should be addressed for each case [65]. For instance, while sterile filtration is a simple method, it cannot be employed for particles larger than 220 nm [66]. Moreover, autoclaving and gamma irradiation could affect sensitive APIs. On the other hand, Tapia-Guerrero

FIGURE 13.8 Factors to be considered for the lyophilization of the NPs. Created with Biorender.com

et al. showed that while gamma irradiation slightly affected the physicochemical properties of polymeric NPs, UV-irradiation did not affect either cell viability or physicochemical properties of the NPs [67], so UV-irradiation can be an alternative green technique as well. In addition to these techniques, aseptic filtration could be an alternative solution to sterilization of NPs as the most continuous manufacturing-friendly practice. However, it also has disadvantages, such as being high-cost and hard to operate [68].

13.5.3 Lyophilization

Lyophilization, also known as freeze-drying, is a dehydration process to improve formulations' (such as NP suspensions) stability, extend their shelf-life, and ease their storage [69] Including lyophilization is a critical step in the scale-up transfer of NPs. Developing a design space for lyophilization processes is essential for successful scale-up and technology-transfer processes and should be a focal point of in-process control strategies [70]. Important factors to be considered for the lyophilization of the NPs could be the appropriate choice of the cryoprotectant, cryoprotectant/surfactant ratio, and extensive comparison of the physicochemical properties of NPs before and after lyophilization (Figure 13.8). For instance, Amis et al. investigated different cryoprotectants for developing lyophilized SLNs. Their study showed that the appropriate choice of cryoprotectant and defining the suitable concentrations were critical, as optimal trehalose concentration prevented significant particle growth after lyophilization [71].

13.6 ASSESSMENT OF RELIABLE RELEASE PROFILE OF NPs

The transition of the NP formulation from bench to bedside also relies on the assessment of reliable preclinical data that resembles future clinical studies. One of the key points to be considered in this transition is the correct assessment of the in vitro release testing of the NPs. While different traditional methods (e.g., dialysis and sample and separate) are available to identify the NPs' release profile [72], these methods could not be sufficient to show a detailed picture of the release profile. For instance, sample and separate methods may result in the undesirable removal of nano and microparticles from the release medium [73]. In addition to this, the dialysis technique suffers from potential loss of the drug encapsulated into NPs and batch-size dependency [74]. Therefore, reliable testing systems such as the NanoDis system are needed. The NanoDis system's filtration principle is based on CFF combined with conventional dissolution apparatus to automate the filtration process (Figure 13.9) [75].

The NanoDis system uses the advantages of CFF to separate the NPs from the dissolution medium. Since CFF is an open filtration system, the formation of a filter cake is prevented, and the NPs are not forced into the filter, which often results in the blockage of filters in a dead-end filtration. Furthermore, the membrane size of the CFF filters can be chosen according to the particle size of the NPs in the formulation. In the NanoDis system, the NPs' dissolution medium is pumped

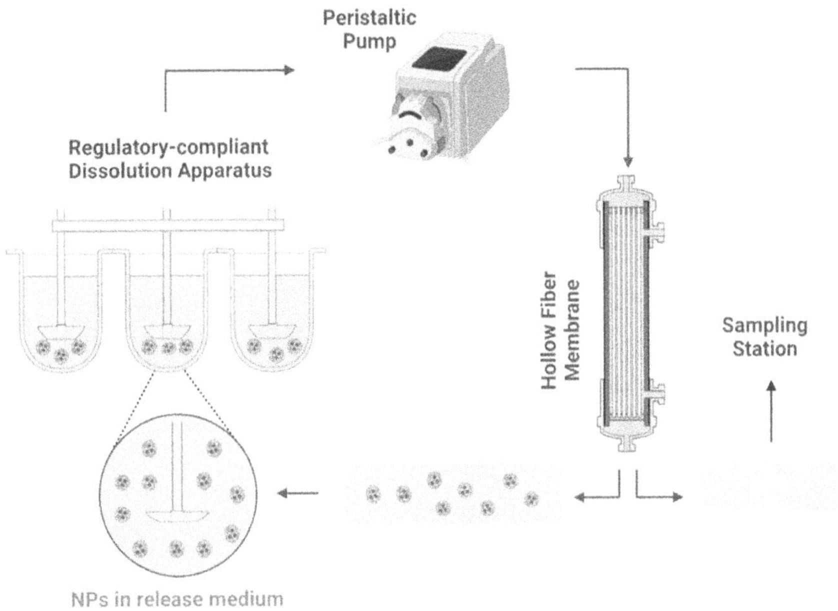

FIGURE 13.9 Principles of the NanoDis system. Created with Biorender.com

through CFF filters where the filtrate, free from particles, is collected with a sample collector. For instance, Lombardo et al. compared the release profiles of polymeric NPs obtained by the dialysis and the NanoDis method. They showed that PLGA nanocapsules (NC) loaded with retinoic acid (RA) showed a burst release when measured by the NanoDis system. In contrast, the traditional dialysis method could not measure the burst release [76].

13.7 GMP-COMPLIANCE

Quality assurance is critical in assuring compliance with the GMP and consistency, quality, and safety for the manufactured drug product. GMP is embedded in a Quality Management System (QMS) to ensure quality in perspective high quality and operation control in processes, procedures, premises and equipment, products, and people (departments) (Figure 13.10). The World Health Organization defines the GMP as an aspect of quality assurance that ensures that medicinal products are consistently produced and controlled to the quality standards appropriate to their intended use and as required by the product specification [77]. The European Medicinal Agency indicates that any manufacturer that intends to European Union (EU) market must comply with the GMP, and according to the GMP, a medicine should meet high quality, comply with the intended use, and comply with the requirements of the marketing and the clinical trial authorization [78].

GMP ensures the product meets certain specifications based on CQAs. Identifying the possible impacts of fluctuations during the processes on the CQAs is essential. Therefore, integrating technologies such as continuous process monitoring is critical to control the processes and minimize the risk of product failure. As mentioned previously, applying the QbD approach is a robust tool for control strategy, providing reassurance about the quality and consistency of the product. While the manufacturing steps should comply with the GMP, laboratory practices, characterization techniques, and scale-up options should meet GMP criteria, and applying QbD from these steps early on can be beneficial in developing nanopharmaceuticals and their industrial development [24]. However, there is still a need for more comprehensive characterization of nanopharmaceuticals by test methods and guidelines, and therefore, not fully addressing the CQAs remains a challenge [79, 80].

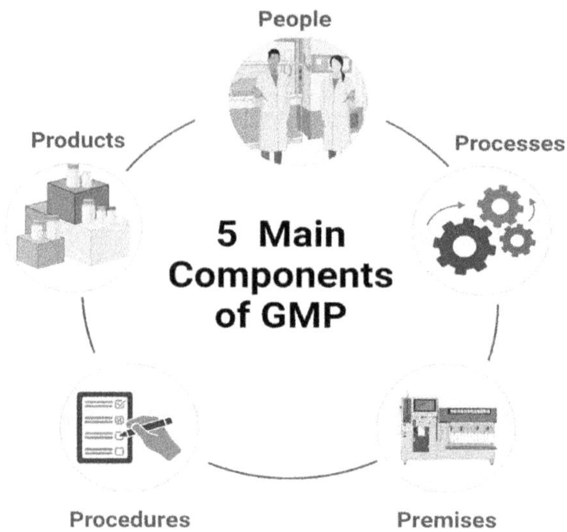

FIGURE 13.10 5 main components of GMP: People, procedure, process, premises, and the product. Created with Biorender.com.

As the challenges of nanopharmaceutical development as DDS are evident from the laboratory scale development and scale-up to the GMP-compliant manufacturing, the importance of single-entry points to offer different services from physicochemical characterization to manufacturing remains significant. To foster innovative approaches such as nanopharmaceuticals in the pharmaceutical industry, PHOENIX Open Innovation Test Bed (OITB), Pharmaceutical Open Innovation Test Bed for Enabling Nano-pharmaceutical Innovative Products, was funded by the EU (Grant agreement ID: 953110). The PHOENIX OITB offers a Single-Entry Point to facilities, technologies, services, and expertise for all technology transfer aspects from characterization, testing, and verification up to scale-up, GMP-compliant manufacturing, and regulatory guidance to offer an easier and cost-friendly transfer of nanopharmaceuticals from bench to bedside [81].

13.8 CONCLUSIONS

DDS have been very promising tools for delivering therapeutic options with high bioavailability, safety, and targeting efficiency. Their application to brain delivery has been highly studied, with different administration routes and approaches, from crossing the BBB to exploiting N2B delivery. From another perspective, their transition from bench to bedside depends on various factors, from manufacturing to compliance with regulatory aspects. Therefore, integrating the quality-by-design approach in the early stages of the DDS development of brain delivery and developing innovative manufacturing techniques allowing easier scale-up of DDS for brain delivery is an essential step forward in their transition from bench to bedside. Moreover, implementing GMP-compliant downstreaming methods, and assuring high quality and operation control in processes, procedures, premises and equipment, products, and people as GMP requires is essential in this transition.

REFERENCES

1. A. Achar, R. Myers, C. Ghosh, Drug delivery challenges in brain disorders across the blood–brain barrier: Novel methods and future considerations for improved therapy, Biomedicines 9 (2021) 1834. https://doi.org/10.3390/biomedicines9121834.

2. W.M. Pardridge, The blood-brain barrier and neurotherapeutics, NeuroRx 2 (2005) 1–2.
3. J.T. Trevino, R.C. Quispe, F. Khan, V. Novak, Non-invasive strategies for nose-to-brain drug delivery, J. Clin. Trials 10 (2020).
4. A. Mishra, R. Kumar, J. Mishra, K. Dutta, P. Ahlawat, A. Kumar, S. Dhanasekaran, A.K. Gupta, S. Sinha, D.K. Bishi, P.K. Gupta, S. Nayak, Strategies facilitating the permeation of nanoparticles through blood-brain barrier: An insight towards the development of brain-targeted drug delivery system, J. Drug Deliv. Sci. Technol. 86 (2023) 104694. https://doi.org/10.1016/j.jddst.2023.104694.
5. M.J. Mitchell, M.M. Billingsley, R.M. Haley, M.E. Wechsler, N.A. Peppas, R. Langer, Engineering precision nanoparticles for drug delivery, Nat. Rev. Drug Discov. 20 (2021) 101–124. https://doi.org/10.1038/s41573-020-0090-8.
6. P. Ghasemiyeh, S. Mohammadi-Samani, Solid lipid nanoparticles and nanostructured lipid carriers as novel drug delivery systems: Applications, advantages and disadvantages, Res. Pharm. Sci. 13 (2018) 288–303. https://doi.org/10.4103/1735-5362.235156.
7. C.-T. Lu, Y.-Z. Zhao, H.L. Wong, J. Cai, L. Peng, X.-Q. Tian, Current approaches to enhance CNS delivery of drugs across the brain barriers, Int. J. Nanomed. 9 (2014) 2241–2257. https://doi.org/10.2147/IJN.S61288.
8. M.A. Beach, U. Nayanathara, Y. Gao, C. Zhang, Y. Xiong, Y. Wang, G.K. Such, Polymeric nanoparticles for drug delivery, Chem. Rev 124 (2024) 5505–5616. https://doi.org/10.1021/acs.chemrev.3c00705.
9. H.O. Alsaab, F.D. Alharbi, A.S. Alhibs, N.B. Alanazi, B.Y. Alshehri, M.A. Saleh, F.S. Alshehri, M.A. Algarni, T. Almugaiteeb, M.N. Uddin, R.M. Alzhrani, PLGA-based nanomedicine: History of advancement and development in clinical applications of multiple diseases, Pharmaceutics 14 (2022) 2728. https://doi.org/10.3390/pharmaceutics14122728.
10. X. Guo, X. Zuo, Z. Zhou, Y. Gu, H. Zheng, X. Wang, G. Wang, C. Xu, F. Wang, PLGA-based micro/nanoparticles: An overview of their applications in respiratory diseases, Int. J. Mol. Sci. 24 (2023) 4333. https://doi.org/10.3390/ijms24054333.
11. C. Vasile, Chapter 1 - polymeric nanomaterials: Recent developments, properties and medical applications, in: C. Vasile (Ed.), Polymeric Nanomaterials for Nanotherapeutics, Elsevier, 2019: pp. 1–66. https://doi.org/10.1016/B978-0-12-813932-5.00001-7.
12. F. Wang, C. Li, J. Cheng, Z. Yuan, Recent advances on inorganic nanoparticle-based cancer therapeutic agents, Int. J. Environ. Res. Public. Health 13 (2016) 1182. https://doi.org/10.3390/ijerph13121182.
13. G. Sanità, B. Carrese, A. Lamberti, Nanoparticle surface functionalization: How to improve biocompatibility and cellular internalization, Front. Mol. Biosci. 7 (2020) 587012. https://doi.org/10.3389/fmolb.2020.587012.
14. W.M. Pardridge, A historical review of brain drug delivery, Pharmaceutics 14 (2022) 1283. https://doi.org/10.3390/pharmaceutics14061283.
15. S.M. Lombardo, M. Schneider, A.E. Türeli, N.G. Türeli, Key for crossing the BBB with nanoparticles: The rational design, Beilstein J. Nanotechnol 11 (2020). https://doi.org/10.3762/BJNANO.11.72.
16. S.K. Niazi, Non-invasive drug delivery across the Blood–Brain barrier: A prospective analysis, Pharmaceutics 15 (2023) 2599. https://doi.org/10.3390/pharmaceutics15112599.
17. A. Burgess, K. Shah, O. Hough, K. Hynynen, Focused ultrasound-mediated drug delivery through the blood-brain barrier, Expert Rev. Neurother. 15 (2015) 477–491. https://doi.org/10.1586/14737175.2015.1028369.
18. S. Gänger, K. Schindowski, Tailoring formulations for intranasal nose-to-brain delivery: A review on architecture, physico-chemical characteristics and mucociliary clearance of the nasal olfactory mucosa, Pharmaceutics 10 (2018). https://doi.org/10.3390/pharmaceutics10030116.
19. S.-H. Jeong, J.-H. Jang, Y.-B. Lee, Drug delivery to the brain via the nasal route of administration: Exploration of key targets and major consideration factors, J. Pharm. Investig. 53 (2023) 119–152. https://doi.org/10.1007/s40005-022-00589-5.
20. J. Wang, Z. Li, M. Pan, M. Fiaz, Y. Hao, Y. Yan, L. Sun, F. Yan, Ultrasound-mediated blood–brain barrier opening: An effective drug delivery system for theranostics of brain diseases, Adv. Drug Deliv. Rev. 190 (2022) 114539. https://doi.org/10.1016/j.addr.2022.114539.
21. E. Perolina, S. Meissner, B. Raos, B. Harland, S. Thakur, D. Svirskis, Translating ultrasound-mediated drug delivery technologies for CNS applications, Adv. Drug Deliv. Rev. 208 (2024) 115274. https://doi.org/10.1016/j.addr.2024.115274.
22. R.K. Thapa, J.O. Kim, Nanomedicine-based commercial formulations: Current developments and future prospects, J. Pharm. Investig. 53 (2023) 19–33. https://doi.org/10.1007/s40005-022-00607-6.
23. R. Paliwal, R.J. Babu, S. Palakurthi, Nanomedicine scale-up technologies: Feasibilities and challenges, AAPS PharmSciTech 15 (2014) 1527–1534. https://doi.org/10.1208/s12249-014-0177-9.

24. N.G. Türeli, A.E. Türeli, Chapter 11 - upscaling and GMP production of pharmaceutical drug delivery systems, in: R. Shegokar (Ed.), Drug Delivery Trends, Elsevier, 2020: pp. 215–229. https://doi.org/10.1016/B978-0-12-817870-6.00011-0.
25. S. Kumar, P. Bhushan, S. Bhattacharya, Fabrication of nanostructures with bottom-up approach and their utility in diagnostics, therapeutics, and others, Environ. Chem. Med. Sens. (2017) 167–198. https://doi.org/10.1007/978-981-10-7751-7_8.
26. N. Günday Türeli, A.E. Türeli, Chapter 16 - industrial perspectives and future of oral drug delivery, in: J.P. Martins, H.A. Santos (Eds.), Nanotechnology for Oral Drug Delivery, Academic Press, 2020: pp. 483–502. https://doi.org/10.1016/B978-0-12-818038-9.00016-8.
27. M.S. El-Eskandarany, A. Al-Hazza, L.A. Al-Hajji, N. Ali, A.A. Al-Duweesh, M. Banyan, F. Al-Ajmi, Mechanical milling: A superior nanotechnological tool for fabrication of nanocrystalline and nanocomposite materials, Nanomaterials 11 (2021) 2484. https://doi.org/10.3390/nano11102484.
28. C. Manoharan, A. Basarkar, J. Singh, Various pharmaceutical disperse systems, in: A.K. Kulshreshtha, O.N. Singh, G.M. Wall (Eds.), Pharmaceutical Suspensions: From Formulation Development to Manufacturing, Springer, New York, NY, 2010: pp. 1–37. https://doi.org/10.1007/978-1-4419-1087-5_1.
29. P. Vinchhi, J.K. Patel, M.M. Patel, High-pressure homogenization techniques for nanoparticles, in: J.K. Patel, Y.V. Pathak (Eds.), Emerging Technologies for Nanoparticle Manufacturing, Springer International Publishing, Cham, 2021: pp. 263–285. https://doi.org/10.1007/978-3-030-50703-9_11.
30. G. Soni, K. Kale, S. Shetty, M.K. Gupta, K.S. Yadav, Quality by design (QbD) approach in processing polymeric nanoparticles loading anticancer drugs by high pressure homogenizer, Heliyon 6 (2020) e03846. https://doi.org/10.1016/j.heliyon.2020.e03846.
31. O.I. Martínez-Muñoz, L.F. Ospina-Giraldo, C.E. Mora-Huertas, O.I. Martínez-Muñoz, L.F. Ospina-Giraldo, C.E. Mora-Huertas, Nanoprecipitation: Applications for entrapping active molecules of interest in pharmaceutics, in: Nano- and Microencapsulation - Techniques and Applications, IntechOpen, 2020. https://doi.org/10.5772/intechopen.93338.
32. S. Salatin, J. Barar, M. Barzegar-Jalali, K. Adibkia, F. Kiafar, M. Jelvehgari, Development of a nanoprecipitation method for the entrapment of a very water soluble drug into Eudragit RL nanoparticles, Res. Pharm. Sci. 12 (2017) 1–14. https://doi.org/10.4103/1735-5362.199041.
33. C. Draheim, F. de Crécy, S. Hansen, E.-M. Collnot, C.-M. Lehr, A design of experiment study of nanoprecipitation and nano spray drying as processes to prepare PLGA nano- and microparticles with defined sizes and size distributions, Pharm. Res. 32 (2015) 2609–2624. https://doi.org/10.1007/s11095-015-1647-9.
34. K. Krishnamoorthy, M. Mahalingam, Selection of a suitable method for the preparation of polymeric nanoparticles: Multi-criteria decision making approach, Adv. Pharm. Bull. 5 (2015) 57–67.
35. A. Zielinska, F. Carreiró, A.M. Oliveira, A. Neves, B. Pires, D. Nagasamy Venkatesh, A. Durazzo, M. Lucarini, P. Eder, A.M. Silva, A. Santini, E.B. Souto, Polymeric nanoparticles: Production, characterization, toxicology and ecotoxicology, Molecules 25 (2020) 3731. https://doi.org/10.3390/molecules25163731.
36. H.K. Makadia, S.J. Siegel, Poly lactic-co-glycolic acid (PLGA) as biodegradable controlled drug delivery carrier, Polymers 3 (2011). https://doi.org/10.3390/polym3031377.
37. Y. Su, B. Zhang, R. Sun, W. Liu, Q. Zhu, X. Zhang, R. Wang, C. Chen, PLGA-based biodegradable microspheres in drug delivery: Recent advances in research and application, Drug Deliv 28 (n.d) 1397–1418. https://doi.org/10.1080/10717544.2021.1938756.
38. K.Y. Hernández-Giottonini, R.J. Rodríguez-Córdova, C.A. Gutiérrez-Valenzuela, O. Peñuñuri-Miranda, P. Zavala-Rivera, P. Guerrero-Germán, A. Lucero-Acuña, PLGA nanoparticle preparations by emulsification and nanoprecipitation techniques: Effects of formulation parameters, RSC Adv. 10 (2020) 4218–4231. https://doi.org/10.1039/c9ra10857b.
39. K. Lim, Z.A.A. Hamid, 10 - polymer nanoparticle carriers in drug delivery systems: Research trend, in: Inamuddin, A.M. Asiri, A. Mohammad (Eds.), Appl. Nanocomposite Mater. Drug Deliv., Woodhead Publishing, 2018: pp. 217–237. https://doi.org/10.1016/B978-0-12-813741-3.00010-8.
40. T. Pulingam, P. Foroozandeh, J.-A. Chuah, K. Sudesh, Exploring various techniques for the chemical and biological synthesis of polymeric nanoparticles, Nanomaterials 12 (2022) 576. https://doi.org/10.3390/nano12030576.
41. S. Rezvantalab, N.I. Drude, M.K. Moraveji, N. Güvener, E.K. Koons, Y. Shi, T. Lammers, F. Kiessling, PLGA-based nanoparticles in cancer treatment, Front. Pharmacol. 9 (2018) 1260. https://doi.org/10.3389/fphar.2018.01260.

42. M. Alvi, A. Yaqoob, K. Rehman, S.M. Shoaib, M.S.H. Akash, PLGA-based nanoparticles for the treatment of cancer: Current strategies and perspectives, AAPS Open 8 (2022) 12. https://doi.org/10.1186/s41120-022-00060-7.
43. K. Huanbutta, K. Suwanpitak, N. Weeranoppanant, P. Sriamornsak, K. Garg, S. Sharma, I. Singh, T. Sangnim, Continuous flow synthesis: A promising platform for the future of nanoparticle-based drug delivery, J. Drug Deliv. Sci. Technol. 91 (2024) 105265. https://doi.org/10.1016/j.jddst.2023.105265.
44. S.J. Shepherd, D. Issadore, M.J. Mitchell, Microfluidic formulation of nanoparticles for biomedical applications, Biomaterials 274 (2021) 120826. https://doi.org/10.1016/j.biomaterials.2021.120826.
45. C.J. Martínez Rivas, M. Tarhini, W. Badri, K. Miladi, H. Greige-Gerges, Q.A. Nazari, S.A. Galindo Rodríguez, R. Román, H. Fessi, A. Elaissari, Nanoprecipitation process: From encapsulation to drug delivery, Int. J. Pharm. 532 (2017) 66–81. https://doi.org/10.1016/j.ijpharm.2017.08.064.
46. N.J. Caggiano, S.K. Nayagam, L.Z. Wang, B.K. Wilson, P. Lewis, S. Jahangir, R.D. Priestley, R.K. Prud'homme, K.D. Ristroph, Sequential flash NanoPrecipitation for the scalable formulation of stable core-shell nanoparticles with core loadings up to 90%, Int. J. Pharm. 640 (2023) 122985. https://doi.org/10.1016/j.ijpharm.2023.122985.
47. X. Zhang, L.Y. Chau, H.W. Chan, J. Weng, K.W. Wong, S.F. Chow, A.H.L. Chow, Physical stability and in vivo brain delivery of polymeric ibuprofen nanoparticles fabricated by flash nanoprecipitation, Int. J. Pharm 598 (2021) 120224. https://doi.org/10.1016/j.ijpharm.2021.120224.
48. A.E. Türeli, Nanoparticle preparation process using novel microjet reactor technology for enhancing dissolution rates of poorly water soluble drugs, Johannes Gutenberg-Universität Mainz, 2015. https://doi.org/10.25358/OPENSCIENCE-1833.
49. N. Günday Türeli, A.E. Türeli, M. Schneider, Optimization of ciprofloxacin complex loaded PLGA nanoparticles for pulmonary treatment of cystic fibrosis infections: Design of experiments approach, Int. J. Pharm. 515 (2016) 343–351. https://doi.org/10.1016/j.ijpharm.2016.10.025.
50. E.L.J. Moya, S.M. Lombardo, E. Vandenhaute, M. Schneider, C. Mysiorek, A.E. Türeli, T. Kanda, F. Shimizu, Y. Sano, N. Maubon, F. Gosselet, N. Günday-Türeli, M.-P. Dehouck, Interaction of surfactant coated PLGA nanoparticles with in vitro human brain-like endothelial cells, Int. J. Pharm. 621 (2022) 121780. https://doi.org/10.1016/j.ijpharm.2022.121780.
51. M.B. Kulkarni, S. Goel, Microfluidic devices for synthesizing nanomaterials—a review, Nano Express 1 (2020) 032004. https://doi.org/10.1088/2632-959X/abcca6.
52. M.I. Teixeira, M.H. Amaral, P.C. Costa, C.M. Lopes, D.A. Lamprou, Recent developments in microfluidic technologies for Central nervous system targeted studies, Pharmaceutics 12 (2020). https://doi.org/10.3390/pharmaceutics12060542.
53. K. Wang, U.S. Kumar, N. Sadeghipour, T.F. Massoud, R. Paulmurugan, A microfluidics-based scalable approach to generate extracellular vesicles with enhanced therapeutic MicroRNA loading for intranasal delivery to mouse glioblastomas, ACS Nano 15 (2021) 18327–18346. https://doi.org/10.1021/acsnano.1c07587.
54. E. Samaridou, H. Walgrave, E. Salta, D.M. Álvarez, V. Castro-López, M. Loza, M.J. Alonso, Nose-to-brain delivery of enveloped RNA - cell permeating peptide nanocomplexes for the treatment of neurodegenerative diseases, Biomaterials 230 (2020) 119657. https://doi.org/10.1016/j.biomaterials.2019.119657.
55. F. Mohebichamkhorami, M. Faizi, M. Mahmoudifard, A. Hajikarim-Hamedani, S.S. Mohseni, A. Heidari, Y. Ghane, M. Khoramjouy, M. Khayati, R. Ghasemi, H. Zali, S. Hosseinzadeh, E. Mostafavi, Microfluidic synthesis of ultrasmall chitosan/graphene quantum dots particles for intranasal delivery in Alzheimer's disease treatment, Small Weinh. Bergstr. Ger. 19 (2023) e2207626. https://doi.org/10.1002/smll.202207626.
56. C. Martins, F. Araújo, M.J. Gomes, C. Fernandes, R. Nunes, W. Li, H.A. Santos, F. Borges, B. Sarmento, Using microfluidic platforms to develop CNS-targeted polymeric nanoparticles for HIV therapy, Eur. J. Pharm. Biopharm. 138 (2019) 111–124. https://doi.org/10.1016/j.ejpb.2018.01.014.
57. F. Sommonte, I. Arduino, R.M. Iacobazzi, L. Laera, T. Silvestri, A.A. Lopedota, A. Castegna, N. Denora, Microfluidic development of brain-derived neurotrophic factor loaded solid lipid nanoparticles: An *in vitro* evaluation in the post-traumatic brain injury neuroinflammation model, J. Drug Deliv. Sci. Technol. 96 (2024) 105699. https://doi.org/10.1016/j.jddst.2024.105699.
58. S. Colombo, M. Beck-Broichsitter, J.P. Bøtker, M. Malmsten, J. Rantanen, A. Bohr, Transforming nanomedicine manufacturing toward quality by design and microfluidics, Adv. Drug Deliv. Rev. 128 (2018) 115–131. https://doi.org/10.1016/j.addr.2018.04.004.

59. L. Gurba-Bryśkiewicz, W. Maruszak, D.A. Smuga, K. Dubiel, M. Wieczorek, Quality by design (QbD) and design of experiments (DOE) as a strategy for tuning lipid nanoparticle formulations for RNA delivery, Biomedicines 11 (2023). https://doi.org/10.3390/biomedicines11102752.
60. C. Vitorino, S. Silva, F. Gouveia, J. Bicker, A. Falcão, A. Fortuna, QbD-driven development of intranasal lipid nanoparticles for depression treatment, Eur. J. Pharm. Biopharm. 153 (2020) 106–120. https://doi.org/10.1016/j.ejpb.2020.04.011.
61. M.C. Operti, A. Bernhardt, V. Sincari, E. Jager, S. Grimm, A. Engel, M. Hruby, C.G. Figdor, O. Tagit, Industrial scale manufacturing and downstream processing of PLGA-based nanomedicines suitable for fully continuous operation, Pharmaceutics 14 (2022) 276. https://doi.org/10.3390/pharmaceutics14020276.
62. S.F. Tehrani, P. Bharadwaj, J. Leblond Chain, V.G. Roullin, Purification processes of polymeric nanoparticles: How to improve their clinical translation? J. Control. Release 360 (2023) 591–612. https://doi.org/10.1016/j.jconrel.2023.06.038.
63. N.K. Shah, R. Ivone, J. Shen, S.A. Meenach, A comparison of centrifugation and tangential flow filtration for nanoparticle purification: A case study on acetalated dextran nanoparticles, Particuology 50 (2020) 189–196. https://doi.org/10.1016/j.partic.2019.06.004.
64. G. Dalwadi, H.A.E. Benson, Y. Chen, Comparison of diafiltration and tangential flow filtration for purification of nanoparticle suspensions, Pharm. Res. 22 (2005) 2152–2162. https://doi.org/10.1007/s11095-005-7781-z.
65. M.A. Vetten, C.S. Yah, T. Singh, M. Gulumian, Challenges facing sterilization and depyrogenation of nanoparticles: Effects on structural stability and biomedical applications, Nanomed. Nanotechnol. Biol. Med. 10 (2014) 1391–1399. https://doi.org/10.1016/j.nano.2014.03.017.
66. S.A. Bernal-Chávez, M.L. Del Prado-Audelo, I.H. Caballero-Florán, D.M. Giraldo-Gomez, G. Figueroa-Gonzalez, O.D. Reyes-Hernandez, M. González-Del Carmen, M. González-Torres, H. Cortés, G. Leyva-Gómez, Insights into terminal sterilization processes of nanoparticles for biomedical applications, Molecules 26 (2021) 2068. https://doi.org/10.3390/molecules26072068.
67. Y.S. Tapia-Guerrero, M.L. Del Prado-Audelo, F.V. Borbolla-Jiménez, D.M. Giraldo Gomez, I. García-Aguirre, C.A. Colín-Castro, J.A. Morales-González, G. Leyva-Gómez, J.J. Magaña, Effect of UV and gamma irradiation sterilization processes in the properties of different polymeric nanoparticles for biomedical applications, Materials 13 (2020) 1090. https://doi.org/10.3390/ma13051090.
68. M.C. Operti, A. Bernhardt, S. Grimm, A. Engel, C.G. Figdor, O. Tagit, PLGA-based nanomedicines manufacturing: Technologies overview and challenges in industrial scale-up, Int. J. Pharm. 605 (2021). https://doi.org/10.1016/j.ijpharm.2021.120807.
69. P. Fonte, S. Reis, B. Sarmento, Facts and evidences on the lyophilization of polymeric nanoparticles for drug delivery, J. Control. Release 225 (2016) 75–86. https://doi.org/10.1016/j.jconrel.2016.01.034.
70. K. Sampathkumar, B.A. Kerwin, Roadmap for drug product development and manufacturing of biologics, J. Pharm. Sci. 113 (2024) 314–331. https://doi.org/10.1016/j.xphs.2023.11.004.
71. T.M. Amis, J. Renukuntla, P.K. Bolla, B.A. Clark, Selection of cryoprotectant in lyophilization of progesterone-loaded stearic acid solid lipid nanoparticles, Pharmaceutics 12 (2020) 892. https://doi.org/10.3390/pharmaceutics12090892.
72. S. D'Souza, A review of in vitro drug release test methods for nano-sized dosage forms, Adv. Pharm. 2014 (2014) 1–12. https://doi.org/10.1155/2014/304757.
73. Y. Kim, E.J. Park, T.W. Kim, D.H. Na, Recent progress in drug release testing methods of biopolymeric particulate system, Pharmaceutics 13 (2021) 1313. https://doi.org/10.3390/pharmaceutics13081313.
74. R. Vargas, M. Romero, T. Berasategui, D.A. Narváez-Narváez, P. Ramirez, A. Nardi-Ricart, E. García-Montoya, P. Pérez-Lozano, J.M. Suñe-Negre, C. Moreno-Castro, C. Hernández-Munain, C. Suñe, M. Suñe-Pou, Dialysis is a key factor modulating interactions between critical process parameters during the microfluidic preparation of lipid nanoparticles, Colloid Interface Sci. Commun. 54 (2023) 100709. https://doi.org/10.1016/j.colcom.2023.100709.
75. E. Türeli, K. Krauel-Göllner, Agilent NanoDis System Method Development Guide, n.d.
76. S.M. Lombardo, N. Günday Türeli, M. Koch, M. Schneider, A.E. Türeli, Reliable release testing for nanoparticles with the NanoDis System, an innovative sample and separate technique, Int. J. Pharm. 609 (2021) 121215. https://doi.org/10.1016/j.ijpharm.2021.121215.
77. Good Manufacturing Practices, World Health Organ. (n.d.). https://www.who.int/teams/health-product-policy-and-standards/standards-and-specifications/gmp (accessed June 28, 2024).
78. Good manufacturing practice, Eur. Med. Agency (n.d.). https://www.ema.europa.eu/en/human-regulatory-overview/research-development/compliance-research-development/good-manufacturing-practice (accessed June 28, 2024).

79. F. Caputo, G. Favre, G. Borchard, L. Calzolai, P. Fisicaro, E. Frejafon, N. Günday-Türeli, D. Koltsov, C. Minelli, B.C. Nelson, J. Parot, A. Prina-Mello, S. Zou, F.-X. Ouf, Toward an international standardisation roadmap for nanomedicine, Drug Deliv. Transl. Res. (2024). https://doi.org/10.1007/s13346-024-01646-2.
80. D.A. Dri, F. Rinaldi, M. Carafa, C. Marianecci, Nanomedicines and nanocarriers in clinical trials: surfing through regulatory requirements and physico-chemical critical quality attributes, Drug Deliv. Transl. Res. 13 (2023) 757–769. https://doi.org/10.1007/s13346-022-01262-y.
81. Pharmaceutical Open Innovation Test Bed for Enabling Nano-pharmaceutical Innovative Products | Phoenix Project | Fact Sheet | H2020, CORDIS Eur. Comm. (n.d.). https://cordis.europa.eu/project/id/953110/de (accessed June 28, 2024).

14 Regulatory Aspects to be Considered in the Development of Drug Delivery Systems to the Brain

Touraj Ehtezazi

14.1 INTRODUCTION

The regulatory approval process for drug delivery systems to the brain is similar to other medicines. The brain can be targeted via different routes: oral, intravenous (i.v.) injection, intraperitoneal injection, intrathecal administration, direct injections into the brain through the skull via a surgery, and nasal cavity. The regulatory approval process will also depend on the route of administration. For example, sterilisation of pharmaceutical products is not required for oral and nasal products. While for systemic routes the pharmaceutical product must be sterilised, and a suitable sterilisation method must be identified. Furthermore, parenteral products should be tested for endotoxin levels as part of the regulatory approval, and accepted levels of endotoxins depend on the route of administration, with intrathecal route being the most stringent one. As well as small and large molecules, nanoparticle formulations have been developed for drug delivery to the brain, which will have their own specific requirements to ensure that the formulation can consistently and safely achieve the therapeutic targets. Therefore, the formulation scientists should consider the specific requirements for brain-targeting drugs/drug delivery systems from the beginning.

Drug delivery to the brain remains a significant challenge due to anatomical and physiological barriers, particularly the protective skull and the blood-brain barrier (BBB). While these protective mechanisms are essential for maintaining optimal neurological functions, they simultaneously impose substantial obstacles to therapeutic interventions. These barriers contribute to the high failure rate of clinical trials aiming to treat neurological disorders. The BBB consists of endothelial cells, which form tight junctions. The BBB allows only selected and small molecules access to the brain. Armstrong *et al.* 1987 (1) demonstrated the size-dependent permeability through comparative studies showing that while sucrose (M_r 340 Da, radius 5 Å) was permeable through the BBB of wild-type rat brain, larger dextran molecules (M_r 200,000 Da, radius 100 Å) exhibited negligible permeation. Therefore, these molecular size limitations should be considered when designing new drug molecules and drug delivery systems targeting the brain. The intranasal delivery offers a way to avoid BBB limitations. Dhuyvetter *et al.* 2020 (2) demonstrated that the absorption of small drug molecules (209 Da $< M_r <$ 512 Da) through the nasal cavity into the brain could depend on the presence of direct transporters and formulation components, such as cyclodextrins which may enhance absorption. However, the presence of the nasal mucus layer may act as a barrier to the absorption of drug molecules into the brain (3, 4).

Nanoparticles (NPs) provide an alternative approach for the delivery of the drug molecules to the brain following systemic (5, 6) or nasal administration (6, 7). It should be noted that BBB dysfunction in the form of leakage and oedema has been demonstrated in several human brain disorders including multiple sclerosis (MS), stroke, and epilepsy (8). Additionally, the basement membrane enlargement was demonstrated in the neurovascular unit of patients with Alzheimer's disease (AD) (Figure 14.1).

DOI: 10.1201/9781003482406-14

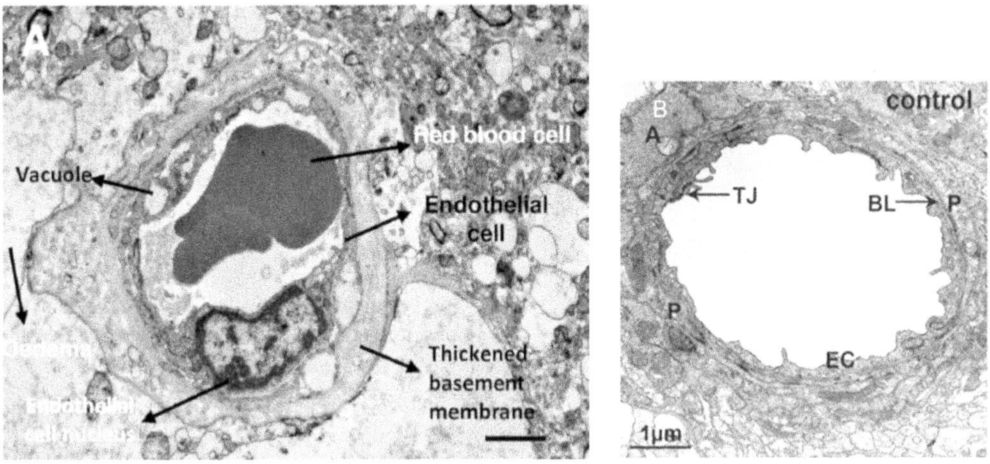

FIGURE 14.1 TEM image of neurovascular unit in (A) patient with Alzheimer's Disease (AD) obtained from the frontal cortex, (B) a healthy control subject. Attention should be given to thickened basement membrane in the AD patient compared to the control subject. In addition, vacuolisation of the endothelial cell, fibrillary deposits, and oedema around brain capillaries can be seen. Reproduced with permission from for panel A (9) and panel B (10).

Therefore, these physiological alterations of the BBB, alongside the chosen administration route must be carefully considered when developing new drugs or formulations for brain diseases.

This chapter examines the drug development process for both small drug molecules and NPs, with emphasis on brain targeting formulations. The regulatory guidelines will be discussed with illustrative examples throughout. Since most brain-targeting formulations require sterility, different sterilisation methods will be reviewed for both conventional drug molecules and NP formulations. The final section addresses the measurement and elimination of bacterial endotoxins from these formulations.

14.2 DRUG DEVELOPMENT PROCESS

The drug development process follows well-established procedures for small drug molecules, which depends on the targeted organ, and it has evolved over the years. These are explained in the following.

14.2.1 Classical Approach

Figure 14.2 illustrates the classical drug discovery process. Stage 1 of the process involves laboratory research to identify and validate a target, which can be a receptor, enzyme, DNA, or RNA. Typically, a construct (such as a small molecule) is identified that shows engagement with the target. This molecule may be considered as the basis and is modified for optimised target engagement in stage 2, which is also known as the "Hit-to-Lead" phase. In this stage, a library of compounds is synthesised by following systematic modifications and rational designing, which are usually synthesised in small quantities. It is crucial to establish validated tests that can be performed for small amounts of new constructs (11). Upon the completion of stage 2, a compound exhibiting relatively favourable effects is designated as the "lead." During stage 3, termed "Lead-optimisation," the chemical structure of the lead compound undergoes a systematic modification, which may yield a small number of molecules highly suitable for the intended therapeutic applications. Optimised lead compounds demonstrate proof-of-concept. Generally, a single optimised lead compound advances to stage 4, selected based on superior bioavailability, physicochemical stability, and feasible formulation processes. This

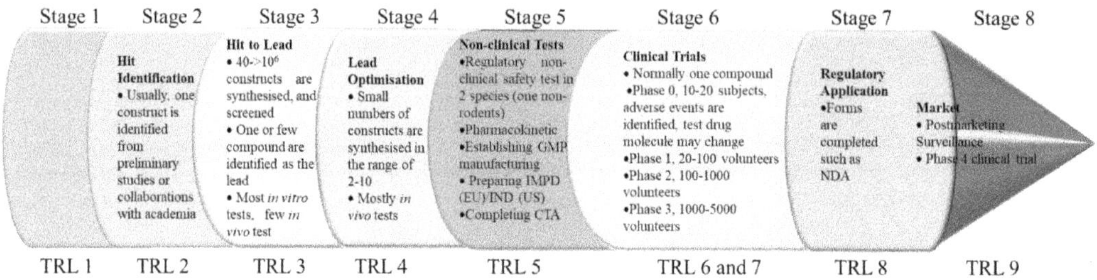

FIGURE 14.2 The schematic presentation of classical drug discovery process. The main stages are represented in a highly simplified manner. The relevant TRL for each stage is presented underneath of each stage. Depending on the disease, the entire process takes about 10–15 years with an approximate cost around US $2.8 billion.

selected compound, designated "Candidate-Drug," undergoes a comprehensive evaluation process in extensive animal studies (12), typically involving two species (one non-rodent, as specified by ICH M3 guidelines) at doses up to 100 times the intended therapeutic dose.

The drug molecules (including degradation products and impurities) may also be investigated for DNA-reactive (mutagenic) impurities to mitigate potential carcinogenic risks as specified in the ICH M7(R2) guideline. Pharmaceutical formulations must also complete genotoxicity testing (ICH S2A and ICH S2B guidelines) and reproductive toxicity assessments (ICH S5(R3) guideline). Stage 5 incorporates comprehensive nonclinical pharmacokinetic studies examining absorption, distribution, metabolism, and excretion of the active ingredient and its metabolites. Additionally, safety pharmacology studies (part of pharmacology studies) evaluate the new drug's effects on the cardiovascular system, the CNS, and respiratory functions, with detailed assessments outlined in the ICH S7A guideline. These safety pharmacology evaluations should preferably comply with Good Laboratory Practice standards. Primary and secondary pharmacodynamics (PD) studies are part of pharmacology studies, with primary pharmacodynamic studies determining the drug substance's mechanisms of action that produce the desired therapeutic effects, while secondary PD studies identify unintended drug mechanisms of action that may lead to side effects.

Manufacturing of pharmaceutical products should be established in compliance with the Good Manufacturing Practice (GMP) in Stage 5 (13), or even from the beginning of the drug development process. Stage 5 would be the latest stage that the manufacturing of the pharmaceutical products should comply with the GMP standards, as these will be tested in humans in clinical trials. The drug development organisation may have its own GMP facilities, however, the GMP activities can also be outsourced. In this scenario, a contract is arranged between the drug development company (Contract Giver) and the GMP pharmaceutical manufacturer (Contract Acceptor). It is part of the regulatory requirements that the Contract Giver is responsible for assessing the legality, suitability, and the competence of the Contract Acceptor to carry out the outsourced activities successfully, as stated in the EU Guidelines for GMP for Medicinal Products for Human and Veterinary Use, Volume 4, Chapter 7.

If a parenteral route of administration is considered for the treatment of a brain disease, then the pharmaceutical product must be sterile for testing in animals at stage 4 (14) and preferably containing acceptable levels of pyrogens (endotoxins): 5 E.U./kg for non-intrathecal parenteral products and 0.2 E.U./kg for intrathecal injections (15). Based on these assumptions, the endotoxin limit is 1.5 E.U./mL for non-intrathecal injections for a mouse with 30 g weight receiving 100 µL of drug formulation per hour. Bacterial endotoxins are found in the serum of all healthy humans in the range of 0.01–1 EU/mL (16). The endotoxin level is reduced to 0.06 E.U./mL per hour for intrathecal injections in both humans and animals (17). This is not only to avoid further harm to animals but also immune reactions to unsterile and pyrogenic parenteral products may mask accurate observations in the safety pharmacology studies. Furthermore, it is recommended that injectable brain drug deliver formulations to be tested in their final (or near final) container and closure systems. This ensures that there are no undesirable byproducts in the formulation during the manufacturing process.

If the candidate drug is found to be promising, then a drug sponsor may decide to test it as a new drug on humans in clinical trials (18). Preparation of an Investigational Medicinal Product Dossier (IMPD) is a regulatory requirement prior to clinical trials application in humans in the EU (Stage 5). The IMPD is part of the Clinical Trial Application (CTA) (19). The equivalent to IMPD in the US is the Investigational New Drug (IND) application, which forms part of the CTA application. The same information is part of the IND application. The IMPD should also be prepared for Phase II and Phase III human clinical trials. The IMPD is not required for an investigational medicinal product (IMP) that already has a marketing authorisation (i.e. a pharmaceutical product). For example, if a pharmaceutical product is going to be investigated in a Phase IV clinical trial, then an IMPD is not required to be submitted as part of the CTA.

The EMA has provided information about the structure and content of an IMPD related to the drug substance and pharmaceutical product through the "Guideline on the requirements for the chemical and pharmaceutical quality documentation concerning investigational medicinal products in clinical trials - Revision 1 (EMA/CHMP/QWP/545525/2017)." This part of the IMPD includes identification, analytical procedures, manufacture, stability studies, packaging of drug substance, medicinal product, and placebo. Nonclinical studies and clinical studies (if available) form the remaining sections of the IMPD. The IMPD follows Common Technical Document (CTD) format (20), and has two main parts: IMPD Quality (IMPD-Q) and IMPD Safety and Efficacy (IMPD-SE). The information in the IMPD should be concise and not unnecessarily voluminous. It is preferable to present data in tabular form, accompanied by brief narrative highlighting the main significant points. The International Council for Harmonisation of Technical Requirements for Pharmaceuticals for Human Use (ICH) provides guidelines for the structure of the CTD in the M4 (R4) (Organisation of Common Technical Document) guideline. Figure 14.3 shows design of the IMPD as well as CTD organisation, which contains 5 modules, with Module 1 being administrative information whose content and format are set by the regulatory authorities.

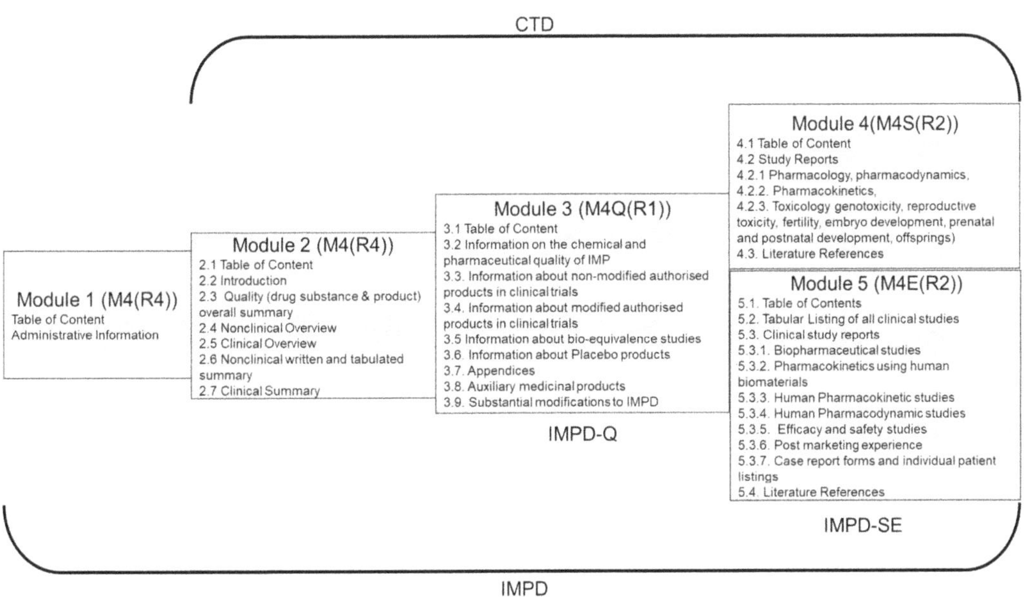

FIGURE 14.3 Design of the IMPD and CTD organisation. The relevant ICH guideline and required information are given for each module.

For example, the following parts should be included in Module 1 for EU application:

- a cover letter,
- a comprehensive table of contents,
- an application form,
- product information documents,
- information on the experts,
- specific requirements for different types of applications (if required),
- an environmental risk assessment,
- information relating to orphan market exclusivity (if required),
- information relating to pharmacovigilance,
- information relating to clinical trials (if required),
- information relating to paediatrics.

The ICH provides the following guidelines for the rest of the modules:

a. The M4Q(R1) guideline (Quality overall summary of Module 2 and Module 3: Quality) related to drug substance and product chemistry, manufacturing, and control,
b. M4 Safety (M 4 S): Nonclinical Overview and Nonclinical Summaries of Module 2 and Organisation of Module 4,
c. M4 Efficacy (M4E(R2)- Clinical overview and Clinical summary of Module 2 and Module 5: Clinical Study reports.

These guidelines are to prepare a dossier, which is well structured and familiar to regulatory reviewers to reduce the time and resources to compile applications. Table 14.1 summarises different parts of a CTA, and an example of a completed application is given by Versyck et al. 2023 (21), which is available online (https://www.clinicaltrialsregister.eu/ctr-search/trial/2019-004813-15/BE).

TABLE 14.1
Summary of the Structure of the Clinical Trial Application in the EU

Title: Request for authorisation of a clinical trial on a medicinal product for human use to the competent authorities and for opinion of the ethics committees in the community

Section	Description
A	Trial Identification • Title of the Clinical Trial, • Allocated trial number such as EduraCT
B	Identification of the Sponsor Responsible for the Request • Sponsor (name and address of the organisation, name of the person to contact) • Legal Representative (Name and Address) • Status of the sponsor (commercial/non-Commercial) • Source(s) of Monetary or Material Support for the clinical trial (organisation that takes responsibility for conducting the trial) • Contact point designated by the sponsor for further information on the trial (such as name and address of Clinical Trial Information Desk)
C	Applicant Identification • Request for the Competent Authority (name and address of the sponsor, registering CTA on EudraCT) • Request for the Ethics Committee (contact details of the principal investigator)

(Continued)

TABLE 14.1 *(Continued)*
Summary of the Structure of the Clinical Trial Application in the EU

Section	Description
D	Information on IMP • IMP Identification (IMP number, IMP used as a comparator) • Status of IMP (does IMP have a marketing authorisation, is the treatment defined only by active substance or a combination of marketed products, status of the IMPD: full or simplified) • IMPD (not required for products that have marketing authorisation) • Somatic Cell Therapy Investigational Medicinal Product (No Genetic Modification) • Gene Therapy Investigational Medicinal Products • Tissue Engineered Product • Products Containing Devices (i.e. Medical Devices, Scaffolds, etc.) • Information on Placebo • Site(S) Where the Qualified Person Certifies Batch Release site and Where the Qualified Person Certifies Batch Release
E	General Information on the Trial • Medical condition or disease under investigation • Objective of the Trial • Principal inclusion criteria • Principal exclusion criteria • End points (primary and secondary outcomes) • Scope of the trial (diagnosis, prophylaxis, therapy, etc) • Trial Type (first administration to human, Phase I, Phase II, Phase III, Phase IV) • Design of the trial (controlled, randomised, crossover, etc.)
F	Population of the trial subjects • Age range • Gender • Group of trial subjects (e.g. healthy volunteers, patients, pregnant women) • Planned number of subjects to be included. • Plans for treatment of care after a subject has ended his/her participation in the trial.
G	Clinical Trial sites/Investigators in the member state concerned by this request • Contact details of a co-ordinating investigator for multi-centre trial and principal investigator for a single-centre trial • Contact details and qualifications of principal investigators responsible for each centre. • Address of laboratories/facilities that will evaluate the efficacy of the treatment and provide outcome measures. • The address and contact details of networks to be involved in the trial such as paediatric networks. Normally these are for Phase II/III clinical trials where a multicentre trial is needed; and it is essential to recruit desired number of patients preferably with different geographical sites. ECMC paediatric network co-ordinates 12 centres across the country in the UK. • The address and contact details of the organisation to whom the sponsor has transferred trial-related duties and functions. The details of tasks are explained in this section for the third party, which is conducting the trial. The third party should follow ICH E6 (R3) guidelines. Any trial-related procedure that is delegated to a service provider should be specified in a written agreement. Where the agreement is between the sponsor and the service provider and involves tasks under investigator's responsibility, it should be clear in the agreement between the sponsor and the service provider as well as in the agreement between the sponsor and the investigator that the investigator maintains adequate control and oversight over the contracted tasks.

(Continued)

TABLE 14.1 *(Continued)*
Summary of the Structure of the Clinical Trial Application in the EU

Section	Description
H	Competent Authority/Ethics Committee in the member state concerned by the request. • Type of application. This part has two options: Competent Authority and Ethics Committee. Completing this part is slightly confusing, as if the application is addressed to the Competent Authority, the Ethics Committee box is ticked and vice versa. • Address and contact details of Competent Authority/Ethics Committee. This is the only part of the application that provides name and contact details of the Ethics Committee if the application is addressed to the Competent Authority. • The details of authorisation outcome, whether the Competent Authority/Ethics Committee has approved conducting the trial. The reasons should be provided if the Competent Authority/Ethics Committee has rejected the application. The anticipated date of resubmission should be stated.
I	Signature of the Applicant in the Member State • Declaration of the accuracy of the provided information • Name, date, and signature of the applicant of the request for the Competent Authority.* • Name, date, and signature of the applicant of the request for the Ethics Committee.

* The Applicant of the Request for Competent Authority: This is the authorised person from the sponsor who requests approval of the clinical trial from the regulatory body in a country that oversees regulation of medicines. The name of this person would be the same as in the section C.1.

Another example can be found in the following link: https://assets.publishing.service.gov.uk/media/6093bd74e90e0726f85de633/PDF_Attachment_5_-_FOI_21-192.pdf.

Here, an interesting example is given for the drug discovery process of Tagamet (cimetidine), which was started in 1964 (22) by Smith Kline & French, a little-known U.S. drug firm in Philadelphia. By the discovery and approval of Tagamet, the company became one of the largest pharmaceutical companies in the world, with $1b Tagamet annual sales in 1986 and $1.13 b in 1987. The development of Tagamet, an H_2-receptor antagonist, started from histamine, an H_2-receptor agonist (Figure 14.4). As histamine was the mediator of gastric acid secretion, then the aim was to modify this molecule and produce a compound that had similar chemical structure but lacked the stimulating effects on gastric acid secretion. The objective was to produce a drug that could be used for the treatment of peptic ulcers. Therefore, histamine was considered as the hit-compound. About 700 compounds were synthesised based on the histamine molecule. It was noticed that all the constructs were still stimulating acid secretion (good H_2 agonists). This was because all the compounds were too similar to histamine, as they all shared similar side chain (Figure 14.4). Therefore, hit-to-lead discovery process involved keeping the histamine ring relatively intact (imidazole) and changing the side chain. In the hit-to-lead process, SK&F 71448 was synthesised, which showed partial histamine antagonistic activities. Therefore, this compound was considered as the "lead" compound (Figure 14.4).

The lead-optimisation involved synthesising a set of compounds that had modified side chains and as a result, burimamide was discovered (Figure 14.4) (23). This compound showed antagonistic activities without histamine agonist activities. Black et al 1972 (from R&D branch of the company in the UK) reported that burimamide inhibited 75% of acid secretion in dogs stimulated by histamine at the dose of 4×10^{-5} mol/kg (21.2 mg/kg); and therefore, the H_2-receptor was identified (23). On the other hand, burimamide lacked oral absorption, and therefore, this compound was modified only on the imidazole ring, and metiamide was synthesised. This compound not only showed good oral bioavailability, but also much stronger H_2-receptor antagonist activities compared to burimamide. As metiamide showed suitable efficacy and safety *in vivo* (24), it was subjected to extensive

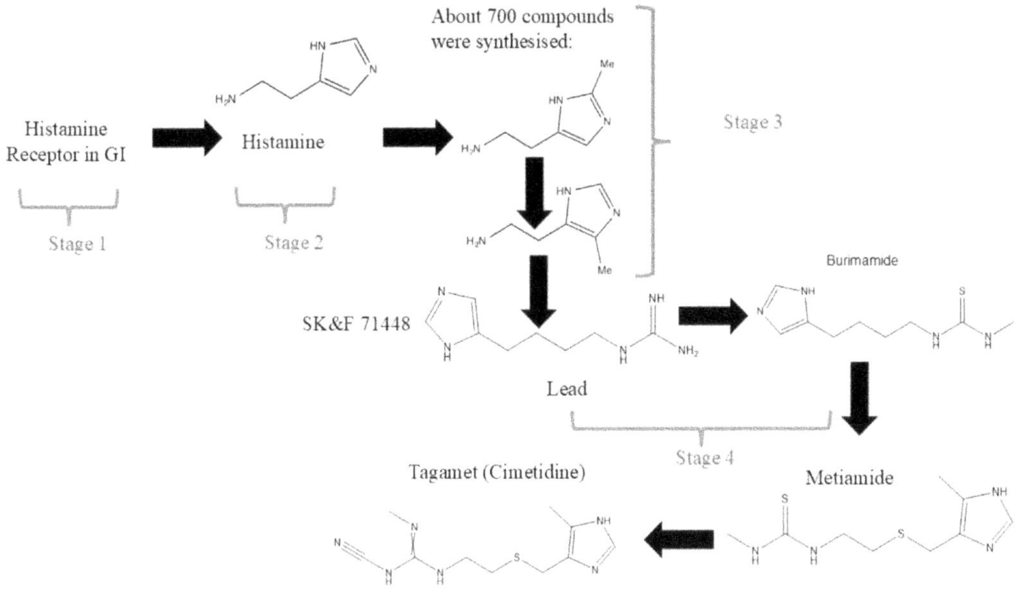

FIGURE 14.4 Molecular structures showing the chemical development pathway from histamine to the final compound Tagamet, illustrating the structural modifications made during drug development process.

animal toxicity studies, which did not show unacceptable toxicity up to 20 times doses of ED_{50}. Metiamide was granted permission to be tested in humans and the results were promising (25, 26). Although favourable results were obtained, there were reports of agranulocytosis in part of patients. This was considered as a serious adverse effect and therefore, metiamide was further optimised and cimetidine was synthesised (Figure 14.4), with no effects on white blood cell count and hemoglobulin. Cimetidine with the brand name of Tagamet was approved in the United Kingdom on the 8th of November 1976. The drug was also approved in the United States in 1977 and then in Europe. James Black was awarded Nobel Prize for medicine or physiology in 1988 for the extensive work that he conducted in identification and antagonism of H_2 receptor in this project.

14.2.2 Novel Techniques Used in the Drug Development Process

To increase the rate of successful drugs for the CNS, brain-on-a-chip (BoC) has been developed to simulate the critical components of the brain's structure. BoC is a microfluidics-based system that may contain human-originated cells such as primary, immortalised, and/or stem cells (27). This is important to replicate human brain *in vitro*. BoC can be 2D (monolayer of cells), or 3D (more than one cell layer), which can mimic better native brain tissues. To address the species difference in the BBB functions, a human brain-chip was developed which comprised human primary astrocytes and pericytes, hiPSC-derived cortical neurons, hiPSC-derived brain microvascular endothelial-like cells, and hiPSC-derived microglia (28). Activation of microglia in the chip increased the number of CD68-positive cells, indicating microglia activation. This would allow *in vitro* simulation of neuroinflammation. Alternatively, an *in vitro* model was developed incorporating the neurovascular unit (astrocytes, pericytes, neurons, and brain micro-vessel endothelial cells) to accelerate drug screening in Stage 1 of the development process for drugs targeting the brain (29).

Receptor-mediated transcytosis is the primary mechanism that BBB endothelial cells internalise extracellular large molecules such as insulin. The expression of certain receptors increases in brain diseases such as the upregulation of basigin in brain endothelial cells under ischemic conditions as well as AD and MS. Therefore, these disease-specific receptors may be considered as targets for the development of novel drugs such as monoclonal antibodies (mAbs) against basigin receptors. Interestingly, internalised basigin mAbs are not sorted to the lysosomal pathway (30).

In another approach, a framework of bidirectional recurrent neural networks was developed to design *de novo* drug molecules for the treatment of brain diseases. The neuronal network (CNSMolGen) was trained by ChemBridge-MPO, an existing database. The model was able to closely predict physicochemical properties of small-molecule drugs penetrating into the CNS (31). Network pharmacology may be applied to tailor or design multi-target drugs for the treatment of complex brain diseases such as AD (32). Artificial Intelligence (AI) has provided possibilities to more effectively develop new drugs (33). AI has potential for application at every stage of the drug discovery and development process. However, the purpose should be defined before using AI to identify a suitable solution. For example, an AI-powered Question & Answer system such as the BioBERT may collect publicly available literature and patent databases to help scientists better understand the biological roles of a protein target or drug-drug interactions (33, 34). Computer-aided drug discovery is becoming more popular due to advances in computational technologies and more data on ligand binding to therapeutic targets (35). *In silico* models have been developed to predict the CNS bioavailability of small drug molecules (36).

Fragment-based drug discovery is a strategy to identify hit constructs for the treatment of brain diseases (37). In this approach small molecular fragments (building blocks) with desired properties are joined to build complex (larger) molecules that have much improved functionality (38). This approach identifies hit-constructs much quicker compared to the traditional high-throughput screening.

14.2.3 Model Informed Drug Development

The efficiency of drug development and the regulatory decision-making process may be enhanced by the application of Model Informed Drug Development (MIDD). MIDD allows for optimising both time and resources used in the early "learning" phases and informing the "confirmatory" phases of development. MIDD is defined as quantitative framework for prediction and extrapolation, centred on knowledge and inference generated from integrated models of compound, mechanisms and disease level data, which aims at improving the quality, efficiency and cost effectiveness of the decision-making process (39). MIDD has been employed in all aspects of the drug development process (39). For example, the application of physiologically-based pharmacokinetic (PBPK) modelling allowed researchers to predict whether a drug molecule such as allopurinol would accumulate in the body before the administration of the next dose (40). MIDD can be employed to design compounds that target a receptor or combinations of receptors to increase the efficacy and also aid in the dose selection (40). For example, a multiscale model was constructed for IL-6 signalling in Crohn's disease. The model contained a simplified signalling network of IL-6-mediated cellular signal transduction. The model also incorporated antibody pharmacokinetic (PK) such as dynamic exchange of free and bound drug between organ compartments, and clearance of free as well as bound antibody. The aim was to examine the role of drug dose and antibody-target affinity in determining drug efficacy (41). The model was validated from previously published data from a pilot trial of a humanised anti–IL-6Rα mAb antibody, tocilizumab, in Crohn's subjects. The optimised model was employed to simulate the effects of anti–IL-6 and anti–IL-6Rα on PD biomarkers including serum CRP, pSTAT3 activity in the GI tract and IL-6 levels in the GI tract and serum after every 4-week i.v. dosing for a 12-week period. In addition, it was assumed that the anti–IL-6 and anti–IL-6Rα antibodies bind only to free IL-6 and IL-6Rα, respectively. The optimised model predicted significant suppression of serum CRP level (almost 100%) when anti-IL6 antibody affinity (Kd) changed from 250 pmol/L to 2.5 pmol/L at 600 mg dose after 12 weeks of treatment. Therefore, this type of simulation may guide drug discovery and development process to effectively and efficiently identify hit or lead constructs. PK and PD models have been shown to be useful in predicting tumour growth in animal models. For example, a PK/PD model was built to link the dosing of anticancer drugs (AZD7762 and irinotecan) to the inhibition of mouse-xenograft tumour growth. The developed models closely simulated the growth of tumours in the animal models and in the control group (vehicles), administrating each anticancer drug and different drug combinations. Therefore, MIDD can be useful in the design of optimal clinical dose, frequency of administration, and suitable combinational therapy (42).

MIDD is a platform to integrate all data available to make critical decisions such as administered dose in Phase III clinical trials. MIDD is employed to predict human doses based on PK and PD data available from nonclinical studies. Furthermore, MIDD has been used to predict the dose escalation in Phase I clinical trials, which normally involves a wide dose range such as Aducanumab, that was tested in doses of 0.3, 1, 3, 10, 20, 30, and 60 mg/kg in a Phase I clinical trial in patients with mild-to moderate AD (43). MIDD (PK/PD) data are used to design the dose escalation in Phase II clinical trials based on the data available from Phase I and nonclinical PK/PD data. For example, Natalizumab was evaluated in Phase I clinical trial in the dose range of 0.03–3 mg/kg (44). Based on the observations, Natalizumab was tested in only two different doses 3 mg/kg and 6 mg/kg, in subsequent clinical studies. The data obtained from Phase I and Phase II clinical trials are used to predict the population PK. MIDD can support go/no-go decisions both in early stages and during clinical trials (40, 45).

14.3 DRUG DEVELOPMENT PROCESS FOR CNS DISEASES

Drug development process for the CNS follows a similar pattern to other diseases, while being more challenging sometimes due to the BBB (46). The market for CNS disorders was over US $80.5 billion in 2010 (47) and it is estimated to grow to US $222.33 billion by 2031 for the pharmaceutical industry. However, there are reports that the failure rate for new drugs targeting CNS diseases is very high relative to most other areas of drug discovery (48). This led to disbanded or significantly reduced CNS programs for the pharmaceutical industry back in 2010 (47), and even recently the closure of GSK Neuroscience R&D site in Shanghai in 2017. Gribkoff and Kaczmarek (2017) suggested better use of animal models such as in stroke and the animals should not have the drug on board at the time of the infarct (49). Table 14.2 shows examples of clinical trial outcomes for pharmaceutical products intended for the treatment of CNS disorders.

14.3.1 Drug Development of Small Drug Molecules for the Brain Diseases

Avagacestat, also known as BMS-708163, was developed by Bristol-Myers Squibb for the treatment of AD. The discovery process of Avagacestat demonstrates the robustness of the classical drug discovery process to achieve a product that delivers intended biological effects but may not produce the desired therapeutic outcomes. Avagacestat is a sulfonamide-based gamma (γ)-secretase inhibitor (Figure 14.5) designed to reduce the formation of Aβ peptides in the brain (50).

The stage 1 of the drug discovery process started from compound **1** (Figure 14.5, BMS 299897) (51), which demonstrated *in vitro* rise (at doses less than 100 nM) and inhibition (at doses of about 1 μM) of Aβ production (52). As a result, a second-generation of γ-secretase inhibitors was initiated with a caprolactam sulfonamide derivative, which was considered the "hit" with IC_{50}=120 nM. The "hit-to-lead" process (structure-activity-relationship approach) resulted in the discovery of constructs containing a caprolactam ring, which exhibited *in vitro* inhibition of $A\beta_{40}$ formation (IC_{50} = 9 nM) (53). Other derivatives (4-methoxy benzyl and N,N-dimethyl analog) showed $brain_{concn}/plasma_{concn}$ ratios in the range of 2.8–7, $A\beta_{40}$ IC_{50}=2.8-11 nM, and $A\beta_{42}$ IC_{50}=2.4–7.8 nM (53). These compounds were approximately 50-fold more potent than the original screening hit, and demonstrated a marked reduction of the concentration of Aβ in the brain and plasma in transgenic Tg2576 mice. On the other hand, caprolactam derivatives were also found to be nanomolar inhibitors of the human CYP450 isoforms 3A4 and 2C19. Further investigations showed that opening the caprolactam ring resulted in a much more potent inhibitor of $A\beta_{40}$ formation ($A\beta_{40}IC_{50}$=0.14 nM) (54). Despite these strong inhibitory effects, plasma and brain $A\beta_{40}$ levels were reduced to

TABLE 14.2
Examples of Clinical Trial Outcomes for Pharmaceutical Products Intended for the Treatment of CNS Disorders

Product	Disease	Sponsor	Phase 2 Outcomes	Phase 3 Outcomes	Approved
Bitopertin	Schizophrenia	Roche	Statistically significant reduction of negative symptoms(124)	Lack of Efficacy(125)	No
Dexmecamylamine	Depression	Targacept/ AstraZeneca	statistically significant results on measures of depression(126)	Lack of Efficacy(127)	No
Long-acting injection Olanzapine Pamoate	Schizophrenia	Eli Lilly	No significant difference with oral formulation(128)	Showed less Efficacy compared to oral formulation(129)	Yes, making product available for patients with the history of non-adherence
Recombinant Factor VIIa	Haemorrhagic stroke	Novo Nordisk	Positive clinical results(130)	Lack of Efficacy and Safety(131)	Yes, because of small benefit for survival during the first 15 days
Semagacestat	AD	Eli Lilly	Positive clinical results(132)	Lack of Efficacy and Safety(133)	No

LoE=Lack of Efficacy
SoE=Showed Efficacy

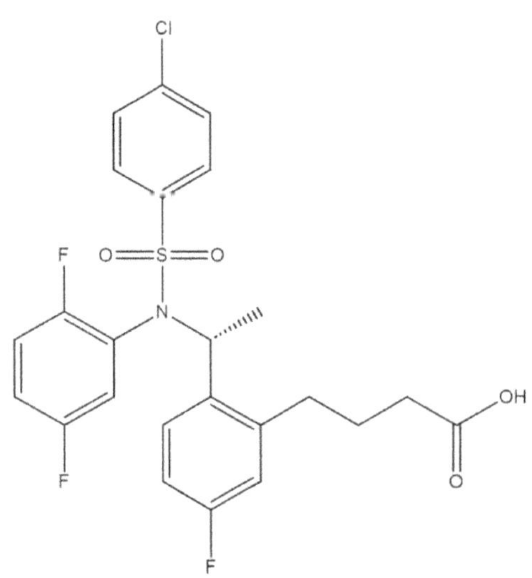

BMS 299897

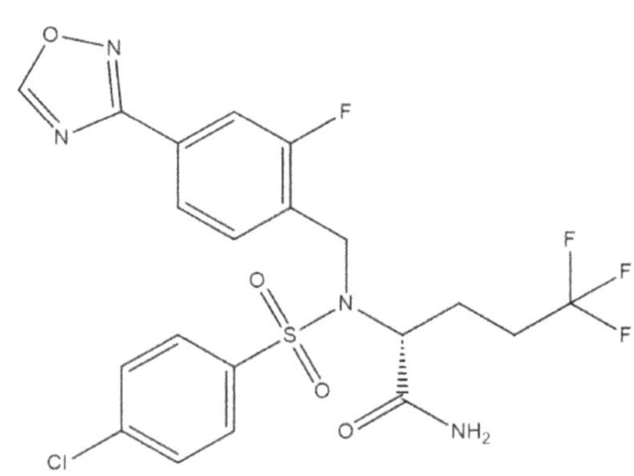

Avagacestat

FIGURE 14.5 Molecular structures of BMS 299897 and avagacestat.

14 ± 21% and 59 ± 12% of control vehicle, respectively, at a very high dose (30 mg/kg) (54). This was attributed to the extremely low $brain_{concn}/plasma_{concn} = 0.01$ of the compound (54). To improve brain-penetration, 1,2,4-oxadiazole derivatives were synthesised, and after a few optimisations BMS-708163 was synthesised. BMS-708163 not only showed weak inhibitors of CPYs (IC_{50} in the range of 20-40 μM), but also a high brain to plasma ratio ($brain_{concn}/plasma_{concn} = 2.4$) in dogs (54). BMS-708163 (Avagacestat) is a small molecule with Mw=520.9 Da, XLog P=4.63, and water solubility < 1 mg/mL. Prior to clinical studies, Avagacestat was subjected to a comprehensive nonclinical toxicology (safety) programme (50). As the drug was intended for continuous administration, repeat-dose toxicity studies were conducted over 6 months in rats and 1-year in dogs. There was a reduction in the number of peripheral B cells at doses ≥ 3 mg/kg/day in dogs after 52 weeks of drug administration, which was related to the interference with notch signalling. There were other toxicological observations such as renal changes, effects on bones, GI toxicity, and ovarian changes both in rats and dogs at doses above human therapeutic exposure levels (1.6x- 12x). It was concluded that Avagacestat demonstrated good oral tolerability in rats and dogs over 1 year at higher human therapeutic exposure doses (50).

Avagacestat was evaluated in a Phase I clinical trial at a single oral dose of 50 mg in both healthy and AD/MCI diagnosed patients. Avagacestat demonstrated favourable safety profiles, with dizziness and headache being the most common adverse events (AEs) (55). Then Avagacestat was evaluated in a Phase II clinical trial, with a total of 209 patients with mild to moderate AD, who were randomised into double-blind treatment and placebo groups at four oral doses of 25, 50, 100, or 125 mg over 24 weeks (56). More AEs were observed with 100 and 125 mg/day doses, including GI AEs, which led to a high discontinuation rate. The number of discontinuations increased with increasing Avagacestat dose. Serious AEs were comparable between treatment groups. Significant decreases were found in cerebrospinal fluid (CSF) biomarkers, as follows: 100 mg dose reduced $A\beta_{38}$, and 125 mg dose reduced $A\beta_{38}$, $A\beta_{40}$, and $A\beta_{42}$ (56). Therefore, Avagacestat met its physiological goals, which were reduction of Aβ peptides in the brains of AD patients. On the other hand, the cognitive function outcomes did not improve compared to placebo, and even the Mini-Mental State Examination (MMSE) score reduced significantly more than placebo for the 100 mg dose after 24 months. There were no significant changes from placebo for 25-mg and 50-mg dose groups. These unfavourable outcomes could be attributed to significant withdrawals from the trial, in particular for higher doses, and unexpected stability of cognitive function in the placebo group. At the end, Bristol-Myers Squibb did not take Avagacestat to Phase III clinical trial, not due to the observed AEs, but because it was clear that reducing Aβ in the brain does not help in maintaining cognitive function. Furthermore, the outcomes of Avagacestat Phase II clinical trial indicated that widespread inhibition of an enzyme in the body could lead to unfavourable side effects. It is the opinion of the author of this chapter that Avagacestat should have been formulated within a brain-targeting drug delivery system. Then, the number of withdrawals might have been reduced due to AEs, and better understanding could have been obtained for the cognitive function effects of Avagacestat at higher doses. Furthermore, although Avagacestat was optimised to cross the BBB, perhaps more optimisations would have been needed to reduce the drug dose.

Despite setbacks in the previous Avagacestat Phase II clinical trial, Bristol-Myers Squibb took Avagacestat to another Phase II clinical trial, but this time for the treatment of prodromal AD, with the aim that early identification of AD and effective early management could have desired therapeutic outcomes (57). This time the treatment group received 50 mg/day for the first 2 weeks and then 125 mg/day. However, to reduce withdrawals from the trial due to AEs, patients were allowed to reduce the dose to 50 mg/day or even 25 mg/day. Even with this treatment protocol, 45 out of 132 patients withdrew due to AEs. The treatment was continued for 104 weeks. The CSF $A\beta_{40}$ reduced between 10%–15% for all dose groups, but this was between 5%–9% for $A\beta_{42}$, which was not significantly different from placebo levels. The MMSE score declined more in the placebo group, but not significantly from the treatment group. Avagacestat trials were not halted due to AEs, but they were halted because stopping beta-amyloid didn't work.

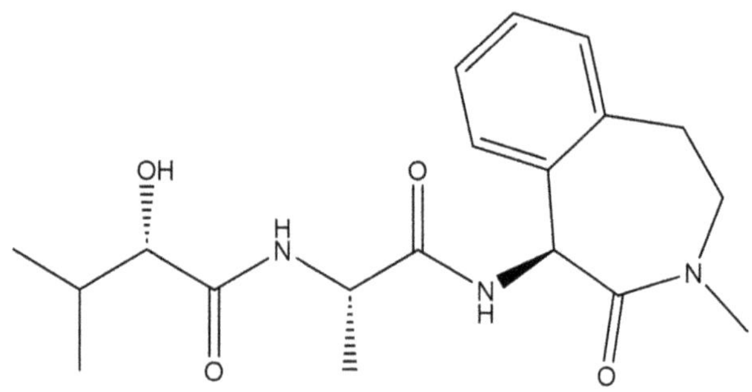

FIGURE 14.6 Molecular structure of semagecestat.

Bristol-Myers Squibb confirmed that conclusion in a 2013 announcement at the Alzheimer's Association International Conference in Paris. Study patients in a midstage trial showed hints of "negative cognitive effects." suggesting that patients get worse on anti-amyloid therapy. Is it possible to conclude, then, that beta-amyloid is protective if reducing it causes patients with beta-amyloid and Alzheimer's to get worse (58)?

Semagacestat (LY450139) was another γ-secretase developed by Eli Lilly & Company (Figure 14.6) (59). Semagacestat is a small drug molecule (Mw= 361.44 Da, XLogP=1.3, <1.2 mg/mL water solubility). Semagacestat inhibited production of $A\beta_{40}$ with IC_{50}=2.7 ± 0.9 μM from APP-C99, and inhibited Notch-100 with IC_{50}= 71.7 ± 13.5 nM (60). Investigating inhibition of Notch was important to ensure reduced drug side effects. Semagecestat (LY450139) demonstrated reduction of hippocampal Aβ burden over 40% with an oral administration dose of 1 mg/kg in PDAPP transgenic mice (61). Chronic administration of LY450139 to PDAPP transgenic mice over 5 months further reduced hippocampal $A\beta_{1\text{-}42}$ (30%–50%) than control groups at the oral dose of 30 mg/kg/day (62). At the time of writing this chapter, a published paper could not been found about nonclinical safety studies of Semagacestat (LY450139); however, this would have been conducted as Semagacestat (LY450139) was tested in three Phase I clinical studies (63–65). Semagacestat (LY450139) was tested in a wide dose range of 5 mg–280 mg with up to 84% Aβ generation inhibition of CNS Aβ production in healthy men (age range, 21-50 years) (65). Notch inhibition was a concern, as in the Phase I clinical trial, women were excluded with possible effects on fertility (65). These studies demonstrated no clinically significant AEs (63). Hence, Semagacestat entered in a Phase II clinical trial with primary outcome measures including AEs, plasma and CSF Aβ levels, vital signs, electrocardiographic data, and laboratory safety tests results with the initial oral dose of 60 mg/day for two weeks then 100 mg/day for 6 additional weeks; then either continuation on 100 mg/day for another 6 weeks, or randomly allocating to 140 mg/day for a 6-week period in patients with mild to moderate AD (66). The total drug exposure period was 14 weeks. Three subjects developed skin rash and another three developed hair-colour change. Although there were reductions of CSF $A\beta_{40}$ and $A\beta_{42}$ between placebo and treatment groups, in particular for $A\beta_{40}$ (20% reduction) in the 100 mg/day treatment group, the difference did not reach the significance level (P=0.15). Surprisingly, the CSF $A\beta_{42}$ levels increased (did not reach the significance level) from the baseline for the 140 mg/day treatment group. Furthermore, patients on 140 mg/day treatment group demonstrated improvements on AD Assessment Scale-Cognitive subscale (ADAS-Cog 11) scores (an increase of 1.66), but this was not significantly different from the placebo group (P=0.36). Eli Lilly & Company conducted Phase III clinical trials. This movement to Phase III clinical trial, without satisfactory outcomes from Phase II clinical trial is known as "overriding" (67). This is based on the fact that the number of recruited patients may not be high enough to demonstrate the medicinal product effects from statistical point of view. Eli Lilly & Company launched three Phase III clinical trials for Semagacestat (LY450139) with the titles of:

a. Effect of γ-Secretase Inhibition on the Progression of Alzheimer's Disease: LY450139 Versus Placebo, which was also called Interrupting Alzheimer's Dementia by Evaluating Treatment of Amyloid Pathology (IDENTITY, Clinical Trial.gov NCT00594568),
b. Effects of LY450139, on the Progression of Alzheimer's Disease as Compared With Placebo (IDENTITY-2, Clinical Trial.gov NCT00762411),
c. A Study of Semagacestat for Alzheimer's Patients (IDENTITY XT, Clinical Trial.gov NCT01035138) (68).

IDENTITY XT was designed for those patients who completed IDENTITY or IDENTITY-2 trials and saw benefits/hoping to see benefits from Semagacestat. Table 14.3 summaries these clinical trials. It can be seen from the inclusion criteria that only women with inactive ovaries were recruited, which was due to the concerns that Elli Lilly & Company had about possible Notch inhibition effects on fertility by using Semagacestat (69).

Then came along Begacestat, a Notch-1-sparing γ-secretase inhibitor for the treatment of AD (73), developed by Wyeth (acquired by Pfizer in 2009 for $68 billion). Begacestat is a hexafluorinated derivative of valine containing a sulfonyl group, with Mw=391.7 Da, XLogP3=3.7, and reported water solubility of 0.23 mg/mL. In the development process of Begacestat, Wyeth company conducted a high-throughput screening (HTS) of the Wyeth and ArQule compound collection (73). The compounds were tested on the hAPPCHO cells, a Chinese hamster ovary cell line engineered to express human APP protein. This investigation led to identifying two hits **1** and **2** (Figure 14.7), with $IC_{50}A\beta_{40}$=5449 nM, and $IC_{50}A\beta_{40}$=2214 nM, respectively.

The hit compounds were not potent γ-secretase inhibitors. Then, structure-activity relationship (SAR) studies were conducted on the hit compounds, which led to the discovery of Begacestat (Figure 14.7). This compound contained trifluoromethyl groups to improve *in vivo* stability of the compound. Begacestat demonstrated potent inhibition of $A\beta_{40}$ ($IC_{50}A\beta_{40}$=15 nM) and Notch-1 sparing selectivity (73). Begacestat reversed cognitive deficits in TG2576 CFC mouse model of AD at a dose of 10 mg/kg orally (73). However, it is believed that Phase I clinical trial of Begacestat (NCT00547560) was discontinued due to the lack of efficacy (74).

14.3.2 Development of Brain-Targeting Antibodies

The main part of this section is provided in Chapter 7 "Targeting the Brain with Monoclonal Antibodies." In this part, only the regulatory aspect will be explained. Regulatory approval rate was 24–30% for non-cancer antibody therapeutics that entered Phase I clinical trial from 2000 to 2019 (75). The number of FDA-approved mAbs was 12 out of 55 in 2023 (76). The number of new chemical entity drugs approved by the FDA was always higher than biologics from 1999 to 2023 (76).

The drug development process for brain-targeting antibodies has steps similar to small molecules, as shown in Figure 14.2. However, there are differences. For example, mAbs have inherent risks for adverse reactions that are mediated by the immune system. These include infusion reaction, anaphylaxis, cytokine storms, immune suppression and developing autoimmune diseases (77, 78). Therefore, it is important that all biological data should be considered during non-clinical evaluations (regulatory safety and toxicity studies) to determine if there are indications for severe immune reactions such as cytokine storms (79). As a result, the ICH guideline S6 (R1) [Preclinical safety evaluation of biotechnology-derived pharmaceuticals] addresses this issue. In addition, safety concerns may arise not due to the mAb, but the presence of impurities derived from animals or cells that mAbs are derived from. Therefore, the purification processes should be optimised to remove impurities and contaminants. As these may not be removed completely, therefore, the product should be characterised in

TABLE 14.3
Semagacestat Phase III Clinical Trials and Outcomes

Identifier	Acronym	Inclusion Criteria	Exclusion Criteria	Recruited	Biomarker Outcome	Cognitive Function Outcome	Ref
NCT00594568	IDENTITY, Alzheimer's disease patients without depression were randomly assigned to receive 100 mg or 140 mg of Semagacestat or placebo daily for 76 weeks.	• Meets criteria for mild to moderate Alzheimer's disease (AD) with MMSE score of 16-26 at Visit 1 • Modified Hachinski Ischemia Scale score of less than or equal to 4 • Ceriatric Depression Scale score of less than or equal to 6 • A magnetic resonance imaging (MRI) or computerised tomography (CT) scan in the last 2 years with no findings inconsistent with a diagnosis of AD • If female, must be without menstruation for at least 12 consecutive months or have had both ovaries removed.	• Is not capable of swallowing whole oral medication • Has serious or unstable illnesses • Does not have a reliable caregiver • Chronic alcohol or drug abuse within the past 5 years • Has ever had active vaccination for AD	1537	• Dose-related reduction in serum uric acid and phosphorus levels • hepatocellular injury • More deaths occurred among patients receiving Semagacestat • Serum $A\beta_{40}$ and $A\beta_{42}$ levels reduced • No significant change in CSF $A\beta_{40}$ and $A\beta_{42}$ levels • More skin cancers • More hair colour change	• MMSE significantly reduced for 140 mg/day dose compared to placebo	(70)
NCT00762411	IDENTITY-2 60 mg of Semagacestat orally once daily for 2 weeks followed by 100 mg orally once daily for 2 weeks, then 140 mg orally once daily until Week 88	Same as above	Same as above	556, but 534 not completed, study was halted by the company	No information available	Slight worsening of ADAS-11 score compared to placebo	(71)

(Continued)

TABLE 14.3 *(Continued)*
Semagacestat Phase III Clinical Trials and Outcomes

Identifier	Acronym	Inclusion Criteria	Exclusion Criteria	Recruited	Biomarker Outcome	Cognitive Function Outcome	Ref
NCT01035138	Identity XT Continuation of study with 140 mg of Semagacestat up to 24 months	• Meets National Institute of Neurological and Communicative Disorders and Stroke/Alzheimer's Disease and Related Disorders Association (NINCDS/ADRDA) criteria for probable Alzheimer's Disease • Completed Semagacestat study IDENTITY or study IDENTITY-2 through 88 weeks • Must continue to have a reliable caregiver • Capable of swallowing whole oral medication • Agrees not to participate in other investigational compounds for the duration of study	Same as above	73, but all terminated the study	Not available	Not available	(72)

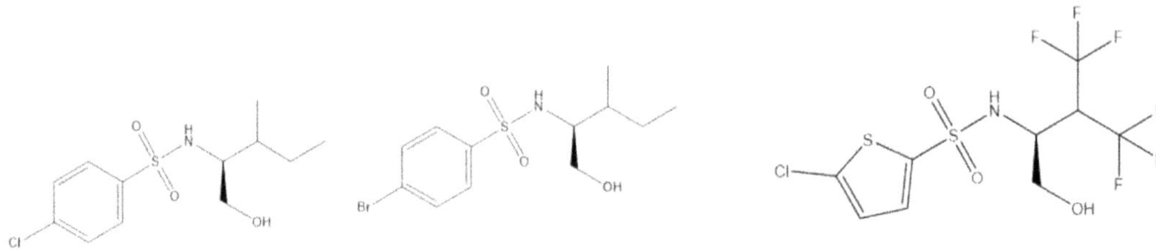

Compound 1 Compound 2 Begacestat

FIGURE 14.7 A diagram showing molecular structures of hit compounds 1 and 2 in the drug development process of Begacestat.

terms of containing these impurities. The ICH S6(R1) suggests that the preclinical safety studies should include toxicities related to the impurities and contaminants, as well as the active ingredient itself.

The ICH guideline indicates that the product tested in non-clinical pharmacological and toxicology studies should be comparable with the product proposed for the initial clinical studies. If there is a significant change to the manufacturing process (such as improving the yield) during the product development programme prior to clinical studies, then additional pharmacokinetics, PD and/or safety studies may be required.

Part of immune reactions in clinical studies can be identified in non-clinical studies. For example, it has been reported that skin rashes were observed both in non-clinical studies carried out in non-human primates and clinical studies for anti-EGFr antibodies (80). To achieve desired therapeutic effects, the mAbs should have high affinity and specificity towards target human proteins. To identify optimal antibodies, high-throughput screening methods have been developed (81, 82).

14.3.2.1 Preclinical Safety Testing for MAbs

Non-clinical (Preclinical) safety studies are conducted to determine the pharmacological and toxicological effects not only for initial human studies, but throughout clinical development. ICH guideline (S6 (R1)) states that when generic mAbs are comparable to a product that is widely used in clinical practice then they may need less extensive toxicity testing. According to ICH S6 (R1) guideline the nonclinical safety testing should consider age, selection of the relevant animal species, physiological state, the manner of delivery, and the stability of the test material under the conditions of use. The selection of the relevant animal species, age, and physiological state are explicitly elaborated below.

a. Selection of the relevant animal species: Non-human primates are the only relevant species for preclinical safety assessment (83), although other species such as rabbits and rats also have been used for single dose toxicity as stated in the summary product characteristics of Synagis (84). While mice, guinea pigs, and *cynomolgus* monkeys were studied in the preclinical studies of natalizumab (85). Natalizumab (TYSABRI®) is a recombinant, humanised anti-human α4 integrin immunoglobulin (Ig) G4k antibody indicated for the treatment of MS and Crohn's disease (86). Although studies in mice did not show growth of melanoma by the administration of Natalizumab, there were cases of melanoma in patients with MS receiving Natalizumab (87, 88).
b. Age/Reproductive and Developmental Toxicity: As part of preclinical studies, Natalizumab was tested on female Hartley guinea pigs ≥12 weeks of age for its effects on fertility (89). These studies were conducted according to ICH guideline S5A (1994): Detection of Toxicity to Reproduction for Medicinal Products. When the female guinea pigs were presumed mated, this was defined as gestational day 0 (GD 0). Natalizumab was tested on 4 groups

of presumed pregnant guinea pigs (30/group), in which three groups received Natalizumab (3, 10, or 30 mg/kg) by bolus i.v. injection every other day from GD 4 to GD 30. This time interval was chosen to ensure the development of the fetal placenta and heart. One group of the presumed pregnant guinea pigs received vehicle only. Natalizumab was also tested from GD 30 of pregnant guinea pigs to GD 30 of their second pregnancy. Seventy pregnant guinea pigs received 30 mg/kg of Natalizumab every other day and 30 pregnant guinea pigs received vehicle. Natalizumab was administered every 4 weeks at a dose of 300 mg, which was about 4 mg/kg. Therefore, the pregnant guinea pigs were exposed to much higher doses compared to humans. The number of animal deaths was similar between control and treatment groups. Foetal histology did not show changes in the heart, thymus, spleen, and intestinal tract of foetuses in the treatment groups. Anti-Natalizumab antibodies were detected in all of the female animals and most of foetuses at 3 mg/kg (89). The levels of anti-Natalizumab antibodies and number of animals (both females and foetuses) presenting anti-Natalizumab antibodies reduced by increasing Natalizumab dose (89).

c. Physiological state: Disease-induced/transgenic animal models may form part of preclinical studies. ICH guideline S6 states that the use of diseased animal models may provide further information about the pharmacological action of the product, as well as pharmacokinetics and safety. For example, Lecanemab is a monoclonal antibody developed for the treatment of early AD (90). Lecanemab is a humanized version of murine antibody mAb158, and this antibody was designed to selectively target Aβ protofibrils. mAb158 was tested in APP transgenic mice with the Arctic (E693G) and Swedish (K670N, M671L) mutations. mAb158 was administered intraperitoneally weekly to four- to five-month-old transgenic mice for 17 weeks at a relatively low dose of 3 mg/kg. The treatment significantly reduced the accumulation of soluble Aβ protofibrils and total Aβ in the brains of 8- to 9-month-old transgenic mice compared to the control transgenic mice (91). mAb158 was further tested in transgenic mice demonstrating selective binding to soluble Aβ protofibrils (92). Additionally, BAN2401 (humanised version of mAb158) targeted soluble Aβ aggregates in the human AD brain extracts (92). The uptake of radiolabelled mAb158 ([^{125}I]mAb158) was demonstrated in transgenic mouse brain, which increased by the age of the mouse (93). It was also found that 0.05%–0.15% of total injected dose reached the brain of the 24-month-old transgenic mice.

14.3.3 Examples of the Development Process for MAbs

Summaries of the development processes for selected mAbs for the treatment of brain disorders are provided below.

14.3.3.1 Lecanemab

BAN2401 (Lecanemab) was tested in a Phase I clinical trial (NCT01230853) in mild to moderate patients with AD (94). The study was split into two parts. A single ascending dose study in which doses of 0.1, 0.3, 1, 3, 10, and 15 mg/kg were tested. A multiple ascending dose study evaluated doses of 0.3, 1, and 3 mg/kg administered every four weeks with a total of four doses over four months and a dose of 10 mg/kg biweekly, with a total of seven doses over four months, all administered as i.v. infusion (94). The study found that BAN2401 (Lecanemab) was well-tolerated for all the treatment regimens. The presence of amyloid-related imaging abnormalities (ARIA, E for oedema, H for hemorrhage) was seen only in 5% of subjects (3/60). The BAN2401 half-life was estimated to be 6.9 days at 10 mg/kg dose, with serum levels of BAN2401 remaining above 10 μg/mL up to 25 days after a single injection. Lecanemab was evaluated in a Phase II clinical trial in early AD, mild cognitive impairment due to AD and mild AD dementia at three doses: 2.5 mg/kg biweekly, 5 mg/kg one group receiving monthly and another group receiving biweekly, and 10 mg/kg one group receiving monthly and another group receiving biweekly (95). At 18 months, 10 mg/kg biweekly Lecanemab reduced brain amyloid while showing a drug-placebo difference in the favour of the active treatment with CSF biomarkers supporting the treatment effect. Lecanemab was evaluated in another Phase II clinical trial with similar dose regimens, but this time the cognitive functions of the patients with early AD were tested. These were: Alzheimer's Disease Composite Score (ADCOMS), Clinical Dementia Rating-Sum-of-Boxes (CDR-SB), and ADAS-Cog14 (96). The 10 mg/kg biweekly dose regimen significantly reduced the decline of cognitive functions compared to the control group (placebo) as measured by the ADCOMS and ADAS-Cog14 (96). Lecanemab was tested in a Phase III clinical trial with a total of 1795 participants who were enrolled to receive Lecanemab (10 mg/kg every two weeks, 898 patients) or placebo (897 subjects) over 18 months. The effects of the treatment were measured by ADCOMS, CDR-SB, ADAS-Cog14, Alzheimer's Disease Cooperative Study–Activities of Daily Living Scale for Mild Cognitive Impairment (ADCS-MCI-ADL), and brain amyloid burden, which was evaluated by positron emission tomography (PET) (90). Lecanemab resulted in moderately but significantly ($P<0.001$) less decline of cognitive function on all measured scales, as well as brain amyloid burden. However, Lecanemab showed AEs, which were infusion-related reactions, but most importantly there were cerebral microhaemorrhages, cerebral macrohaemorrhages, or superficial siderosis (ARIA-H; 17.3% with Lecanemab and 9.0% with placebo) (90). FDA approved the use of Lecanemab to treat adult patients with Alzheimer's Disease on 6 July 2023. However, the European Medicines Agency (EMA) refused the approval of Lecanemab in a letter issued on 26 July 2024, on the grounds that the observed effects of Lecanemab on delaying cognitive decline would not counterbalance the risk of serious AEs associated with the medicine. The frequent occurrence of ARIA was considered the most important safety concern with Lecanemab (Leqembi), an AE that was detected by brain imaging. ARIA indicates swelling of the brain tissue and risks of bleeding in the brain. Although most cases of ARIA in the main study were not serious and did not involve symptoms, some patients had serious AEs, including large bleeds in the brain, which required hospitalisation. The seriousness of this side effect should be considered in the context of the small effect seen with the medicine.

14.3.3.2 Aducanumab

Aducanumab is another antibody that was developed to selectively target aggregated Aβ plaques. Aducanumab was administered intraperitoneally at a dose of 30 mg/kg to 22-month-old female Tg2576 transgenic mice (97). Preclinical studies discovered a half-life of 2.5 days for Aducanumab with a brain C_{max} of 0.1062 μg/g and plasma C_{max} of 181 μg/mL (approximate brain/plasma concentration of 0.0006). The brain/plasma AUC ratio was 1.3%. These data demonstrated penetration of the antibody into the brains of the transgenic mice, although not in high quantities. Aducanumab reduced the Aβ burden in the brains of transgenic mice potentially through the recruitment and phagocytosis by microglia. Aducanumab (BIIB037) was tested for the first time in humans through a randomized double-blind, placebo-controlled single ascending-dose study evaluating the safety, tolerability, and measuring PK parameters of Aducanumab in patients with mild-to moderate AD (43). Aducanumab was tested in doses of 0.3, 1, 3, 10, 20, 30, and 60 mg/kg. There were three patients that developed serious AEs who received 60 mg/kg of the drug dose. These were related to the development of ARIA-E with one patient developing ARIA-H too. ARIA-related AEs were not observed at Aducanumab doses ≤ 30 mg/kg. The terminal elimination half-life showed an increase from 326 h (13.6 days) at 0.3 mg/kg dose to 606 h (25.3 days) at 60 mg/kg dose. This was 571 h (23.8 days) for 30 mg/kg dose. Therefore, in the subsequent Phase Ib clinical trial, Aducanumab was tested in patients with prodromal or mild AD with doses up to 10 mg/kg each month administered intravenously between October 2012 and January 2014 (97). The ARIA incidents increased with the treatment dose and 47% of patients showed this ARIA with 10 mg/kg dose. CDR-SB scores were significantly lower compared to the placebo group in week 54 of the treatment with 10 mg/kg dose for 23 patients. The number of withdrawals due to AEs increased and reached 10 patients (out of 32) for 10 mg/kg dose. The median plasma half-life was determined to be 21 days (about 10 times more than the transgenic mice), which was consistent with the previous Phase I clinical trial measurements. The results were considered promising (98), and Biogen conducted two Phase III clinical trials (EMERGE and ENGAGE) to evaluate Aducanumab in early AD (99) without sponsoring a Phase II clinical trial, which is known as a bypass of Phase II clinical trial (67). Aducanumab was administered at a low dose of 3 mg/kg to ApoE ε4+ patients or 6 mg/kg to ApoE ε4− patients, and a high dose of 10 mg/kg to ApoE ε4− or 6 mg/kg to ApoE ε4+patients. The drug was administered via saline-diluted i.v. infusions every 4 weeks over 76 weeks (20 doses in total). Aducanumab significantly reduced the brain amyloid load compared to placebo group at all doses by week 78. Aducanumab resulted in 22% statistically significant reduction in the decline of CDR-SB, 18% reduction in decline of MMSE, and 27% reduction in the decline of the ADCS-MCI-ADL scores for the 10 mg/kg dose regimen in the EMERGE cohort. The incidences of ARIA-E were much higher for both doses in both cohorts (99). As the EMERGE trial met its primary outcome and the ENGAGE trial did not, then Biogen submitted the data to the U.S. Food and Drug Administration (FDA) for review and possible marketing approval. Following reviewing the data, an FDA Advisory Committee voted (10 opposed, 0 in favour, 1 undecided) to recommend that the Agency not approve Aducanumab based on the current data. A decision that was later deemed to reflect a perspective that was too narrow, ignoring important scientific findings and clinically important outcomes (98). However, the FDA granted approval to Aducanumab in June 2021, despite the lack of clear clinical evidence for the drug's cognitive benefits (100).

Although not part of the regulatory approval process, the manufacturer initially priced the drug at an astonishing $56,000 per year, a price that was later reduced to $28,200. The unclear approval process of the FDA and the drug's price prompted criticism and led Congress and the FDA to launch investigations into relations between Biogen and key figures at the regulatory agency (101). The House Oversight Committee's investigation found that the FDA worked unusually closely with Biogen staff in a "collaborative workstream" that massaged the poor trial results and changed study endpoints. The FDA considered only the reduction of brain amyloid load as the evidence of clinical benefit, while the plaque formation and disease progression have never been connected. FDA statisticians never accepted non-significant clinical benefits, and as a result they were excluded from key meetings. It was further discovered that part of Biogen's presentations were prepared by the FDA for the advisory committee. Also, the FDA's own investigations found the collaborations between the drug company and the regulatory agency "exceeded the norm." The House report also identified that the initial price tag did not represent the drug's research and production costs, and unfortunately was driven by the company's "ambition to make history," as one of the corporate documents revealed. Furthermore, a group of international researchers, clinicians, and policy experts met on 15 December 2021, and they voted unanimously to recommend that the FDA withdraw its approval for Aducanumab (102). In light of these investigations, high price tag, refusal of insurance companies to cover the cost of the treatment using Aducanumab, and global recommendations Biogen decided to withdraw its application for a marketing authorisation of Aducanumab in Europe on 20 April 2022. Biogen stated in a letter to the EMA that the decision was based on the fact that the data provided thus far would not be sufficient to support a positive opinion on the marketing authorisation of Aduhelm (Aducanumab). Furthermore, Biogen announced in January 2024 that Aduhelm would be discontinued starting in November 2024.

14.3.3.3 Solanezumab

m266 is a murine version of Solanezumab (103), which demonstrated a reduction in the Aβ load in the brains of PDAPP transgenic mouse models of AD via the sink biological process (104). This study showed that m266 formed complexes with all free Aβ species in the plasma, and as a result the equilibrium between blood Aβ levels and CNS Aβ levels shifted to compensate for the loss of Aβ in the blood, leading to the reduction of the brain Aβ load (104). Records of preclinical studies are not available in the public domain for the immunogenicity of Solanezumab due to the repeated drug administrations. These would have been conducted prior to obtaining approval for clinical trials. Solanezumab (LY2062430) entered a Phase I clinical trial in patients with mild to moderate

AD testing single doses of 0.5, 1.5, 4.0, and 10.0 mg/kg (105). The outcomes were that a single dose of Solanezumab was generally well tolerated, and infusion reactions occurred only at higher doses. Dose-dependent changes were observed in the plasma and CSF Aβ levels. Solanezumab was tested in a Phase II clinical trial, in which patients with mild-to-moderate AD again well tolerated i.v. administration of the drug doses up to 400 mg weekly (106). Solanezumab was evaluated in two Phase III clinical trials EXPEDITION 1 and EXPEDITION 2 in mild to moderate AD patients, with ClinicalTrials.gov numbers of NCT00905372 and NCT00904683, respectively (107). Solanezumab was administered at a dose of 400 mg every 4 weeks for 18 months. The primary outcomes of the EXPEDITION 1 trial were the changes from baseline on the ADAS-Cog 11 score (range, 0 to 70, with higher scores indicating greater cognitive impairment) and ADCS-ADL scale (range, 0 to 78, with lower scores indicating worse functioning) at week 80 of the treatment. For EXPEDITION 2, the primary outcome was change from baseline on the ADAS-Cog 14 score (range, 0 to 90, with higher scores indicating greater impairment). The values for the plasma and CSF levels of Aβ and CSF levels of tau and phospho-tau formed part of the secondary outcomes. Although the primary outcomes were not met in either EXPEDITION 1 or 2 trials, the free and total $A\beta_{40}$ as well as total $A\beta_{42}$ levels significantly decreased in the CSF compared to the placebo group in week 80 of the treatment. Solanezumab Phase III clinical trials demonstrated a significant decrease in the total $A\beta_{42}$ CSF levels compared to placebo, supporting the sink effects (107). Most importantly, ARIA-E occurred only in nine subjects of the treatment group and in four subjects of the control group. ARIA-H was less frequent in the treatment group (50) compared to the placebo group (57). These observations suggested that lack of or minimal Solanezumab penetration into the brain might have been associated with a low incidence of ARIA-E or ARIA-H for this immunotherapy method. Solanezumab was tested in other trials for the treatment of AD (108, 109), but did not meet the primary outcomes. As a result of these negative outcomes, Eli Lilly & Company announced the end of development for Solanezumab on 8 March, 2023.

14.3.3.4 Bapineuzumab

Bapineuzumab is an anti-amyloid antibody that targets all forms of Aβ plaques. The murine version of Bapineuzumab (antibody 3D6) was tested on the PDAPP transgenic mouse models of AD (110). 3D6 reduced the amyloid burden in the brain of the animal models more effectively compared to m266. Furthermore, 3D6 antibody was tested in the PDAPP transgenic mice to investigate potential brain microhaemorrhage (111). 3D6 was able to completely clear vascular Aβ, which causes cerebral amyloid angiopathy (CAA), at 3 mg/kg dose administered intraperitoneally once weekly. However, microhaemorrhage was observed in a subset of vessels for all doses (0.1–3 mg/kg). The incidence and severity of CAA-associated microhaemorrhage increased when PDAPP transgenic mice were treated with the N-terminally directed 3D6 antibody, which bound to the CAA associated with the microvessels (112). 3D6 antibody was tested in a large number of PDAPP mice (345 animals) at a dose of 3 mg/kg for 36 weeks (113). Antibody 3D6 transiently increased the occurrence of microhaemorrhages in leptomeningeal vessels of PDAPP mice. However, this was reduced by week 36. It was concluded that vascular leakage events, such as microhaemorrhage, might be related to the removal of vascular Aβ. However, with continued treatment, the initial susceptibility period was followed by restoration of vascular morphology and reduced vulnerability to further vascular leakage incidents. Further information about the safety of antibody 3D6 could not be found in the literature and it would be assumed that the evidence was gathered to obtain regulatory approval of Bapineuzumab to be tested in humans.

Bapineuzumab was tested in a Phase I clinical trial, and a half-life of 21 to 26 days was measured in patients with mild-to-moderate AD (114). This finding supported a 13-week dosing interval for Bapineuzumab. The antibody was tested in a Phase II clinical trial with the doses of 0.5, 1, or 2 mg/kg over 78 weeks in patients with AD, who received the medicine every 13 weeks (115). Carbon-11-labelled Pittsburgh compound B (^{11}C-PiB) PET is a marker of cortical fibrillar amyloid-β load in the brain and this was used to investigate whether Bapineuzumab would reduce cortical fibrillar amyloid-β load in the patients. The outcomes confirmed that Bapineuzumab significantly reduced the brain Aβ load compared to control subjects but not significant reductions in CSF biomarkers such as CSF tau and clinical endpoints such as ADAS-cog 11 scores. Phase III clinical trials of Bapineuzumab were conducted, despite not completely satisfactory outcomes from Phase II trials. This may be considered overriding Phase II clinical trial outcomes. Salloway et al. 2014 conducted two double-blind, randomized, placebo-controlled, Phase III trials evaluating Bapineuzumab by recruiting patients with mild-to-moderate Alzheimer's disease — one involving 1121 carriers of the apolipoprotein E (APOE) ε4 allele and the other involving 1331 noncarriers (116). Bapineuzumab was administered via i.v. infusions every 13 weeks for 78 weeks. CSF phospho-tau concentrations significantly decreased at the 0.5 mg/kg dose for APOE ε4 allele carriers and 1 mg/kg dose for noncarriers. Only 0.5 mg/kg dose showed slight but not significant less decline in the Disability Assessment for Dementia (DAD), with scores ranging from 0 to 100 and higher scores indicating less impairment, score at week 78 of the treatment for the noncarrier patients. ARIA was observed in 14 patients receiving 0.5 mg/kg dose, 31 patients receiving 1 mg/kg dose, and only in 1 patient in the placebo group. Due to the lack of Bapineuzumab efficacy and serious AEs in the clinical trials (117), Pfizer and Johnson & Johnson announced they were stopping the development of Bapineuzumab on 7 August, 2012.

There were follow-up studies to better understand the lack of efficacy and serious AEs observed for Bapineuzumab. Antibody 3D6 was tested in the TG-ArcSwe mouse models of AD and wild-type mice and resulted in slight penetration of large molecules through the BBB (118). This could be an explanation for the observed prevalence of ARIA during treatment with Bapineuzumab. In addition, the ARIA AEs could be due to the activation of perivascular macrophages and CAA resulting from the deposition of Aβ plaques in meningeal and cortical (parenchymal) arteries, arterioles, capillaries, and veins (119). The lack of Bapineuzumab efficacy was attributed to poor target engagement in plaques, as Bapineuzumab targets only the first 5 amino acids of Aβ (both soluble and insoluble) with high affinity of 3–5 nM. Then, the truncation of the Aβ at the first amino acids by amino-terminal proteases would reduce the affinity of Bapineuzumab towards Aβ (120). Ideally, antibodies should target the $A\beta_{4\text{-}10}$ amino acid sequence (i.e. FRHDSGY) (121).

Hypersensitivity reactions to mAbs may occur, which include flushing, chills, fever, and tachycardia. These are classified as immunogenicity (122). A desensitisation protocol may be applied, which consists of gradually exposing the patient to mAbs. This approach reduced type I immune reactions (consisting of the activation of mononuclear phagocytes (123)) from 63% of patients to 32% in patients (122).

14.4 REGULATORY ASPECTS OF BLOCK COPOLYMER MICELLES AND LIPOSOMES WITH EMPHASIS ON FORMULATIONS FOR TARGETING THE BRAIN

Nanotechnology has earned significant attention for the treatment of neurological disorders. NPs are materials with overall dimensions in the range of 10 to 1000 nm(132) with different shapes such as spheres (133), rods (134), fibres (135), or irregular (136). NPs can have filled or core-shell structures (137). Several types of NPs have been developed for drug delivery to the brain, such as polymeric NPs (138), liposomes (139), exosomes (140), and peptide-based NPs (141). There are several advantages of using NPs for drug delivery to the brain:

- Facilitating crossing the BBB
- Sustained release of active ingredients
- Protection of sensitive drugs
- Reduced side effects
- Improved solubility and stability
- Potential for non-invasive administration
- Multifunctional capabilities

Until 2012, there was a lack of officially published regulatory standards in the evaluation of nanomedicines. The regulatory agencies examined NP formulations case-by-case (142). To provide a greater clarity, FDA issued six final guidance documents for industry on the use of nanotechnology (https://www.fda.gov/science-research/nanotechnology-programs-fda/nanotechnology-guidance-documents) in June 2014: (143) as follows:

1. Assessing the Effects of Significant Manufacturing Process Changes, Including Emerging Technologies, on the Safety and Regulatory Status of Food Ingredients and Food Contact Substances, Including Food Ingredients that Are Colour Additives,
2. Considering Whether an FDA-Regulated Product Involves the Application of Nanotechnology,
3. Liposome Drug Products: Chemistry, Manufacturing, and Controls; Human Pharmacokinetics and Bioavailability; and Labelling Documentation,
4. Safety of Nanomaterials in Cosmetic Products,
5. Use of Nanomaterials in Food for Animals,
6. Drug Products, Including Biological Products, that Contain Nanomaterials Guidance for Industry.

FDA considers nanoscale in the size range of 1–100 nm. FDA issued a final guidance for industry related to liposome drug products in April 2018: Liposome Drug Products Chemistry, Manufacturing, and Controls; Human Pharmacokinetics and Bioavailability; and Labeling Documentation (https://www.fda.gov/media/70837/download). The EMA has also issued four scientific guidelines on nanomedicines to help medicine developers in preparing marketing authorisation applications for human medicines (https://www.ema.europa.eu/en/human-regulatory-overview/research-and-development/scientific-guidelines/multidisciplinary-guidelines/multidisciplinary-nanomedicines) as follows:

1. Reflection paper on the data requirements for intravenous iron-based nano-colloidal products developed with reference to an innovator medicinal product (26 March 2015)
2. Reflection paper on the data requirements for intravenous liposomal products developed with reference to an innovator liposomal product (21 Feb 2013)
3. Joint Japanese Ministry of Health, Labour and Welfare of Japan (MHLW) and EMA reflection paper on the development of block copolymer micelle medicinal products (19 December 2013)
4. Reflection paper on surface coatings: general issues for consideration regarding parenteral administration of coated nanomedicine products (22 May 2013).

MHLW also issued a detailed guideline about the development of liposome medicinal products (Guideline for the Development of Liposome Drug Products https://www.nihs.go.jp/drug/section4/160328_MHLW_liposome_guideline.pdf) in March 2016. All three guidelines (FDA, EMA, and MHLW) demonstrate the requirements of highly detailed documents for the approval of liposome formulations. A recent publication listed regulatory-approved nanomedicines and those in clinical trials (144).

14.4.1 Block Copolymer Micelle Products

A large number of block copolymer micelle products have been in clinical trials (145). Micelles can be formed by block copolymers or peptides. There are several regulatory-approved micelle formulations such as Genexol-PM (Samyang Co., Seoul, Korea), sold as Cynviloq or IG-001 (145). Therefore, micelle formulations can reach the market; hence, regulatory guidelines have been prepared for them. Perhaps regulatory experience in processing these formulations helped in the preparation of the guidelines. In the following parts, regulatory guidelines are stated, and examples are provided related to polymeric micelles. The regulatory requirements are expressed in italic font. The regulatory requirement "Joint MHLW/EMA reflection paper on the development of block copolymer micelle medicinal products Final" has been the centre of attention in this chapter. It should be noted that the document contains "General considerations" in the "Non-clinical studies," followed by more specifications. Therefore, a sense of repetition may be noticed for this part. More attention has been paid to drug delivery to the brain by block copolymer micelles, wherever possible.

14.4.1.1 Properties Related to the Block Copolymer Micelle

Micelle formulations may have different properties, but normally they share the quality characterisations listed below:

- *Size*: This normally can be measured by dynamic light scattering (146).
- *Morphology*: This typically is achieved by transmission electron microscopy or scanning electron microscopy (147).
- *Zeta potential*: This is usually determined by dynamic light scattering method (148).
- *Other surface properties (e.g. targeting ligand)*: This criterion could imply that the presence of targeting ligands should be proven on the surface of the micelles. This is a challenging criterion, but use of scrambled peptides could be one way of presenting desired surface properties of micelle NPs crossing the BBB (149), formation of desired NP structure such as formation of nanofibres (150), or entering target cells such as glioblastoma cells (151).
- *Association number*: Amphiphilic compounds self-assemble into micelles in aqueous solutions at concentrations above the critical micelle concentration (CMC). The micelle aggregation (also called association) number is defined as the number of amphiphilic molecules that form a micelle. Typical aggregation number varies between 20 and 118 for surfactants (152). An aggregation number of 91 has been reported for brain-targeting peptide micelles (141). The aggregation number may be determined by employing static light scattering technique (153) or by fluorescence quenching methods (152) through the calculation of the ratio of the micelle weight-average to the molecular weight of each amphiphilic compound molecule.
- *Determination of CMC*: The CMC of brain-targeting amphiphilic compounds can be determined by spectrophotometric methods such as employing 1-(2-pyridylazo)-2-naphthol (PAN)(146) and pyrene as probes (141, 146), tensiometry (154), or conductometry (154). Using the PAN procedure, a serial dilution of the amphiphilic peptide is prepared typically with 20 ml each, and 500 μL of PAN solution in pentane at 1.6 mM concentration is added to each peptide solution. The PAN/amphiphilic compound solutions are left for a short duration (30 minutes) for the evaporation of pentane. The UV absorptions of solutions are measured at 470 nm. The CMC values are determined from a sudden increase in the UV absorption (Figure 14.8).

  For the tensiometry method, the CMC is determined by employing the Du Nouy ring platinum detachment method. Serial dilutions of the amphiphilic compound are prepared inside beakers with diameters much larger than the platinum ring. The surface tension is plotted against the amphiphilic compound concentrations, and the CMC of the micellar solutions is determined from the breakpoint of the surface tension (similar to Figure 14.8). Static light scattering has also been employed to determine the CMC of surfactants (153). In this approach, the intensity of the scattered light is plotted against the amphiphilic compound concentrations, and the CMC value is determined from the sudden increase in the scattered light intensity (similar to Figure 14.8).
- *Drug loading*: Micelles can be employed for the delivery of APIs such as siRNA or microRNA to the brain. Two parameters should be determined: loading capacity (LC) and loading efficiency (LE). LC is defined as the ratio of encapsulated or loaded drug mass

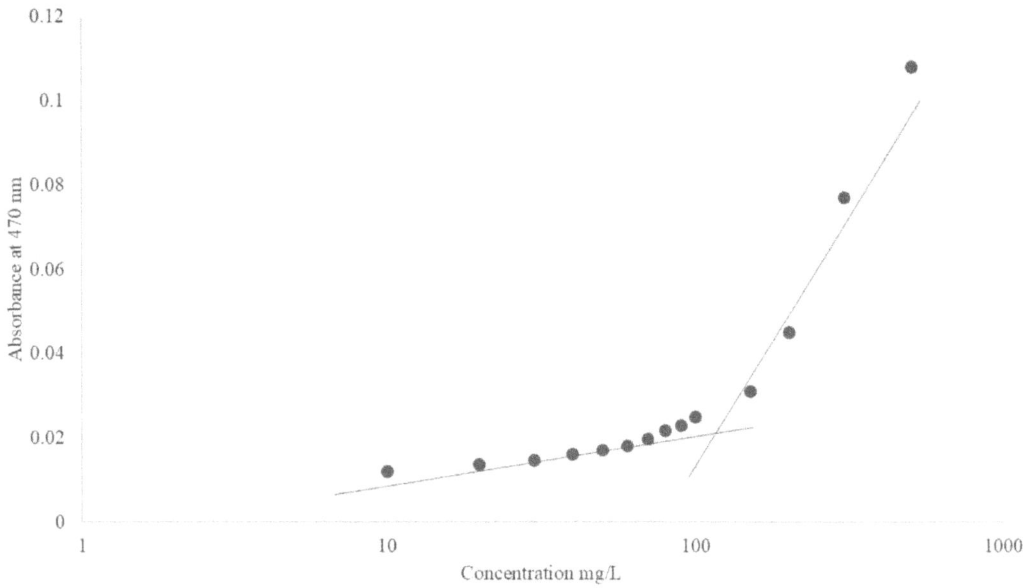

FIGURE 14.8 The plot of PAN absorbance against PAN concentration, with best-fit lines to the horizontal and vertical parts of the data. The crossing point of the best-fit lines corresponds to CMC, which is 114 mg/L in this example.

(M_{drug}) over the total mass of the carrier (total amphiphilic compound), i.e. (M_{drug}+$M_{carrier}$), as in Equation 14.1:

- $$LC = \frac{M_{drug}}{M_{drug} + M_{Carrier}} \times 100 \quad (14.1)$$
- While LE is defined as the ratio of encapsulated or loaded drug mass (M_{drug}) over the total mass of the drug added at the beginning of the formulation process (*M added drug*), as in Equation 14.2
- $$LE = \frac{M_{drug}}{M_{added\ drug}} \times 100 \quad (14.2)$$
- LC can be determined by excessive dilution of micelles to decrease the amphiphilic compound concentration below the CMC, leading to micelle disintegration and release of active ingredient. The drug content then may be determined by applying high-pressure liquid chromatography (HPLC) (148). The LC can also be determined indirectly. In this approach, the solution containing micelles is transferred to a centrifugal cellulose filter tubes with a cut-off diameter that allows crossing of only free API. The filter unit containing micelle solution is centrifuged at a typical speed of 10,000 rpm. The filter retains the micelles, but free API crosses the membrane. Then, the amount of the API is determined in the filtrate (146). Validation studies should be conducted to ensure not retention of the API by the cellulose filter.

- *Chemical Structure*: This is relevant to the determination of the monomer numbers for each block of a copolymer, as well as identification of the molecular structure of the block copolymer. For example, a molecular structure of poly [polyethylene glycol methacrylate-co-2-(acetoacetoxy) ethyl methacrylate]-*b*-poly [2-(dimethylamino) ethyl methacrylate] may be represented by [P($PEGMA_9$-co-$AEMA_6$)-*b*-$PDMAEMA_{10}$] or [P($PEGMA_m$-co-$AEMA_n$)-*b*-$PDMAEMA_p$] with the values of *m*, *n*, and *p* being indicated (155). It is also important to determine the polydispersity index (PDI), which indicates the variations in the M_W of each block (156). The broader Mw leads to larger PDIs. ^{1}H NMR (400 MHz) spectrum has been employed to determine the degree of polymerisation for each block of copolymer. Gel permeation chromatography was employed to determine the molecular weight distribution of block copolymers (155).

- *Physical state of the active substance*: The active substance typically is in the form of a powder such as paclitaxel (PTX), which may have limited water solubility (157). Cremophor® EL (a non-ionic surfactant) was added to increase the aqueous solubility.
- *Viscosity*: The viscosity of a micellar solution may be influenced by the shape of micelles. For example, the formation of worm-like micelles increased the viscoelastic properties of the media compared to spherical micelles (158). Therefore, measuring the product viscosity may be a quality control checkpoint to ensure desired micellar shapes (159).
- *In vitro stability of the block copolymer micelle in plasma and/or relevant media*: The *in vitro* stability of micelles can be determined by asymmetric flow field-flow fractionation technology (AF4)(160) in combination with the dynamic light scattering (DLS) technique (161). In this method, NPs are separated by two opposing flows and a cross flow, in a flat thin channel with a semi-permissive bottom wall membrane. The AF4 allows separation of micelle NPs based on their hydrodynamic sizes with NPs of smaller hydrodynamic size and higher diffusion coefficient values being eluted at an earlier time point (shorter retention time). NPs with a large hydrodynamic size and a low diffusion coefficient value elute later (longer retention time). Size-based separated NPs exit the channel, and a detector (such as refractive index) is used at the end of the channel to determine the presence and concentration of NPs. Then AF4 fractograms are plotted, which are refractive index versus retention time. When micelle NPs are incubated with serum, the AF4 separates micelle NPs from the serum proteins. In this procedure, serum proteins exit earlier than NPs, and also smaller NPs exit earlier than larger NPs (160). The change in the particle size of micelle NPs is evidence of micelle instability in serum. It was shown that covalently cross-linked poly(ε-caprolactone)-co-poly(1,2-dithiolane-carbonate)-*b*-poly (ethylene glycol) (p(CL-co-DTC)-PEG) showed excellent stability in human plasma. While non-covalently crosslinked micelles composed of poly(ε-caprolactone)-b-poly (ethylene glycol) (PCL-PEG) showed an increase in NP size when incubated with human plasma, which suggested instability of the micelles (161). It appears AF4 would be able to better detect subtle changes in particle size than the DLS.
- *In vitro release of the active substance from the block copolymer micelle product in plasma and/or relevant media*: Slow or controlled release from micelles is required, in particular outside of the target area. Typically, phosphate buffer has been used as the release media (162, 163). Rapid release of block-copolymer micelle cargo has been reported in the circulation after i.v. administration (164). This rapid *in vivo* release was in line with the *in vitro* release studies in the serum and plasma, suggesting the importance of pre-testing of the *in vitro* release behaviour in the plasma for predicting the fate of a delivery system *in vivo* (164). Switching from 1,2-distearoyl-*sn*-glycero-3-phosphoethanolamine-N-[amino(polyethylene glycol)-5000] (PEG-DSPE) micelles to PEG block-poly(N-hexyl stearate L-aspartamide) micelles improved serum stability of the micelle formulation (165). The guideline does not address the effects of plasma protein corona on the size of NPs. It has been shown that the formation of protein corona increased the average NP size, which was detected by the DLS technique (166).
- *In vitro degradation of the block copolymer in plasma and/or relevant media*: The degradation of block copolymers in plasma is important mainly to prevent premature drug release due to the early degradation of the delivery system despite the fact that block copolymers should be biodegradable and biocompatible. To meet this regulatory requirement, poly(2-deoxy-2-methacrylamido-D-glucose) [PMAG]-*b*-poly amino acid [PAA] (PMAG-*b*-PAA) NPs were incubated with enzyme solutions or the plasma at 37°C. At pre-determined time

intervals, a sample of particle-free supernatant was obtained, and the amounts of amino acid (isoleucine in this case) were determined by applying a HPLC method (167). It was found that the biodegradation rate was low in human blood plasma.

14.4.1.2 Properties Related to the Manufacturing Process

The following points should be addressed in relation to the manufacturing process:

- *Validated process for reconstitution*: Block copolymers may be freeze-dried (lyophilised), such as poly(ethylene oxide)-block-poly(β-benzyl-L-aspartate) (PEO-*b*-PBLA) loaded with doxorubicin (DOX) (168). The weight average diameters of DOX-loaded micelles were 37 nm (determined by the DLS method) before freeze-drying, and 28 nm after reconstitution. Inspecting particle size distribution suggested the secondary association of NPs leading to the formation of larger sizes (120 nm) as well as formation of smaller size micelles (possibly loss of DOX during freeze-drying process). Stern-Volmer plots of the fluorescence quenching suggested that PEO-*b*-PBLA micelles loaded with DOX were largely unaffected by freeze-drying and reconstitution. In other words, DOX remained inside the micelles during the freeze-drying process and after the reconstitution step. When the word "validated" appears in a regulatory guidance, it implies the collection and evaluation of data. This is from the process design stage through commercial production, which establishes scientific evidence that a process is capable of consistently delivering quality product. Therefore, the collected data should demonstrate not only the reproducibility of the reconstitution process, but also the reliability of the collected information. A successful validated process should include:
  - Understanding the sources of variation,
  - Detecting the presence and degree of variation,
  - Understanding the impact of variation on the process and ultimately on product attributes,
  - Controlling the variation in a manner appropriate with the risk it represents to the process and product.

  Therefore, if cryoprotectants are added to micelle solutions prior to freeze-drying process, their effects and impurities should be investigated on the reconstituted product.
- *Validated process for ensuring sterility*: Microfiltration has been suggested for the sterilisation of block copolymers (169), and it appears this technique was applied for the sterilisation of Genexol®-PM polymeric micellar formulation (170) and carfilzomib-loaded polymeric micelles (171). The sterilisation of NPs will be elaborated further in section 14.5.1.

14.4.1.3 Properties Related to the *In Vivo* Behaviour

The following requirements are addressed in relation to *in vivo* stability of polymeric micelles:

- *Osmolarity*: In this section, osmolality (number of fully dissociated substances in one kilogram of solvent, normally water) and osmolarity (number of fully dissociated substances in one litre of solvent, normally water) have been used interchangeably. Osmolality is one of the key formulation parameters for products to be instilled into the eye or injected. Normal human serum osmolality is within the narrow range of 280 to 295 mOsm/kg (172, 173). The average osmolarity value is 302.2 ± 8.3 mOsm/L for normal human tears (174). When only NPs are dispersed in water, the osmolality of the final solution may be very low (37 mOsm/kg), but potentially may cause eye irritations for the ophthalmic formulations due to hypo-osmolarity (175). Therefore, NaCl may be added to match tear osmolality. On the other hand, the osmolarity was in the range of 304–313 mOsm/L for an ophthalmic formulation that contained poly

(N-palmitoyl-N-monomethyl-N,N-dimethyl-N,N,N-trimethyl-6-O-glycolchitosan) micelles as well as active ingredient (cyclosporine) and glycerol. The osmolarity was determined by using Roebling Milliosmol Osmometer (Geminibv, Apeldoorn, The Netherlands) coupled with a digital display and a freezing needle (171). Measuring osmolarity was considered as part of the storage stability studies. Osmolarity/Osmolality is known to be one of the key formulation parameters contributing to the tolerability of injectable products for both systemic and local administrations (176). Extremely hypertonic injectable products (>8000 mOsm/kg) have been reported for marketed drugs (177). Intraperitoneal administration of hypertonic solutions (0.95 M NaCl plus 0.05 M $NaHCO_3$ resulting in 1900 mOsm/L osmolarity) increased cerebral blood flow by 50%, causing the BBB disruption and potential demyelination (178). Administration of >1000 mOsm/L significantly increased the chance of developing phlebitis or infiltration in patients receiving peripheral parenteral nutrition (179). For i.v. administrations, osmolality in the range of 300 ± 30mOsm/kg is recommended (176). I.v. injection of hypertonic solutions caused bruising of the tail in mice and subcutaneous injections in rabbits (180). The solutions were 7.5% hypertonic saline (2564 mOsm/L) or hypertonic saline-dextran (7.5% NaCl and 6% Dextran-70). It is recommended to inject hypertonic solutions with osmolality greater than 600 mOsm/kg into central veins of animals to avoid vascular damage such as thrombophlebitis (181).

- *Fraction of active substance that is surface-associated*: Two methods for physical loading of hydrophobic drugs into polymeric micelles have been developed: (1) Dialysis and (2) O/W emulsion (182). In the dialysis method, both drug and copolymer are dissolved in a water-miscible solvent (such as N,N-dimethylacetamide or water itself) and the resulting mixture is dialysed against water using a dialysis membrane with MWCO 12-14 kDa (183). This process removes the organic solvent, untrapped drug molecules, and also part of encapsulated drug molecules (due to the release from NPs). A drug entrapment of 20.4% (w/w) was reported for poly-(ethylene oxide)-poly(β-benzyl L-aspartate) (PEO-PBLA) micelles loaded with indomethacin (183). In the O/W emulsion method, the drug is dissolved in an organic solvent such as chloroform, and the copolymer aqueous solution is added, which leads to the formation of an oil/water emulsion. The drug gets incorporated into the micelles as the solvent evaporates (184). The solution is filtered to remove unbound drug molecules and other impurities (183). An entrapment of 22.1% (w/w) was reported for the encapsulation of indomethacin in PEO-PBLA micelles by this method (183). The O/W emulsion method was more efficient for loading DOX into PEO-PBLA micelles (65% O/W vs 8% dialysis) (182). In this method, most of the drug molecules are incorporated inside the micelles, such as 80% of loaded drug (182), and potentially 20% of drug molecules attaching to the surface of NPs. The surface-bound drug may lead to rapid drug release, leading to the accumulation of hydrophobic drugs in the vascular system. Therefore, it is essential to determine the fraction of the active ingredient trapped inside micelles and the fraction that is surface-bound. It was found that methoxy polyethylene glycol-block-poly(d,l-lactide) micelles encapsulating PTX accumulated 85% of loaded PTX in the hydrophobic core of micelles (185), suggesting 15% of PTX was surface bound. This was determined by mild centrifugation conditions (2000 × g) at 4 °C, and by measuring PTX in the supernatant (185). Burst PTX release in the range of 20%–40% of loaded drug from triblock copolymers P($MeOx_{33}$-*b*-$BuOx_{26}$-*b*-$MeOx_{45}$) micelles could account for surface-bound PTX molecules (163). It may be hypothesised that the location of the drug within polymeric micelles depends on the physicochemical properties of the drug and the length of the polymer chain of the amphiphilic molecule, with more hydrophilic drugs being closer to the hydrophilic surface of the micelle (186). The application of the DEEP-STD NMR allowed researchers to determine the location of fluconazole and indomethacin within micellar surfactants such as Pluronic®F-127 (187). This information may be used to explain the drug release behaviour of the investigated system.

- *Release rate and place of active substance release*: The release rate and place of active substance release have been demonstrated for Genexol®-P (170). To achieve this, mice bearing murine B16 melanoma tumour were administered intravenously with Genexol®-PM at a dose of 50 mg/kg, and four mice were sacrificed at 3 min, 30 min, 1, 2, 3, 4, 8, 12, and 24 h post drug administration. To determine the drug release rate at the target site, the amounts of PTX were determined in the samples of the tumours at the sampling time intervals (170). A mixed polymeric micelle system was employed to load lurasidone hydrochloride (LH), a medicine that is used for the treatment of bipolar disorders such as schizophrenia (188). The LH-loaded mixed micelles (LHMM) were administered intravenously (a dose equivalent to 1 mg/kg of LH) to Wistar male rats, and the brains were harvested at 0.5, 1, 1.5, 2, 3, 4, 6, 10, and 24 h post-drug administration. Then the amounts of the active ingredient were measured in brain samples. The study demonstrated the release of the active ingredient at the target site. It should be noted that micelles may migrate to off-targets, and therefore, the time course of the drug levels may be obtained in other organs as well (164).
- *Block copolymer degradation rate and place of degradation*: The block-copolymer degradation rate has been determined by manufacturing implants followed by surgery to insert the implants under the skin (intradermal). The inserted samples were extracted from the skin at predetermined time intervals, and the differences in the weights of the samples were determined before and after insertion (189). The low concentrations of polymers in biological samples make the quantification of the polymers challenging. Therefore, a triple-quadrupole mass spectrometer-based method was developed to achieve detection limits in the range of nanogram to picogram per mL (190). This method was applied to determine the concentration of oleyl hyaluronan (HA-C18:1) polymeric micelles encapsulating DOX. Following i.v. administration of the DOX-loaded polymeric micelles, homogenates of the liver from animals were obtained at predetermined time intervals. The samples were fractionated by size exclusion chromatography, and each size fraction was degraded by Hyaluronan lyase from *Streptococcus pneumoniae* and analysed for ΔHA2 and ΔHA2C18:1 (digestion products) by LC-MS/MS. It was shown that HA-C18:1 was already partly cleaved after 30 min of administration and the resulting fragments exhibited different ΔHA2/ΔHA2-C18:1 ratios from the original HAC18:1 chain, and the liver was the main compartment for the elimination of oleyl-modified HA (190).

14.4.1.4 Stability

The conventional stability studies should be conducted for block copolymer micelles encapsulating the active substance according to the ICH Q1A(R2) guideline. The stability studies may not be required for a drug substance if regulatory approval already exists, and GMP-grade materials can be obtained. Formal stability studies should be designed and conducted for copolymer micelles. This includes conducting stability studies of the copolymer micelles in the container closure system proposed for marketing. It may be suggested to pack freeze-dried copolymer micelles in impermeable containers such as glass vials for Genexol®-PM or CPC634 (docetaxel-entrapped core-cross-linked polymeric micelles) (191). The stability studies should cover both physical and chemical stabilities. The physical stability studies should include average NP size, the release of physically or chemically entrapped active substance, and secondary aggregation. Methoxy poly(ethylene glycol)-*b*-(N-(2-benzoyloxypropyl) methacrylamide (mPEG-*b*-p(HPMAm-Bz)) micelles were loaded with PTX via a nanoprecipitation method (192). The micelle formulations were stored frozen at –20°C for 180 days and average micelle size was determined after thawing the samples at different time points (0, 15, 30, 90, 180 days post-production). The average micelle size remained within 10% deviations (68-83 nm) from the original average micelle size (75 nm on day 0). Furthermore, the release profiles of PTX from the micelles were comparable on day 0 (freshly prepared) and day 180 (stored frozen at –20°C) (192). The secondary aggregation of micelles was defined as the formation of large units from the aggregation of small micelles (193). Figure 14.9 demonstrates a TEM image

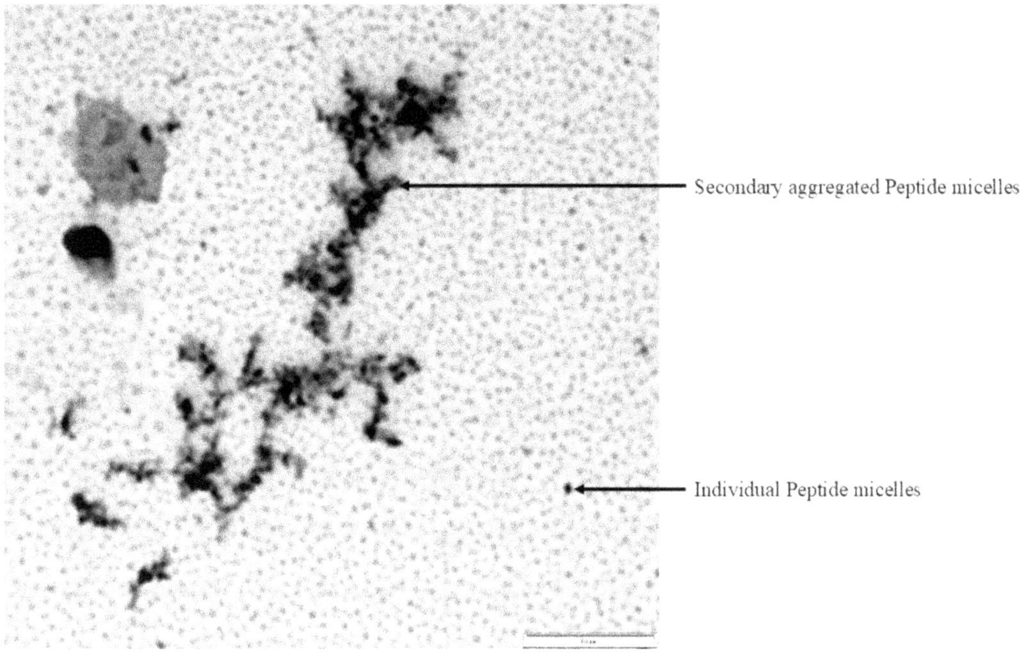

FIGURE 14.9 The aggregation of peptide micelles, individual micelles had diameters in the range of 20–30 nm, scale bar = 200 nm (from Fattahi et al unpublished data).

of brain-targeting peptide micelles encapsulating miR-34a (Author's unpublished data), exhibiting individual micelles, as well as aggregated micelles (secondary aggregation).

Figure 14.10 presents the size distribution of peptide micelles measured by DLS, which confirms the secondary aggregation observed in the matching TEM image. However, the individual NP size cannot be detected in the DLS data. The secondary aggregation was also reported by DLS data for Pluronic 123 micelles encapsulating curcumin (194).

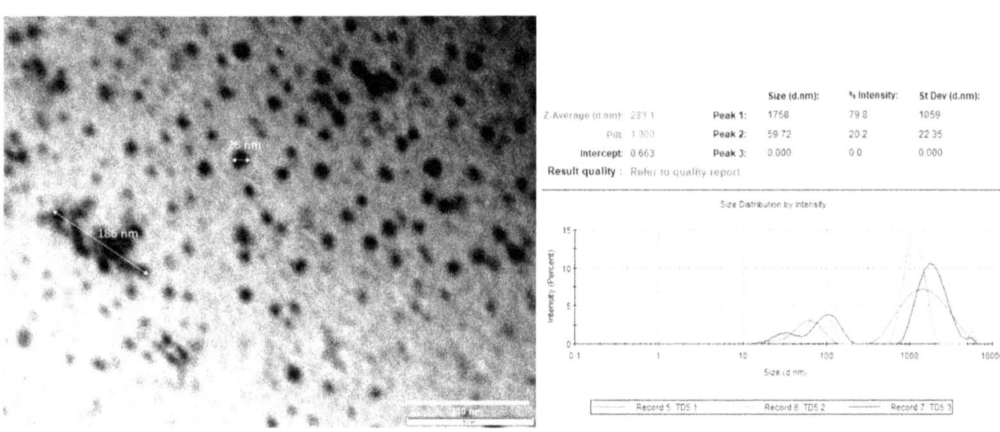

FIGURE 14.10 TEM image and corresponding DLS size distribution of peptide micelles showing both individual micelles and aggregates (scale bar = 200 nm). The DLS size distribution reveals multiple peaks corresponding to different particle populations: Individual NPs and their aggregates. Representative size measurements are overlaid on the TEM image using the scale bar dimensions, demonstrating good agreement between the measured particle sizes and the peaks observed in the DLS size distribution analysis (from Fattahi et al unpublished data).

According to ICH Q1A (R2), the stability studies should also be conducted for reconstituted formulations, such as lyophilised polymeric micelles, in relation to the active substance stability under recommended storage conditions. For example, the chemical stability of PTX was shown in mPEG-*b*-p (HPMAm-Bz) micelles over 6 months at –20°C (192). It should be noted that the ICH Q1A(R2) guideline recommends 12 months storage at –20°C ± 5°C for drug products intended for storage in a freezer.

Studies on the chemical stability of the block copolymer should also be conducted. The chemical stabilities were tested for poly (ε-caprolactone), poly (D,L-lactide), and poly (D,L-lactide-co-glycolide) NPs stored at different temperatures and in different media (195). It was found that average molecular weight of poly (ε-caprolactone) remained relatively unchanged over 12 months stored in water and at room temperature or 4°C. If an active ingredient is chemically conjugated to a block copolymer, then the stability of the chemical bond should also be demonstrated. For example, core-crosslinked polymeric micelles (CCPM), also known as CPC634, encapsulated docetaxel (DTX) within the core of the micelles via a hydrolytically sensitive ester bond (191). Due to the hydrolysis of the ester bond between DTX and the CCPM during storage at room temperature or even fridge, this led to the premature release of the drug (191). Therefore, the formulations were stored at −60°C, which demonstrated a good stability for a minimum period of 5 years. This led to the use of a cold chain for the supply of CPC634 for clinical trials. Therefore, it was needed for shipment on dry ice, continuous temperature logging, and use of −60°C freezers at clinical sites. Expectedly, these all contributed to a very expensive model of transportation. To reduce the costs, a lyophilised form was developed (191). It was found that critical quality attributes (CQAs) such as size, morphology, drug retention, and drug release profiles of freeze-dried product (containing trehalose as cryoprotectant) were identical to those of non-freeze-dried CPC634. Similar comparisons should be made for long-term stability studies over a 12-month storage at room temperature (25°C or 30°C), 5°C, or –20°C, depending on the recommended storage conditions. Photostability studies are also required according to ICH Q1B, if the nanomedicine contains photosensitive ingredients (196).

Micelles typically load small hydrophobic molecules; however, if polymeric micelles contain biotechnological products such as antibodies, then the ICH Q5C stability guideline should be followed. For example, brain-targeting PEG-poly-L-lysine (PEG-PLL) micelles were developed for the treatment of glioblastoma. These micelles encapsulated an IgG antibody via cross-linking to PEG-PLL with angiopep 2 as brain targeting ligand on the surface of the micelles co-encapsulating PTX through electrostatic interaction with lysine (197).

14.4.1.5 *In Vitro* Methods, Using Conditions Relevant to the Proposed Use

As part of the stability programme, suitable *in vitro* methods should be used to determine:

- *The Release Rate of the Active Substance Entrapped in the Block Copolymer Micelles*
- *The Rate of Release of Active Substance Chemically Bound to Block Copolymer Micelles.*

The above requirements are to demonstrate that upon storage the release rate is not affected when micelles are not intended to release their cargo in the plasma/serum. For example, PBS containing albumin at physiological concentration (40 g/L) was used to simulate the plasma as release media to investigate drug release from mPEG-*b*-p(HPMA-Bz) micelles (198). Similar studies should be conducted to evaluate the drug release upon storage under recommended conditions, for example, over 12 months for long-term stability.

The joint MHLW/EMA reflection paper guideline requires the development of discriminating, bio-relevant *in vitro* release methods for the purpose of:

- *Defining the release of active substance or block-copolymer substance conjugate from the block copolymer micelle when in the circulation.*

    - Bio-relevant *in vitro* release study is defined as using a release media that simulates *in vivo* conditions. For example, adding albumin at the concentration of 40 g/L to the phosphate buffer saline (PBS) may provide some degree of bio-relevance for the blood circulation (198, 199). The aim of bio-relevant *in vitro* release studies is to adequately predict the *in vivo* performance of a polymeric micelle drug product (200).
- *Defining the release of the active substance or block copolymer-active substance conjugate from the block copolymer micelle at the targeted site of action. The proposed media should reflect the physiological environment of the block copolymer micelle in use*:
    - For the purpose of drug delivery to the brain, CSF may be considered as the target release medium. As 99% of CSF is comprised of water with a pH value of 7.3–7.4 (201) therefore 0.01 M sodium phosphate buffer containing 150 mol/L NaCl (PBS), pH 7.4 (202) may be considered as a relevant physiological environment for the *in vitro* drug release from the brain-targeting NPs/micelles.
- *Defining the stability in storage.*
    - The bio-relevant release profiles should be reproducible upon storage at the recommended conditions.

14.4.1.6 Manufacturing Process and Process Control

It is required that the molecular structure of amphiphilic compounds be consistent from batch to batch. It was shown that variation in polymerisation of block copolymers can affect the micelle size (203). For example, $mPEG_{5k}$-*b*-p(HPMA-Bz)$_{18.5k}$ micelles had 14.1 nm (radius) in THF, and this was reduced to 4.6 nm (radius) for $mPEG_{5k}$-*b*-p(HPMA-Bz)$_{2.2k}$ block copolymer in the same solvent (203). Interestingly, the polymers of $mPEG_{5k}$-*b*-p(HPMA-Bz)$_{18.5k}$ formed micelles with a diameter of 80 nm and polydispersity index (related to NP size distribution) of less than 0.1 following the addition of a THF polymer content solution (203). This study found that increasing polymerisation variation (shown by polydispersity index) increased NP polydispersity index. Therefore, if NP polydispersity index of less than 0.1 is desired, then perhaps polymer polydispersity index should be $\leq$ 1.12. Furthermore, a well-defined manufacturing process is required to ensure that acceptable product is produced on a consistent basis. For example, it was shown that increasing the addition rate of an organic solution of $mPEG_{5k}$-*b*-p(HPMA-Bz)$_{18.5k}$ into an aqueous solution from 0.15 mL/min to 1.5 mL/min reduced average particle size from 82 nm to 62 nm (203). As well as recommendations related to block copolymers, the ICHQ8(R2) guideline (Pharmaceutical Development) should be followed. It is required to provide a detailed description of the synthesis process (204), extraction (205), and purification (206). The starting materials should be specified, as well as macromolecular reaction by-products. For example, polymers may be synthesised by atomic transfer radical polymerisation (ATRP), in which polymer chain ends equipped with an alkyl halide react with a transitional metal ligand, leading to the formation of a polymer and halide-transitional metal, which is a by-product (207). There could be side reactions (208), which produce unwanted compounds (209).

14.4.1.7 Non-Clinical Studies

14.4.1.7.1 General Considerations

It is required to commence conducting all non-clinical safety/PKPD studies when block copolymers are well characterised. The product stability (including upon storage) and dissociation of the micelles should also be determined and established in a physiologically relevant environment (*in vitro/in vivo*) (164, 210) prior to PK studies. The PK studies should address the following points:

a. *The rate of clearance of the block copolymer micelle containing entrapped or chemically bound active substance.* Clarithromycin was loaded into PLGA-PEG-phosphatidylethanolamine (PE) micelles and was intravenously injected into male Sprague-Dawley rats with an equivalent clarithromycin dose of 22 mg/kg (211). Micelles had 16% loading content and showed lower clearance, higher plasma concentration, and lower volume distribution than free clarithromycin i.v. administrations (211).

b. *The rate of dissociation of the block copolymer micelle. This may lead to release of block copolymer unimers (with or without bound active substance) that would have lower molecular weight (smaller size characteristics) and may display different clearance characteristics.* The size of clarithromycin-loaded PLGA-PEG-PE micelles increased by incubating in 10% FBS (211). NP size affected circulation half-life in humans, with 12 h circulation half-life (212) for Genexol-PM NPs with an average size of 23 nm (162) and 33 h (213) for CPC634 NPs with typical size of 65 nm (191).
c. *The rate of release of entrapped active substance from the block copolymer micelles.* The *in vitro* PTX release was slower from Pluronic P123 micelles compared to Taxol, and this was reflected as a higher mean plasma concentration-time profile for PTX delivered by Pluronic P123 micelles compared to Taxol in rats (214).
d. *The rate of release of active substance chemically bound to the block copolymer unimer.* Drug-amphiphilic conjugates have been developed to prevent premature drug release before reaching pathological sites. Camptothecin was chemically conjugated via a disulfide bond to a poly caprolactone-PEG-folate co-polymer. The breakage of the disulfide bond occurred in the presence of glutathione. I.v. injections of free camptothecin resulted in higher plasma concentration at 15 min post injection compared to camptothecin-conjugated formulation. However, the trend was reversed at 12 h postinjection. Physical encapsulation of camptothecin within micelles showed plasma concentrations of camptothecin between free camptothecin injections and camptothecin-conjugated formulations (215).
e. *The rate of degradation of the block copolymer.* Typically, copolymers are developed for a long degradation time in the body. For example, PLGA-PEG diblock copolymers formed NPs with detectable degradation up to 220 h (216). Therefore, the effects of copolymer degradation on the PK properties may not be detected for slow-biodegradable copolymers (217).
f. *Clearance and metabolism of free active substance.* This requirement may be addressed by administering free drugs to animal models for evaluating the PK properties. For example, PTX was intravenously administered as free drug, which also contained small amounts of ^{3}HPTX to CD1 mice (217). Blood samples were collected and PK parameters were determined. In parallel, ^{14}C labelled PTX was loaded to ^{3}H labelled methoxy poly(ethylene glycol)-block-poly(ε-caprolactone) (MePEG-*b*-PCL) copolymers and the micelles were intravenously administered to the mice. Employing radiolabelled PTX and block copolymers allowed investigation of the simultaneous distribution of PTX and the copolymer in the body following i.v. administration in mice. Unexpectedly, the $T_{1/2}$ (the amount of time required for the drug concentration in plasma to be reduced to exactly half of its starting concentration) of free PTX decreased from 6.45 h to 2.63 h using MePEG$_{114}$-*b*-PCL$_{104}$ micelles (217). The results suggested rapid partitioning out of the drug from the micelles in the blood circulation and perhaps another mechanism of elimination associated exclusively with the micelles. The metabolism of the active ingredient may already have been known for the active ingredient such as PTX, which is mainly metabolised by the liver's cytochromes P450 2C8 and 3A4 (218). It is important to determine if polymorphism of metabolising enzymes would have effects on the toxicity and efficacy of the active ingredient.
g. *The distribution of the block copolymer micelles:* DOX was loaded to lactose-modified poly(L-phenylalanine-co-γ-benzyl-L-lysine) (P(LP$_{10}$-co-ZLL$_{11}$)) micelles (GPM/DOX) (219), and the aim of lactose-conjugation was to develop micelles surface-decorated with lactose. There were two aims for the addition of lactose on the micelle surface: firstly, to prevent hydrolysis of the peptides by proteases, and secondly, to improve targeting of tumour cells. GPM/DOX micelles had an average diameter of 54 nm, and the semi-quantitative organ distribution was obtained by measuring the fluorescence signals related to the

fluorescence activity of DOX from different organs. Three formulations were evaluated: GPM/DOX, free DOX, and normal saline (as control) in H22-xenografted Kunming mice. The fluorescence activities were measured at 12 h post i.v. injection for the heart, liver, spleen, lung, kidneys, and tumour itself. GPM/DOX micelles accumulated more in the tumour than free DOX. It should be noted that organ distribution of NPs changes with time. For example, organ distribution of cetuximab-conjugated docetaxel-loaded poly(γ-glutamic acid) (γ-PGA) NPs significantly changed from 30 min compared to 2 h, day 1, and day 4 post i.v. injection in the MKN-28 xenograft gastric cancer nude mice model (220). A similar trend was also observed for poly(ethylene glycol)-block-poly(2-methyl-2-carboxyl-propylene carbonate-graft-dodecanol-graft-tetraethylenepentamine) (PEG-*b*-PCC-g-DC-g-TEPA) micelles loaded with GDC-0449 (hedgehog signalling inhibitor) and let-7b (a tumour suppressor miRNA) in orthotopic pancreatic tumour-bearing mice (221). The organ distribution of PEG-*b*-PCC-g-DC-g-TEPA micelles reached a maximum 0.25 h post i.v. injection in the kidneys and liver, but reached maximum in the tumour 4 h post injection. Organ content of PEG-*b*-PCC-g-DC-g-TEPA micelles significantly reduced 24h post-injection. For drug delivery to the brain, a time duration should be considered for NPs crossing the BBB and reaching the brain. A two-hour time was sufficient to observe strong signals in the brains of Sprague-Dawley rats from intravenously injected DiR-loaded Pluronic F127 and D-α-tocopheryl polyethylene glycol succinate mixed micelles (222). The micelles were also distributed to other organs such as the heart, liver, lungs, spleen, and kidneys by that time. However, as little as 0.5 h was sufficient to detect the amounts of PTX loaded to methoxy poly(ethylene glycol)$_{3000}$-poly(ε-caprolactone)$_{20000}$ copolymer micelles within the tumours of mice bearing C6 glioma xenograft (223). The micelles were decorated with an MMP2/MMP9 (overexpressed in the glioma tumours) cleavable peptide (ALMWP:E10PLGLAGVSRRRRRRGGRRRRC), which was selectively cleaved by MMPs and exposed the glioma penetrating peptide (CVSRRRRRRGGRRRR) (223). PTX amounts were also quantified in other organs (the liver, heart, spleen, kidneys, lungs) at the time intervals of 0.5 h, 1 h, 3 h, 6 h, 12 h, 24 h post i.v. injection of different NP formulations and free PTX (Taxol) (223).

h. *The interaction of the block copolymer micelle with plasma or serum proteins or blood cells.* It would be expected that the interaction of block copolymer micelles with serum proteins or blood cells would reduce the clearance of NPs from the blood stream and affect the drug organ distribution. It was shown that coating polymeric micelles with cell membrane mimetic polymers reduced the uptake of the NPs by macrophages, leading to higher blood concentrations in male New Zealand rabbits compared to uncoated NPs, as well as modified drug organ distribution (224). The uncoated NPs accumulated more in the heart, liver, spleen, lungs, and kidneys, compared to NPs coated with cell membrane mimetic polymers (224).

The regulatory guideline indicates that the rate and location of *in vivo* active ingredient release are to be determined. This has been considered as a crucial parameter which often determines the toxicity and efficacy of the active ingredient. This requirement is different from measuring the organ distribution of the active ingredient. For drug delivery to the brain, if the active ingredient is a nucleotide such as siRNA or miRNA, then NPs should deliver the cargo to neurons or glia cells and the expression of proteins would be measured (225, 226). This evaluation should also involve evidence of NPs reaching the brain, downregulation/upregulation of gene expressions both at mRNA and protein levels, as well as observing significant therapeutic effects. The experiments would include the use of negative control nucleotides to establish that the observed therapeutic effects are due to the successful delivery of the active nucleotides to the target cells in the brain. However, when the release of an active ingredient is required in the brain, then the release of active ingredient from the NPs may be measured by applying the microdialysis technique (227, 228). This method was applied to investigate the release of a model drug FITC-dextran MW 4000 (FD4) loaded into copolymeric micelles of methoxy-polyethylene glycol-poly lactic acid surface decorated with CD71 monoclonal antibody. The FD4-loaded NPs crossed the BBB following i.v. administration and the amounts of free FD4 were measured by microdialysis implanted in the ventral hippocampus of male Sprague Dawley rats (229). FD4 free concentration gradually decreased in the brain over three days.

To summarise, prior to non-clinical studies, the copolymer micelles should be well characterised in terms of chemical structure, composition, stability under storage conditions, micelle dissociation in the blood (also explained in the following sections), NP size, charge, stability of active-ingredient-copolymer conjugates, optimised manufacturing process with *in vitro* product consistency performance, and known levels of impurities.

14.4.1.8 Non-Clinical Pharmacokinetics

14.4.1.8.1 Analytical Methods

The analytical methods should be able to determine the concentrations of the active ingredient both in total and in free form in the blood, plasma, or serum, as well as the total concentration of active substance in organs and/or tissues. The analytical methods should also be validated (ICH Q2 (R1) Validation of Analytical Procedures: Text and Methodology). When the active ingredient is released from a micelle in the blood, it can be in free form, bound to plasma proteins, and red blood cells (230). Therefore, the total amount of the drug in the blood is the sum of entrapped drug amounts in micelles, amounts of bound drug to proteins/cells, and unbound (free) drug amounts. Ultrafiltration may be used to determine free drug concentration in plasma (231, 232). Ultrafiltration may be considered as a high-throughput assay, since it is straightforward and fast to conduct (ca. 30 min). However, it suffers from the leakage of peptides and proteins through the pores of the semipermeable membrane. Therefore, careful selections are required for molecular cut-off and pore size of the semipermeable membrane (230). The protein-bound drug fraction may be interpolated from an established correlation between free (ultrafilterable) and protein-bound drug concentrations. Then, the encapsulated drug concentration may be calculated by subtracting the ultrafilterable and protein-bound drug concentrations from the total drug concentration (233).

14.4.1.8.2 Pharmacokinetics

Suitable species and models should be employed for the measuring of *in vivo* copolymer micelle PK parameters. DSPE-PEG$_{2000}$-Angiopep-2 copolymer micelles were loaded with isoliquiritigenin to target the brain for the treatment of middle cerebral artery occlusion (MCAO) ischemic stroke (234). The mean plasma concentrations of isoliquiritigenin-loaded micelles were higher than free isoliquiritigenin administration following i.v. injection into Sprague Dawley male rats. Furthermore, the circulation time was longer for the drug molecules loaded into the micelles compared to the free drug administration. A similar pattern was also observed for isoliquiritigenin-loaded micelles and free isoliquiritigenin in the brains of MCAO mouse model following i.v. administration (234).

As indicated, suitable species and models should be selected for the PK studies. For the purpose of brain drug delivery, humans have the highest cardiac output (over 10%) to the brain compared to mice, dogs, and rats (less than 5%) (235). Humans also have a higher blood flow to the brain than non-human primates (236). However, non-human primates have been considered vital models for understanding human brain functions (237) and diseases such as non-human primate model of AD, which was developed by overexpressing human tau in the bilateral hippocampi of adult rhesus macaque monkeys (238). When direct delivery to the brain is required, larger-sized brains are more challenging and invasive to work with than small-sized brains (235). Perhaps due to the diffusional difficulties in larger brains, the iron oxide NPs stayed longer in the brains of macaques compared to mice following i.v. administration of the NPs (239). Rodents have been considered as an inexpensive animal model for developing new therapeutic techniques (240), however overreliance on rodents (mouse trap) may not lead to the development of effective treatments for humans (241). Therefore, it is essential to employ non-human primates in validating potential drugs for the treatment of human brain diseases (242, 243). The squirrel monkey has been considered as an animal model to fill the gap between humans and rodents in the translational research to humans (244). Non-human primates have been used as animal models for drug delivery to the brain by NPs (245). The pattern of organ distribution significantly changed for poly(ethylene glycol)-*b*-poly(D,L-lactide) micelles between rats and non-human primates for targeting the brain (245). When possible, the PK parameters of copolymeric micelles should also be determined in diseased animal models (221). However, the PK parameters have been determined in healthy animals for drug delivery to the brain (234). Perhaps this could be due to the technical challenges in developing neurologically diseased animal models. For example, a surgery is required to insert a monofilament into the cerebral middle artery to induce ischemic stroke, then after one hour from the operation, the removal of the monofilament is needed. To measure and calculate PK parameters, four to six animals may be required for each timepoint, which would require many diseased animals. When transgenic mouse models are employed, then the diseased animals may also be included to determine PK parameters (246, 247). PK models have been developed to compare the brain-therapeutic index of unformulated drug, NP-encapsulated drug, and drug in brain-targeting NPs (248). As expected, the brain-therapeutic index of brain-targeting NPs was much higher (159 times) than regular NPs (248).

The composition of the block copolymer micelles should be justified when conducting PK studies. For example, the use of PEG in block copolymers may be justified to enhance the circulation time of the drug-loaded micelles. In addition, the release of the active substance should be justified in the target tissue. For drug delivery to the brain, the release of active substances such as isoliquiritigenin may be justified in the brain to exert anti-inflammatory effects for the treatment of ischemic stroke (234).

The regulatory guideline requires justification of the product specification in terms of NP size, charge, and morphology by conducting organ distribution studies and applying variations in the physiochemical characteristics of the NPs (e.g. size, charge, shape). It is recommended to follow the ICH S3A (Toxicokinetics: A Guidance for Assessing Systemic Exposure in Toxicology Studies), ICH S3B (Pharmacokinetics: Repeated Dose Tissue Distribution Studies), ICH S6 (R1) (Preclinical safety evaluation of biotechnology-derived pharmaceuticals), and ICH M3 (R2) (non-clinical safety studies for the conduct of human clinical trials and marketing authorisation for pharmaceuticals).

The PK studies should also provide $C_{\max}$, half-life ($T_{1/2}$), and AUC of the block copolymer micelle product both for total active ingredient (encapsulated +free) and free active ingredient in the blood, plasma, or serum. PK studies have compared the PK parameters between micelle-loaded drug and free drug administration, for (p(CL-co-DTC)-PEG) micelles loaded with mTHPG (161), Pluronic 123 micelles loaded with PTX (214), and DOX loaded into PEG-PLA micelles (249). Furthermore, this regulatory requirement addresses distinguishing between free drug and encapsulated drug in the blood, following administration of the polymeric micelle formulations. As explained above, ultrafiltration may be used to determine free drug concentration in the plasma (231, 232). It should be noted that C_{max} is the same as the blood/serum/plasma concentration of the drug at t=0 following an i.v. administration. While $C_{\max}$ should be determined following other routes such as an oral administration, as shown for a mixed polymeric micelle formulation of itraconazole (ITZ-PM) (250).

Pharmacokinetic parameters should be measured at different dose levels and at appropriate time points. A micellar formulation containing monomethoxy poly (ethylene glycol)-*b*-poly (lactic acid) and poly(lactic acid) as the drug carrier material with encapsulating itraconazole (ITZ-PM) was injected intravenously at the doses of 5, 7.5, and 10 mg/kg/day for single dose and repeated daily dose for 28 days to male and female dogs, and PK parameters were determined (250). For a single dose administration, the blood samples were obtained at 30 min after starting the infusion, and 0, 5, 15, and 30 min, 1, 2, 3, 6, 12, 24, 36, 48, and 72 h after the end of the drug infusion. AUC_{24h} increased from 3.8 ± 0.3 μg.h/mL to 9.5 ± 2.1 μg.h/mL by doubling the dose after a single injection on day 1 for the male dogs. While AUC_{24h} increased about five-fold (from 29.4 ± 15.3 μg.h/mL to 162.4 ± 82.3 μg.h/mL) by doubling the intravenously administered dose on day 28, after 28 repeated dose administrations to the male dogs. These observations suggested that doubling the dose led to accumulation of the drug in the body.

Distribution of the block copolymer micelle products in organs and/or tissues relevant to proposed clinical use and route of administration. This was addressed in the above under "The distribution of the block copolymer micelles."

Specifically total amounts of active substance may be required. A distribution time profile should be obtained using multiple time points with justification of the time course of the study. This was explained in the above under "The distribution of the block copolymer micelles."

Sampling time points and sampling duration should be carefully selected so as to accurately quantify the time course of the concentrations of active substance both in free form and metabolites in blood, plasma, or serum, and the total concentrations of active substance and metabolites in organs and/or tissues. Sampling time and duration were addressed above for itraconazole and its major metabolite hydroxy itraconazole delivered by ITZ-PM (250). The organ distribution of the active ingredient is addressed in the section "The distribution of the block copolymer micelles" for poly(ethylene glycol)$_{3000}$-poly(ε-caprolactone)$_{20000}$ copolymer micelles (223). Metabolite distribution of ^{13}N[amide-^{13}N]nicotinamide in organs is another example, which was injected into the tail veins of C3H mice. One of the metabolites, [^{13}N]nicotinamide adenine dinucleotide, accumulated in the intestine due to the bile duct excretion. The metabolite could not cross the BBB due to low lipophilicity (251). It should be noted that most of NP organ distribution studies focused on the measurements of active ingredients, not metabolites.

In particular, samples taken in the initial distribution phase (e.g. < 15 min) are considered very informative for calculating the distribution volume to estimate the stability of block copolymer micelles in the blood circulation. Blood samples were taken at 5 min, 30 min, 1, 2, 3, 6, 9, and 24 h following i.v. administration of ITZ-PM to rats for calculating apparent volume of distribution during terminal phase (Vz) (250). Dose, AUC_{last} (AUC up to last blood concentration measurement), $T_{1/2}$, and first-order terminal rate constant were required to calculate (Vz). This regulatory requirement addresses identifying micelle stability in serum by observing differences between the distribution volumes of the free drug and micelle-encapsulated drug. This was shown for covalently crosslinked p(CL-co-DTC)-PEG micelles loading meta-tetra (hydroxyphenyl) chlorin (mTHPC), for which the distribution volume was compared with non-covalently crosslinked micelles of PCL-PEG loading mTHPC, and free mTHPC. The distribution volumes were 120 mL/kg, 125 mL/kg, and 60 mL/kg for free mTHPC, PCL-PEG, and p(CL-co-DTC)-PEG, respectively (161). The instability of PCL-PEG micelles in the blood was concluded from the proximity of the PCL-PEG distribution volume to free mTHPC distribution volume.

Measurement of active substance metabolites in blood, plasma, or serum, and maybe organs and/or tissue, is especially important when the metabolite is acknowledged to be the primary active compound. If one or more metabolites have substantial clinical activity then it might be necessary to compare their kinetics, and where necessary, and where necessary, toxicokinetics, to determine accumulation following multiple doses. PTX has low water solubility (0.34 mg/L or 0.34 μg/mL) (252), which limits its clinical applications using aqueous formulation for i.v. administrations. A prodrug of PTX, oligo (lactic acid)$_8$-PTX (o(LA)$_8$-PTX), was synthesised to create a hydrophobic derivative of PTX to enhance drug loading into PEG-*b*-PLA micelles (253). The o(LA)$_8$-PTX-PEG-*b*-PLA micelles were hydrolysed in the plasma and formed bioactive metabolites o(LA)$_1$PTX and o(LA)$_2$PTX as well as PTX free molecule. Plasma concentrations were obtained in Sprague–Dawley rats, and average $T_{1/2}$ were 0.98, 3.82, and 6.88 h for o(LA)$_2$-PTX, o(LA)$_1$-PTX, and PTX, respectively (253). As part of toxicokinetic studies, the accumulation of itraconazole and its metabolite (hydroxy-itraconazole) was determined for PEG-*b*-PLA micelles that were administered to Sprague–Dawley rats and Beagle dogs following multiple dose i.v. injections over 28 days at the doses of 5, 7.5, and 10 mg/kg/day (250). The itraconazole accumulation ratios increased from

2.1 to 3.8 by increasing the doses from 5 to 10 mg/kg/day, but hydroxy-itraconazole accumulation ratio increased from 7.7 to 17.1 for 5 and 10 mg/kg/day doses in male dogs (250).

Comparing the PK of the block copolymer micelle product and the active substance administered by itself is recommended. Such comparative studies are also considered useful to demonstrate a claimed pharmacokinetic advantage of the block copolymer product against the active substance administered by itself. The PK parameters were compared for free Dapoxetine (DPX) and DPX-loaded PEG-PLGA polymeric micelles for the treatment of premature ejaculation in Wistar rats following oral administration (254). T_{max}, $AUC_{0\text{-inf-obs}}$, and mean residence time (MRT) were increased for DPX encapsulated in the polymeric micelles compared to free DPX. Additionally, T_{max}, C_{max}, $AUC_{0\text{-inf-obs}}$ and MRT increased in the brain for DPX-PEG-PLGA micelles compared to free DPX (254). These observations demonstrated the advantages of using polymeric micelles for brain drug delivery. Similar PK improvements were observed for encapsulated PTX (223) and mTHPC (161) in polymeric micelles.

It may also be important to consider the protein and cellular interaction of block copolymers micelles administered intravenously as these factors are known to have potential to influence the distribution, stability, and safety of nanomedicines. This was shown for poly methoxypolyethyleneglycol cyanoacrylate-co-*n*-hexadecyl cyanoacrylate (PEG-PHDCA) polymeric micelles loading recombinant human TNF-α (labelled with ^{125}I) (255). The interaction of NPs with serum proteins was investigated by Coomassie brilliant blue staining of sodium dodecyl sulfate-polyacrylamide gel electrophoresis. NP size and chemical structure affected the amounts of adsorbed proteins, with PEG_{5000}-PHDCA NPs (80 nm) having only 6% protein but PHDCA NPs (242 nm) having 58% protein adsorption. Consequently, PEG_{5000}-PHDCA NPs (80 nm) showed higher blood concentrations compared to PHDCA NPs (242 nm). Furthermore, it was found that reducing particle size of PEG_{5000}-PHDCA NPs from 243 nm to 80 nm increased average $AUC_{0\text{-}24}$ from 1611 h cpm (count per minute)/μL to 6303 h cpm/μL. The stability of polymeric micelles in serum was also investigated for pCL-PEG micelles (161), and benzyl-poly(ε-caprolactone)-*b*-methoxy poly(ethylene glycol) (Ben-PCL-mPEG) micelles both loaded with mTHPC (164). PK data demonstrated the instability of the micelles in the blood circulation. Regarding the safety aspects of protein adsorption on NPs (known as protein corona), reducing NP size of PEG_{5000}-PHDCA micelles reduced protein adsorption and also reduced the uptake of the micelles by macrophages (255). Furthermore, PHDCA micelles showed much higher uptake by macrophages compared to PEG_{5000}-PHDCA micelles, which could explain much shorter MRT for PHDCA micelles (0.43 h) compared to PEG_{5000}-PHDCA micelles (4.67 h) with similar NP size (255). The variations in the protein corona composition of gold NPs affected Wnt/β-Catenin signalling activity leading to variations in gene expressions and cancer staging (256). The formation of protein corona affected the cytotoxicity of gold NPs, with reduced surface-protein adsorption demonstrating higher cytotoxicity in HeLa cells (257).

Some factors should be considered for the sampling schedules, for example, the block copolymer micelle stability after administration, and the profile of localisation to specific organs and/or tissues. The copolymer micelle stability in serum was shown for Ben-PCL-mPEG micelles loading mTHPC (164). The micelle tissue localisation was demonstrated for poly(ethylene glycol)$_{3000}$-poly(ε-caprolactone)$_{20000}$ copolymer micelles by measuring the localisation and time course of NP-PTX in glioma tumours (223).

The metabolic and excretion pathways of the active substance should be determined and fully characterised after administration of the block copolymer micelle products. Furthermore, the metabolic and excretion pathways of the micelle constituents are by themselves of interest. Their detailed characterisation is needed unless otherwise justified. The plasma metabolites were determined for o(LA)$_8$-PTX-PEG-*b*-PLA micelles following i.v. injections to Sprague–Dawley rats (253). Metabolites were also excreted into bile (253). The metabolism and excretion of the free active ingredient itself may serve as a guide to determine the excretion pathways of the active ingredient loaded into micelles. For example, excretion of free PTX and its metabolites has been reported into the faeces (56% of administered dose through biliary secretion) following i.v. administration in humans (258). These observations helped in measuring o(LA)$_8$-PTX-PEG-*b*-PLA micelle metabolites in the bile (253). After administration of free PTX into humans, an average of 5% of administered dose was excreted into the urine in the form of free PTX and its metabolites (6α-hydroxypaclitaxel, 3'-*p*-hydroxypaclitaxel), with other undefined compounds (259). Similarly, urinary excretion was observed in humans when PTX was loaded into polymeric micelles (NK105) but with less excretion rate (5%) compared to conventional PTX formulation (9%) (260). However, another elimination route was identified when PTX was administered loaded into mPEG-PDLLA micelles, which was through the raised antibodies against PEG (185). The raised antibodies reduced $AUC_{0\text{-}48}$ of mPEG-PDLLA micelles by 50% (185). In terms of excretion of micelles and micelle constituents, purified Poloxamer 188 was mainly (84%) eliminated from humans through urine (261). Gel permeation chromatographic analysis of plasma, urine, and faecal samples for radiolabelled P188 indicated the presence of intact polymer, suggesting limited metabolism of the polymer in the body (261). When radiolabelled Pluronic P85 was administered to mice, it was found that the total clearance of P85 did not change by the formation of micelles (through self-assembly). This suggested excretion of unimers mainly through renal elimination. While the AUC/dose of P85 decreased in the liver by increasing the administered dose from 100 μL of 0.02% (w/v) to 100 μL of 0.2% (262). Furthermore, the total clearance value (1 mL/h) independent of administered dose and absence of P85 protein binding suggested re-absorption of the unimers in the kidney tubules back to the blood (262).

If there is concern that components of the block copolymer micelle drug products may cause drug-drug interactions, for example by modulating membrane transporters such as p-glycoprotein, an appropriate evaluation should be carefully undertaken. Poloxamer P85 was reported to inhibit p-glycoprotein (P-gp) (263), a transporter that can be found in the BBB, and affect drug efflux across the BBB (264). P-gp is thought to act as a defence mechanism by returning back those lipophilic drugs that attempt to cross the BBB and reach the brain. On the other hand, Poloxamer P85 may inhibit the penetration of drug molecules into the brain at concentrations above CMC (265).

14.4.1.9 Non-clinical Pharmacodynamics

The non-clinical pharmacodynamic studies should include demonstration of pharmacodynamic response in appropriately justified in vitro (where possible) and in vivo models. This regulatory point addresses whether encapsulation of an API affects the time course of a clinical response. Nanoemulsion formulation of travoprost maintained lower intraocular pressure in rabbit models of ocular hypertension for a longer period compared to an aqueous formulation of the free drug (266). Also, NK911 (polymeric micelles containing DOX) remarkably reduced tumour markers (CEA and CA 19-9) in the serum over time in a patient with pancreatic cancer (267).

In vivo evaluation should involve an appropriate route of administration, justified dose levels and a justified dosing regimen depending on proposed clinical application. For brain drug delivery, the electroencephalogram (EEG) can be used as a PD endpoint (268). This end point was used to evaluate the performance of polymeric micelles formed by ferrocene (Fc)-conjugated D-a-tocopherol polyethylene glycol succinate (TPGS) polymer and amphiphilic copolymer Poloxamer 407, which were loaded with anti-epileptic drugs such as phenytoin. (269). Phenytoin loaded micelles (TFP@PHT) were administered at two doses of 10 or 20 mg/kg intraperitoneally to mouse models of epilepsy prior to the general seizure induction and behavioural measurements (269). The behavioural assessments indicated similar results for TFP@PHT at 10 mg/kg and 20 mg/kg doses. Therefore, EEG was measured only for the 10 mg/kg TFP@PHT dose (269). In another approach, carfilzomib (CFZ) was loaded into PEG-PCL micelles containing deoxycholic acid (CFZ-PM), which were administered intravenously at the doses of 3 mg/kg and 6 mg/kg (two consecutive days per week for 3 weeks) to mice bearing H460 xenograft tumours (270). The tumour volume was measured over 16 days of drug administration. The tumour volume was slightly smaller for the CFZ-PM formulation at 6 mg/kg dose compared to 3 mg/kg dose (270).

Appropriateness of the pharmacological model should be discussed in respect of the PK of the block copolymer micelle product, and the PD and PK of the active substance when administered by itself. This point was addressed above for the evaluations of the CFZ-PM on tumour volume over 16 days of treatment, by comparing the results with a clinically approved CFZ formulation (cyclodextrin-based CFZ) (270).

The chemical composition and physicochemical properties (including size and surface charge, and the rate of release of the active substance) of block copolymer micelle product affect pharmacodynamic properties. As addressed before, published research studies mainly focus on the proof-of-concept, which includes the design, development, *in vitro* evaluations with limited *in vivo* tests for NPs (271), and this is also the case for polymeric micelles. Therefore, little attention has been given to the effects of micelle size and charge on the PD properties, while the effect of NP size has been reported on the blood PK profile and biodistribution (272). The size of micelles depends on the molecular structure of the polymer, with only certain structures leading to reproducible stable (upon storage) NP sizes (273–275). A bimodal size distribution has been reported for polymeric micelles (269). Therefore, it is essential to determine the pharmacological response to the polymeric micelles when the NP size changes from one batch to another, or even upon storage (274). Perhaps it should be added that ideally the PD properties should be evaluated in relevant diseased animal models. However, developing animal models itself is challenging, for example, only 6 out of 27 mice developed tumours for studying the anticancer properties of (mPEG-*b*-p(HPMA-Bz) micelles loaded with curcumin (276). Furthermore, micelle size may change during the scale-up process depending on the operational parameters (277). Perhaps, then, the effects of micelle physicochemical properties on PD should be investigated, when the formulation is obtained by a large-scale production process.

Some important factors to consider when designing studies to discuss the mechanism of action include:

- *The fate of active substance (the location and rate of in vivo substance release)*
    - The fate of polymeric micelles is defined as the journey of the micelles as a whole in the body (278). As well as organ distribution, the term includes uptake of micelles by blood macrophages, distribution into the target tissue (for example, tumour), and excretion. The *in vivo* active substance release was measured in different organs and tumours of mice bearing murine B16 melanoma for Genexol-PM micelles loaded with PTX at a dose of 50 mg/kg injected intravenously (170). It was found that the amount of PTX reached maximum in the tumour 1 h post injection. Furthermore, Fluorescence Resonance Energy Transfer, also known as Förster resonance energy transfer (FRET) has been employed as a non-invasive method to demonstrate or even semi-quantitatively measure drug release in the target site or whole animal model (279, 280).
- *The fate of the micelles (block copolymers or other stabilising components) following administration and/or cellular entry by endocytosis or other mechanisms.* The FRET technique may be applied to demonstrate the cellular uptake of polymeric micelles as well as the release of the active ingredient in the cells (279). The fate of the micelles may be addressed by employing radiolabelled polymers such as $_3$H-labelled MePEG$_{5000}$-*b*-PCL$_{5000}$ (281). This technique allowed researchers to determine the biodistribution of the micelles as well as the unimers at different dose levels over different time points. The blood and tissue samples were analysed by scintillation counting (281).

The PD effect of the micelles should be assessed using in vitro and in vivo pharmacodynamic models. Failure to use both in vitro and in vivo models to assess the PD effects of the micelles should be extensively justified by the applicant. PD in conjunction with PK modelling aims to determine optimum dose or dose range. On the PK side, the time course of drug concentrations in different tissues including the target tissue is determined. This helps to determine the effects of drug dose on the drug concentrations in the target tissue as well as off-target organs. Meanwhile, PD modelling aims to determine the relationship between the drug concentration at the target tissue and therapeutic/pharmacological responses. The combined knowledge helps to determine optimal dosing regimens (282). An *in vitro* PK/PD model incorporates three levels of pharmacological response: (1) maximum, (2) 50%, and (3) when no drug is used, to relate the pharmacological response to the drug dose (exposure) (282). Additionally, *in vitro* PK modelling may predict drug concentrations in different tissues such as adipose tissue (283) and the brain (284, 285) based on the drug's physicochemical properties. The ultimate goal of *in vitro* PK/PD modelling is to simulate and predict pharmacological/therapeutic responses before moving to *in vivo* studies not only to observe therapeutic effects but also to reduce animal usage (286). The active ingredient release should be known/measured in the target tissue from the micelles to perform PK/PD modelling. Therefore, an *in vitro* model employing *ex-vivo* ovarian tumour tissue was developed to quantify the tumour inflow and outflow of DOX encapsulated in pH-responsive polyethylene glycol-poly (aspartate-hydrazone-DOX/phenylalanine)$_n$ (PPDF-Hyd-DOX) micelles (287). The tumour tissue was suspended in a biologically mimicking solution (Hanks' balanced salt solution+BSA) with polyether tubes cannulated into the tumour vessels to supply the drug formulation. A microdialysis probe was inserted into the tumour to measure DOX released from the micelles. DOX concentrations were measured in the perfusate as well as inside the tumour released from the micelles. The tumour volume was measured over time for the PD aspect of the study. Then, a compartmental PK-PD analysis was developed for curve fitting of the DOX concentrations in the tumour (released from the micelles) and the tumour volume over time (287). The *ex-vivo-in vitro* PK-PD model allowed measuring the uptake of micelles by the tumour, DOX released from micelles inside the tumour, and amounts of DOX that escaped from the tumour. Measuring these amounts would have been challenging *in vivo*. Furthermore, the *ex-vivo* model provided key PK rate constants in relation to drug concentration in the tumour (287).

14.4.1.10 Safety Pharmacology

When applicable (e.g. for block copolymer micelle drug out of the scope of ICH S9) the core battery of safety pharmacology studies should be conducted, in accordance with ICH M3 (R2), ICH S7A and ICH S7B. Safety pharmacology aims to determine the unexpected/undesired modification of normal physiological functions in the manner of PD effects from a drug/drug delivery system, which is administered within or above the therapeutic doses (288). Safety pharmacology aims to characterise the PK/PD relationship of a drug's adverse effects. A drug/drug delivery system may have receptors in the off-target organs and orchestrate a series of physiological responses, which could lead to undesired adverse effects. Ultimately, safety pharmacology seeks to predict whether a drug/drug delivery system is safe to be administered to patients. The term safe here also means "lack of rare and potentially lethal adverse effects." The difference between pre-clinical toxicity studies and safety pharmacology is that a small number of animals are tested against high chronic doses for a short period of time. While rare adverse effects/potentially lethal effects may not occur in a small number of animals to draw the attention of the investigators. Safety pharmacology includes three Core Battery Studies, which are CNS, respiratory, and cardiovascular. There are parameters recommended to be measured for each part as indicated in Figure 14.11.

Neurological evaluations have been reported for undesired effects of drugs on CNS as part of safety pharmacology studies (289). These include measuring body temperature and giving scores to neurological behaviours such as fear, awareness, and spontaneous locomotor activities. For cardiovascular effects, changes in the blood pressure, heart rate, and QA interval of ECG may be monitored over time (290). The safety pharmacology of siponimod was conducted on cardiovascular, respiratory, and CNS functions in four GLP-compliant studies in accordance with the ICH S7A and 7B guidelines (https://www.ema.europa.eu/en/documents/assessment-report/mayzent-epar-public-assessment-report_en.pdf). These were also supported by an extensive non-GLP cardiovascular safety pharmacology package. Siponimod did not show significant effects on the QT interval in rats, rabbits, guinea pigs, and monkeys. However, siponimod induced bradycardia in association with atrioventricular (AV) dissociation and second-degree AV block in rats, rabbits, guinea pigs, and

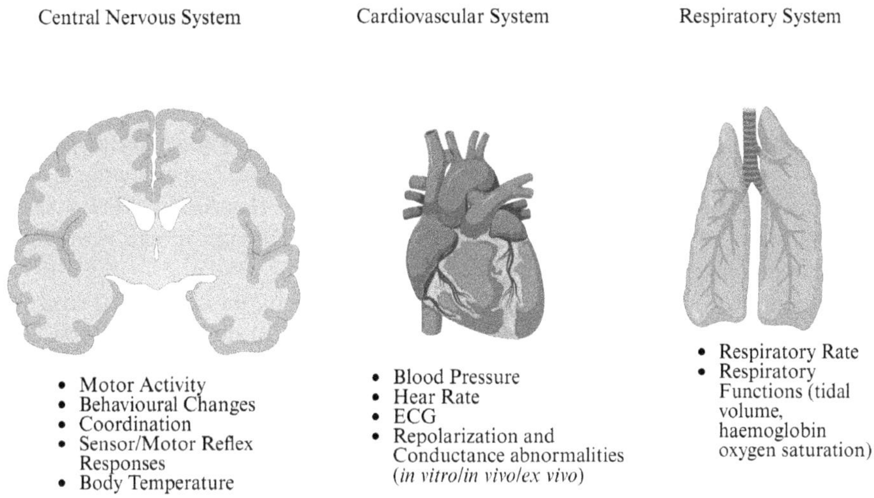

FIGURE 14.11 Core battery studies for safety pharmacology as outlined in the ICH S7A guideline. There are three primary physiological systems that must be evaluated for test substance effects: Central nervous system, cardiovascular system, and respiratory system. Specific responses and functions requiring assessment are detailed for each system.

monkeys. Siponimod slightly increased respiratory tidal volume from baseline. Using the Irwin test (291), no effects on the CNS were observed from siponimod. A safety pharmacology study has been reported for self-assembled micelle inhibitory RNA-targeting amphiregulin (SAMiRNA-AREG) formulations (292). The formulations were administered intravenously to mice at the escalating doses of 0, 100, 200, and 300 mg/kg meanwhile monkeys received i.v. injections of the escalating doses of 0, 25, 50, and 100 mg/kg, where each monkey had a transmitter surgically inserted to measure ECG parameters such as QT, PR, and RR intervals. The safety pharmacology tests covered all three core battery studies, as well as inhibition of human ether-a-go-go (hERG) gene (related to cardiac QT prolongation (293)). Chinese hamster ovary (CHO) cells were used to assay hERG *in vitro* at a formulation concentration of 200 μg/mL. Neurobehavior evaluations were conducted only in mice (both sexes), while cardiovascular evaluations were conducted in monkeys. Mice did not show treatment-related neurobehavioral changes, which were including death, startle reflex, locomotion, respiratory rate, tail elevation, eyelid reflex (checked for involuntary blink or eyelid twitch), piloerection, exophthalmos, righting reflex, skin coloration, pinna reflex, abdominal tone, salivation, lacrimation, tremor, convulsion, diarrhoea, catalepsy, and traction (292). On the other hand, there were transient changes in the body temperature for both sexes at a dose of 300 mg/kg and 0.5 h post injection. There were no changes in the respiratory rate and tidal volume. Also, there were no changes in the cardiovascular system.

14.4.1.11 Toxicology

The non-clinical evaluations of a block copolymer micelle product toxicity are similar to other medicinal products and should follow the ICH S4, S6(R1), S9, and M3 (R2) guidelines. Further relevant toxicity studies should be conducted based on the therapeutic effects of the polymeric micelle product following the ICH safety batch guidelines (ICH S1 to ICH S13).

- *Toxicokinetics:* The primary objective of a toxicokinetic study is to determine the relationship between administered dose levels by exposing the organs to different levels of the investigating drug product dose and observing toxicological effects in animal models. For example, animals may be exposed to escalating doses of 20 mg/kg, 150 mg/kg, and

500 mg/kg with the 500 mg/kg dose producing toxicity symptoms (294). In another example, a protein (protein A or micellar formulation) may be administered to animals at escalating doses of 5 mg/kg, 50 mg/kg, and 150 mg/kg every day, and plasma concentrations of protein A or the micellar formulation are determined. It may be observed that the plasma concentrations of protein A or the micellar formulation may drop down to zero at the doses of 5 mg/kg and 50 mg/kg on day 28 due to antibodies produced against protein A or the micellar formulation (294). As part of a toxicokinetic study, four different micellar formulations (PEG-PG-PCL, PEEP-PCL, PEG-PCL, PEG-DSPE) with similar hydrophilic and hydrophobic moieties were intravenously administered at a dose of 8 mg/kg every other day for eight days (receiving four doses in total). Therefore, the organs were exposed to repeated doses (although at a single dose level). Histological observations showed slight liver damage by PEG-PCL micelles. (295). NP-uptake and excretion formed part of a toxicokinetic study, when animals were exposed to different NP sizes and compositions (296). As mentioned above, the active substance concentration/amounts should be measured in blood, plasma, or serum, target tissues, and toxicologically relevant organs, which include CNS, peripheral nervous system, cardiovascular, respiratory, digestive, hepatic, renal, and endocrine systems (294, 297–299). Histological examinations have been conducted on the heart, liver spleen, kidney, and brain as part of NP-toxicokinetic studies for two different doses (3.6 mg/kg and 36 mg/kg) over 28 days of exposure with twice-weekly oral dose administration (299).

- *Additional Studies:* Additional studies are required if excipients, block copolymers, and/or manufacturing process may affect the function of the target organ. For example, Pluronic surfactants affected P-gp transporters on the BBB (263). Furthermore, NP formulations may present infusion-related reactions (300, 301), complement activation, hematotoxicity, antigenicity, and/or immunotoxicity as suppression or enhancement of immune responses, which are all addressed in the ICH S8 guideline.
  - Infusion-related reactions (IRRs) may appear as anaphylaxis, anaphylactoid reactions, and cytokine release syndrome (300). Acute lower back pain has also been categorised as an IRR (302). Thrombocytopenia, nausea, fatigue, and anorexia have been IRRs that were observed in a clinical trial for cisplatin-incorporated micelles (NC-6004) (301). Non-clinical studies should be conducted for identifying possible IRRs, although these may not predict human incidence or severity (303).
  - Polymeric micelles have been investigated for complement activation. Interestingly, PEGylated micelles induced *in vivo* complement activation, while micelles containing a hydrophilic sulfoxide did not show induction of the complement activation *in vivo* (304).
  - Hematotoxicity is defined as adverse effects created by an agent on the blood or blood-forming tissues (305). The hematotoxicity evaluations include histological examination of the bone marrow and performing bone marrow cytology (305). Injury to peripheral blood cells may occur by drug/micelles/xenobiotics (305). Haemolysis (306) and platelet activation/aggregation (307) are typical tests for hematotoxicity of polymeric micelles.
  - Antigenicity is defined as the ability of a foreign material (antigen) to bind/interact with the products of the final cell-mediated response such as T-cell or B-cell receptors (308). The antibodies or immunoglobulins form part of B-cell transmembrane receptors, which interact with certain structural designs of foreign materials. A foreign material is considered immunogenic if, upon this interaction, the B-cells differentiate to antibody-producing plasma cells (309). Therefore, antigenicity of a foreign material may lead to the formation of host antibodies against the foreign material, for example, the formation of antibodies against therapeutic monoclonal antibodies (310). Immunogenic materials initiate an immune response, first by triggering the innate immune response and subsequently by the adaptive (acquired) immune response, sensitising the body to foreign antigens (308). As a result, all immunogenic materials are antigenic, but the reverse is

not true. Therefore, there are molecules such as metal ions that are antigenic but not immunogenic. The innate immune responses were measured for polymeric micelles by the release of proinflammatory cytokines (IL-6 and TNF-α) from human monocytes (311). A minimum of seven amino acid residues is required for synthetic peptides to induce an immune response (312).

- Immunotoxicity is defined as unintended immunosuppression or enhancement. Immunosuppression is defined as the decrease in host resistance to infectious agents or tumour cells. On the other hand, enhancing the immune response may trigger autoimmune diseases. Standard toxicity studies may hint at the immunotoxicity of the polymeric micelles. These include changes in the number/ratios of immune cells in the blood (haematological changes), changes in the size/weight of immune system organs (e.g. spleen), changes in the serum immunoglobulin levels, increased incidence of infections, and increased occurrence of tumours without clear reasons. Additionally, histological studies should be conducted in the following tissues: the thymus, the spleen, draining lymph node, at least one additional lymph node, bone marrow, and Peyer's patch. Bronchus-associated lymphoid tissues and nasal-associated lymphoid tissues should also be examined for inhalation or nasal administration only. Additional immunotoxicity studies may be conducted, such as T-cell-dependent antibody response (TDAR). Host resistance studies also form part of the additional immunotoxicity studies. In these tests, rodents are treated with different doses of the test compound with varying concentrations of pathogens (bacteria, fungal, viral, parasitic) or tumour cells. The test compound is compared against a control vehicle to determine if the infectivity of the pathogens or tumour cells is changed by the test compound, leading to the alteration of host resistance. There are other immunotoxicity studies listed in the ICH S8 guideline: immunophenotyping, natural killer cell activity assays, macrophage/neutrophil function, and cell-mediated immunity assays. The study design involves using rodents, of both sexes, for 28 days, receiving consecutive daily dosing. The doses should be above the no-observed-adverse-effect-level (NOAEL) but below levels inducing stress on the immune system. Furthermore, multiple dose levels are advised to determine a dose-response relationship leading to the identification of a non-immunotoxicity dose. The immunotoxicity studies should be completed before Phase III clinical trials, when a large population of patients is exposed to the drug. TDAR and immunophenotyping tests were conducted as part of immunotoxicity studies for Siponimod (https://www.ema.europa.eu/en/documents/assessment-report/mayzent-epar-public-assessment-report_en.pdf). TDAR to keyhole limpet hemocyanin (KLH) was conducted for silver NPs in rats for 28-day repeated dose i.v. administration with daily doses of 0, 0.0082, 0.025, 0.074, 0.22, 0.67, 2, and 6 mg/kg with each dose forming one group of animals (five animals per group). The rats were immunised with KLH via subcutaneous injections on days 14 and 28 post silver NP administration. Then serum was collected from blood samples obtained on day 49 (post NP administration) and analysed for KLH-specific IgG. The reduction of KLH-specific IgG levels was observed in the serum at a 0.40 mg/kg/day dose of silver NPs, which indicated the induction of suppression of the functional immune system (313). Reduced thymus weight and increased spleen weight were also observed for silver NPs (313). The TDAR limit of detection was improved to 113.43 ng/mL by covalently bonding the KLH antigen to cytometric microspheres (314).

14.4.1.12 Considerations for First-In-Human Studies

Key pre-clinical pharmacokinetic parameters (C_{max}, $T_{1/2}$, AUC) should be determined for total active substance (encapsulated plus released drug) and for free active substance in the blood/serum/plasma, prior to first-in-human studies. If analytical techniques provide only total drug concentrations in the blood (free+ encapsulated) (249), then ultrafiltration could be used to determine free drug concentration in the plasma (231, 232). Subsequently, encapsulated drug concentration may be calculated by the subtraction of free drug concentration from total blood/serum/plasma concentration. Measuring the time course of metabolites is required (250). Sufficient, early, and long enough blood samples are advised to be obtained. Blood samples were taken at 0 (pre-dose), 0.083, 0.25, 0.5 min, 1, 2, 3, 4, 6, and 8 h from Sprague-Dawley rats after i.v. administration for PTX loaded into Pluronic P123 micelles (214). The time profile of organ distribution is required for the block copolymer micelles including the target organ in relation to total amounts of the active substance. This was shown for the PEG-PCL micelles loaded with PTX for the treatment of glioblastoma. NP-organ distributions were obtained as early as 0.5 h including the brain tumour over adequate time points (0.5 h, 1 h, 3 h, 6 h, 12 h, 24 h) (223). First human dose should be calculated by considering all related non-clinical data and following the ICH M3 (R2) guideline. Product critical attributes (e.g. average molecular weight and polydispersity of polymers and co-polymers, micelle size), stability, and batch-to-batch consistency should be achieved and established prior to the first-in-human test. Furthermore, it is required to establish and ensure the stability of the block copolymer micelles throughout the first-in-human studies. The quality attributes of the product should be consistent between the non-clinical studies and those intended for the first-in-human test. Test procedures should be established for the determination of quality attributes prior to first-in-human studies to ensure consistency of product batches. If the manufacturing process is changed for the preparation of block copolymer micelle product for first-in-human tests compared to non-clinical studies, then suitable comparability studies should be demonstrated or justified. For example, premature drug release was observed due to the hydrolysis of the ester bond between DTX and the CCPM during storage at room temperature or even in the fridge (191). As a result, the formulations were stored at −60°C. Due to the high costs, a lyophilised form was developed (191), and CQAs such as size, morphology, drug retention, and drug release profiles of the freeze-dried product (containing trehalose as cryoprotectant) were comparable to those of non-freeze-dried formulations (191). Potential hypersensitivity reactions should be considered for polymeric micelles, and these may not be dose dependent and may not be similar to non-clinical data. Also, Phase I clinical trial/first-in-human use studies provide the opportunity to determine the drug dose that cannot be used due to the severity of the observed toxicities, which are known as the dose-limiting toxicities (DLTs) (315). Traditionally, DLT is defined as grade 3 or 4 of non-haematological toxicity (47 categories including vomiting, myalgia, renal and hepatic toxicities), and grade 4 of haematological toxicities (9 categories including neutropenia and thrombocytopenia) (316). The grading scale is: 1=mild, 2=moderate, 3=severe, 4=life threatening, 5= death (317). The procedure for determining DLTs for polymeric micelles may follow similar methods as those used for conventional drugs. For example, the dose escalation and observing DLTs may follow the standard 3+3 design for polymeric micelles. According to this design, at least 3 patients are treated at each dose level by applying the following rules: 1. dose escalation to the next level if there are no DLTs, 2. if 1 out of 3 patients develops a DLT, then treat 3 more patients at the same dose level. When there are no observed DLTs in the second group of 3, then the dose can be escalated. 3. Stop dose escalation if ≥ 2 out of 3 or ≥ 2 out of 6 patients experience a DLT (212, 213). Based on nonclinical safety/efficacy, a dose-escalation regimen was planned from 135 mg/m^2 to 390 mg/m^2 for the Phase I clinical trial of Genexol-PM, to determine DLTs and maximum tolerated dose (MTD) (212). Three to six patients were allocated to each dose level, and the treatment was repeated every 3 weeks for a total of cycles between 10 and 24. As the treatment dose increased, more DLTs were observed, and the number of cycles decreased. At the MTD of 390 mg/m^2, a range of DLTs were observed, which included neuropathy, myalgia, and neutropenia. PK studies may also be conducted by taking blood samples before and after drug administration. For example, Genexol-PM was administered by continuous infusion over 3 h, and blood samples were taken prior to infusion and at 1.5 h into infusion, 0 (post infusion completion), 0.25, 0.5, 1, 2, 3, 4, 6, 8, 10, 24, 34, and 48 h after infusion (212).

14.4.2 Liposomes

Liposomes have accumulated a relatively large number of regulatory-approved products, which contributed to issuing guidelines about liposomes by the FDA, EMA, MHLW, and Chinese National Medical Products Administration (318). EMA requires addressing the following parameters for the liposomal formulations at the time of submission for liposome pharmaceutical products. Each parameter has been expanded wherever possible.

a. *Critical discussion of the lipidic components (description, source and characterisation, manufacture, assay, impurity profile, isomers and stability characteristics);*
b. *Quality, purity, and stability characteristics of other critical excipients;*
    i. For example, impurities as the result of cholesterol oxidation in the starting material and during manufacturing process (319, 320).
c. *Identification and control of key intermediates in the manufacturing process;*
    i. This can refer to the lipid thin membranes or lipid suspensions as the manufacturing intermediates. It is important to ensure the homogeneity and consistency of these intermediates through maintaining consistency in the process parameters. Critical process parameters can be hydration time, mixing speed, and temperature of the hydration process. For example, vigorous shaking or bath sonication may be applied during manufacturing of liposomes by the hydration of thin lipid membranes (321). Different sonication methods and duration could lead to liposomes with different sizes, with longer sonication periods (10 s on-off and 30 min duration) making smaller size liposomes (321) compared to a shorter sonication time (6 s on, 4 s off, and 10 min duration) (322).
d. *Active substance/lipidic moiety ratio at relevant manufacturing steps to be within acceptable range to ensure consistent formulation performance;*
    i. This could be attributed to the oxidation of phospholipids due to the heat generated during the manufacturing process when sonication is applied (323, 324).
e. *Liposome morphology, mean size and size distribution, aggregates;*
    i. The liposome size and size distribution may be measured by the DLS method (325), phospholipid aggregation by an analytical method (326), and morphology by a cryo-TEM technique (327).
f. *Fraction of encapsulated active substance (amount of free/entrapped);*
    i. The fraction of encapsulated active substance is the ratio of the active ingredient inside the liposomes to the total amount of the active ingredient added at the start of formulation process. The fraction of the encapsulated active substance may be represented as encapsulation efficiency, which could depend on the preparation method (328).
g. *Stability of the active substance, lipids, and functional excipients in the finished product, including quantification of critical degradation products (e.g. lyso phosphatidylcholine, oxidated/hydrolytic moieties);*
    i. The stability of liposomal formulation and approved kit with drug and blank liposomes should be conducted in separate containers under recommended storage conditions (329). Furthermore, the stability studies should be conducted for the reconstituted liposome formulations (kit products) under in-use conditions and during the in-use period (329). Oxidative and hydrolytic degradation pathways have been identified for phospholipids (330). Phospholipid oxidation could change the zeta potential of liposomes (331), which plays an important role in cellular uptake (332). The active ingredient may degrade in liposome formulations depending on the microenvironmental pH (333), which also plays a key role in the lipid hydrolysis (334). Freeze drying may be applied to improve the chemical stability of liposomal formulations. (329).

h. *In vitro drug substance release rate from the liposome in physiologically/clinically relevant media. Reliable and discriminating validated in-vitro release methods should be developed to monitor the simulated release of the active substance from the liposomes in physiologically/clinically relevant media. If justified, an in vitro leakage test in relevant media under multiple conditions (e.g. range of temperatures and pH values) could be appropriate.*
   i. Therefore, drug release studies should be conducted from liposomes at physiological pH (335), in cell culture media (336), in the serum (337), in the plasma (338), in the blood (*in vivo*) (339), and at gastric pH for oral formulations (340).
i. *Monitor stability on storage, and be sensitive to ensure batch-to-batch consistency.*
   i. It has been shown that liposome size may change during storage (341).
j. *Stability studies under proposed in-use conditions;*
   i. Liposomes may be administered by reconstitution in large volume parenteral products before administration. Particle size may change slightly (342). However, liposomal rupture may occur due to osmotic influx of solvent (diluent) into the highly loaded liposomes (343). Furthermore, this regulatory requirement would address evaluations of liposomal stability in blood circulation, which can be related to the initial burst release from liposomes (339) or leakage of active ingredient from liposomes in the serum (344).
k. *Robustness of process for reconstitution and/or pharmacy preparation.*
   i. Liposomal formulations may be freeze-dried to improve their stability, and the reconstitution process should be applicable in the clinic. For example, water-for-injection should be added aseptically into the vial containing the freeze-dried powder, and the powder should dissolve/disperse completely by vortexing the vial within three minutes (345). The effects of the reconstitution process may be evaluated, as faster reconstitution may lead to smaller-sized liposomes (345).

14.4.2.1 Pharmaceutical Development

The manufacturing process should be sufficiently equipped with process controls to assure products with acceptable consistency (346, 347). There could be changes in the manufacturing process, which may result in changes in the liposomal products, significantly affecting their performance. Therefore, it is essential to determine critical process parameters such as heating temperature or aspirator rate during spray drying (347). If there are changes in the key product parameters such as particle size and charge during the manufacturing process, then *in vivo* studies may be necessary to demonstrate that those changes do not affect the safety and efficacy profile of the product. For example, increasing liposome sizes from 194 (± 65) nm to a broader size of 338 nm increased spleen uptake from 10% to 40% of the injected dose, while the spleen uptake significantly decreased for liposomes < 70 nm compared to liposomes with the size of 194 (± 65) nm (348).

14.4.2.2 Toxicological Studies

The toxicological studies of liposomal formulations depend on the active substance and liposomal formulation. If regulatory approval of the active substance (active pharmaceutical ingredient) already exists, then toxicity studies may not be needed for the active substance, as stated in the EMA guideline. For example, doxorubicin was approved in 1974, while liposomal formulation (Doxil®) was approved later in 1995. In this case, the toxicity studies will focus on the liposomal formulation containing the active substance. As part of toxicological studies, organ function tests should be conducted such as toxicity to the heart (reviewed in ref (349) and clinical observation in ref (350)) or the brain (351). In addition, the toxicity and toxicokinetic data should compare the liposomal formulation with the active substance administered by itself. When the active substance is novel, then the toxicity and toxicokinetic studies should be conducted for both the active ingredient and drug-loaded liposomes. If it is certain that the active substance does not leak from the liposomal formulations, then the toxicity and toxicokinetic studies may be performed only on the liposomal formulation. The toxicity studies should also be conducted on the blank liposomes (not loaded with the drug) and other excipients. Liposomes may also initiate immune reactions when they are introduced into the bloodstream. Liposomes may stimulate the release of cytokines such as TNF-α and IL-1β from human peripheral blood cells, with larger liposome size leading to higher levels of cytokine induction (352). Immunotoxicity studies should be conducted following the ICH S8 guideline. The other regulatory requirements are briefly explained in the following, which have been addressed in more detail above for block copolymer micelles.

14.4.2.3 Methods of Analysis and Non-clinical Studies

The analytical methods should be validated to allow quantification of both encapsulated and unencapsulated drug in the blood/plasma and tissues. Non-clinical PK and PD studies should be conducted. Suitable animal models should be selected and justified for PK studies. It is likely that *in vivo* tests will be required for elucidating PD properties of the liposomal formulations.

14.4.2.4 Clinical Studies

The PK behaviour of liposomal formulation should be compared with the reference unencapsulated drug (353, 354). The PK parameters may be obtained only for liposomal formulation in a Phase I clinical trial (355). When the active substance cannot be tolerated by healthy volunteers, the PK studies may be performed in patients. It is recommended to measure the blood/serum levels of a metabolite of the active ingredient in the clinical PK studies such as doxorubicinol, which is a major metabolite of DOX and can be measured using a validated reverse-phase high-performance liquid chromatography (355). The conventional PK metrics should be measured such as AUC, $T_{1/2}$, C_{max}, clearance, and volume of distribution (355). Additional PK parameters should be measured related to the excretion of the active ingredient, such as urinary excretion (353). It is recommended to take early and sufficient blood sampling time points for a liposomal formulation such as prior to dosing, at the end of infusion (t=0 h), and 1, 4, 8, 24, 34, 48, 96, 168, and 240 h after completion of the infusion (355). Similar to block copolymer micelles, liposomal formulations may present infusion-related reactions (356), and it is recommended re-formulation of the product if the risk of acute infusion reaction rates increases compared to a reference product.

14.5 STERILISATION PROCESS AND PYROGEN LEVELS OF CNS MEDICINES

14.5.1 Sterilisation of Pharmaceutical Products

Sterilisation is defined as the killing or removal of all living organisms. There are four methods of sterilisation: heat (dry heat or moist heat), irradiation, gas, and aseptic processing including microfiltration. All pharmaceutical products must be sterile when they are injected into the body. Staff training forms a key part in the manufacture of sterile products. Ten suitable GMP rules are identified for staff training (357), which are listed in Table 14.4A. The author (TE) has summarised these rules in Table 14.4B in the form of the TEN COMMANDMENTS.

14.5.1.1 Sterilisation by Heat

Moist heat sterilisation is the method of choice, as there is less chance of microbial contamination after sealing the product. Typical moist heat sterilisation conditions are 121°C for 15 minutes. This is to achieve the sterility assurance level (SAL) of 10^{-6}, i.e., the probability of a product remaining microbially contaminated or non-sterile after sterilisation process. As a typical batch size is less than 1,000,000, then there will be no microbially contaminated product after the moist heat sterilisation process. The typical 15-min duration process is an overkill, and typically the SAL is achieved within 5 or 10 min of the product reaching 121°C. However, this may change due to the heat resistance of the product microbial bioburden (358). Moist heat sterilisation is a commonly used method

TABLE 14.4A
Ten GMP Rules that Should Be Considered in the Staff Training.

Number	Rule
1	Confirm trained is carried out and correct written instructions have been received before starting any job
2	Errors and bad practices should be reported immediately
3	All instructions must be followed exactly
4	It must be ensured that all correct materials are provided.
5	Correct equipment must be used for the job, and its status and cleanliness should be confirmed.
6	Good segregation should be maintained with protection against contamination
7	Work should be done accurately, precisely, and methodically
8	Good standards of cleanliness and tidiness must be maintained
9	Changes must be pre-approved (through the change control system).
10	Assumptions must not be made, and always confirmations should be obtained

for heat-stable products. There are several advantages to moist heat sterilisation such as fast and non-toxic (358); however, autoclaves require the initial capital investment and routine repair costs (359). Conventional moist heat sterilisation (121°C for 15 min) of solid lipid nanoparticles (SLNPs) increased average NP diameter from 78.5 nm to 120.0 nm when the NPs were made from stearic acid (360). While the average NP size only increased from 70.0 nm to 75.5 nm for Acidan N12 NPs (360). Similarly, SLNPs stabilised with soy lecithin were sterilised by the moist heat sterilisation method (121°C for 15 min) (361). Chemical modifications may occur during moist heat sterilisation. Our investigations also found that moist heat sterilisation of the 100 nm polystyrene NPs did not change the size distribution significantly before (Figure 14.12A) and after (Figure 14.12B) the moist heat sterilisation process, but the zeta potential significantly reduced from −1 mV (Figure 14.13A) to −27 mV (Figure 14.13B).

For the purpose of brain drug delivery, vesicular phospholipid gels were formulated with a size range of 20 to 100 nm (average ± SD: 58.3 ± 6.5 nm) for injection into the created cavity in the brain as an implant following tumour resection (362). The vesicles formed a gel-like semisolid dispersion suitable for injection as an implant. The NPs were sterilised by moist heat sterilisation at 121°C for

TABLE 14.4B
The TEN COMMANDMENTS of Good Manufacturing Practice

Number	Commandment
I	THOU SHALT CONFIRM THOU ART TRAINED AND HAST CORRECT WRITTEN INSTRUCTIONS BEFORE COMMENCING ANY LABOR
II	THOU SHALT FOLLOW INSTRUCTIONS WITH EXACTITUDE
III	THOU SHALT REPORT ERRORS AND UNSEEMLY PRACTICES FORTHWITH
IV	THOU SHALT ENSURE THE RIGHT MATERIALS ARE GATHERED BEFORE THY LABOR BEGINS
V	THOU SHALT USE THE CORRECT EQUIPMENT FOR THY LABOR, AND VERIFY ITS CONDITION AND CLEANLINESS
VI	THOU SHALT MAINTAIN PROPER SEGREGATION AND GUARD AGAINST CONTAMINATION
VII	THOU SHALT WORK WITH ACCURACY, PRECISION, AND METHOD
VIII	THOU SHALT UPHOLD RIGHTEOUS STANDARDS OF CLEANLINESS AND ORDER
IX	THOU SHALT ENSURE ALL CHANGES RECEIVE PRIOR APPROVAL THROUGH THE APPOINTED SYSTEM OF CONTROL
X	THOU SHALT NOT MAKE ASSUMPTIONS—VERIFY ALL THINGS

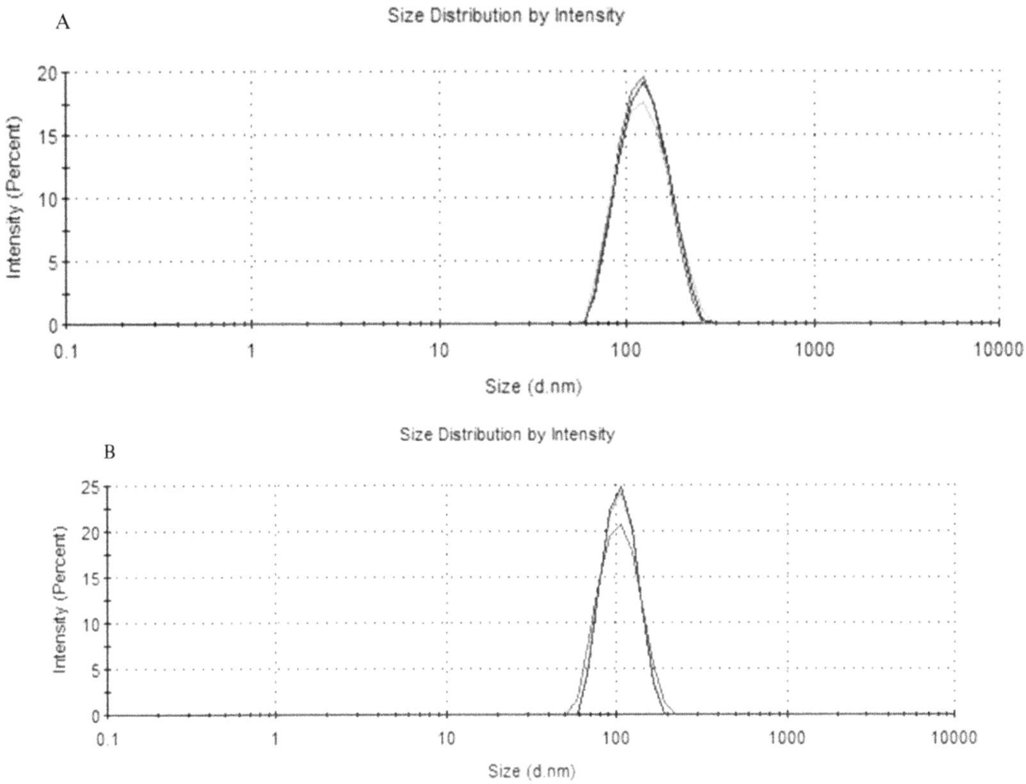

FIGURE 14.12 The size distribution of 100 nm polystyrene NPs (A) before, and (B) after moist heat sterilisation at 121°c for 15 min. The comparison shows a slight reduction in the particle size following the sterilisation process.

15 min. The NP size slightly increased to 76.7 ± 8.1 nm, but the drug release rate decreased, with remarkable drug (cytarabine) degradation (about 27% loss). The size distribution of self-assembled peptide NPs did not significantly change upon sterilisation by moist heat at 121°C for 15 min (363). The moist heat sterilisation method was applicable for chitosan NPs that formed by self-assembly (364). Liposomes have been successfully sterilised by autoclaving (365). Also, archaeosomes made of PLFE lipids demonstrated vesicle integrity (stability of NP size) after multiple autoclaving cycles between pH 4–10 (366). The removal of air from a malonaldehyde bis(dimethyl acetate) liposome formulation prevented lipid oxidation during the conventional autoclaving process (121°C, 15 min) (367). The removal of air from a liposomal formulation may be achieved by applying vacuum (368).

14.5.1.2 Sterilisation by Irradiation

Sterilisation by irradiation is a terminal process and there are two types: ionising and non-ionising radiation (369). Gamma rays, electron beam, and X-rays are categorised as ionising radiation, and ultraviolet (UV) radiation is an example of non-ionising radiation. Radiation may have two forms: particles or waves (electromagnetic). Gamma rays are high-energy electromagnetic waves similar to light but with much shorter wavelength (0.1 Å), which allows them to penetrate into pharmaceutical products. Particle radiation is generated by the disintegration of an unstable atom. The particles have mass, energy, and may have charge. Alpha, β, and neutron particles form part of ionising radiation. Alpha particles are large and positively charged but with low penetrating power. On the other hand, β particles are high-energy electrons (negatively charged) or positrons (positively charged). Neutron particles are neutral with

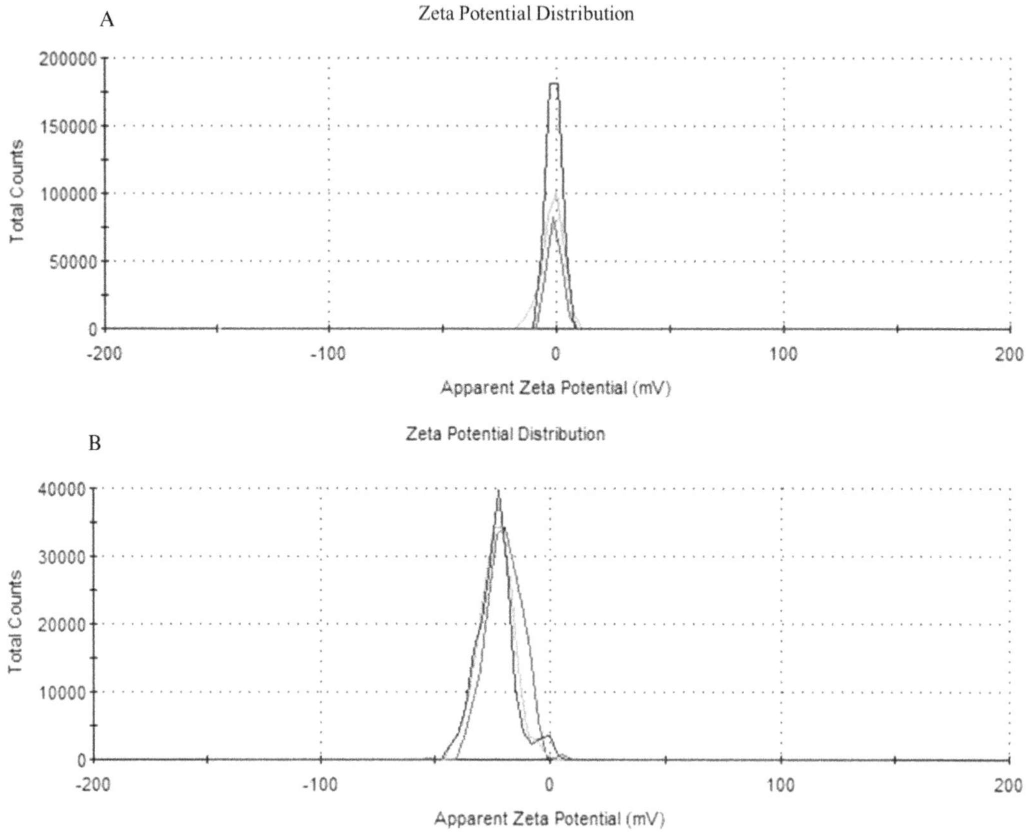

FIGURE 14.13 The zeta potential distribution of 100 nm polystyrene NPs (A) before, and (B) after moist heat sterilisation at 121°c for 15 min. The comparison shows a significant shift toward more negative zeta potential following the sterilisation process.

potential application in cancer treatment (370). β particles have been used for the sterilisation of pharmaceutical products. Gamma irradiation has been a popular method for terminal sterilisation of pharmaceutical ingredients and APIs (358). This is due to the associated lower costs in terms of capital investment, and it can be adapted to small-volume batches. Gamma irradiation has been employed for sterilisation of NPs (371–374). The typical gamma irradiation dose has been in the range of 15-50 kGy (371, 372). Gamma irradiation increased the Mw of NPs (372). β-irradiation did not change the average size of nanostructured lipid carriers (NLCs) loaded with platelet-derived growth factor BB (PDGF-BB) at 12 kGy, but reduced it significantly at 25 kGy dose (374). On the other hand, β irradiation did not change the average size of NLCs loaded with vascular endothelial growth factor 165 (VEGF165) at either 12 kGy or 25 kGy doses (374). However, gamma irradiation significantly reduced the average size of NLCs loaded with PDGF-BB or VEGF 165 at a 25 kGy dose (374). Both β and gamma irradiations have been applied to the brain-targeting PLGA NPs loaded with DOX and coated with poloxamer 188 at dose ranges of 10-15 kGy (375). β irradiation did not significantly change the NP size, but gamma irradiation at a dose of 25 kGy significantly increased average NP size from 127 nm to 564 nm (375). UV irradiation has been employed for the sterilisation of PCL and PLGA NPs with 2 h exposure time at a dose of 100 μJ/Cm^2 using a UVC 500 ultraviolet crosslinker (376). The average NP size did not change with UV irradiation, while gamma irradiation significantly reduced average NP size at both doses of 5 and 10 kGy (376).

14.5.1.3 Sterilisation by Filtration

Sterilisation by filtration physically and completely removes microorganisms (bacteria, yeasts, and molds, but not viruses) from the product (358). The typical pore size of the filters is 0.22 μm. This is a widely used method for NPs with average sizes less than 200 nm (245, 375, 377). Normally, NP size does not change during filtration (375). Cellulose acetate (372) and polyethersulfone (378, 379) have been polymers used in filters for sterilisation. Filter fouling occurred slowly for sterile filtration of 200 nm NPs using 0.22 μm membrane filters, but significant filter retention was observed for 400 nm size NPs (380). When sterilisation by filtration is not achievable due to large NP size, then other methods such as UV irradiation may be applied (376). Filtration has been employed for the sterilisation of self-assembled NPs either with pore sizes of 0.45 μm (381) or 0.22 μm (382).

The sterilisation by filtration should be conducted within an aseptic environment, ideally ISO 5 or grade A air. The procedures are normally conducted inside a disinfected laminar flow cabinet or sterilised isolator. This is to prevent potential contamination of the product by microorganisms in the air. Microbial sterility testing may be conducted following sterilisation by filtration to demonstrate the sterility of final NP formulation (379). To conduct the microbial sterility testing, the final NP product (sterilised by filtration) should be passed through a 0.22 μm sterile membrane filter housed inside a filter holder to capture contaminating microorganisms in the product. Then, the membrane filter (containing possible microorganisms) is incubated inside a culture medium chosen as a testing media. Different culture media should be used suitable for the growth of aerobic, anaerobic, and fungal organisms. These operations should be conducted again within an aseptic environment. The sterility testing should be validated to demonstrate that if there are contaminating microorganisms in the final product, they are able to flourish inside the microbial culture media and form colony-forming units or turn clear liquid culture media turbid. To achieve this, the same procedure for sterility testing is conducted, but at the end microorganisms are added to the microbial culture media. The microorganisms may include *Staphylococcus aureus* (aerobic), *Clostridium sporogenes* (anaerobic), *and Candida albicans* (fungi). These culture media (control) plus the initial culture media from the sterility testing (test) are incubated at suitable temperatures. It would be expected that liquid culture media for test bottles remain clear, but the culture media containing the microorganisms (control) will turn turbid. If the bottles of the control culture media remain clear, then there is a possibility of antibacterial/antifungal or bacteriostatic/fungistatic activity of the final product. Therefore, suitable procedures should be taken to inactivate antibacterial/antifungal or bacteriostatic/fungistatic ingredients.

It is recommended to conduct a prefiltration step for the removal of any debris, mainly to avoid clogging of the final 0.22 μm membrane filter. The FDA has produced guidance to help manufacturers meet the requirements in the Agency's current good manufacturing practice (cGMP) regulations (21 CFR parts 210 and 211) when manufacturing sterile drug and biological products using aseptic processing (383). The EMA also produced a guidance to help the industry on the selection of appropriate sterilisation techniques for sterile products (384). The EMA states that terminal sterilisation is the method of choice. The aim of the guidance is to provide information on when other terminal sterilisation processes, sterilisation by filtration, or aseptic processing (either alone or when combined with a terminal heat treatment) could be accepted as an alternative. Sterilisation by filtration within an aseptic processing is acceptable for highly sensitive products such as most biological products. The EMA guidance recommends that all sterilisation procedures should be carried out according to the instructions given by the Ph. Eur., and any deviations should be justified. Filter integrity tests should be conducted pre- and post-filtration to ensure the ability of the filter to retain microorganisms. The bubble point test is the conventional method for performing filter integrity tests (358).

As stated above, sterilisation by filtration does not remove viruses by size exclusion mechanism, as pores of filters typically have sizes greater than 100 nm (385). Regulatory notes on virus validation studies state that viral contamination is a feature common to all biologicals whose production involves the use of material of animal or human origin (386). This document suggests that testing of starting materials is key to minimise viral contamination, as viral testing normally can detect many virus species but not all known viruses. Therefore, it should be demonstrated that the manufacturing process is capable of removing or inactivating viruses (386). Filters may be optimised to remove viruses by mechanisms such as electrostatic interactions. For example, hydrophilic nylon filters with a pore size of 0.2 μm became slightly positively charged at pH 5 and were effective in the removal of MS2, Qβ, phiX174, and phi6 viruses with hydrodynamic diameters in the range of 25–85 nm (385). The filter did not remove the viruses at pH 7.

14.5.1.4 Sterilisation by Gas

Sterilisation by gas is normally employed for heat and radiation-sensitive materials. In addition, it could be an alternative method for NP formulations that cannot be sterilised by filtration. Ethylene oxide (EO) is widely used for sterilisation by gas. It is a highly reactive agent and alkylates cellular components such as nucleic acids and proteins. EO alkylates carboxyl, hydroxyl, sulfhydryl, and amino groups ($-NH_2$) of proteins and nucleic acids. Its high diffusivity into the cells contributes to the microbial inactivation. EO is considered toxic to humans as well. EO is classified as a known human carcinogen agent, and epidemiological data demonstrated a higher risk of developing cancer for workers exposed to EO (387). Therefore, EMA issued a note of guidance limiting the use of EO for sterilisation of pharmaceutical products to cases when its use is absolutely necessary, with EO residual not exceeding 1 ppm (388). Therefore, residues of EO must be removed from the product by procedures such as gas washing. Furthermore, EO is highly explosive, and therefore, certain safety precautions should be in place (389). EO was employed for sterilisation of PEG-coated gold NPs (390). The sterilisation method was considered appropriate, as the EO sterilisation procedure did not induce cell cytotoxicity, intracellular reactive oxygen species (ROS), morphological changes or aggregation of particles, or damage to the PEG shell around gold NPs. The sterilisation process was applied to a solution of the gold NPs (390).

14.5.2 Bacterial Endotoxin Levels in CNS Pharmaceutical Formulations

Bacterial endotoxins are a type of lipopolysaccharides (LPS) with four main parts connected sequentially in the order of lipid A, inner core, outer core, and O-antigen. Lipid A contains hydrocarbon chains, but the rest are polymers of sugars (391). The bacterial endotoxin is embedded on the lipid side into the cell membrane and with sugar molecules stretching towards the outside of the cell (391). The endotoxins are major components of the cell walls of Gram-negative bacteria. Bacterial endotoxins have large molecular weights in the range of 10,000 to 1,000,000 Da (392). Bacterial endotoxins belong to a group of non-viable materials/particles which are called pyrogens (393). A pyrogen is defined as a substance that causes a rise in body temperature (394). Pyrogens are categorised into two classes: (1) Exogenous, (2) Endogenous. Exogenous substances such as bacterial endotoxins are introduced into the body by passing body's natural defence mechanism, such as parenteral administration. Following the entrance of exogenous pyrogens, the body responses through the release of pro-inflammatory cytokines such as TNF-α, and IL-1β (endogenous pyrogens) by macrophages causing a rise in the body's temperature. The presence of bacterial endotoxins in the blood circulation may cause development of seizures (395) or even death (396).

The British Pharmacopeia recognises the limulus amoebocyte lysate (LAL) and recombinant factor C assays to detect and quantify levels of endotoxins from Gram-negative bacteria (397). In addition, the rabbit pyrogen test may be applied to determine whether a sterile solution for injection contains sufficiently high levels of pyrogens to raise temperature in rabbits (398). Endotoxins are typically measured by using a LAL technique and it is expressed in endotoxin units (EU). The LAL assay contains several inactive proteins (Factor C, Factor B, and proclotting enzyme), but these become activated and converted to enzymes as a result of interaction with bacterial endotoxins. Initially, bacterial endotoxin converts Factor C to activated Factor C, which converts Factor B to activated Factor B. This enzyme activates the proclotting enzyme, which cleaves coagulogen to coagulin. This is a fibrinogen-like protein that forms a coagulin gel. The LAL test can be conducted through three methods: (1) the gel-clot technique, (2) the turbidimetric technique, and (3) the chromogenic technique (Figure 14.14A). The gel-clot technique is a popular method that is based on coagulin gel formation. The turbidimetric technique is based on the development of turbidity as a result of coagulin formation. The solution turbidity is measured at 340 nm. For the chromogenic method, a chromogenic peptide is added to the LAL proteins, which replaces coagulogen. Therefore, when bacterial endotoxin reacts with the LAL proteins and a proclotting enzyme is finally formed, this enzyme cleaves the chromogenic peptide and produces para-nitro aniline, which has a yellow colour with an absorbance at 405 nm (399). The flowchart of LAL assays is demonstrated in Figure 14.14A. The recombinant Factor C test is an alternative to the LAL test, which adopts a sustainable method for the detection and quantification of bacterial endotoxin levels. The recombinant Factor C assay works based on a single enzymatic step in which a fluorogenic substrate (Boc-Val-Pro-Arg-7-amido-4-methylcoumarin) is converted to a fluorescence product (amino-methylcoumarin), as a

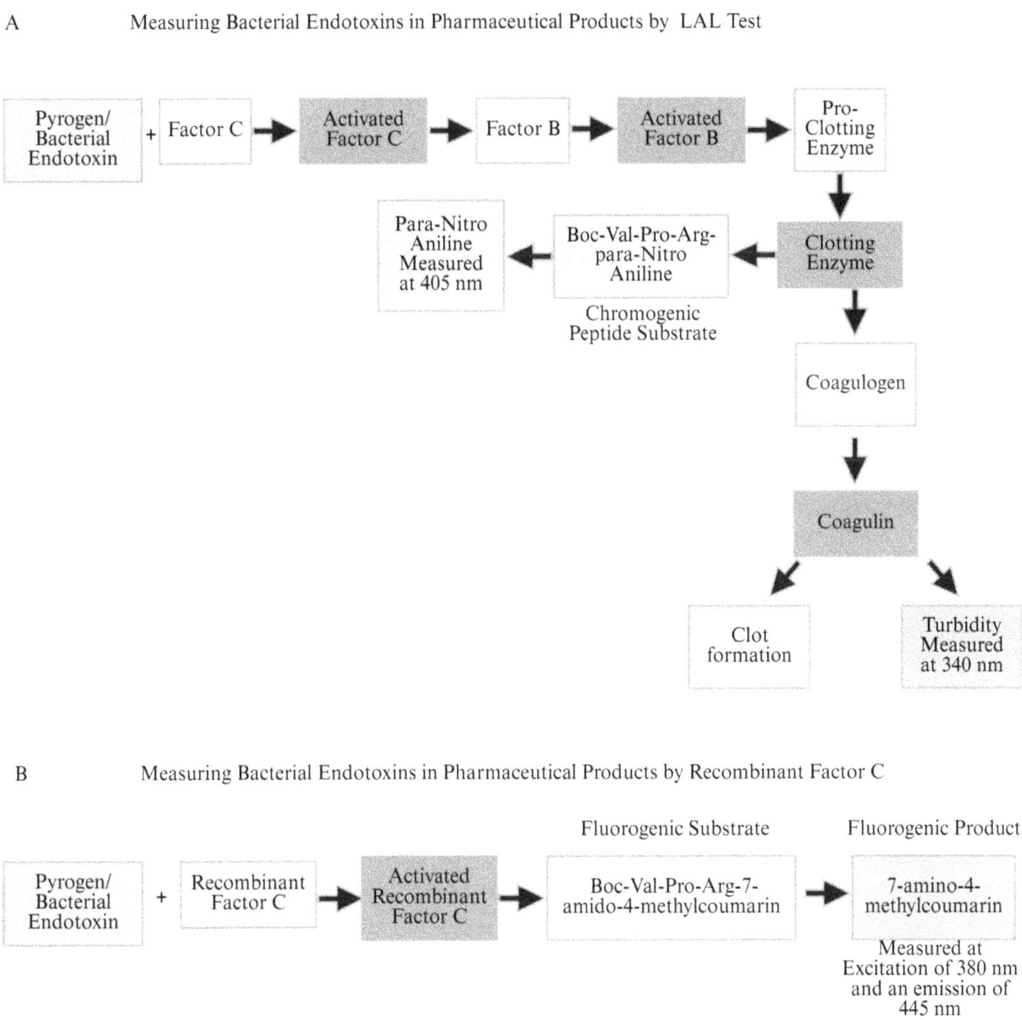

FIGURE 14.14 Quantification of bacterial endotoxin/pyrogen levels by (A) the LAL assay, (B) the recombinant Factor C assay, and (C) detection of bacterial endotoxins/pyrogens by the rabbit pyrogen test method. (*Continued*)

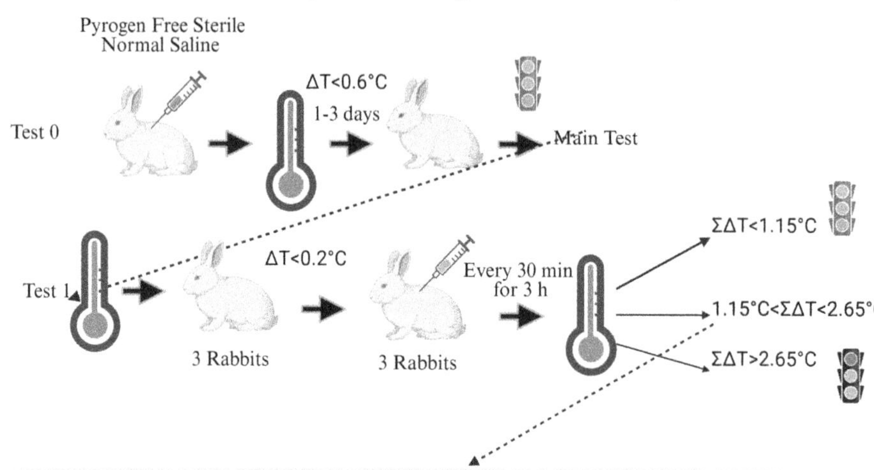

Total No of Rabbits	Product passes if summed response does not exceed	Product fails if summed response exceeds
6	2.80°C	4.30°C
9	4.45°C	5.95°C
12	6.60°C	6.60°C

FIGURE 14.14 *(Continued)*

result of the bacterial endotoxin interaction with the recombinant Factor C (Figure 14.14B) (400–402). There are attempts to replace the conventional LAL methods with the recombinant Factor C assay, which eases the pressure on the horseshoe crab (403). The rabbit pyrogen test is schematically presented in Figure 14.14C, based on the method described in the British Pharmacopoeia (398).

For the rabbit pyrogen tests, selected rabbits must be healthy, mature animals weighing at least 1.5 kg and maintained on a balanced diet. Prior to the main test, all animals undergo examination to assess their suitability for the pyrogen test. This preliminary assessment, known as the sham test (394) or Test 0 (Figure 14.14C), involves injecting rabbits with 10 mL/kg of pyrogen-free 9 g/L sodium chloride R solution or another suitable liquid. Animals are monitored for three days, and any rabbit exhibiting temperature variation greater than 0.6°C is excluded from the main test. For the main test, animals are grouped into sets of three rabbits per group. Body temperature is monitored for 90 min with measurements taken every 30 min. Only rabbits showing less than 0.2°C body temperature variation are selected for testing. Suitable rabbits then receive injections of the test product at volumes ranging from 0.5 to 10 mL/kg, and temperature rise is monitored for 3 h. The test interpretation follows a sequential protocol based on cumulative temperature rise. For the first group of three rabbits, the product passes if the sum of temperature rise does not exceed 1.15°C and fails if it exceeds 2.65°C. If the sum falls between 1.15–2.65°C, testing proceeds to a second group of three rabbits. With six rabbits in total, the product passes if the summed temperature rise does not exceed 2.80°C and fails if it exceeds 4.30°C. If the sum falls between 2.80–4.30°C, testing continues with a third group. With nine rabbits in total, the product passes if the cumulative temperature rise does not exceed 4.45°C and fails if it exceeds 5.95°C. If the sum falls between 4.45–5.95°C, a final group of three rabbits is tested. With all twelve rabbits, the product passes if the cumulative temperature rise does not exceed 6.60°C and fails if it exceeds this threshold. The popularity of this method is declining, but it is recommended when other *in vitro* tests are not conclusive (404).

The pharmaceutical products may be administered intravenously or directly into the CNS for the treatment of CNS disorders. These products are not only sterile, but also have low levels of bacterial endotoxin with 0.06 EU/mL for intrathecal administrations and 0.5 EU/mL for i.v. products (17). Pharmaceutical products can be administered into the CNS via intrathecal lumbar puncture, lateral ventricle of the brain, or convection-enhanced delivery into brain tumours. CNS direct therapeutic delivery has been employed for the treatment of lymphoma (405), glioblastoma (406), and pain relief (407).

The bacterial endotoxins may be removed from the pharmaceutical products by methods such as ultrafiltration and ion-exchange chromatography (403). Ultrafiltration was applied for BMS-753493 with molecular weight of 1570 Da using 10 kDa cut-off regenerated cellulose membrane filters at 5°C (408). The choice of cut-off diameter was based on the LPS molecular weight, which is in the range of 10-20 kDa. Ion-exchange chromatography can be used to separate negatively charged endotoxin molecules from neutral/positively charged drug molecules. Filters with larger cut-offs such as 100 kDa may be used for the removal of bacterial endotoxin from NPs. This is because NPs typically would have much larger sizes than LPS molecules, and NPs are retained by the filter, but bacterial endotoxins will pass through the filter. However, bacterial endotoxins may aggregate and form larger particles (409). To avoid this, surfactants such as cholate or deoxycholate (409), Triton™X-100 (410), or Deviron™C-16 as an ecofriendly replacement for Triton™X-100 may be added to the formulation (410). It was found that a significant portion of NPs (about 50%) was lost during an ultrafiltration process using 300 kDa Mw cut-off filters to remove bacterial endotoxins (411).

NPs may interfere with the detection/quantification of bacterial endotoxins using assays such as LAL (412). For example, purple/red coloured 30 nm gold NPs interfered with both chromogenic and gel-clot LAL assays (413). Therefore, a decision tree has been proposed for the determination of bacterial endotoxin levels using LAL assays for NPs to overcome interferences (414). In this approach, two non-interfering LAL assays are recommended for the NP formulation with the results should not differing more than 25%. When the LAL assays present differences larger than 25% differences, then the rabbit pyrogen test should be conducted. Diluting NP solutions may be applied to overcome the NP interference with the LAL assays (415). However, the dilution should not be beyond the maximum valid dilution (MVD). The MVD is defined as the dilution of a product at which the endotoxin limit would not be detected using the LAL assays. The FDA recommends that, ideally, the undiluted products should be screened using the LAL tests, as long as there is no enhancement/inhibition of LAL assay components. During product development, the interference with LAL assays should be identified, and it is recommended that the lowest product dilution be determined that neutralises the interference. An example is given for the calculation of MVD for NP solutions by the National Cancer Laboratories (416).

There are alternative techniques to quantify endotoxins in NP formulations. For example, surface-enhanced Raman spectroscopy was applied to quantitatively determine the amounts of LPS on gold NPs with a limit of detection to be 2.6 ± 0.1 ng/mL (417). Due to the challenges of detecting and removing of endotoxins, it may be recommended to avoid/minimise contamination of the formulation with endotoxins by following aseptic techniques (wearing gloves), using pyrogen-free water, pyrogen-free solvents, pyrogen-free glassware, and conducting synthesis in a laminar flow cabinet (418).

14.6 CONCLUSIONS

The classical drug development process is well established for small drug molecules. For drug delivery to the brain, targeting this organ may be considered during product development and suitable alterations may be made to the drug molecule structure to improve $\text{brain}_{\text{concentration}}/\text{plasma}_{\text{concentration}}$. The drug development process for brain-targeting monoclonal antibodies has steps similar to those for small molecules, with a few differences. For example, monoclonal antibodies have inherent risks for adverse reactions that are mediated by the immune system. Liposome and block copolymer micelle products are becoming widely investigated, and therefore, regulatory guidelines have been issued. These are key documents that formulation scientists should be familiar with prior to the development of formulations for the treatment of brain diseases. For example, a well-defined manufacturing process should be designed with its associated process controls to ensure that therapeutically consistent products are manufactured. Clear distinctions should be made between pharmacokinetic and PD studies. In particular, more attention should be given to the effects of NP physicochemical properties on the PD outcomes. Since most pharmaceutical formulations are injectables for the treatment of neurodegenerative diseases, sterilisation of the formulation and removal of pyrogens may be considered during the early stages of product development.

The diseased brain is different from a healthy brain in terms of physiology and even morphology (9). Therefore, in the development of new drugs for the treatment of CNS disorders, it is required to have better hypotheses and predict the efficacy/effectiveness of the new drug by considering the latest understanding of the disease pathophysiology in the brain. The researcher should always be prepared for the possibility that the primary outcomes of CNS clinical trials may not be favourable due to our lack of knowledge of disease progression. It is also stated in an FDA publication examining 22 case studies where Phase 2 and Phase 3 trials had divergent results that a medical product's apparent mechanism of action does not automatically predict clinical outcomes (419). Furthermore, the clinical trial outcomes presented in this chapter demonstrate that a "one therapeutic strategy" approach may not be appropriate for treating neurodegenerative diseases. Given the complexity of CNS diseases, the pharmaceutical industry has traditionally pursued a paradigm where one drug targets one specific neurodegenerative condition. However, this approach may be inadequate for addressing the multifaceted nature of CNS diseases.

REFERENCES

1. Armstrong BK, Robinson PJ, Rapoport SI. Size-dependent blood-brain barrier opening demonstrated with [14C]sucrose and a 200,000-da [3H]dextran. Exp Neurol. 1987;97(3):686–96.
2. Dhuyvetter D, Tekle F, Nazarov M, Vreeken RJ, Borghys H, Rombouts F et al. Direct nose to brain delivery of small molecules: Critical analysis of data from a standardized in vivo screening model in rats. Drug Deliv. 2020;27(1):1597–607.
3. Fortuna A, Alves G, Serralheiro A, Sousa J, Falcão A. Intranasal delivery of systemic-acting drugs: Small-molecules and biomacromolecules. Eur J Pharm Biopharm. 2014;88(1):8–27.
4. Larhed AW, Artursson P, Gråsjö J, Björk E. Diffusion of drugs in native and purified gastrointestinal mucus. J Pharm Sci. 1997;86(6):660–5.
5. Islam Y, Leach AG, Smith J, Pluchino S, Coxon CR, Sivakumaran M et al. Physiological and pathological factors affecting drug delivery to the brain by nanoparticles. Adv Sci (Weinh). 2021;8(11):e2002085.
6. Yousfan A, Al Rahwanji MJ, Hanano A, Al-Obaidi H. A comprehensive study on nanoparticle drug delivery to the brain: Application of machine learning techniques. Mol Pharm. 2024;21(1):333–45.
7. Lee D, Minko T. Nanotherapeutics for nose-to-brain drug delivery: An approach to bypass the blood brain barrier. Pharmaceutics. 2021;13(12):2049.
8. Profaci CP, Munji RN, Pulido RS, Daneman R. The blood-brain barrier in health and disease: Important unanswered questions. J Exp Med. 2020;217(4):e20190062.
9. Boyanova ST, Lloyd-Morris E, Corpe C, Rahman KM, Farag DB, Page LK et al. Interaction of amisulpride with GLUT1 at the blood-brain barrier. Relevance to Alzheimer's disease. PLoS One. 2023;18(10):e0286278.
10. Noell S, Wolburg-Buchholz K, Mack AF, Wolburg H, Fallier-Becker P. The Blood-Brain Barrier in Brain Tumours. In: Garami M, editor. Management of CNS Tumors 2011. p. 77–96.
11. Teleki A, Nylander O, Bergström CAS. Intrinsic dissolution rate profiling of poorly water-soluble compounds in biorelevant dissolution media. Pharmaceutics. 2020;12(6):493.
12. Törnqvist E, Annas A, Granath B, Jalkesten E, Cotgreave I, Öberg M. Strategic focus on 3R principles reveals major reductions in the use of animals in pharmaceutical toxicity testing. PLoS One. 2014;9(7):e101638.
13. Litten RZ, Egli M, Heilig M, Cui C, Fertig JB, Ryan ML et al. Medications development to treat alcohol dependence: A vision for the next decade. Addict Biol. 2012;17(3):513–27.

14. Turner PV, Brabb T, Pekow C, Vasbinder MA. Administration of substances to laboratory animals: Routes of administration and factors to consider. J Am Assoc Lab Anim Sci. 2011;50(5):600–13.
15. Malyala P, Singh M. Endotoxin limits in formulations for preclinical research. J Pharm Sci. 2008;97(6):2041–4.
16. Nádházi Z, Takáts A, Offenmüller K, Bertók L. Plasma endotoxin level of healthy donors. Acta Microbiol Immunol Hung. 2002;49(1):151–7.
17. FDA. Pyrogen and Endotoxins Testing; Questions and Answers 2012.
18. US-GAO. NEW DRUG DEVELOPMENT:Science, Business, Regulatory, and Intellectual Property Issues Cited as Hampering Drug Development Efforts. US: Scholar's Choice 2006.
19. Nugent P, Duncan JN, Colagiovanni DB. Chapter 8 - Preparation of a preclinical dossier to support an investigational new drug (IND) application and first-in-human clinical trial. In: Faqi AS, editor. A Comprehensive Guide to Toxicology in Nonclinical Drug Development (Second Edition). Boston: Academic Press; 2017. p. 189–213.
20. Gastelurrutia P, Prat-Vidal C, Vives J, Coll R, Bayes-Genis A, Gálvez-Montón C. Transitioning from preclinical evidence to advanced therapy medicinal product: A Spanish experience. Front Cardiovasc Med. 2021;8:604434.
21. Versyck B, Vermeylen K, Willemse J, van Geffen GJ, Leunen I, Soetens F et al. Serum concentrations of local anesthetics after unilateral interpectoral-pectoserratus plane block in breast cancer surgery: A pharmacokinetic study. Reg Anesth Pain Med. 2023;48(8):399–402.
22. Molinder HK. The development of cimetidine: 1964-1976. A human story. J Clin Gastroenterol. 1994;19(3):248–54.
23. Black JW, Duncan WA, Durant CJ, Ganellin CR, Parsons EM. Definition and antagonism of histamine H2-receptors. Nature. 1972;236(5347):385–90.
24. Black JW, Duncan WA, Emmett JC, Ganellin CR, Hesselbo T, Parsons ME et al. Metiamide–an orally active histamine H2-receptor antagonist. Agents Actions. 1973;3(3):133–7.
25. Wyllie JH, Ealding WD, Hesselbo T, Plack JW. Inhibition of gastric secretion in man by metiamide: A new, orally active histamine H2-receptor antagonist. Gut. 1973;14(5):424.
26. Burland WL. Treatment of duodenal ulcer by metiamide: A multicentre trial. The Lancet. 1975;306(7939):779–81.
27. Cecen B, Saygili E, Zare I, Nejati O, Khorsandi D, Zarepour A et al. Biosensor integrated brain-on-a-chip platforms: Progress and prospects in clinical translation. Biosens Bioelectron. 2023;225:115100.
28. Chim SM, Howell K, Kokkosis A, Zambrowicz B, Karalis K, Pavlopoulos E. A human brain-chip for modeling brain pathologies and screening blood–brain barrier crossing therapeutic strategies. Pharmaceutics. 2024;16(10):1314.
29. Lubin KE, Xu Z, Sluka JP, Knipp GT. Development of an in vitro model of the neurovascular unit for BBB permeability-linked neuroactivity screening. Med Chem Res. 2024;33(8):1450–63.
30. Rennie K, Yogi A, Costain WJ. Mechanisms and methods for evaluating drug delivery via transcytosis to the brain. In: Talevi A, editor. CNS Drug Development and Delivery: Concepts and Applications. Cham: Springer Nature Switzerland; 2024. p. 31–68.
31. Gou R, Yang J, Guo M, Chen Y, Xue W. CNSMolGen: A bidirectional recurrent neural network-based generative model for de novo Central nervous system drug design. J Chem Inf Model. 2024;64(10):4059–70.
32. Talevi A. Network-pharmacology and Central nervous system drug development. In: Talevi A, editor. CNS Drug Development and Delivery: Concepts and Applications. Cham: Springer Nature Switzerland; 2024. p. 213–27.
33. Liu Z, Roberts RA, Lal-Nag M, Chen X, Huang R, Tong W. AI-based language models powering drug discovery and development. Drug Discov Today. 2021;26(11):2593–607.
34. Zhu Y, Li L, Lu H, Zhou A, Qin X. Extracting drug-drug interactions from texts with BioBERT and multiple entity-aware attentions. J Biomed Inform. 2020;106:103451.
35. Sadybekov AV, Katritch V. Computational approaches streamlining drug discovery. Nature. 2023; 616(7958):673–85.
36. Talevi A. In silico prediction of CNS bioavailability. In: Talevi A, editor. CNS Drug Development and Delivery: Concepts and Applications. Cham: Springer Nature Switzerland; 2024. p. 93–112.
37. Chan B, Lynch NB, Tran W, Joyce JM, Savage GP, Meutermans W et al. Fragment-based drug discovery for disorders of the central nervous system: Designing better drugs piece by piece. Front Chem. 2024;12:1379518.
38. Petros AM, Korepanova A, Jakob CG, Qiu W, Panchal SC, Wang J et al. Fragment-based discovery of an apolipoprotein E4 (apoE4) stabilizer. J Med Chem. 2019;62(8):4120–30.

39. Marshall SF, Burghaus R, Cosson V, Cheung SY, Chenel M, DellaPasqua O et al. Good practices in model-informed drug discovery and development: Practice, application, and documentation. CPT Pharmacometrics Syst Pharmacol. 2016;5(3):93–122.
40. Krauss M, Schaller S, Borchers S, Findeisen R, Lippert J, Kuepfer L. Integrating cellular metabolism into a multiscale whole-body model. PLOS Comput Biol. 2012;8(10):e1002750.
41. Dwivedi G, Fitz L, Hegen M, Martin S, Harrold J, Heatherington A et al. A multiscale model of interleukin-6–Mediated immune regulation in Crohn's disease and its application in drug discovery and development. CPT: Pharmacometrics Syst Pharmacol. 2014;3(1):89.
42. Goteti K, Garner CE, Utley L, Dai J, Ashwell S, Moustakas DT et al. Preclinical pharmacokinetic/pharmacodynamic models to predict synergistic effects of co-administered anti-cancer agents. Cancer Chemother Pharmacol. 2010;66(2):245–54.
43. Ferrero J, Williams L, Stella H, Leitermann K, Mikulskis A, O'Gorman J et al. First-in-human, double-blind, placebo-controlled, single-dose escalation study of aducanumab (BIIB037) in mild-to-moderate Alzheimer's disease. Alzheimers Dement (N Y). 2016;2(3):169–76.
44. Sheremata WA, Vollmer TL, Stone LA, Willmer-Hulme AJ, Koller M. A safety and pharmacokinetic study of intravenous natalizumab in patients with MS. Neurology. 1999;52(5):1072–4.
45. Goteti K, Edwin Garner C, Utley L, Dai J, Ashwell S, Moustakas DT et al. Preclinical pharmacokinetic/pharmacodynamic models to predict synergistic effects of co-administered anti-cancer agents. Cancer Chemother Pharmacol. 2010;66(2):245–54.
46. Morofuji Y, Nakagawa S. drug development for Central nervous system diseases using in vitro blood-brain barrier models and drug repositioning. Curr Pharm Des. 2020;26(13):1466–85.
47. Abbott A. Novartis to shut brain research facility. Nature. 2011;480(7376):161–2.
48. Miller G. Is pharma running out of brainy ideas? Science. 2010;329(5991):502–4.
49. Gribkoff VK, Kaczmarek LK. The need for new approaches in CNS drug discovery: Why drugs have failed, and what can be done to improve outcomes. Neuropharmacology. 2017;120:11–9.
50. Simutis FJ, Sanderson TP, Pilcher GD, Graziano MJ. Nonclinical safety assessment of the γ-secretase inhibitor avagacestat. Toxicol Sci. 2018;163(2):525–42.
51. Anderson JJ, Holtz G, Baskin PP, Turner M, Rowe B, Wang B et al. Reductions in β-amyloid concentrations in vivo by the γ-secretase inhibitors BMS-289948 and BMS-299897. Biochem Pharmacol. 2005;69(4):689–98.
52. Burton CR, Meredith JE, Barten DM, Goldstein ME, Krause CM, Kieras CJ et al. The amyloid-β rise and γ-secretase inhibitor potency depend on the level of substrate Expression*. J Biol Chem. 2008;283(34):22992–3003.
53. Parker MF, Bronson JJ, Barten DM, Corsa JA, Du W, Felsenstein KM et al. Amino-caprolactam derivatives as γ-secretase inhibitors. Bioorg Med Chem Lett. 2007;17(21):5790–5.
54. Gillman KW, Starrett JE Jr., Parker MF, Xie K, Bronson JJ, Marcin LR et al. Discovery and evaluation of BMS-708163, a potent, selective and orally bioavailable γ-secretase inhibitor. ACS Med Chem Lett. 2010;1(3):120–4.
55. Tong G, Wang JS, Sverdlov O, Huang SP, Slemmon R, Croop R et al. A contrast in safety, pharmacokinetics and pharmacodynamics across age groups after a single 50 mg oral dose of the γ-secretase inhibitor avagacestat. Br J Clin Pharmacol. 2013;75(1):136–45.
56. Coric V, van Dyck CH, Salloway S, Andreasen N, Brody M, Richter RW et al. Safety and tolerability of the γ-secretase inhibitor avagacestat in a phase 2 study of mild to moderate Alzheimer's disease. Archives Neurol. 2012;69(11):1430–40.
57. Coric V, Salloway S, van Dyck CH, Dubois B, Andreasen N, Brody M et al. Targeting prodromal alzheimer disease with avagacestat: A randomized clinical trial. JAMA Neurol. 2015;72(11):1324–33.
58. Lewis TJ, Trempe CL. Chapter 2 - the amyloid Cascade hypothesis. In: Lewis TJ, Trempe CL, editors. The End of Alzheimer's (Second Edition): Academic Press; 2017. p. 26–50.
59. Henley DB, May PC, Dean RA, Siemers ER. Development of semagacestat (LY450139), a functional gamma-secretase inhibitor, for the treatment of Alzheimer's disease. Expert Opin Pharmacother. 2009;10(10):1657–64.
60. Yang G, Zhou R, Guo X, Yan C, Lei J, Shi Y. Structural basis of γ-secretase inhibition and modulation by small molecule drugs. Cell. 2021;184(2):521–33.e14.
61. Boggs LN, Fuson KS, Gitter BD, Czilli DL, Hyslop PA, Bender MH et al. P1-419 in vivo characterization of LY450139, a novel, stereoselective, functional gamma-secretase inhibitor. Neurobiol Aging. 2004;25:S218.
62. Ness DK, Boggs LN, Hepburn DL, Gitter B, Long GG, May PC et al. P2-053 reduced β-amyloid burden, increased c-99 concentrations and evaluation of neuropathology in the brains of PDAPP mice given LY450139 dihydrate daily by gavage for 5 months. Neurobiology of Aging. 2004;25:S238–S9.

63. Siemers ER, Dean RA, Friedrich S, Ferguson-Sells L, Gonzales C, Farlow MR et al. Safety, tolerability, and effects on plasma and cerebrospinal fluid amyloid-beta after inhibition of gamma-secretase. Clin Neuropharmacol. 2007;30(6):317–25.
64. Siemers E, Skinner M, Dean RA, Gonzales C, Satterwhite J, Farlow M et al. Safety, tolerability, and changes in amyloid beta concentrations after administration of a gamma-secretase inhibitor in volunteers. Clin Neuropharmacol. 2005;28(3):126–32.
65. Bateman RJ, Siemers ER, Mawuenyega KG, Wen G, Browning KR, Sigurdson WC et al. A gamma-secretase inhibitor decreases amyloid-beta production in the central nervous system. Ann Neurol. 2009;66(1):48–54.
66. Fleisher AS, Raman R, Siemers ER, Becerra L, Clark CM, Dean RA et al. Phase 2 safety trial targeting amyloid beta production with a gamma-secretase inhibitor in alzheimer disease. Arch Neurol. 2008;65(8):1031–8.
67. Moyer H, Mellett R, Vigneault K, McKeown M, Karlawish J, Augustine E et al. Prevalence and impact of bypassing or overriding phase 2 trials in neurologic drug development. Neurology. 2024; 103(1):e209533.
68. Panza F, Frisardi V, Imbimbo BP, Capurso C, Logroscino G, Sancarlo D, et al. REVIEW: γ-Secretase inhibitors for the treatment of Alzheimer's disease: The current state. CNS Neurosci Ther. 2010; 16(5):272–84.
69. Henley DB, Sundell KL, Sethuraman G, Dowsett SA, May PC. Safety profile of semagacestat, a gamma-secretase inhibitor: IDENTITY trial findings. Curr Med Res Opin. 2014;30(10):2021–32.
70. Doody RS, Raman R, Farlow M, Iwatsubo T, Vellas B, Joffe S et al. A phase 3 trial of semagacestat for treatment of Alzheimer's disease. N Engl J Med. 2013;369(4):341–50.
71. Henley DB, Dowsett SA, Chen YF, Liu-Seifert H, Grill JD, Doody RS et al. Alzheimer's disease progression by geographical region in a clinical trial setting. Alzheimer's Res Ther. 2015;7(1):43.
72. Eli-lilly&Company. A Study of Semagacestat for Alzheimer's Patients (Identity XT). https://clinicaltrialsgov/study/NCT01035138?term=NCT01035138&rank=1. 2014.
73. Mayer SC, Kreft AF, Harrison B, Abou-Gharbia M, Antane M, Aschmies S et al. Discovery of begacestat, a notch-1-sparing γ-secretase inhibitor for the treatment of Alzheimer's disease. J Med Chem. 2008;51(23):7348–51.
74. Zhao J, Liu X, Xia W, Zhang Y, Wang C. Targeting amyloidogenic processing of APP in Alzheimer's disease. Front Mol Neurosci. 2020;13:137.
75. Crescioli S, Kaplon H, Chenoweth A, Wang L, Visweswaraiah J, Reichert JM. Antibodies to watch in 2024. MAbs. 2024;16(1):2297450.
76. de la Torre BG, Albericio F. The pharmaceutical industry in 2023: An analysis of FDA drug approvals from the perspective of molecules. Molecules. 2024;29(3):585.
77. Mould DR, Meibohm B. Drug development of therapeutic monoclonal antibodies. BioDrugs. 2016;30(4):275–93.
78. Hansel TT, Kropshofer H, Singer T, Mitchell JA, George AJT. The safety and side effects of monoclonal antibodies. Nat Rev Drug Discov. 2010;9(4):325–38.
79. Suntharalingam G, Perry MR, Ward S, Brett SJ, Castello-Cortes A, Brunner MD et al. Cytokine storm in a phase 1 trial of the anti-CD28 monoclonal antibody TGN1412. N Engl J Med. 2006; 355(10):1018–28.
80. Tabrizi MA, Roskos LK. Preclinical and clinical safety of monoclonal antibodies. Drug Discov Today. 2007;12(13–14):540–7.
81. Leonard P, Säfsten P, Hearty S, McDonnell B, Finlay W, O'Kennedy R. High throughput ranking of recombinant avian scFv antibody fragments from crude lysates using the Biacore A100. J ImmunolMethods. 2007;323(2):172–9.
82. Canziani GA, Klakamp S, Myszka DG. Kinetic screening of antibodies from crude hybridoma samples using Biacore. Anal Biochem. 2004;325(2):301–7.
83. Chapman K, Pullen N, Coney L, Dempster M, Andrews L, Bajramovic J et al. Preclinical development of monoclonal antibodies: Considerations for the use of non-human primates. MAbs. 2009; 1(5):505–16.
84. EMA. Synagis SUMMARY OF PRODUCT CHARACTERISTICS, 2013. https://www.ema.europa.eu/en/documents/product–information/synagis–epar–product–information_en.pdf.
85. EMA. Tysabri SUMMARY OF PRODUCT CHARACTERISTICS. 2005. https://www.ema.europa.eu/en/documents/product–information/tysabri–epar–product–information_en.pdf.
86. Wehner NG, Gasper C, Shopp G, Nelson J, Draper K, Parker S et al. Immunotoxicity profile of natalizumab. J Immunotoxicol. 2009;6(2):115–29.

87. Mullen JT, Vartanian TK, Atkins MB. Melanoma complicating treatment with natalizumab for multiple sclerosis. N Engl J Med. 2008;358(6):647–8.
88. Sabol RA, Noxon V, Sartor O, Berger JR, Qureshi Z, Raisch DW et al. Melanoma complicating treatment with natalizumab for multiple sclerosis: A report from the Southern network on adverse reactions (SONAR). Cancer Med. 2017;6(7):1541–51.
89. Wehner NG, Shopp G, Rocca MS, Clarke J. Effects of natalizumab, an alpha4 integrin inhibitor, on the development of Hartley Guinea pigs. Birth Defects Res B Dev Reprod Toxicol. 2009;86(2):98–107.
90. Dyck CH, Swanson CJ, Aisen P, Bateman RJ, Chen C, Gee M et al. Lecanemab in early Alzheimer's disease. N Engl J Med. 2023;388(1):9–21.
91. Lord A, Gumucio A, Englund H, Sehlin D, Sundquist VS, Söderberg L et al. An amyloid-β protofibril-selective antibody prevents amyloid formation in a mouse model of Alzheimer's disease. Neurobio Disease. 2009;36(3):425–34.
92. Tucker S, Möller C, Tegerstedt K, Lord A, Laudon H, Sjödahl J et al. The murine version of BAN2401 (mAb158) selectively reduces amyloid-β protofibrils in brain and cerebrospinal fluid of tg-ArcSwe mice. J Alzheimers Dis. 2015;43(2):575–88.
93. Magnusson K, Sehlin D, Syvänen S, Svedberg MM, Philipson O, Söderberg L et al. Specific uptake of an amyloid-β protofibril-binding antibody-tracer in aβpp transgenic mouse brain. J Alzheimers Dis. 2013;37(1):29–40.
94. Logovinsky V, Satlin A, Lai R, Swanson C, Kaplow J, Osswald G et al. Safety and tolerability of BAN2401 - a clinical study in Alzheimer's disease with a protofibril selective aβ antibody. Alzheimer's Res Therapy. 2016;8(1):14.
95. Swanson CJ, Zhang Y, Dhadda S, Wang J, Kaplow J, Lai RYK et al. A randomized, double-blind, phase 2b proof-of-concept clinical trial in early Alzheimer's disease with lecanemab, an anti-aβ protofibril antibody. Alzheimer's Res Ther. 2021;13(1):80.
96. Dhadda S, Kanekiyo M, Li D, Swanson CJ, Irizarry M, Berry S et al. Consistency of efficacy results across various clinical measures and statistical methods in the lecanemab phase 2 trial of early Alzheimer's disease. Alzheimer's Res Ther. 2022;14(1):182.
97. Sevigny J, Chiao P, Bussière T, Weinreb PH, Williams L, Maier M et al. The antibody aducanumab reduces aβ plaques in Alzheimer's disease. Nature. 2016;537(7618):50–6.
98. Cummings J, Aisen P, Lemere C, Atri A, Sabbagh M, Salloway S. Aducanumab produced a clinically meaningful benefit in association with amyloid lowering. Alzheimer's Res Ther. 2021;13(1):98.
99. Budd Haeberlein S, Aisen PS, Barkhof F, Chalkias S, Chen T, Cohen S et al. Two randomized phase 3 studies of aducanumab in early Alzheimer's disease. J Prev Alzheimers Dis. 2022;9(2):197–210.
100. Brockmann R, Nixon J, Love BL, Yunusa I. Impacts of FDA approval and medicare restriction on anti-amyloid therapies for Alzheimer's disease: Patient outcomes, healthcare costs, and drug development. Lancet Reg Health Am. 2023;20:100467.
101. Dyer O. Aduhelm: Biogen abandons Alzheimer's drug after controversial approval left it unfunded by medicare. Bmj. 2024;384:q281.
102. Whitehouse P, Gandy S, Saini V, George DR, Larson EB, Alexander GC et al. Making the case for accelerated withdrawal of aducanumab. J Alzheimers Dis. 2022;87(3):1003–7.
103. Mably AJ, Liu W, Mc Donald JM, Dodart J-C, Bard F, Lemere CA et al. Anti-aβ antibodies incapable of reducing cerebral aβ oligomers fail to attenuate spatial reference memory deficits in J20 mice. Neurobiol Dis. 2015;82:372–84.
104. DeMattos RB, Bales KR, Cummins DJ, Dodart JC, Paul SM, Holtzman DM. Peripheral anti-Abeta antibody alters CNS and plasma A beta clearance and decreases brain A beta burden in a mouse model of Alzheimer's disease. Proc Natl Acad Sci U S A. 2001;98(15):8850–5.
105. Siemers ER, Friedrich S, Dean RA, Gonzales CR, Farlow MR, Paul SM et al. Safety and changes in plasma and cerebrospinal fluid amyloid beta after a single administration of an amyloid beta monoclonal antibody in subjects with alzheimer disease. Clin Neuropharmacol. 2010;33(2):67–73.
106. Farlow M, Arnold SE, van Dyck CH, Aisen PS, Snider BJ, Porsteinsson AP et al. Safety and biomarker effects of solanezumab in patients with Alzheimer's disease. Alzheimer's Dement. 2012;8(4):261–71.
107. Doody RS, Thomas RG, Farlow M, Iwatsubo T, Vellas B, Joffe S et al. Phase 3 trials of solanezumab for mild-to-moderate Alzheimer's disease. N Engl J Med. 2014;370(4):311–21.
108. Honig LS, Vellas B, Woodward M, Boada M, Bullock R, Borrie M et al. Trial of solanezumab for mild dementia due to Alzheimer's disease. N Engl J Med. 2018;378(4):321–30.
109. Sperling RA, Donohue MC, Raman R, Rafii MS, Johnson K, Masters CL et al. Trial of solanezumab in preclinical Alzheimer's disease. N Engl J Med. 2023;389(12):1096–107.

110. Seubert P, Barbour R, Khan K, Motter R, Tang P, Kholodenko D et al. Antibody capture of soluble abeta does not reduce cortical abeta amyloidosis in the PDAPP mouse. Neurodegener Dis. 2008;5(2):65–71.
111. Schroeter S, Khan K, Barbour R, Doan M, Chen M, Guido T et al. Immunotherapy reduces vascular amyloid-beta in PDAPP mice. J Neurosci. 2008;28(27):6787–93.
112. Racke MM, Boone LI, Hepburn DL, Parsadainian M, Bryan MT, Ness DK et al. Exacerbation of cerebral amyloid angiopathy-associated microhemorrhage in amyloid precursor protein transgenic mice by immunotherapy is dependent on antibody recognition of deposited forms of amyloid beta. J Neurosci. 2005;25(3):629–36.
113. Zago W, Schroeter S, Guido T, Khan K, Seubert P, Yednock T et al. Vascular alterations in PDAPP mice after anti-aβ immunotherapy: Implications for amyloid-related imaging abnormalities. Alzheimer's Dement. 2013;9(5 Suppl):S105–15.
114. Black RS, Sperling RA, Safirstein B, Motter RN, Pallay A, Nichols A et al. A single ascending dose study of bapineuzumab in patients with alzheimer disease. Alzheimer Dis Assoc Disord. 2010;24(2):198–203.
115. Rinne JO, Brooks DJ, Rossor MN, Fox NC, Bullock R, Klunk WE et al. 11C-PiB PET assessment of change in fibrillar amyloid-beta load in patients with Alzheimer's disease treated with bapineuzumab: A phase 2, double-blind, placebo-controlled, ascending-dose study. Lancet Neurol. 2010;9(4):363–72.
116. Salloway S, Sperling R, Fox NC, Blennow K, Klunk W, Raskind M et al. Two phase 3 trials of bapineuzumab in mild-to-moderate Alzheimer's disease. N Engl J Med. 2014;370(4):322–33.
117. Abushouk AI, Elmaraezy A, Aglan A, Salama R, Fouda S, Fouda R et al. Bapineuzumab for mild to moderate Alzheimer's disease: A meta-analysis of randomized controlled trials. BMC Neurol. 2017;17(1):66.
118. Gustafsson S, Gustavsson T, Roshanbin S, Hultqvist G, Hammarlund-Udenaes M, Sehlin D et al. Blood-brain barrier integrity in a mouse model of Alzheimer's disease with or without acute 3D6 immunotherapy. Neuropharmacology. 2018;143:1–9.
119. Taylor X, Clark IM, Fitzgerald GJ, Oluoch H, Hole JT, DeMattos RB et al. Amyloid-β (aβ) immunotherapy induced microhemorrhages are associated with activated perivascular macrophages and peripheral monocyte recruitment in Alzheimer's disease mice. Mol Neurodegener. 2023;18(1):59.
120. Demattos RB, Lu J, Tang Y, Racke MM, Delong CA, Tzaferis JA et al. A plaque-specific antibody clears existing β-amyloid plaques in Alzheimer's disease mice. Neuron. 2012;76(5):908–20.
121. McLaurin J, Cecal R, Kierstead ME, Tian X, Phinney AL, Manea M et al. Therapeutically effective antibodies against amyloid-beta peptide target amyloid-beta residues 4-10 and inhibit cytotoxicity and fibrillogenesis. Nat Med. 2002;8(11):1263–9.
122. Isabwe GAC, Garcia Neuer M, de Las Vecillas Sanchez L, Lynch DM, Marquis K, Castells M. Hypersensitivity reactions to therapeutic monoclonal antibodies: Phenotypes and endotypes. J Allergy Clin Immunol. 2018;142(1):159–70.e2.
123. Annunziato F, Romagnani C, Romagnani S. The 3 major types of innate and adaptive cell-mediated effector immunity. J Allergy and Clinical Immunol. 2015;135(3):626–35.
124. Umbricht D, Alberati D, Martin-Facklam M, Borroni E, Youssef EA, Ostland M et al. Effect of bitopertin, a Glycine reuptake inhibitor, on negative symptoms of schizophrenia: A randomized, double-blind, proof-of-concept study. JAMA Psychiatry. 2014;71(6):637–46.
125. Bugarski-Kirola D, Wang A, Abi-Saab D, Blättler T. A phase II/III trial of bitopertin monotherapy compared with placebo in patients with an acute exacerbation of schizophrenia - results from the CandleLyte study. Eur Neuropsychopharmacol. 2014;24(7):1024–36.
126. Tummala R, Desai D, Szamosi J, Wilson E, Hosford D, Dunbar G et al. Safety and tolerability of dexmecamylamine (TC-5214) adjunct to ongoing antidepressant therapy in patients with major depressive disorder and an inadequate response to antidepressant therapy: Results of a long-term study. J Clin Psychopharmacol. 2015;35(1):77–81.
127. Vieta E, Thase ME, Naber D, D'Souza B, Rancans E, Lepola U et al. Efficacy and tolerability of flexibly-dosed adjunct TC-5214 (dexmecamylamine) in patients with major depressive disorder and inadequate response to prior antidepressant. Eur Neuropsychopharmacol. 2014;24(4):564–74.
128. Kane JM, Detke HC, Naber D, Sethuraman G, Lin DY, Bergstrom RF et al. Olanzapine long-acting injection: A 24-week, randomized, double-blind trial of maintenance treatment in patients with schizophrenia. Am J Psychiatry. 2010;167(2):181–9.
129. McDonnell DP, Kryzhanovskaya LA, Zhao F, Detke HC, Feldman PD. Comparison of metabolic changes in patients with schizophrenia during randomized treatment with intramuscular olanzapine long-acting injection versus oral olanzapine. Hum Psychopharmacol. 2011;26(6):422–33.
130. Mayer SA, Brun NC, Begtrup K, Broderick J, Davis S, Diringer MN et al. Recombinant activated factor VII for acute intracerebral hemorrhage. N Engl J Med. 2005;352(8):777–85.

131. Mayer SA, Brun NC, Begtrup K, Broderick J, Davis S, Diringer MN et al. Efficacy and safety of recombinant activated factor VII for acute intracerebral hemorrhage. N Engl J Med. 2008;358(20):2127–37.
132. Khan I, Saeed K, Khan I. Nanoparticles: Properties, applications and toxicities. Arabian J Chem. 2019;12(7):908–31.
133. Wang Y, Xia Y. Bottom-Up and top-down approaches to the synthesis of monodispersed spherical colloids of low melting-point metals. Nano Letters. 2004;4(10):2047–50.
134. Nusz GJ, Marinakos SM, Curry AC, Dahlin A, Höök F, Wax A et al. Label-free plasmonic detection of biomolecular binding by a single gold nanorod. Analytical Chem. 2008;80(4):984–9.
135. Xue J, Wu T, Dai Y, Xia Y. Electrospinning and electrospun nanofibers: Methods, materials, and applications. Chem Rev. 2019;119(8):5298–415.
136. Jangir H, Bharadwaj A, Srivastava G, Das M. Fertilizer-free cultivation of wheat in nutrient-deficient soil by treating the seeds with nanopyrite. Nanotechnol Environmental Eng. 2020;5(1):9.
137. Park JH, Jin S-M, Lee E, Ahn HS. Electrochemical synthesis of core–shell nanoparticles by seed-mediated selective deposition. Chem Sci. 2021;12(40):13557–63.
138. Zhang W, Chen H, Ding L, Gong J, Zhang M, Guo W et al. Trojan horse delivery of 4,4'-dimethoxychalcone for Parkinsonian neuroprotection. Adv Sci (Weinh). 2021;8(9):2004555.
139. Johnsen KB, Burkhart A, Melander F, Kempen PJ, Vejlebo JB, Siupka P et al. Targeting transferrin receptors at the blood-brain barrier improves the uptake of immunoliposomes and subsequent cargo transport into the brain parenchyma. Sci Rep. 2017;7(1):10396.
140. Alvarez-Erviti L, Seow Y, Yin H, Betts C, Lakhal S, Wood MJA. Delivery of siRNA to the mouse brain by systemic injection of targeted exosomes. Nat Biotech. 2011;29(4):341–5.
141. Liu L, Xu K, Wang H, Tan PK, Fan W, Venkatraman SS et al. Self-assembled cationic peptide nanoparticles as an efficient antimicrobial agent. Nature Nanotechnol. 2009;4(7):457–63.
142. Desai N. Challenges in development of nanoparticle-based therapeutics. AAPS J. 2012;14(2):282–95.
143. Havel H, Finch G, Strode P, Wolfgang M, Zale S, Bobe I et al. Nanomedicines: From bench to bedside and beyond. AAPS J. 2016;18(6):1373–8.
144. Halwani AA. Development of pharmaceutical nanomedicines: From the bench to the market. Pharmaceutics. 2022;14(1).
145. Zheng X, Xie J, Zhang X, Sun W, Zhao H, Li Y et al. An overview of polymeric nanomicelles in clinical trials and on the market. Chinese Chem Lett. 2021;32(1):243–57.
146. Islam Y, Ehtezazi P, Cashmore A, Marinsalda E, Leach AG, Coxon CR et al. The inclusion of a matrix metalloproteinase-9 responsive sequence in self-assembled peptide-based brain-targeting nanoparticles improves the efficiency of nanoparticles crossing the blood-brain barrier at elevated MMP-9 levels. J Pharm Sci. 2021;110(3):1349–64.
147. Bonafé Allende JC, Ambrosioni F, Ruiz Moreno FN, Marin C, Romero VL, Virgolini MB et al. Pyrogallol-rich supramolecular hydrogels with enzyme-sensitive microdomains for controlled topical delivery of hydrophobic drugs. Biomat Adv. 2025;166:214075.
148. Yazdimamaghani M, Kolupaev OV, Lim C, Hwang D, Laurie SJ, Perou CM et al. Tumor microenvironment immunomodulation by nanoformulated TLR 7/8 agonist and PI3k delta inhibitor enhances therapeutic benefits of radiotherapy. Biomaterials. 2025;312:122750.
149. Wu LP, Ahmadvand D, Su J, Hall A, Tan X, Farhangrazi ZS et al. Crossing the blood-brain-barrier with nanoligand drug carriers self-assembled from a phage display peptide. Nat Commun. 2019; 10(1):4635.
150. Alnemeh-Al Ali H, Griveau A, Artzner F, Dupont A, Lautram N, Jourdain MA et al. Investigation on the self-assembly of the NFL-TBS.40-63 peptide and its interaction with gold nanoparticles as a delivery agent for glioblastoma. Int J Pharm: X. 2022;4:100128.
151. Berges R, Balzeau J, Takahashi M, Prevost C, Eyer J. Structure-function analysis of the glioma targeting NFL-TBS.40-63 peptide corresponding to the tubulin-binding site on the light neurofilament subunit. PLoS One. 2012;7(11):e49436.
152. Alargova RG, Kochijashky II, Sierra ML, Zana R. Micelle aggregation numbers of surfactants in aqueous solutions: A comparison between the results from steady-state and time-resolved fluorescence quenching. Langmuir. 1998;14(19):5412–8.
153. Kroll P, Benke J, Enders S, Brandenbusch C, Sadowski G. Influence of temperature and concentration on the self-assembly of nonionic CiEj surfactants: A light scattering study. ACS Omega. 2022; 7(8):7057–65.
154. Vasileva L, Gaynanova G, Valeeva F, Romanova E, Pavlov R, Kuznetsov D et al. Synthesis, properties, and biomedical application of dicationic gemini surfactants with dodecane spacer and carbamate fragments. Int J Mol Sci. 2023;24(15):12312.

155. Tan L, Fan J, Zhou Y, Xiong D, Duan M, Hu D et al. Preparation of reversible cross-linked amphiphilic polymeric micelles with pH-responsive behavior for smart drug delivery. RSC Adv. 2023;13(40):28165–78.
156. Shrivastava A. 2 - polymerization. In: Shrivastava A, editor. Introduction to Plastics Engineering: William Andrew Publishing; 2018. p. 17–48.
157. Zhang JA, Anyarambhatla G, Ma L, Ugwu S, Xuan T, Sardone T et al. Development and characterization of a novel cremophor® EL free liposome-based paclitaxel (LEP-ETU) formulation. Eur J Pharm Biopharm. 2005;59(1):177–87.
158. Bao X-N, Zhang W-D, Gang H-Z, Yang S-Z, Li Y-C, Mu B-Z. Formation of viscoelastic micellar solutions by a novel cationic surfactant and anionic salt system. Colloids and Surf A: Physicochem Eng Asp. 2021;611:125795.
159. Lavagnini E, Booth J, Helm K, El-Benni F, Warren PB, Bray DJ et al. Simulating micelle self-assembly to assess potential for viscosity build in surfactant formulations. Front Soft Matter. 2024;4.
160. Zhang H, Lyden D. Asymmetric-flow field-flow fractionation technology for exomere and small extracellular vesicle separation and characterization. Nature Protocols. 2019;14(4):1027–53.
161. Liu Y, Fens MHAM, Capomaccio RB, Mehn D, Scrivano L, Kok RJ et al. Correlation between in vitro stability and pharmacokinetics of poly(ε-caprolactone)-based micelles loaded with a photosensitizer. J Control Release. 2020;328:942–51.
162. Werner ME, Cummings ND, Sethi M, Wang EC, Sukumar R, Moore DT et al. Preclinical evaluation of genexol-PM, a nanoparticle formulation of paclitaxel, as a novel radiosensitizer for the treatment of non-small cell lung cancer. Int J Radiation Oncol*Biol*Phys. 2013;86(3):463–8.
163. He Z, Wan X, Schulz A, Bludau H, Dobrovolskaia MA, Stern ST et al. A high capacity polymeric micelle of paclitaxel: Implication of high dose drug therapy to safety and in vivo anti-cancer activity. Biomaterials. 2016;101:296–309.
164. Wennink JWH, Liu Y, Mäkinen PI, Setaro F, de la Escosura A, Bourajjaj M et al. Macrophage selective photodynamic therapy by meta-tetra(hydroxyphenyl)chlorin loaded polymeric micelles: A possible treatment for cardiovascular diseases. Eur J Pharm Sci. 2017;107:112–25.
165. Diezi TA, Bae Y, Kwon GS. Enhanced stability of PEG-block-poly(N-hexyl stearate l-aspartamide) micelles in the presence of serum proteins. Mol Pharm. 2010;7(4):1355–60.
166. Wu W, Wu Q, Liu Q-H, Li Y-Y, Ren P, Wu Y-Y et al. Identification and characterization of soft protein corona absorbed on iron oxide nanoparticles. Chin J Anal Chem. 2023;51(5):100246.
167. Levit M, Zashikhina N, Vdovchenko A, Dobrodumov A, Zakharova N, Kashina A et al. Bio-inspired amphiphilic block-copolymers based on synthetic glycopolymer and poly(Amino acid) as potential drug delivery systems. Polymers (Basel). 2020;12(1).
168. Kwon G, Naito M, Yokoyama M, Okano T, Sakurai Y, Kataoka K. Block copolymer micelles for drug delivery: Loading and release of doxorubicin. J Control Release. 1997;48(2):195–201.
169. Kulthe SS, Choudhari YM, Inamdar NN, Mourya V. Polymeric micelles: Authoritative aspects for drug delivery. Designed Monomers and Polymers. 2012;15(5):465–521.
170. Kim SC, Kim DW, Shim YH, Bang JS, Oh HS, Kim SW et al. In vivo evaluation of polymeric micellar paclitaxel formulation: Toxicity and efficacy. J Control Release. 2001;72(1):191–202.
171. Uchegbu IF, Breznikar J, Zaffalon A, Odunze U, Schätzlein AG. Polymeric micelles for the enhanced deposition of hydrophobic drugs into ocular tissues, without plasma exposure. Pharmaceutics. 2021;13(5).
172. Kache S, Ferry RJ. CHAPTER 15 - diabetes insipidus. In: Moshang T, editor. Pediatric Endocrinology: Mosby; 2005. p. 257–67.
173. Meesaragandla B, Komaragiri Y, Schlüter R, Otto O, Delcea M. The impact of cell culture media on the interaction of biopolymer-functionalized gold nanoparticles with cells: Mechanical and toxicological properties. Sci Rep. 2022;12(1):16643.
174. Sullivan BD, Whitmer D, Nichols KK, Tomlinson A, Foulks GN, Geerling G et al. An objective approach to dry eye disease severity. Invest Ophthalmol Vis Sci. 2010;51(12):6125–30.
175. Zingale E, Rizzo S, Bonaccorso A, Consoli V, Vanella L, Musumeci T et al. Optimization of lipid nanoparticles by response surface methodology to improve the ocular delivery of diosmin: Characterization and in-vitro anti-inflammatory assessment. Pharmaceutics. 2022;14(9):1961.
176. Wang W. Tolerability of hypertonic injectables. Int J Pharm. 2015;490(1):308–15.
177. Pereira-da-Silva L, Henriques G, Videira-Amaral JM, Rodrigues R, Ribeiro L, Virella D. Osmolality of solutions, emulsions and drugs that may have a high osmolality: Aspects of their use in neonatal care. J Matern Fetal Neonatal Med. 2002;11(5):333–8.
178. Adler S, Verbalis JG, Meyers S, Simplaceanu E, Williams DS. Changes in cerebral blood flow and distribution associated with acute increases in plasma sodium and osmolality of chronic hyponatremic rats. Experimental Neurol. 2000;163(1):63–71.

179. Dugan S, Le J, Jew RK. Maximum tolerated osmolarity for peripheral administration of parenteral nutrition in pediatric patients. J Parenteral Enteral Nutr. 2014;38(7):847–51.
180. Dubick MA, Wade CE. Evaluation of the local irritation potential of hypertonic saline-dextran (HSD) in mice and rabbits. J Appl Toxicol. 2004;24(6):409–13.
181. Hansen B. Chapter 15 - technical aspects of fluid therapy. In: DiBartola SP, editor. Fluid, Electrolyte, and Acid-Base Disorders in Small Animal Practice (Fourth Edition). Saint Louis: W.B. Saunders; 2012. p. 351–85.
182. Kwon GS, Okano T. Polymeric micelles as new drug carriers. Adv Drug Deliv Rev. 1996;21(2):107–16.
183. La SB, Okano T, Kataoka K. Preparation and characterization of the micelle-forming polymeric drug indomethacin-lncorporated polyfethylene oxide)-poly(β-benzyl l-aspartate) block copolymer micelles. J Pharm Sci. 1996;85(1):85–90.
184. Allen C, Maysinger D, Eisenberg A. Nano-engineering block copolymer aggregates for drug delivery. Colloids and Surf B: Biointerfaces. 1999;16(1):3–27.
185. Lin S, Yu Y, Wu E, Ding T, Chu Y, Pan F et al. Reexamining in vivo fate of paclitaxel-loaded polymeric micelles. Nano Today. 2024;56:102255.
186. Negut I, Bita B. Polymeric micellar systems—A special emphasis on “Smart. Pharmaceutics. 2023;15(3):976.
187. Malec K, Monaco S, Delso I, Nestorowicz J, Kozakiewicz-Latała M, Karolewicz B et al. Unravelling the mechanisms of drugs partitioning phenomena in micellar systems via NMR spectroscopy. J Colloid and Interface Sci. 2023;638:135–48.
188. Pokharkar V, Suryawanshi S, Dhapte-Pawar V. Exploring micellar-based polymeric systems for effective nose-to-brain drug delivery as potential neurotherapeutics. Drug Deliv Transl Res. 2020;10(4):1019–31.
189. Yoon SK, Chung DJ. In vivo degradation studies of PGA-PLA block copolymer and their histochemical analysis for spinal-fixing application. Polymers (Basel). 2022;14(16).
190. Šimek M, Hermannová M, Šmejkalová D, Foglová T, Souček K, Binó L et al. LC–MS/MS study of in vivo fate of hyaluronan polymeric micelles carrying doxorubicin. Carbohydrate Polym. 2019;209:181–9.
191. Ojha T, Hu Q, Colombo C, Wit J, van Geijn M, van Steenbergen MJ et al. Lyophilization stabilizes clinical-stage core-crosslinked polymeric micelles to overcome cold chain supply challenges. Biotechnol J. 2021;16(6):2000212.
192. Mihyar R, Shalmani AA, Wildt V, Sheybanifard M, Wang A, May J-N et al. Microfluidic formulation, cryoprotection and long-term stability of paclitaxel-loaded π electron-stabilized polymeric micelles. J Control Release. 2024;375:614–26.
193. Staples EJ, Tiddy GJT. Nuclear magnetic resonance technique to distinguish between micelle size changes and secondary aggregation in anionic and nonionic surfactant solutions. J Chem Soc, Faraday Trans 1: Physical Chemistry in Condensed Phases. 1978;74(0):2530–41.
194. Tănase MA, Soare AC, Diţu LM, Nistor CL, Mihaescu CI, Gifu IC et al. Influence of the hydrophobicity of pluronic micelles encapsulating curcumin on the membrane permeability and enhancement of photoinduced antibacterial activity. Pharmaceutics. 2022;14(10):2137.
195. Lemoine D, Francois C, Kedzierewicz F, Preat V, Hoffman M, Maincent P. Stability study of nanoparticles of poly(ε-caprolactone), poly(d,l-lactide) and poly(d,l-lactide-co-glycolide). Biomaterials. 1996;17(22):2191–7.
196. Muthu MS, Feng S-S. Pharmaceutical stability aspects of nanomedicines. Nanomedicine. 2009;4(8):857–60.
197. Zhang Z, Xu X, Du J, Chen X, Xue Y, Zhang J et al. Redox-responsive polymer micelles co-encapsulating immune checkpoint inhibitors and chemotherapeutic agents for glioblastoma therapy. Nat Commun. 2024;15(1):1118.
198. Varela-Moreira A, van Leur H, Krijgsman D, Ecker V, Braun M, Buchner M et al. Utilizing in vitro drug release assays to predict in vivo drug retention in micelles. Int J Pharm. 2022;618:121638.
199. Østergaard J, Mertz N, Gancho V, Le A, Barber M, Bezawada P et al. Development of a bio-relevant in vitro release testing method for subcutaneous and intramuscular oil depot formulations. J Drug Deliv Sci Technol. 2024;102:106412.
200. Klein S. The use of biorelevant dissolution media to forecast the in vivo performance of a drug. AAPS J. 2010;12(3):397–406.
201. Shende P, Trivedi R. Biofluidic material-based carriers: Potential systems for crossing cellular barriers. J Control Release. 2021;329:858–70.
202. Belyaeva AA, Averchuk AS, Rozanova NA, Alexandrova OP, Solomakha OA, Nashchekina YA et al. Thermosensitive injectable fibrillar gels based on cellulose nanocrystals grafted with poly(n-isopropylacrylamide) as biocompatible brain implants. Carbohydrate Polym. 2024;346:122596.

203. Bagheri M, Bresseleers J, Varela-Moreira A, Sandre O, Meeuwissen SA, Schiffelers RM et al. Effect of formulation and processing parameters on the size of mPEG- b-p(HPMA-bz) polymeric micelles. Langmuir. 2018;34(50):15495–506.
204. Salomatina E, Pikulin A, Semenov D, Grishin I, Kirillova N, Smirnova L. Synthesis of block copolymers by radical polymerization in presence of SI-based chain transfer agent. Polym Chem. 2023;14(28):3294–301.
205. Zhang H, Marin V, Fijten MWM, Schubert US. High-throughput experimentation in atom transfer radical polymerization: A general approach toward a directed design and understanding of optimal catalytic systems. J Polym Sc Part A: Polym Chem. 2004;42(8):1876–85.
206. Schuett T, Kimmig J, Zechel S, Schubert US. Fully automated multi-step synthesis of block copolymers. Polymers (Basel). 2022;14(2):292.
207. Feng H, Lu X, Wang W, Kang NG, Mays JW. Block copolymers: Synthesis, self-assembly, and applications. Polymers (Basel). 2017;9(10):494.
208. Yousi Z, Jian L, Rongchuan Z, Jianliang Y, Lizong D, Lansun Z. Synthesis of block copolymer from dissimilar vinyl monomer by stable free radical polymerization. Macromolecules. 2000;33(13):4745–9.
209. On byproducts and Side products. Organic Process Research & Development. 2012;16(12):1877–.
210. Zhang H, He Z, Fu C, Pan P, Zhu Y, Xu M et al. Dissociation of polymeric micelle under hemodynamic shearing. Nano Today. 2022;45:101517.
211. Wen S-N, Chu C-H, Wang Y-C, Huang H-Y, Wang Y-J, Lin J-Y et al. Polymer-stabilized micelles reduce the drug rapid clearance in vivo. J Nanomater. 2018;2018(1):5818592.
212. Kim T-Y, Kim D-W, Chung J-Y, Shin SG, Kim S-C, Heo DS et al. Phase I and pharmacokinetic study of Genexol-PM, a cremophor-free, polymeric micelle-formulated paclitaxel, in patients with advanced malignancies. Clin Cancer Res. 2004;10(11):3708–16.
213. Atrafi F, Dumez H, Mathijssen RHJ, Menke van der Houven van Oordt CW, Rijcken CJF, Hanssen R et al. A phase I dose-escalation and pharmacokinetic study of a micellar nanoparticle with entrapped docetaxel (CPC634) in patients with advanced solid tumours. J Control Release. 2020;325:191–7.
214. Han LM, Guo J, Zhang LJ, Wang QS, Fang XL. Pharmacokinetics and biodistribution of polymeric micelles of paclitaxel with Pluronic P123. Acta Pharmacol Sin. 2006;27(6):747–53.
215. Luo X, Chen M, Zhang Y, Chen Z, Li X. Pharmacokinetics and antitumor efficacy of micelles assembled from multiarmed amphiphilic copolymers with drug conjugates in comparison with drug-encapsulated micelles. Eur J Pharm Biopharm. 2016;98:9–19.
216. Zhu Q, Zhang B, Wang Y, Liu X, Li W, Su F et al. Self-assembled micelles prepared from poly(d,L-lactide-co-glycolide)-poly(ethylene glycol) block copolymers for sustained release of valsartan. Polym Adv Technol. 2021;32(3):1262–71.
217. Letchford K, Burt HM. Copolymer micelles and nanospheres with different in vitro stability demonstrate similar paclitaxel pharmacokinetics. Mol Pharm. 2012;9(2):248–60.
218. Spratlin J, Sawyer MB. Pharmacogenetics of paclitaxel metabolism. Crit Rev Oncol/Hematol. 2007;61(3):222–9.
219. Li P, Han J, Li D, Chen J, Wang W, Xu W. Synthetic glycopolypeptide micelle for targeted drug delivery to hepatic carcinoma. Polymers (Basel). 2018;10(6).
220. Sreeranganathan M, Uthaman S, Sarmento B, Mohan CG, Park I-K, Jayakumar R. In vivo evaluation of cetuximab-conjugated poly(γ-glutamic acid)-docetaxel nanomedicines in EGFR-overexpressing gastric cancer xenografts. Int J Nanomed. 2017;12(null):7165–82.
221. Kumar V, Mundra V, Peng Y, Wang Y, Tan C, Mahato RI. Pharmacokinetics and biodistribution of polymeric micelles containing miRNA and small-molecule drug in orthotopic pancreatic tumor-bearing mice. Theranostics. 2018;8(15):4033–49.
222. Meng X, Liu J, Yu X, Li J, Lu X, Shen T. Pluronic F127 and d-α-tocopheryl polyethylene glycol succinate (TPGS) mixed micelles for targeting drug delivery across the blood brain barrier. Sci Rep. 2017;7(1):2964.
223. Gu G, Xia H, Hu Q, Liu Z, Jiang M, Kang T et al. PEG-co-PCL nanoparticles modified with MMP-2/9 activatable low molecular weight protamine for enhanced targeted glioblastoma therapy. Biomaterials. 2013;34(1):196–208.
224. Zhao J, Chai Y-D, Zhang J, Huang P-F, Nakashima K, Gong Y-K. Long circulating micelles of an amphiphilic random copolymer bearing cell outer membrane phosphorylcholine zwitterions. Acta Biomaterialia. 2015;16:94–102.
225. Liu C, Wen J, Li D, Qi H, Nih L, Zhu J et al. Systemic delivery of microRNA for treatment of brain ischemia. Nano Res. 2021;14(9):3319–28.

226. Jiang T, Qiao Y, Ruan W, Zhang D, Yang Q, Wang G et al. Cation-free siRNA micelles as effective drug delivery platform and potent RNAi nanomedicines for glioblastoma therapy. Adv Mater. 2021;33(45):e2104779.
227. Kreuter J. Drug delivery to the central nervous system by polymeric nanoparticles: What do we know? Adv Drug Deliv Rev. 2014;71:2–14.
228. Rognin E, Willis-Fox N, Daly R. Quantitative monitoring and modelling of retrodialysis drug delivery in a brain phantom. Sci Rep. 2023;13(1):1900.
229. Bommana MM, Kirthivasan B, Squillante E. In vivo brain microdialysis to evaluate FITC-dextran encapsulated immunopegylated nanoparticles. Drug Deliver. 2012;19(6):298–306.
230. Hann E, Malagu K, Stott A, Vater H. Chapter three - the importance of plasma protein and tissue binding in a drug discovery program to successfully deliver a preclinical candidate. In: Witty DR, Cox B, editors. Progress in Medicinal Chemistry. 61: Elsevier; 2022. p. 163–214.
231. Skoczen S, McNeil SE, Stern ST. Stable isotope method to measure drug release from nanomedicines. J Control Release. 2015;220:169–74.
232. Bulitta JB, Zhao P, Arnold RD, Kessler DR, Daifuku R, Pratt J et al. Mechanistic population pharmacokinetics of total and unbound paclitaxel for a new nanodroplet formulation versus Taxol in cancer patients. Cancer Chemother Pharmacol. 2009;63(6):1049–63.
233. Bekersky I, Fielding RM, Dressler DE, Lee JW, Buell DN, Walsh TJ. Plasma protein binding of amphotericin B and pharmacokinetics of bound versus unbound amphotericin B after administration of intravenous liposomal amphotericin B (AmBisome) and amphotericin B deoxycholate. Antimicrob Agents Chemother. 2002;46(3):834–40.
234. Song W, Bai L, Yang Y, Wang Y, Xu P, Zhao Y et al. Long-circulation and brain targeted isoliquiritigenin micelle nanoparticles: Formation, characterization, tissue distribution, pharmacokinetics and effects for ischemic stroke. Int J Nanomed. 2022;17:3655–70.
235. Martinez MN. Factors influencing the use and interpretation of animal models in the development of parenteral drug delivery systems. AAPS J. 2011;13(4):632–49.
236. Seymour RS, Bosiocic V, Snelling EP, Chikezie PC, Hu Q, Nelson TJ et al. Cerebral blood flow rates in recent great apes are greater than in *Australopithecus* species that had equal or larger brains. Proc Royal Soc B: Biol Sci. 2019;286(1915):20192208.
237. Mitchell AS, Thiele A, Petkov CI, Roberts A, Robbins TW, Schultz W et al. Continued need for nonhuman primate neuroscience research. Current Biol. 2018;28(20):R1186–R7.
238. Jiang Z, Wang J, Qin Y, Liu S, Luo B, Bai F et al. A nonhuman primate model with Alzheimer's disease-like pathology induced by hippocampal overexpression of human tau. Alzheimer's Res Ther. 2024;16(1):22.
239. Chiarelli PA, Revia RA, Stephen ZR, Wang K, Jeon M, Nelson V et al. Nanoparticle biokinetics in mice and nonhuman primates. ACS Nano. 2017;11(9):9514–24.
240. Elsea SH, Lucas RE. The mousetrap: What we can learn when the mouse model does not mimic the human disease. ILAR J. 2002;43(2):66–79.
241. Huang L, Merson TD, Bourne JA. In vivo whole brain, cellular and molecular imaging in nonhuman primate models of neuropathology. Neuroscience & Biobehavioral Reviews. 2016;66:104–18.
242. Kaushik A, Rodriguez J, Rothen D, Bhardwaj V, Jayant RD, Pattany P et al. MRI-guided, noninvasive delivery of magneto-electric drug nanocarriers to the brain in a nonhuman primate. ACS Appl Bio Mater. 2019;2(11):4826–36.
243. Xiya L, Jinsheng Z. Why non-human primates are needed in stroke preclinical research. Stroke Vascular Neurol. 2025;10(2):e003504.
244. Royo J, Forkel SJ, Pouget P, Thiebaut de Schotten M. The squirrel monkey model in clinical neuroscience. Neurosci Biobehavioral Rev. 2021;128:152–64.
245. Wilson MG, Webb TD, Odéen H, Kubanek J. Remotely controlled drug release in deep brain regions of non-human primates. J Control Release. 2024;369:775–85.
246. Quitschke WW, Steinhauff N, Rooney J. The effect of cyclodextrin-solubilized curcuminoids on amyloid plaques in alzheimer transgenic mice: Brain uptake and metabolism after intravenous and subcutaneous injection. Alzheimer's Res Ther. 2013;5(2):16.
247. Li Y, Chen Z, Lu Z, Yang Q, Liu L, Jiang Z et al. Cell-addictive" dual-target traceable nanodrug for Parkinson's disease treatment via flotillins pathway. Theranostics. 2018;8(19):5469–81.
248. Hu Y, Hammarlund-Udenaes M, Fridén M. Understanding the influence of nanocarrier-mediated brain delivery on therapeutic performance through pharmacokinetic-pharmacodynamic modeling. J Pharm Sci. 2019;108(10):3425–33.
249. Jiang Y, Lee J, Seo J-M, Davaa E, Shin K-J, Yang S-G. Enhanced thermodynamic, pharmacokinetic and theranostic properties of polymeric micelles via hydrophobic core-clustering of superparamagnetic iron oxide nanoparticles. Biomater Res. 2022;26(1):8.

250. Yi Y, Yoon HJ, Kim BO, Shim M, Kim S-O, Hwang S-J et al. A mixed polymeric micellar formulation of itraconazole: Characteristics, toxicity and pharmacokinetics. Journal of Controlled Release. 2007;117(1):59–67.
251. Tominaga T, Inoue O, Suzuki K, Yamasaki T, Hirobe M. Organ distribution and metabolism of [13N]nicotinamide in mice. Int JRadiation Appl Instrumentation Part B Nuclear Med d Biology. 1987;14(5):491–6.
252. Hamada H, Ishihara K, Masuoka N, Mikuni K, Nakajima N. Enhancement of water-solubility and bioactivity of paclitaxel using modified cyclodextrins. J Biosci Bioeng. 2006;102(4):369–71.
253. Repp L, Skoczen SL, Rasoulianboroujeni M, Stern ST, Kwon GS. plasma stability and plasma metabolite concentration-time profiles of oligo(Lactic acid)(8)-paclitaxel prodrug loaded polymeric micelles. AAPS J. 2023;25(3):39.
254. Abourehab MA, Ahmed OA, Balata GF, Almalki WH. Self-assembled biodegradable polymeric micelles to improve dapoxetine delivery across the blood-brain barrier. Int J Nanomed. 2018;13:3679–87.
255. Fang C, Shi B, Pei Y-Y, Hong M-H, Wu J, Chen H-Z. In vivo tumor targeting of tumor necrosis factor-α-loaded stealth nanoparticles: Effect of MePEG molecular weight and particle size. Eur Journal Pharm Sci. 2006;27(1):27–36.
256. Wang C, Quan Y, Jiang J, Yu H, Liu J, Tang W et al. Protein coronation-induced cancer staging-dependent multilevel cytotoxicity: An all-humanized study in blood vessel organoids. ACS Nano. 2025;19(1):345–68.
257. Terracciano R, Carcamo-Bahena Y, Butler EB, Demarchi D, Grattoni A, Filgueira CS. Hyaluronate-thiol passivation enhances gold nanoparticle peritumoral distribution when administered intratumorally in lung cancer. Biomedicines. 2021;9(11).
258. Malingré MM, Schellens JH, van Tellingen O, Rosing H, Koopman FJ, Duchin K et al. Metabolism and excretion of paclitaxel after oral Administration in combination with cyclosporin A and after i.v. Administration. Anticancer Drugs. 2000;11(10):813–20.
259. Panday VRN, Rosing H, Huizing MT, Koopman FJ, van Warmerdam LJ, ten Bokkel Huinink WW et al. Urinary excretion of paclitaxel and metabolites in a large series study. J Oncol Pharmacy Practice. 1998;4(3):159–64.
260. Hamaguchi T, Kato K, Yasui H, Morizane C, Ikeda M, Ueno H et al. A phase I and pharmacokinetic study of NK105, a paclitaxel-incorporating micellar nanoparticle formulation. Br J Cancer. 2007;97(2):170–6.
261. Michael Grindel J, Jaworski T, Piraner O, Martin Emanuele R, Balasubramanian M. Distribution, metabolism, and excretion of a novel surface-active agent, purified poloxamer 188, in rats, dogs, and humans. J Pharm Sci. 2002;91(9):1936–47.
262. Batrakova EV, Li S, Li Y, Alakhov VY, Elmquist WF, Kabanov AV. Distribution kinetics of a micelle-forming block copolymer Pluronic P85. J Control Release. 2004;100(3):389–97.
263. Alakhov VY, Moskaleva EY, Batrakova EV, Kabanov AV. Hypersensitization of multidrug resistant human ovarian carcinoma cells by Pluronic P85 block copolymer. Bioconjugate Chem. 1996;7(2):209–16.
264. Kabanov AV, Batrakova EV, Miller DW. Pluronic® block copolymers as modulators of drug efflux transporter activity in the blood–brain barrier. Adv Drug Deliv Rev. 2003;55(1):151–64.
265. Fang Z, Cao P, Pan N, Lu H. Pluronic P85 decreases the delivery of phenytoin to the brain in drug-resistant rats with p-glycoprotein overexpressed chronic mesial temporal lobe epilepsy. IBRO Neurosci Rep. 2023;15:100–6.
266. Ismail A, Nasr M, Sammour O. Nanoemulsion as a feasible and biocompatible carrier for ocular delivery of travoprost: Improved pharmacokinetic/pharmacodynamic properties. Int J Pharm. 2020;583:119402.
267. Matsumura Y, Hamaguchi T, Ura T, Muro K, Yamada Y, Shimada Y et al. Phase I clinical trial and pharmacokinetic evaluation of NK911, a micelle-encapsulated doxorubicin. Br J Cancer. 2004;91(10):1775–81.
268. de Lange EC, Ravenstijn PG, Groenendaal D, van Steeg TJ. Toward the prediction of CNS drug-effect profiles in physiological and pathological conditions using microdialysis and mechanism-based pharmacokinetic-pharmacodynamic modeling. AAPS J. 2005;7(3):E532–43.
269. Zhang Q, Yang L, Zheng Y, Wu X, Chen X, Fei F et al. Electro-responsive micelle-based universal drug delivery system for on-demand therapy in epilepsy. J Control Release. 2023;360:759–71.
270. Park JE, Chun S-E, Reichel D, Min JS, Lee S-C, Han S et al. Polymer micelle formulation for the proteasome inhibitor drug carfilzomib: Anticancer efficacy and pharmacokinetic studies in mice. PLoS One. 2017;12(3):e0173247.
271. Mukherjee B, Satapathy BS, Bhattacharya S, Chakraborty R, Mishra VP. Chapter 19 - pharmacokinetic and pharmacodynamic modulations of therapeutically active constituents from orally administered nanocarriers along with a glimpse of their advantages and limitations. In: Grumezescu AM, editor. Nano- and Microscale Drug Delivery Systems: Elsevier; 2017. p. 357–75.

272. Fan W, Yu Z, Peng H, He H, Lu Y, Qi J et al. Effect of particle size on the pharmacokinetics and biodistribution of parenteral nanoemulsions. Int J Pharm. 2020;586:119551.
273. Carstens MG, Bevernage JJL, van Nostrum CF, van Steenbergen MJ, Flesch FM, Verrijk R et al. Small oligomeric micelles based on end group modified mPEG–Oligocaprolactone with monodisperse hydrophobic blocks. Macromolecules. 2007;40(1):116–22.
274. Liu Y, Fens M, Lou B, van Kronenburg NCH, Maas-Bakker RFM, Kok RJ et al. π-π-stacked poly(ε-caprolactone)-b-poly(ethylene glycol) micelles loaded with a photosensitizer for photodynamic therapy. Pharmaceutics. 2020;12(4).
275. Liu Y, Scrivano L, Peterson JD, Fens MHAM, Hernández IB, Mesquita B et al. EGFR-targeted nanobody functionalized polymeric micelles loaded with mTHPC for selective photodynamic therapy. Mol Pharm. 2020;17(4):1276–92.
276. Bagheri M, Fens MH, Kleijn TG, Capomaccio RB, Mehn D, Krawczyk PM et al. in vitro and in vivo studies on HPMA-based polymeric micelles loaded with curcumin. Mol Pharm. 2021;18(3):1247–63.
277. Gupta A, Costa AP, Xu X, Burgess DJ. Continuous processing of paclitaxel polymeric micelles. Int J Pharm. 2021;607:120946.
278. Cai Y, Qi J, Lu Y, He H, Wu W. The in vivo fate of polymeric micelles. Advanced Drug Delivery Reviews. 2022;188:114463.
279. Guo X, Wang L, Duval K, Fan J, Zhou S, Chen Z. Dimeric drug polymeric micelles with acid-active tumor targeting and FRET-traceable drug release. Adv Mater. 2018;30(3).
280. Zou P, Chen H, Paholak HJ, Sun D. Noninvasive fluorescence resonance energy transfer imaging of in vivo premature drug release from polymeric nanoparticles. Molecular Pharmaceutics. 2013;10(11):4185–94.
281. Liu J, Zeng F, Allen C. In vivo fate of unimers and micelles of a poly(ethylene glycol)-block-poly(caprolactone) copolymer in mice following intravenous administration. European Journal of Pharmaceutics and Biopharmaceutics. 2007;65(3):309–19.
282. Derendorf H, Lesko LJ, Chaikin P, Colburn WA, Lee P, Miller R et al. Pharmacokinetic/pharmacodynamic modeling in drug research and development. J Clin Pharmacol. 2000;40(12 Pt 2):1399–418.
283. Poulin P, Theil FP. Prediction of pharmacokinetics prior to in vivo studies. II. Generic physiologically based pharmacokinetic models of drug disposition. J Pharm Sci. 2002;91(5):1358–70.
284. Ball K, Bouzom F, Scherrmann JM, Walther B, Declèves X. Physiologically based pharmacokinetic modelling of drug penetration across the blood-brain barrier–towards a mechanistic IVIVE-based approach. AAPS J. 2013;15(4):913–32.
285. Hansen DK, Scott DO, Otis KW, Lunte SM. Comparison of in vitro BBMEC permeability and in vivo CNS uptake by microdialysis sampling. J Pharm Biomed Anal. 2002;27(6):945–58.
286. Bouhaddou M, Yu LJ, Lunardi S, Stamatelos SK, Mack F, Gallo JM et al. Predicting in vivo efficacy from in vitro data: Quantitative systems pharmacology modeling for an epigenetic modifier drug in cancer. Clin Transl Sci. 2020;13(2):419–29.
287. Yamashita S, Kimura S, Kiriyama A. Tumoral pharmacokinetic-pharmacodynamic simulation of doxorubicin-encapsulated polymeric micelles using a tissue-isolated tumor perfusion system. Mol Pharm. 2024;21(11):5736–48.
288. Pugsley MK, Authier S, Curtis MJ. Principles of safety pharmacology. Br J Pharmacol. 2008;154(7):1382–99.
289. Moscardo E, Maurin A, Dorigatti R, Champeroux P, Richard S. An optimised methodology for the neurobehavioural assessment in rodents. Journal of Pharmacological and Toxicological Methods. 2007;56(2):239–55.
290. Tontodonati M, Fasdelli N, Moscardo E, Giarola A, Dorigatti R. A canine model used to simultaneously assess potential neurobehavioural and cardiovascular effects of candidate drugs. Journal of Pharmacological and Toxicological Methods. 2007;56(2):265–75.
291. Irwin S. Comprehensive observational assessment: Ia. A systematic, quantitative procedure for assessing the behavioral and physiologic state of the mouse. Psychopharmacologia. 1968;13(3):222–57.
292. Kim TR, Kim H-Y, Kim I-H, Kim KC, Ko Y, Park JH et al. Safety pharmacology of self-assembled-micelle inhibitory RNA-targeting amphiregulin (SAMiRNA-AREG), a novel siRNA nanoparticle platform. Toxicol Rep. 2021;8:839–45.
293. Schneider J, Hauser R, Andreas J-O, Linz K, Jahnel U. Differential effects of human ether-a-go-go-related gene (HERG) blocking agents on QT duration variability in conscious dogs. Eur J Pharmacol. 2005;512(1):53–60.
294. Horii I. Advantages of toxicokinetics in new drug development. Toxicol Lett. 1998;102-103:657–64.
295. Zhao B, Wang X-Q, Wang X-Y, Zhang H, Dai W-B, Wang J et al. Nanotoxicity comparison of four amphiphilic polymeric micelles with similar hydrophilic or hydrophobic structure. Particle and Fibre Toxicology. 2013;10(1):47.

296. Silva PV, Santos CSA, Papadiamantis AG, Gonçalves SF, Prodana M, Verweij RA et al. Toxicokinetics of silver and silver sulfide nanoparticles in Chironomus riparius under different exposure routes. Sci Total Environ. 2023;865:161087.
297. Welling PG. Differences between pharmacokinetics and toxicokinetics. Toxicol Pathol. 1995;23(2):143–7.
298. Chen Q, Riviere JE, Lin Z. Toxicokinetics, dose–response, and risk assessment of nanomaterials: Methodology, challenges, and future perspectives. WIREs Nanomed Nanobiotechnol. 2022;14(6):e1808.
299. Qu J, Wang J, Zhang H, Wu J, Ma X, Wang S et al. Toxicokinetics and systematic responses of differently sized indium tin oxide (ITO) particles in mice via oropharyngeal aspiration exposure. Env Pollut. 2021;290:117993.
300. Doessegger L, Banholzer ML. Clinical development methodology for infusion-related reactions with monoclonal antibodies. Clin Transl Immunology. 2015;4(7):e39.
301. Plummer R, Wilson RH, Calvert H, Boddy AV, Griffin M, Sludden J et al. A phase I clinical study of cisplatin-incorporated polymeric micelles (NC-6004) in patients with solid tumours. Br J Cancer. 2011;104(4):593–8.
302. Majenka P, Loquai C, Schöning T, Enk A, Hassel J. Acute low back pain as infusion-related reaction to monoclonal antibodies. Front Oncol. 2023;13.
303. Mease KM, Kimzey AL, Lansita JA. Biomarkers for nonclinical infusion reactions in marketed biotherapeutics and considerations for study design. Curr Opin Toxicol. 2017;4:1–15.
304. Wang Y, Wang J, Li J, Mu Y, Ying J, Liu Z et al. Sulfoxide-containing polymers conjugated prodrug micelles with enhanced anticancer activity and reduced intestinal toxicity. J Control Release. 2024;371:313–23.
305. Ramaiah L, Bounous DI, Elmore SA. Chapter 50 - hematopoietic system. In: Haschek WM, Rousseaux CG, Wallig MA, editors. Haschek and Rousseaux's Handbook of Toxicologic Pathology (Third Edition). Boston: Academic Press; 2013. p. 1863–933.
306. Ali A, Bhadane R, Asl AA, Wilén C-E, Salo-Ahen O, Rosenholm JM et al. Functional block copolymer micelles based on poly (jasmine lactone) for improving the loading efficiency of weakly basic drugs. RSC Advances. 2022;12(41):26763–75.
307. Hou Z, Zhou W, Guo X, Zhong R, Wang A, Li J et al. Poly(ε-caprolactone)-methoxypolyethylene glycol (PCL-MPEG)-based micelles for drug-delivery: The effect of PCL chain length on blood components, phagocytosis, and biodistribution. Int J Nanomed. 2022;17:1613–32.
308. Fishman JM, Wiles K, Wood KJ. Chapter 8 - the acquired immune system response to biomaterials, including both naturally occurring and synthetic biomaterials. In: Badylak SF, editor. Host Response to Biomaterials. Oxford: Academic Press; 2015. p. 151–87.
309. Gold MR. To make antibodies or not: Signaling by the b-cell antigen receptor. Trends Pharmacol Sci. 2002;23(7):316–24.
310. Olsson PG, Hammarström L, Edvard Smith CI. Antigenicity of mouse monoclonal antibodies. A study on the variable region of the heavy chain. JTheor Bio. 1991;151(1):111–22.
311. Gardey E, Sobotta FH, Hoeppener S, Bruns T, Stallmach A, Brendel JC. Influence of core cross-linking and shell composition of polymeric micelles on immune response and their interaction with human monocytes. Biomacromolecules. 2020;21(4):1393–406.
312. Welling GW, Fries H. Choice of peptide and peptide length for the generation of antibodies reactive with the intact protein. FEBS Lett. 1985;182(1):81–4.
313. Vandebriel RJ, Tonk EC, de la Fonteyne-Blankestijn LJ, Gremmer ER, Verharen HW, van der Ven LT et al. Immunotoxicity of silver nanoparticles in an intravenous 28-day repeated-dose toxicity study in rats. Part Fibre Toxicol. 2014;11:21.
314. Zhong W, Penghuan C, Lianfang G, Lifan Z, Yang Z. A t-cell-dependent antibody response (TDAR) method in BALB/c mice based on a cytometric bead array. J Immunotoxicol. 2022;19(1):34–40.
315. Paoletti X, Le Tourneau C, Verweij J, Siu LL, Seymour L, Postel-Vinay S et al. Defining dose-limiting toxicity for phase 1 trials of molecularly targeted agents: Results of a DLT-TARGETT international survey. Eur J Cancer. 2014;50(12):2050–6.
316. Le Tourneau C, Razak ARA, Gan HK, Pop S, Diéras V, Tresca P et al. Heterogeneity in the definition of dose-limiting toxicity in phase I cancer clinical trials of molecularly targeted agents: A review of the literature. Eur J Cancer. 2011;47(10):1468–75.
317. Potter DM. Phase i studies of chemotherapeutic agents in cancer patients: A review of the designs. J Biopharm Stat. 2006;16(5):579–604.
318. Liu P, Chen G, Zhang J. A review of liposomes as a drug delivery system: Current status of approved products, regulatory environments, and future perspectives. Molecules. 2022;27(4):1372.
319. Wang C, Gamage PL, Jiang W, Mudalige T. Excipient-related impurities in liposome drug products. International Journal of Pharmaceutics. 2024;657:124164.

320. Wang C, Siriwardane DA, Jiang W, Mudalige T. Quantitative analysis of cholesterol oxidation products and desmosterol in parenteral liposomal pharmaceutical formulations. Int J Pharm. 2019;569:118576.
321. Jovanović AA, Balanč BD, Ota A, Ahlin Grabnar P, Djordjević VB, Šavikin KP et al. Comparative effects of cholesterol and β-sitosterol on the liposome membrane characteristics. Eur J Lipid Sci Technol. 2018;120(9):1800039.
322. Hadian Z, Sahari MA, Moghimi HR, Barzegar M. Formulation, characterization and optimization of liposomes containing eicosapentaenoic and docosahexaenoic acids; A methodology approach. Iran J Pharm Res. 2014;13(2):393–404.
323. Shah S, Dhawan V, Holm R, Nagarsenker MS, Perrie Y. Liposomes: Advancements and innovation in the manufacturing process. Adv Drug Deliv Rev. 2020;154–155:102–22.
324. Almog R, Forward R, Samsonoff C. Stability of sonicated aqueous suspensions of phospholipids under air. Chem Physics Lipids. 1991;60(2):93–9.
325. Lubitz LJ, Rieger H, Leneweit G. Laminar and turbulent flow effects in high-pressure homogenization of liposomes and perfluorocarbon nanoemulsions. Sci Rep. 2024;14(1):27856.
326. Oshima G, Nagasawa K. Binding and aggregate formation of phospholipid liposomes containing phosphatidic acid with ovalbumin. Biochim Biophys Acta. 1973;291(1):1–14.
327. Zhigaltsev IV, Cullis PR. Morphological behavior of liposomes and lipid nanoparticles. Langmuir. 2023;39(9):3185–93.
328. Gonzalez Gomez A, Syed S, Marshall K, Hosseinidoust Z. Liposomal nanovesicles for efficient encapsulation of staphylococcal antibiotics. ACS Omega. 2019;4(6):10866–76.
329. Sainaga Jyothi VGS, Bulusu R, Venkata Krishna Rao B, Pranothi M, Banda S, Kumar Bolla P et al. Stability characterization for pharmaceutical liposome product development with focus on regulatory considerations: An update. Int J Pharm. 2022;624:122022.
330. Grit M, Crommelin DJA. Chemical stability of liposomes: Implications for their physical stability. Chem Physics Lipids. 1993;64(1):3–18.
331. Mosca M, Ceglie A, Ambrosone L. Effect of membrane composition on lipid oxidation in liposomes. Chem and Physics Lipids. 2011;164(2):158–65.
332. Miller CR, Bondurant B, McLean SD, McGovern KA, O'Brien DF. liposome–Cell interactions in vitro: Effect of liposome surface charge on the binding and endocytosis of conventional and sterically stabilized liposomes. Biochemistry. 1998;37(37):12875–83.
333. Shao X-R, Wei X-Q, Zhang S, Fu N, Lin Y-F, Cai X-X et al. Effects of micro-environmental pH of liposome on chemical stability of loaded drug. Nanoscale Res Lett. 2017;12(1):504.
334. Allen Zhang J-a, Pawelchak J. Effect of pH, ionic strength and oxygen burden on the chemical stability of EPC/cholesterol liposomes under accelerated conditions: Part 1: Lipid hydrolysis. Eur J Pharm Biopharm. 2000;50(3):357–64.
335. Ramana LN, Sethuraman S, Ranga U, Krishnan UM. Development of a liposomal nanodelivery system for nevirapine. J Biomed Sci. 2010;17(1):57.
336. Deng W, Chen W, Clement S, Guller A, Zhao Z, Engel A et al. Controlled gene and drug release from a liposomal delivery platform triggered by X-ray radiation. Nat Commun. 2018;9(1):2713.
337. Cipolla D, Wu H, Eastman S, Redelmeier T, Gonda I, Chan HK. Development and characterization of an in vitro release assay for liposomal ciprofloxacin for inhalation. J Pharm Sci. 2014;103(1):314–27.
338. Yamamoto E, Hyodo K, Ohnishi N, Suzuki T, Ishihara H, Kikuchi H et al. Direct, simultaneous measurement of liposome-encapsulated and released drugs in plasma by on-line SPE–SPE–HPLC. J Chromatogr B. 2011;879(30):3620–5.
339. Shabbits JA, Chiu GNC, Mayer LD. Development of an in vitro drug release assay that accurately predicts in vivo drug retention for liposome-based delivery systems. J Control Release. 2002;84(3):161–70.
340. Taira MC, Chiaramoni NS, Pecuch KM, Alonso-Romanowski S. Stability of liposomal formulations in physiological conditions for oral drug delivery. Drug Deliv. 2004;11(2):123–8.
341. Isailović BD, Kostić IT, Zvonar A, Đorđević VB, Gašperlin M, Nedović VA et al. Resveratrol loaded liposomes produced by different techniques. Innov Food Sci Emerg Technol. 2013;19:181–9.
342. Urey C, Weiss VU, Gondikas A, von der Kammer F, Hofmann T, Marchetti-Deschmann M et al. Combining gas-phase electrophoretic mobility molecular analysis (GEMMA), light scattering, field flow fractionation and cryo electron microscopy in a multidimensional approach to characterize liposomal carrier vesicles. Int J Pharm. 2016;513(1):309–18.
343. Idiart MA, Levin Y. Rupture of a liposomal vesicle. Phys Rev E Stat Nonlin Soft Matter Phys. 2004;69(6 Pt 1):061922.

344. Allen TM, Cleland LG. Serum-induced leakage of liposome contents. Biochimica Et Biophysica Acta (BBA) – Biomembranes. 1980;597(2):418–26.
345. Benjakul R, Panyarachun B, Sarisuta N. Preparation of dry reconstituted liposomal powder by freeze-drying at room temperature. J Liposome Res. 2011;21(1):28–37.
346. Mozafari MR. Liposomes: An overview of manufacturing techniques. Cell Mol Biol Lett. 2005;10(4):711–9.
347. Liu H, Rivnay B, Avery K, Myung JH, Kozak D, Landrau N et al. Optimization of the manufacturing process of a complex amphotericin B liposomal formulation using quality by design approach. Int J Pharm. 2020;585:119473.
348. Litzinger DC, Buiting AMJ, van Rooijen N, Huang L. Effect of liposome size on the circulation time and intraorgan distribution of amphipathic poly(ethylene glycol)-containing liposomes. Biochimica Et Biophysica Acta (BBA) - Biomembranes. 1994;1190(1):99–107.
349. Li X-R, Cheng X-H, Zhang G-N, Wang X-X, Huang J-M. Cardiac safety analysis of first-line chemotherapy drug pegylated liposomal doxorubicin in ovarian cancer. J Ovarian Res. 2022;15(1):96.
350. Dioun SM, Vilardo N, Goldberg GL, Gressel GM. Necessity of routine cardiac evaluation in patients receiving pegylated liposomal doxorubicin for gynecologic cancer. Gynecol Oncol. 2019;155(2):301–4.
351. Adams DH, Joyce G, Richardson VJ, Ryman BE, Wiśniewski HM. Liposome toxicity in the mouse central nervous system. J Neurolo Sci. 1977;31(2):173–9.
352. Yamamoto S, Ishida T, Inoue A, Mikami J, Muraguchi M, Ohmoto Y et al. HEPC-based liposomes trigger cytokine release from peripheral blood cells: Effects of liposomal size, dose and lipid composition. Int J Pharm. 2002;236(1–2):125–33.
353. Rahman A, Treat J, Roh JK, Potkul LA, Alvord WG, Forst D et al. A phase I clinical trial and pharmacokinetic evaluation of liposome-encapsulated doxorubicin. J Clin Oncol. 1990;8(6):1093–100.
354. Veal GJ, Griffin MJ, Price E, Parry A, Dick GS, Little MA et al. A phase I study in paediatric patients to evaluate the safety and pharmacokinetics of SPI-77, a liposome encapsulated formulation of cisplatin. Br J Cancer. 2001;84(8):1029–35.
355. Fujisaka Y, Horiike A, Shimizu T, Yamamoto N, Yamada Y, Tamura T. Phase 1 clinical study of pegylated liposomal doxorubicin (JNS002) in Japanese patients with solid tumors. Jpn J Clin Oncol. 2006;36(12):768–74.
356. Kanhai KMS, Zuiker R, Stavrakaki I, Gladdines W, Gaillard PJ, Klaassen ES et al. Glutathione-PEGylated liposomal methylprednisolone in comparison to free methylprednisolone: Slow release characteristics and prolonged lymphocyte depression in a first-in-human study. Br J Clin Pharmacol. 2018;84(5):1020–8.
357. Sandle T. 3 - GMP and regulations. In: Sandle T, editor. Pharmaceutical Microbiology. Oxford: Woodhead Publishing; 2016. p. 25–34.
358. Sharma A, Anup N, Tekade RK. Chapter 20 - achieving sterility in biomedical and pharmaceutical products (part I): Thermal, chemical, and filtration sterilization. In: Tekade RK, editor. The Future of Pharmaceutical Product Development and Research: Academic Press; 2020. p. 695–788.
359. McCreanor V, Graves N. An economic analysis of the benefits of sterilizing medical instruments in low-temperature systems instead of steam. Am J Infect Control. 2017;45(7):756–60.
360. Cavalli R, Caputo O, Carlotti ME, Trotta M, Scarnecchia C, Gasco MR. Sterilization and freeze-drying of drug-free and drug-loaded solid lipid nanoparticles. Int J Pharm. 1997;148(1):47–54.
361. Schwarz C, Mehnert W, Lucks JS, Müller RH. Solid lipid nanoparticles (SLN) for controlled drug delivery. I. Production, characterization and sterilization. J Control Release. 1994;30(1):83–96.
362. Qi N, Tang X, Lin X, Gu P, Cai C, Xu H et al. Sterilization stability of vesicular phospholipid gels loaded with cytarabine for brain implant. Int J Pharm. 2012;427(2):234–41.
363. Matsusaki M, Masahiro M, Tomonori W, Akashi M. Self-assembled structure of peptide nanospheres induces high stability against hydrolysis and sterilization. J Biomater Sci, Polymer Edition. 2011;22(8):1035–48.
364. Fülöp Z, Balogh A, Saokham P, Jansook P, Loftsson T. Formation and stability assessment of self-assembled nanoparticles from large mw chitosan and sulfobutylether-β-cyclodextrin. J Drug Deliv Sci Technol. 2015;30:478–85.
365. Zuidam NJ, Lee SS, Crommelin DJ. Sterilization of liposomes by heat treatment. Pharm Res. 1993;10(11):1591–6.
366. Brown DA, Venegas B, Cooke PH, English V, Chong PL. Bipolar tetraether archaeosomes exhibit unusual stability against autoclaving as studied by dynamic light scattering and electron microscopy. Chem Phys Lipids. 2009;159(2):95–103.

367. Lukyanov AN, Torchilin VP. Autoclaving of liposomes. J Microencapsul. 1994;11(6):669–72.
368. Voss OH, Lee HN, Tian L, Krzewski K, Coligan JE. Liposome preparation for the analysis of lipid-receptor interaction and efferocytosis. Curr Protoc Immunol. 2018;120(14.44.1–14.44.21.
369. Sharma A, Anup N, Tekade RK. Chapter 21 - achieving sterility in biomedical and pharmaceutical products (part-II): Radiation sterilization. In: Tekade RK, editor. The Future of Pharmaceutical Product Development and Research: Academic Press; 2020. p. 789–848.
370. Kalet IJ. Chapter 6 - biomedical data, knowledge, and systems in context. In: Kalet IJ, editor. Principles of Biomedical Informatics (Second Edition). San Diego: Academic Press; 2014. p. 579–602.
371. Zheng J, Clogston JD, Patri AK, Dobrovolskaia MA, McNeil SE. Sterilization of silver nanoparticles using standard gamma irradiation procedure affects particle integrity and biocompatibility. J Nanomed Nanotechnol. 2011;2011(Suppl 5):001.
372. Masson V, Maurin F, Fessi H, Devissaguet JP. Influence of sterilization processes on poly(ε-caprolactone) nanospheres. Biomaterials. 1997;18(4):327–35.
373. Bozdag S, Dillen K, Vandervoort J, Ludwig A. The effect of freeze-drying with different cryoprotectants and gamma-irradiation sterilization on the characteristics of ciprofloxacin HCl-loaded poly(d,L-lactide-glycolide) nanoparticles. J Pharm Pharmacol. 2005;57(6):699–707.
374. Ordoyo-Pascual J, Ruiz-Alonso S, Gallego I, Saenz-del-Burgo L, Pedraz JL. Effects of beta And gamma radiation sterilization on growth factor-loaded nanoparticles: An innovative approach for osteoarticular disorders treatment. Drug Deliv Transl Res. 2025;15(10):3716–3736.
375. Maksimenko O, Malinovskaya J, Shipulo E, Osipova N, Razzhivina V, Arantseva D et al. Doxorubicin-loaded PLGA nanoparticles for the chemotherapy of glioblastoma: Towards the pharmaceutical development. Int J Pharm. 2019;572:118733.
376. Tapia-Guerrero YS, Del Prado-Audelo ML, Borbolla-Jiménez FV, Giraldo Gomez DM, García-Aguirre I, Colín-Castro CA et al. Effect of UV and gamma irradiation sterilization processes in the properties of different polymeric nanoparticles for biomedical applications. Materials. 2020;13(5):1090.
377. Patel M, Panja S, Zaman LA, Yeapuri P, Bhattarai S, Gorantla S et al. CCR5-ligand decorated rilpivirine lipid-based nanoparticles for sustained antiretroviral responses. Nat Commun. 2025;16(1):513.
378. Konan YN, Gurny R, Allémann E. Preparation and characterization of sterile and freeze-dried sub-200 nm nanoparticles. Int J Pharm. 2002;233(1):239–52.
379. Konan YN, Cerny R, Favet J, Berton M, Gurny R, Allémann E. Preparation and characterization of sterile sub-200 nm meso-tetra(4-hydroxylphenyl)porphyrin-loaded nanoparticles for photodynamic therapy. Eur J Pharm Biopharm. 2003;55(1):115–24.
380. Taylor N, Ma WJ, Kristopeit A, Wang S-c, Zydney AL. Retention characteristics of sterile filters – Effect of pore size and structure. J Membrane Sci. 2021;635:119436.
381. Salvage JP, Rose SF, Phillips GJ, Hanlon GW, Lloyd AW, Ma IY et al. Novel biocompatible phosphorylcholine-based self-assembled nanoparticles for drug delivery. J Control Release. 2005;104(2):259–70.
382. Salvage JP, Thom C, Lewis AL, Phillips GJ, Lloyd AW. Nanoprecipitation of polymeric nanoparticle micelles based on 2-methacryloyloxyethyl phosphorylcholine (MPC) with 2-(diisopropylamino)ethyl methacrylate (DPA), for intracellular delivery applications. J Mater Sci: Mater Med. 2015;26(3):150.
383. FDA. Guidance for Industry Sterile Drug Products Produced by Aseptic Processing —Current Good Manufacturing Practice 2004.
384. EMA. Guideline on the sterilisation of the medicinal product, active substance, excipient and primary container. 2019.
385. Yasui M, Wang X, Tarabara VV, Katayama H. Virus removal by microfiltration: Effects of electrostatic and hydrophobic interactions. Sep Purif Technol. 2024;350:127902.
386. EMA. Note for guidance on virus validation studies: The design, contribution and interpretation of studies validating the inactivation and removal of viruses. 1996.
387. Vincent MJ, Kozal JS, Thompson WJ, Maier A, Dotson GS, Best EA et al. Ethylene oxide: Cancer evidence integration and dose-response implications. Dose Response. 2019;17(4):1559325819888317.
388. EMA. Note for guidance on limitations to the use of ethylene oxide in the manufacture of medicinal products. 2001.
389. NIOSH. Preventing Worker Injuries and Deaths from Explosions in Industrial Ethylene Oxide Sterilization Facilities (Revised Edition). 2007.
390. França A, Pelaz B, Moros M, Sánchez-Espinel C, Hernández A, Fernández-López C et al. Sterilization matters: Consequences of different sterilization techniques on gold nanoparticles. Small. 2010;6(1):89–95.

391. Thakur M, Dan A. Poly-l-lysine-functionalized green-light-emitting carbon dots as a fluorescence turn-on sensor for ultrasensitive detection of endotoxin. ACS Appl Bio Mater. 2021;4(4):3410–22.
392. Demmer W, Fischer-Frühholz S, Nußbaumer D, Melzner D. Chapter 14 - economic production of biopharmaceuticals by high-speed membrane adsorbers. In: Bhattacharyya D, Butterfield DA, editors. Membrane Science and Technology. 8: Elsevier; 2003. p. 283–95.
393. McCormick K, Sanders JH. Chapter 13 - pharmaceutical microbiology. In: McCormick K, Sanders JH, editors. Quality (Second Edition): Butterworth-Heinemann; 2022. p. 299–325.
394. Shanmugam PST, Sampath T, Jagadeeswaran I, Thamizharasan S, Fathima S. Chapter 4 - material-mediated pyrogenicity. In: Timiri Shanmugam PS, Sampath T, Jagadeeswaran I, editors. Biocompatibility Protocols for Medical Devices and Materials: Academic Press; 2023. p. 55–66.
395. Sayyah M, Javad-Pour M, Ghazi-Khansari M. The bacterial endotoxin lipopolysaccharide enhances seizure susceptibility in mice: Involvement of proinflammatory factors: Nitric oxide and prostaglandins. Neuroscience. 2003;122(4):1073–80.
396. Kellum JA, Ronco C. The role of endotoxin in septic shock. Cri Care. 2023;27(1):400.
397. MHRA. Appendix XIV C. Test for Bacterial Endotoxins. The British Pharmacopoeia 2025.
398. MHRA. Appendix XIV D. Test for Pyrogens. The British Pahrmacopoeia 2025.
399. Sandle T. 11 - endotoxin and pyrogen testing. In: Sandle T, editor. Pharmaceutical Microbiology. Woodhead Publishing; 2016. p. 131–45.
400. Maloney T, Phelan R, Simmons N. Saving the horseshoe crab: A synthetic alternative to horseshoe crab blood for endotoxin detection. PLoS Biol. 2018;16(10):e2006607.
401. Ding JL, Ho B. A new era in pyrogen testing. Trends Biotechnol. 2001;19(8):277–81.
402. Kang DH, Yun SY, Eum S, Yoon KE, Ryu SR, Lee C et al. A study on the application of recombinant factor c (rFC) assay using biopharmaceuticals. Microorganisms. 2024;12(3):516.
403. Schneier M, Razdan S, Miller AM, Briceno ME, Barua S. Current technologies to endotoxin detection and removal for biopharmaceutical purification. Biotechnol Bioeng. 2020;117(8):2588–609.
404. Vipond C, Findlay L, Feavers I, Care R. Limitations of the rabbit pyrogen test for assessing meningococcal OMV based vaccines. Altex. 2016;33(1):47–53.
405. Ribera JM, García O, Grande C, Esteve J, Oriol A, Bergua J et al. Dose-intensive chemotherapy including rituximab in Burkitt's leukemia or lymphoma regardless of human immunodeficiency virus infection status: Final results of a phase 2 study (Burkimab). Cancer. 2013;119(9):1660–8.
406. Mueller S, Kline C, Stoller S, Lundy S, Christopher L, Reddy AT et al. PNOC015: Repeated convection-enhanced delivery of MTX110 (aqueous panobinostat) in children with newly diagnosed diffuse intrinsic pontine glioma. Neuro Oncol. 2023;25(11):2074–86.
407. Gonvers E, El-Boghdadly K, Grape S, Albrecht E. Efficacy and safety of intrathecal morphine for analgesia after lower joint arthroplasty: A systematic review and meta-analysis with meta-regression and trial sequential analysis. Anaesthesia. 2021;76(12):1648–58.
408. de Mas N, Kientzler DC, Kleindienst D. Endotoxin removal from a small-molecule aqueous drug substance using ultrafiltration: A case study. Org Process Res Dev. 2015;19(9):1293–8.
409. Sweadner KJ, Forte M, Nelsen LL. Filtration removal of endotoxin (pyrogens) in solution in different states of aggregation. Appl Environ Microbiol. 1977;34(4):382–5.
410. Kiesewetter A, Gupta A, Heinen-Kreuzig A, Greenhalgh T, Stein A. Improved endotoxin removal using ecofriendly detergents for intensified plasmid capture. Biotechnol Prog. 2023;39(6):e3375.
411. Vannefors C. Detection and removal of endotoxin in nanomaterial preparations: Mater thesis, UMEA University; 2022.
412. Mangini M, Verde A, Boraschi D, Puntes VF, Italiani P, De Luca AC. Interaction of nanoparticles with endotoxin importance in nanosafety testing and exploitation for endotoxin binding. Nanotoxicology. 2021;15(4):558–76.
413. Dobrovolskaia MA, Neun BW, Clogston JD, Ding H, Ljubimova J, McNeil SE. Ambiguities in applying traditional limulus amebocyte lysate tests to quantify endotoxin in nanoparticle formulations. Nanomedicine (Lond). 2010;5(4):555–62.
414. Dobrovolskaia MA, Germolec DR, Weaver JL. Evaluation of nanoparticle immunotoxicity. Nat Nanotechnol. 2009;4(7):411–4.
415. Hannon G, Prina-Mello A. Endotoxin contamination of engineered nanomaterials: Overcoming the hurdles associated with endotoxin testing. WIREs Nanomed Nanobiotechnol. 2021;13(6):e1738.
416. Neun BW, Dobrovolskaia MA. Detection and Quantification of Gram Negative Bacterial Endotoxin Contamination in Nanoparticle Formulations by End Point Chromogenic LAL Assay: Version 3.1. 2024 Jul. In: National Cancer Institute's Nanotechnology Characterization Laboratory Assay Cascade Protocols. Bethesda (MD): National Cancer Institute (US); 2005 May 1. NCL Method STE-1.1..

417. Verde A, Mangini M, Managò S, Tramontano C, Rea I, Boraschi D et al. SERS sensing of bacterial endotoxin on gold nanoparticles. Front Immunol. 2021;12:758410.
418. Dobrovolskaia MA. Endotoxin & Depyrogenation Guide: Version 4. National Cancer Institute's Nanotechnology Characterization Laboratory Assay Cascade Protocols. Bethesda (MD): National Cancer Institute (US); 2005.
419. FDA. 22 Case Studies Where Phase 2 and Phase 3 Trials Had Divergent Results. https://www.fda.gov/about-fda/reports/22-case-studies-where-phase-2-and-phase-3-trials-had-divergent-results. 2017.

Index

Note: Pages in **bold** refer to tables. Pages in *Italics* refer to figures.

For Product Safety Concerns and Information please contact our EU
representative GPSR@taylorandfrancis.com
Taylor & Francis Verlag GmbH, Kaufingerstraße 24, 80331 München, Germany

www.ingramcontent.com/pod-product-compliance
Ingram Content Group UK Ltd.
Pitfield, Milton Keynes, MK11 3LW, UK
UKHW052218190926
475702UK00006B/1082

* 9 7 8 1 0 3 2 7 7 1 9 1 5 *